HIGH RESOLUTION SOLAR PHYSICS: THEORY, OBSERVATIONS, AND TECHNIQUES

COVER ILLUSTRATION: Image of solar granulation at 571 nm, The image shows an individual granule surrounded by small bright (possibly magnetic) features. The image was taken with a 10 second exposure time using the NSO adaptive optics system. The smallest structures seen in this image are of size 0.16 inch, the diffraction limit of the Dunn Solar Telescope.

A SERIES OF BOOKS ON RECENT DEVELOPMENTS IN ASTRONOMY AND ASTROPHYSICS

Printed by Sheridan Books, Inc.

First published 1999

Library of Congress Catalog Card Number: 99-64739
ISBN 1-58381-009-9

Please contact proper address for information on:

PUBLISHING:
Managing Editor
PO Box 24463
211 KMB
Brigham Young University
Provo, UT 84602-4463 USA

Phone: 801-378-2298
Fax: 801-378-4049
E-mail: pasp@astro.byu.edu

ORDERING BOOKS:
Astronomical Society of the Pacific
CONFERENCE SERIES
390 Ashton Avenue
San Francisco, CA 94112 - 1722 USA

Phone: 415-337-2624
Fax: 415-337-5205
E-mail: catalog@aspsky.org
Web Site: http://www.aspsky.org

A SERIES OF BOOKS ON RECENT DEVELOPMENTS IN ASTRONOMY AND ASTROPHYSICS

Vol. 1-Progress and Opportunities in Southern Hemisphere Optical Astronomy: CTIO 25th Anniversary Symposium
ed. V. M. Blanco and M. M. Phillips ISBN 0-937707-18-X

Vol. 2-Proceedings of a Workshop on Optical Surveys for Quasars
ed. P. S. Osmer, A. C. Porter, R. F. Green, and C. B. Foltz ISBN 0-937707-19-8

Vol. 3-Fiber Optics in Astronomy
ed. S. C. Barden ISBN 0-937707-20-1

Vol. 4-The Extragalactic Distance Scale: Proceedings of the ASP 100th Anniversary Symposium
ed. S. van den Bergh and C. J. Pritchet ISBN 0-937707-21-X

Vol. 5-The Minnesota Lectures on Clusters of Galaxies and Large-Scale Structure
ed. J. M. Dickey ISBN 0-937707-22-8

Vol. 6-Synthesis Imaging in Radio Astronomy: A Collection of Lectures from the Third NRAO Synthesis Imaging Summer School
ed. R. A. Perley, F. R. Schwab, and A. H. Bridle ISBN 0-937707-23-6

Vol. 7-Properties of Hot Luminous Stars: Boulder-Munich Workshop
ed. C. D. Garmany ISBN 0-937707-24-4

Vol. 8-CCDs in Astronomy
ed. G. H. Jacoby ISBN 0-937707-25-2

Vol. 9-Cool Stars, Stellar Systems, and the Sun. Sixth Cambridge Workshop
ed. G. Wallerstein ISBN 0-937707-27-9

Vol. 10-Evolution of the Universe of Galaxies: Edwin Hubble Centennial Symposium
ed. R. G. Kron ISBN 0-937707-28-7

Vol. 11-Confrontation Between Stellar Pulsation and Evolution
ed. C. Cacciari and G. Clementini ISBN 0-937707-30-9

Vol. 12-The Evolution of the Interstellar Medium
ed. L. Blitz ISBN 0-937707-31-7

Vol. 13-The Formation and Evolution of Star Clusters
ed. K. Janes ISBN 0-937707-32-5

Vol. 14-Astrophysics with Infrared Arrays
ed. R. Elston ISBN 0-937707-33-3

Vol. 15-Large-Scale Structures and Peculiar Motions in the Universe
ed. D. W. Latham and L. A. N. da Costa ISBN 0-937707-34-1

Vol. 16-Proceedings of the 3rd Haystack Observatory Conference on Atoms, Ions and Molecules: New Results in Spectral Line Astrophysics
ed. A. D. Haschick and P. T. P. Ho ISBN 0-937707-35-X

Vol. 17-Light Pollution, Radio Interference, and Space Debris
ed. D. L. Crawford ISBN 0-937707-36-8

Vol. 18-The Interpretation of Modern Synthesis Observations of Spiral Galaxies
ed. N. Duric and P. C. Crane ISBN 0-937707-37-6

Vol. 19-Radio Interferometry: Theory, Techniques, and Applications, IAU Colloquium 131
ed. T. J. Cornwell and R. A. Perley ISBN 0-937707-38-4

Vol. 20-Frontiers of Stellar Evolution: 50th Anniversary McDonald Observatory (1939-1989)
ed. D. L. Lambert ISBN 0-937707-39-2

Vol. 21-The Space Distribution of Quasars
ed. D. Crampton ISBN 0-937707-40-6

Vol. 22-Nonisotropic and Variable Outflows from Stars
ed. L. Drissen, C. Leitherer, and A. Nota ISBN 0-937707-41-4

Vol. 23-Astronomical CCD Observing and Reduction Techniques
ed. S. B. Howell ISBN 0-937707-42-4

Vol. 24-Cosmology and Large-Scale Structure in the Universe
ed. R. R. de Carvalho ISBN 0-937707-43-0

Vol. 25-Astronomical Data Analysis Software and Systems I
ed. D. M. Worrall, C. Biemesderfer, and J. Barnes ISBN 0-937707-44-9

Vol. 26-Cool Stars, Stellar Systems, and the Sun, Seventh Cambridge Workshop
ed. M. S. Giampapa and J. A. Bookbinder ISBN 0-937707-45-7

Vol. 27-The Solar Cycle: Proceedings of the National Solar Observatory/Sacramento Peak 12th Summer Workshop
ed. K. L. Harvey ISBN 0-937707-46-5

Vol. 28-Automated Telescopes for Photometry and Imaging
ed. S. J. Adelman, R. J. Dukes, Jr., and C. J. Adelman ISBN 0-937707-47-3

Vol. 29-Viña Del Mar Workshop on Catacysmic Variable Stars
ed. N. Vogt ISBN 0-937707-48-1

Vol. 30-Variable Stars and Galaxies
ed. B. Warner ISBN 0-937707-49-X

Vol. 31-Relationships Between Active Galactic Nuclei and Starburst Galaxies
ed. A. V. Filippenko ISBN 0-937707-50-3

Vol. 32-Complementary Approaches to Double and Multiple Star Research, IAU Collouquium 135
ed. H. A. McAlister and W. I. Hartkopf ISBN 0-937707-51-1

Vol. 33-Research Amateur Astronomy
ed. S. J. Edberg ISBN 0-937707-52-X

Vol. 34-Robotic Telescopes in the 1990s
ed. A. V. Filippenko ISBN 0-937707-53-8

Vol. 35-Massive Stars: Their Lives in the Interstellar Medium
ed. J. P. Cassinelli and E. B. Churchwell ISBN 0-937707-54-6

Vol. 36-Planets Around Pulsars
ed. J. A. Phillips, S. E. Thorsett, and S. R. Kulkarni ISBN 0-937707-55-4

Vol. 37-Fiber Optics in Astronomy II
ed. P. M. Gray ISBN 0-937707-56-2

Vol. 38-New Frontiers in Binary Star Research: Pacific Rim Colloquium
ed. K. C. Leung and I.-S. Nha ISBN 0-937707-57-0

Vol. 39-The Minnesota Lectures on the Structure and Dynamics of the Milky Way
ed. Roberta M. Humphreys ISBN 0-937707-58-9

Vol. 40-Inside the Stars, IAU Colloquium 137
ed. Werner W. Weiss and Annie Baglin ISBN 0-937707-59-7

Vol. 41-Astronomical Infrared Spectroscopy: Future Observational Directions
ed. Sun Kwok ISBN 0-937707-60-0

Vol. 42-GONG 1992: Seismic Investigation of the Sun and Stars
ed. Timothy M. Brown ISBN 0-937707-61-9

Vol. 43-Sky Surveys: Protostars to Protogalaxies
ed. B. T. Soifer ISBN 0-937707-62-7

Vol. 44-Peculiar Versus Normal Phenomena in A-Type and Related Stars, IAU Colloquium 138
ed. M. M. Dworetsky, F. Castelli, and R. Faraggiana ISBN 0-937707-63-5

Vol. 45-Luminous High-Latitude Stars
ed. D. D. Sasselov ISBN 0-937707-64-3

Vol. 46-The Magnetic and Velocity Fields of Solar Active Regions, IAU Colloquium 141
ed. H. Zirin, G. Ai, and H. Wang ISBN 0-937707-65-1

Vol. 47-Third Decennial US-USSR Conference on SETI
ed. G. Seth Shostak ISBN 0-937707-66-X

Vol. 48-The Globular Cluster-Galaxy Connection
ed. Graeme H. Smith and Jean P. Brodie ISBN 0-937707-67-8

Vol. 49-Galaxy Evolution: The Milky Way Perspective
ed. Steven R. Majewski ISBN 0-937707-68-6

Vol. 50-Structure and Dynamics of Globular Clusters
ed. S. G. Djorgovski and G. Meylan ISBN 0-937707-69-4

Vol. 51-Observational Cosmology
ed. G. Chincarini, A. Iovino, T. Maccacaro, and D. Maccagni ISBN 0-937707-70-8

Vol. 52-Astronomical Data Analysis Software and Systems II
ed. R. J. Hanisch, R. J. V. Brissenden, and Jeannette Barnes ISBN 0-937707-71-6

Vol. 53-Blue Stragglers
ed. Rex A. Saffer ISBN 0-937707-72-4

Vol. 54-The First Stromlo Symposium: The Physics of Active Galaxies
ed. Geoffrey V. Bicknell, Michael A. Dopita, and Peter J. Quinn ISBN 0-937707-73-2

Vol. 55-Optical Astronomy from the Earth and Moon
ed. Diane M. Pyper and Ronald J. Angione ISBN 0-937707-74-0

Vol. 56-Interacting Binary Stars
ed. Allen W. Shafter ISBN 0-937707-75-9

Vol. 57-Stellar and Circumstellar Astrophysics
ed. George Wallerstein and Alberto Noriega-Crespo ISBN 0-937707-76-7

Vol. 58-The First Symposium on the Infrared Cirrus and Diffuse Interstellar Clouds
ed. Roc M. Cutri and William B. Latter ISBN 0-937707-77-5

Vol. 59-Astronomy with Millimeter and Submillimeter Wave Interferometry, IAU Colloquium 140
ed. M. Ishiguro and Wm. J. Welch ISBN 0-937707-78-3

Vol. 60-The MK Process at 50 Years: A Powerful Tool for Astrophysical Insight: A Workshop of the Vatican Observatory
ed. C. J. Corbally, R. O. Gray, and R. F. Garrison ISBN 0-937707-79-1

Vol. 61-Astronomical Data Analysis Software and Systems III
ed. Dennis R. Crabtree, R. J. Hanisch, and Jeannette Barnes ISBN 0-937707-80-5

Vol. 62-The Nature and Evolutionary Status of Herbig Ae / Be Stars
ed. P. S. Thé, M. R. Pérez, and E. P. J. van den Heuvel ISBN 0-937707-81-3

Vol. 63-Seventy-Five Years of Hirayama Asteroid Families: The Role of Collisions in the Solar System History
ed. Y. Kozai, R. P. Binzel, and T. Hirayama ISBN 0-937707-82-1

Vol. 64-Cool Stars, Stellar Systems, and the Sun, Eighth Cambridge Workshop
ed. Jean-Pierre Caillault ISBN 0-937707-83-X

Vol. 65-Clouds, Cores, and Low Mass Stars
ed. Dan P. Clemens and Richard Barvainis ISBN 0-937707-84-8

Vol. 66- Physics of the Gaseous and Stellar Disks of the Galaxy
ed. Ivan R. King ISBN 0-937707-85-6

Vol. 67-Unveiling Large-Scale Structures Behind the Milky Way
ed. C. Balkowski and R. C. Kraan-Korteweg ISBN 0-937707-86-4

Vol. 68-Solar Active Region Evolution: Comparing Models with Observations
ed. K. S. Balasubramaniam and George W. Simon ISBN 0-937707-87-2

Vol. 69-Reverberation Mapping of the Broad-Line Region in Active Galactic Nuclei
ed. P. M. Gondhalekar, K. Horne, and B. M. Peterson ISBN 0-937707-88-0

Vol. 70-Groups of Galaxies
ed. Otto G. Richter and Kirk Borne ISBN 0-937707-89-9

Vol. 71-Tridimensional Optical Spectroscopic Methods in Astrophysics, IAU Colloquium 149
ed. G. Comte and M. Marcelin ISBN 0-937707-90-2

Vol. 72-Millisecond Pulsars: A Decade of Surprise.
ed. A. A. Fruchter, M. Tavani, and D. C. Backer ISBN 0-937707-91-0

Vol. 73-Airborne Astronomy Symposium on the Galactic Ecosystem: From Gas to Stars to Dust
ed. M. R. Haas, J. A. Davidson, and E. F. Erickson ISBN 0-937707-92-9

Vol. 74-Progress in the Search for Extraterrestrial Life: 1993 Bioastronomy Symposium
ed. G. Seth Shostak ISBN 0-937707-93-7

Vol. 75-Multi-Feed Systems for Radio Telescopes
ed. D. T. Emerson and J. M. Payne ISBN 0-937707-94-5

Vol. 76-GONG '94: Helio- and Astero-Seismology from the Earth and Space
ed. Roger K. Ulrich, Edward J. Rhodes, Jr., and Werner Däppen ISBN 0-937707-95-3

Vol. 77-Astronomical Data Analysis Software and System IV
ed. R. A. Shaw, H. E. Payne, and J. J. E. Hayes ISBN 0-937707-96-1

Vol. 78-Astrophysical Applications of Powerful New Databases: Joint Discussion No. 16 of the 22nd General Assembly of the IAU
ed. S. J. Adelman and W. L. Wiese ISBN 0-937707-97-X

Vol. 79-Robotic Telescopes: Current Capabilities, Present Developments, and Future Prospects for Automated Astronomy
ed. Gregory W. Henry and Joel A. Eaton ISBN 0-937707-98-8

Vol. 80-The Physics of the Interstellar Medium and Intergalactic Medium
ed. A. Ferrara, C. F. McKee, C. Heiles, and P. R. Shapiro ISBN 0-937707-99-6

Vol. 81-Laboratory and Astronomical High Resolution Spectra
ed. A. J. Sauval, R. Blomme, and N. Grevesse ISBN 1-886733-01-5

Vol. 82-Very Long Baseline Interferometry and the VLBA
ed. J. A. Zensus, P. J. Diamond, and P. J. Napier ISBN 1-886733-02-3

Vol. 83-Astrophysical Applications of Stellar Pulsation. IAU Colloquium 155
ed. R. S. Stobie and P. A. Whitelock ISBN 1-886733-03-1

Vol. 84-The Future Utilisation of Schmidt Telescopes, IAU Colloquium 148
ed. Jessica Chapman, Russell Cannon, Sandra Harrison, and Bambang Hidayat ISBN 1-886733-05-8

Vol. 85-Cape Workshop on Magnetic Cataclysmic Variables
ed. D. A. H. Buckley and B. Warner ISBN 1-886733-06-6

Vol. 86-Fresh Views of Elliptical Galaxies
ed. Alberto Buzzoni, Alvio Renzini, and Alfonso Serrano ISBN 1-886733-07-4

Vol. 87-New Observing Modes for the Next Century
ed. Todd Boroson, John Davies, and Ian Robson ISBN 1-886733-08-2

Vol. 88- Clusters, Lensing, and the Future of the Universe
ed. Virginia Trimble and Andreas Reisenegger ISBN 1-886733-09-0

Vol. 89-Astronomy Education: Current Developments, Future Coordination
ed.John R. Percy ISBN 1-886733-10-4

Vol. 90-The Origins, Evolution, and Destinies of Binary Stars in Clusters
ed. E. F. Milone and J. C. Mermilliod ISBN 1-886733-11-2

Vol. 91-Barred Galaxies, IAU Colloquium 157
ed. R. Buta, D. A. Crocker, and B. G. Elmegreen ISBN 1-886733-12-0

Vol. 92-Formation of the Galactic Halo–Inside and Out
ed. H. L. Morrison and A. Sarajedini ISBN 1-886733-13-9

Vol. 93-Radio Emission from the Stars and the Sun
ed. A. R. Taylor and J. M. Paredes ISBN 1-886733-14-7

Vol. 94-Mapping, Measuring, and Modelling the Universe
ed. Peter Coles, Vincent Martinez, and Maria-Jesus Pons-Borderia ISBN 1-886733-15-5

Vol. 95-Solar Drivers of Interplanetary and Terrestrial Disturbances: Proceedings of 16th International Workshop, National Solar Observatory/Sacramento Peak
ed. K. S. Balasubramaniam, S. L. Keil, and R. N. Smartt ISBN 1-886733-16-3

Vol. 96- Hydrogen-Deficient Stars
ed. C. S. Jeffery and U. Heber ISBN 1-886733-17-1

Vol. 97-Polarimetry of the Interstellar Medium
ed.W. G. Roberge and D. C. B. Whittet ISBN 1-886733-18-X

Vol. 98-From Stars to Galaxies: The Impact of Stellar Physics on Galaxy Evolution
ed. Claus Leitherer, Uta Fritze-von Alvensleben, and John Huchra ISBN 1-886733-19-8

Vol. 99-Cosmic Abundances: Proceedings of the 6th Annual October Astrophysics Conference
ed. Stephen S. Holt and Geroge Sonneborn ISBN 1-886733-20-1

Vol. 100-Energy Transport in Radio Galaxies and Quasars
ed. P. E. Hardee, A. H. Bridle, and J. A. Zensus ISBN 1-886733-21-X

Vol. 101-Astronomical Data Analysis Software and Systems V
ed. George H. Jacoby and Jeannette Barnes ISSN 1080-7926

Vol. 102-The Galactic Center, 4th ESO/CTIO Workshop
ed. Roland Gredel ISBN 1-886733-22-8

Vol. 103-The Physics of Liners in View of Recent Observations
ed. M. Eracleous, A. Koratkar, C. Leitherer, and L. Ho ISBN 1-886733-23-6

Vol. 104-Physics, Chemistry, and Dynamics of Interplanetary Dust, IAU Colloquium 150
ed. Bo A. S. Gustafson and Martha S. Hanner ISBN 1-886733-24-4

Vol. 105-Pulsars: Problems and Progress, IAU Colloquium 160
ed. M. Bailes, S. Johnston, and M. A. Walker ISBN 1-886733-25-2

Vol. 106-Minnesota Lectures on Extragalactic Neutral Hydrogen
ed. Evan D. Skillman ISBN 1-886733-26-0

Vol. 107-Completing the Inventory of the Solar System: A Symposium held in conjuunction with the 106th Annual Meeting of the ASP
ed. Terrence W. Rettig and Joseph M. Hahn ISBN 1-886733-27-9

Vol. 108-M. A. S. S. Model Atmospheres and Spectrum Synthesis: 5th Vienna Workshop
ed. S. J. Adelman, F. Kupka, and W. W. Weiss ISBN 1-886733-28-7

Vol. 109-Cool Stars, Stellar Systems, and the Sun, Ninth Cambridge Workshop
ed. Roberto Pallavicini and Andrea K. Dupree ISBN 1-886733-29-5

Vol. 110-Blazar Continuum Variability
ed.H. R. Miller, J. R. Webb, and J. C. Noble ISBN 1-886733-30-9

Vol. 111-Magnetic Reconnection in the Solar Atmosphere: Proceedings of a Yohkoh Conference
ed. R. D. Bentley and J. T. Mariska ISBN 1-886733-31-7

Vol. 112-The History of the Milky Way and Its Satellite System
ed. A. Burkert, D. H. Hartmann, and S. R. Majewski ISBN 1-886733-32-5

Vol. 113-Emission Lines in Active Galaxies: New Methods and Techniques, IAU Colloquium 159
ed. B. M. Peterson, F. Z. Cheng, and A. S. Wilson ISBN 1-886733-33-3

Vol. 114-Young Galaxies and QSO Absorption-Line Systems
ed. Sueli M. Viegas, Ruth Gruenwald, and Reinaldo R. de Carvalho ISBN 1-886733-34-1

Vol. 115-Galactic and Cluster Cooling Flows
ed. Noam Soker ISBN 1-886733-35-X

Vol. 116-The Second Stromlo Symposium: The Nature of Elliptical Galaxies
ed. M. Arnaboldi, G. S. Da Costa, and P. Saha ISBN 1-886733-36-8

Vol. 117- Dark and Visible Matter in Galaxies
ed. Massimo Persic and Paolo Salucci ISBN 1-886733-37-6

Vol. 118-First Advances in Solar Physics Euroconference: Advances in the Physics of Sunspots
ed. B. Schmieder, J. C. del Toro Iniesta, and M. Vázquez ISBN 1-886733-38-4

Vol. 119-Planets Beyond the Solar System and the Next Generation of Space Missions
ed. David R. Soderblom ISBN 1-886733-39-2

Vol. 120-Luminous Blue Variables: Massive Stars in Transition
ed. Antonella Nota and Henny J. G. L. M. Lamers ISBN 1-886733-40-6

Vol. 121-Accretion Phenomena and Related Outflows, IAU Colloquium 163
ed. D. T. Wickramasinghe, G. V. Bicknell and L. Ferrario ISBN 1-886733-41-4

Vol. 122-From Stardust to Planetesimals: Symposium held as part of the 108th Annual Meeting of the ASP
ed. Yvonne J. Pendleton and A. G. G. M. Tielens ISBN 1-886733-42-2

Vol. 123-The 12th 'Kingston Meeting': Computational Astrophysics
ed. David A. Clarke and Michael J. West ISBN 1-886733-43-0

Vol. 124-Diffuse Infrared Radiation and the IRTS
ed. Haruyuki Okuda, Toshio Matsumoto, and Thomas L. Roellig ISBN 1-886733-44-9

Vol. 125- Astronomical Data Analysis Software and Systems VI
ed. Gareth Hunt and H. E. Payne ISBN 1-886733-45-7

Vol. 126-From Quantum Fluctuations to Cosmological Structures
ed. D. Vallis-Gabaud, M. A. Hendry, P. Molaro, and K. Chamcham ISBN 1-886733-46-5

Vol. 127-Proper Motions and Galactic Astronomy
ed. Roberta M. Humphreys ISBN 1-886733-47-3

Vol. 128- Mass Ejection from AGN (Active Galactic Nuclei)
ed. N. Arav, I. Shlosman, and R. J. Weymann ISBN 1-886733-48-1

Vol. 129-The George Gamow Symposium
ed. E. Harper, W. C. Parke, and G. D. Anderson ISBN 1-886733-49-X

Vol. 130-The Third Pacfic Rim Conference on Recent Development on Binary Star Research
ed. Kam-Ching Leung ISBN 1-886733-50-3

Vol. 131-Boulder-Munich II: Properties of Hot, Luminous Stars
ed. Ian D. Howarth ISBN 1-886733-51-1

Vol. 132-Star Formation with the Infrared Space Observatory (ISO)
ed. João L. Yun and René Liseau ISBN 1-886733-53-X

Vol. 133-Science with the NGST
ed. Eric P. Smith and Anuradha Koratkar ISBN 1-886733-53-8

Vol. 134-Brown Dwarfs and Extrasolar Planets
ed. Rafael Rebolo, Eduardo L. Martin,
and Maria Rosa Zapatero Osorio ISBN 1-886733-54-6

Vol. 135-A Half Century of Stellar Pulsation Interpretations: A Tribute to Arthur N. Cox
ed. P. A Bradley and J. A. Guzik ISBN 1-886733-55-4

Vol. 136- Galactic Halos: A UC Santa Cruz Workshop
ed. Dennis Zaritsky ISBN 1-886733-56-2

Vol. 137-Wild Stars in the Old West: Proceedings of the 13th North American Workshop
on Cataclysmic Variables and Related Objects
ed. S. Howell, E.Kuulkers, and C. Woodward ISBN 1-886733-57-0

Vol. 138-1997 Pacific Rim Conference on Stellar Astrophysics
ed. Kwing L. Chan, K. S. Cheng, and Harinder P. Singh ISBN 1-886733-58-9

Vol. 139-Preserving the Astronomical Windows, Proceedings of Joint Discussion
No. 5 of the 23rd General Assembly of the IAU
ed. Syuzo Isobe and Tomohiro Hirayama ISBN 1-886733-59-7

Vol. 140-Synoptic Solar Physics – 18th NSO/Sacramento Peak Summer Workshop
ed. K. S. Balasubramaniam, J. W. Harvey, and D. M. Rabin ISBN 1-886733-60-0

Vol. 141-Astrophysics from Antarctica
A Symposium held as part of the 109th Annual Meeting of the ASP
ed. Giles Novak and Randall H. Landsberg ISBN 1-886733-61-9

Vol. 142-The Stellar Initial Mass Function, 38th Herstmonceux Conference
ed. Gerry Gilmore and Debbie Howell ISBN 1-886733-62-7

Vol. 143-The Scientific Impact of the Goddard High Resolution Spectrograph
ed. John C. Brandt, Thomas B. Ake III, and Carolyn Collins Petersen ISBN 1-886733-63-5

Vol. 144- Radio Emission from Galactic and Extragalactic Compact Sources,
IAU Colloquium 164
ed. J. Anton Zensus, G. B. Taylor, and J. M. Wrobel ISBN 1-886733-64-3

Vol. 145-Astronomical Data Analysis Software and Systems VII
ed. Rudolf Albrecht, Richard N. Hook, and Howard A. Bushouse ISBN 1-886733-65-1

Vol. 146-The Young Universe: Galaxy Formation and Evolution at
Intermediate and High Redshift
ed. S. D'Odorico, A. Fontana, and E. Giallongo ISBN 1-886733-66-X

Vol. 147-Abundance Profiles: Diagnostic Tools for Galaxy History
ed. Daniel Friedli, Mike Edmunds, Carmelle Robert,
and Laurent Drissen ISBN 1-886733-67-8

Vol. 148-Origins
ed. Charles E. Woodward, J. Michael Shull,
and Harley A. Thronson, Jr. ISBN 1-886733-68-6

Vol. 149-Solar System Formation and Evolution
ed. D. Lazzaro, R. Vieira Martins, S. Ferraz-Mello,
J. Fernández, and C. Beaugé ISBN 1-886733-69-4

Vol. 150-New Perspectives on Solar Prominences, IAU Colloquium 167
ed. David Webb, David Rust, and Brigitte Schmieder ISBN 1-886733-70-8

Vol. 151-Cosmic Microwave Background and Large Scale Structure of the Universe
ed. Yong-Ik Byun and Kin-Wang Ng ISBN 1-886733-71-6

Vol. 152-Fiber Optics in Astronomy III
ed. S. Arribas, E. Mediavilla, and F. Watson ISBN 1-886733-72-4

Vol. 153-Library and Information Services in Astronomy III, (LISA III)
ed. Uta Grothkopf, Heinz Andernach, Sarah Stevens-Rayburn,
and Monique Gomez ISBN 1-886733-73-2

Vol. 154-Cool Stars, Stellar Systems, and the Sun, Tenth Cambridge Workshop
ed. Robert A. Donahue and Jay A. Bookbinder ISBN 1-886733-74-0

Vol. 155-Second Advances in Solar Physics Euroconference:
Three-Dimensional Structure of Solar Active Regions
ed. Costas E. Alissandrakis and Brigitte Schmieder ISBN 1-886733-75-9

Vol. 156-Highly Redshifted Radio Lines
ed. C. L. Carilli, S. J. E. Radford, K. M. Menten and
G. I. Langston ISBN 1-886733-76-7

Vol. 157-Annapolis Workshop on Magnetic Cataclysmic Variables
ed. Coel Hellier and Koji Mukai ISBN 1-886733-77-5

Vol. 158-Solar and Stellar Activity: Similarities and Differences
ed. C. J. Butler and J. G. Doyle ISBN 1-886733-78-3

Vol. 159-BL Lac Phenomenon
ed. Leo O. Takalo and Aimo Sillanpää ISBN 1-886733-79-1

Vol. 160-Astrophysical Discs, An EC Summer School
ed. J. A. Sellwood and Jeremy Goodman ISBN 1-886733-80-5

Vol. 161-High Energy Processes in Accreting Black Holes
ed. Juri Poutanen and Roland Svensson ISBN 1-886733-81-3

Vol. 162-Quasars and Cosmology
ed. Gary Ferland and Jack Baldwin ISBN 1-886733-83-X

Vol. 163-Star Formation in Early Type Galaxies
ed. Jordi Cepa and Patricia Carral ISBN 1-886733-84-8

Vol. 164-Ultraviolet–Optical Space Astronomy Beyond HST
ed. Jon A. Morse, J. Michael Shull and Anne L. Kinney ISBN 1-886733-85-6

Vol. 165-The Third Stromlo Symposium: The Galactic Halo
ed. Brad K. Gibson, Tim S. Axelrod, and Mary E. Putman ISBN 1-886733-86-4

Vol. 166-Stromlo Workshop on High-Velocity Clouds
ed. Brad K. Gibson and Mary E. Putman ISBN 1-886733-87-2

Vol. 167-Harmonizing Cosmic Distance Scales in a Post-HIPPARCOS Era
ed. Daniel Egret and André Heck ISBN 1-886733-88-0

Vol. 168-New Perspectives on the Interstellar Medium
ed. A. R. Taylor, T. L. Landecker, and G. Joncas ISBN 1-886733-89-9

Vol. 169-11th European Workshop on White Dwarfs
ed. J.-E. Solheim and E. G. Meistas ISBN 1-886733-91-0

Vol. 170-The Low Surface Brightness Universe, IAU Colloquium 171
ed. J. I. Davies, C. Impey, and S. Phillipps ISBN 1-886733-92-9

Vol. 171-LiBeB, Cosmic Rays, and Related X-and Gamma-Rays
ed. Reuven Ramaty, Elisabeth Vangioni-Flam,
Michel Cassé, and Keith Olive ISBN 1-886733-93-7

Vol. 172-Astronomical Data Analysis Software and Systems VIII
ed. David M. Mehringer, Raymond L. Plante, and
Douglas A. Roberts ISBN 1-886733-94-5

Vol. 173-Theory and Tests of Convection In Stellar Structure
ed. Álvaro Giménez, Edward F. Guinan, and Benjamín Montesinos ISBN 1-886733-95-3

Vol. 174-Catching the Perfect Wave: Adaptive Optics and Optical Interferometry for the 21st Century, A Symposium held as a part of the 110th Annual Meeting of the ASP
ed. Sergio R. Restaino, William Junor, Nebojsa Duric ISBN 1-886733-96-1

Vol. 175-Structure and Kinematics of Quasar Broad Line Regions
ed. C.M. Gaskell, W. N. Brandt, M. Dietrich, D. Dultzin-Hacyan,
and M. Eracleous ISBN 1-886733-97-X

Vol. 176-Observational Cosmology: The Development of Galaxy Systems
ed. Giuliano Giuricin, Marino Mezzetti, and Paolo Salucci ISBN 1-58381-000-5

Vol. 177-Astrophysics with Infrared Surveys: A Prelude to SIRTF
ed. Michael D. Bicay, Roc M. Cutri, and Barry F. Madore ISBN 1-58381-001-3

Vol. 178-Stellar Dynamos: Nonlinearity and Chaotic Flows
ed. Manuel Núñez and Antonio Ferriz-Mas ISBN 1-58381-002-1

Vol. 179-Eta Carinae At The Millennium
ed. Jon A. Morse, Roberta M. Humphreys, and Augusto Damineli ISBN 1-58381-003-X

Vol. 180-Synthesis Imaging in Radio Astronomy II
ed. G. B. Taylor, C. L. Carilli, and R. A. Perley ISBN 1-58381-005-6

Vol. 181-Microwave Foregrounds
ed. Angelica de Oliveira-Costa and Max Tegmark ISBN 1-58381-006-4

Vol. 182-Galaxy Dynamics, A Rutgers Symposium
ed. David Merritt, J. A. Sellwood, and Monica Valluri ISBN 1-58381-007-2

Vol. 183-High Resolutin Solar Physics: Theory, Observations and Techniques
ed. T. R. Rimmele, K. S. Balasubramaniam, and R. R. Radick ISBN 1-58381-009-9

Book orders or inquiries concerning these volumes should be directed to the:

Astronomical Society of the Pacific Conference Series
390 Ashton Avenue
San Francisco, CA 94112-1722 USA

Phone: 415-337-2126
Fax: 415-337-5205
E-mail: catalog@aspsky.org
Web Site: http://www.aspsky.org

ASTRONOMICAL SOCIETY OF THE PACIFIC
CONFERENCE SERIES

Volume 183

HIGH RESOLUTION SOLAR PHYSICS: THEORY, OBSERVATIONS, AND TECHNIQUES

Proceedings of the 19th Sacramento Peak Summer Workshop held at National Solar Observatory, Sunspot, New Mexico, USA
28 September - 2 October 1998

Edited by

T. R. Rimmele, K. S. Balasubramaniam
National Solar Observatory, Sunspot, New Mexico, USA

and

R. R. Radick
Air Force Research Laboratory, Sunspot, New Mexico, USA

Table of Contents

Part 4. Quiet Sun Dynamics - Flows, Waves, Oscillations

Part 5. Active Regions and Flares

High Resolution Solar Physics: Theory, Observations, and Techniques
ASP Conference Series, Vol. 183, 1999
T.R. Rimmele, K. S. Balasubramaniam, R.R. Radick, eds.

PREFACE

The 19th NSO/SP International Workshop on High Resolution Solar Physics: Theory, Observations, and Techniques was held in Sunspot, New Mexico, USA from 28 September to 2 October 1998.

High resolution angular, spectral, and temporal observations of the Sun are required in order to answer many important questions in solar and stellar physics. Furthermore, the Sun is the only star that can be spatially resolved well enough to allow detailed study of fundamental astrophysical processes such as magneto-convection.

Previous NSO/SP summer workshops on high resolution solar observations were held in 1980, 1988 and 1992. It was evident during the 1998 workshop that substantial progress continues to be made in the solar astronomer's quest for observations of ever higher angular resolution, but that much remains to be accomplished. Exciting new observations of the dynamic solar atmosphere have resulted from space missions such as SOHO and TRACE. New instrumentation and observing techniques have greatly improved the scientific productivity of ground based facilities and have advanced our understanding of the physics of the solar atmosphere. Major new facilities in space (e.g. Solar-B) and on the ground, utilizing adaptive optics to achieve diffraction limited observations, promise a breakthrough in the understanding of the fundamental physics of solar finestructure in the near future. The proceedings of the 1998 workshop review the current state of high resolution solar physics from the theoretical, observational, and modeling perspectives, and the state-of-the-art of instrumentation used to achieve high resolution solar observations.

The workshop was dedicated to Dr. Richard B. Dunn, who recently retired from the NSO. Dr. Dunn's pioneering contributions in engineering, design, and observation have done much to lay the foundations of present-day high resolution solar physics. A highlight of the workshop was the ceremony during which the Vacuum Tower Telescope at Sacramento Peak, one of the finest high resolution solar telescopes in the world, was renamed the Richard. B. Dunn Solar Telescope after its builder.

The workshop was jointly sponsored and funded by the National Solar Observatory, NOAO[1], the National Optical Astronomy Observatories[1], the Divisions of Astronomy and Atmospheric Sciences of the National Science Foundation (NSF), the Office of Space Science of the National Aeronautics and Space Administration (NASA), and the Air Force Office of Scientific Research (AFOSR), including the European Office of Aerospace Research and Development (AFOSR/EOARD), the Asian Office of Aerospace Research and Development (AFOSR/AOARD) and the Space Hazards Branch of the Air Force Research Labs. It is a pleasure to thank the following individuals for supporting

[1]The National Optical Astronomy Observatories are sponsored by the National Science Foundation

the goals of the workshop and facilitating its funding: Sidney Wolff (NOAO), Jacques Beckers (NSO), Ken Schatten and Ben Snavely (NSF), Bill Wagner (NASA), Kent Miller and Stephanie Brewis (EOARD), Koto White and Michele Gaudreault (AOARD).

The Scientific Organizing Committee (SOC) members were: Thomas Rimmele (Chair, NSO), K. S. Balasubramaniam (NSO and AFRL/VSBS), W.C. Livingston (NSO), R. R. Radick (AFRL/VSBS), P. Goode (NJIT/BBSO), T. Bogdan (HAO), T. Berger (LMSAL), O. von der Luehe (KIS).

We thank the members of the SOC for extensive help in organizing the meeting and reviewing the papers. Last but not least, we thank the session chairs and all the participating scientists for creating a stimulating week of ideas and discussion.

Rebecca Coleman and the entire administrative and facilities maintenance staff at NSO/SP played a major role in local organization and support. Brenda Ramos typed all the discussion material. Robert McGraw played a key role in computing support before, during and after the workshop. Finally, we are grateful to the scientific staff of NSO/SP and AFRL/VSBS for their local support and help in organizing this workshop.

T.R. Rimmele, K. S. Balasubramaniam, R. R. Radick
19th NSO/SP Summer Workshop
National Solar Observatory

High Resolution Solar Physics: Theory, Observations, and Techniques
ASP Conference Series, Vol. 183, 1999
T.R. Rimmele, K. S. Balasubramaniam, R.R. Radick, eds.

Workshop Participants

Dick Altrock, Air Force Research Labs/VSBS and National Solar Observatory, Sunspot, NM 88349, USA
altrock@sunspot.noao.edu

Thomas R. Ayres, Univ. of Colorado, Boulder
Campus Box 389 (CASA), Boulder, CO 80309,USA
ayres@vulcan.colorado.edu

K.S. Balasubramaniam Air Force Research Labs/VSBS and National Solar Observatory,
Sunspot, NM 88349, USA
altrock@sunspot.noao.edu

Horst Balthasar, Astrophysikalisches Institut Potsdam,
Telegrafenberg, D-14473 Potsdam (Germany)
hbalthasar@aip.de

Dipankar Banerjee, Armagh Observatory,
Armagh, BT61 9DG, N. Ireland dipu@star.arm.ac.uk

J. Beckers, National Solar Observatory,
P. O. Box 62, Sunspot, NM 88349, USA
jbeckers@noao.edu

Luis R. Bellot Rubio, Instituto de Astrofisica de Canarias
38200-La Laguna, Tenerife, Spain
lbellot@iac.es

Thomas Berger, Lockheed Martin Solar and Astrophysics Lab
O/H1-12 B/252 3251 Hanover St., Palo Alto, Ca. 94304
tberger@solar.stanford.edu

Thomas J. Bogdan, HAO/NCAR, P.O. Box 3000
Boulder, CO 80307-3000, USA
tom@hao.ucar.edu

Paal Brekke, Institute of Theoretical Astrophysics
P.O. Box 1029, Blindern, N-0315 Oslo, NORWAY
paalb@astro.uio.no

Jo H.M.J. Bruls, Kiepenheuer-Institut fuer Sonnenphysik
Schoeneckstrasse 6, D-79104 Freiburg, Germany
bruls@kis.uni-freiburg.de

Andrei V. Boulatov, Institute of Solar-Terrestrial Physics SD RAS,
P.O.Box 4026, Irkutsk-33, 664033, Russia
bulat@iszf.irk.ru

Jongchul Chae, Big Bear Solar Observatory
40386 North Shore Lane, Big Bear City, CA 92314-9672, USA
chae@bbso.njit.edu

Guennadi Chernov, 142092 Troitsk, Moscow region, IZMIRAN, RUSSIA
gchernov@izmiran.rssi.ru

Carsten Denker, Big Bear Solar Observatory
40386 North Shore Lane, Big Bear City, CA 92314-9672, USA
cdenker@bbso.njit.edu

Richard B. Dunn, National Solar Observatory,
P. O. Box 62, Sunspot, NM 88349, USA
rdunn@sunspot.noao.edu

Peter T. Gallagher, Dept. of Pure and App. Physics
The Queen's University of Belfast
Belfast, BT7 1NN,N. Ireland
P.Gallagher@qub.ac.uk

Dan Gezari, NASA/Goddard Space Flight Center
Infrared Astrophysics Branch
Code 685, Greenbelt, MD 20771
gezari@astrolab.gsfc.nasa.gov

Phil Goode, Big Bear Solar Observatory
40386 North Shore Lane, Big Bear City, CA 92314-9672
pgoode@bbso.njit.edu

Arnold Hanslmeier, Institut fuer Astronomie,
Univ.-Platz 5, A-8010 Graz, Austria
arh@bimgs1.kfunigraz.ac.at

Johann Hirzberger, Institut fuer Astronomie
Univ.-Platz 5, A-8010 Graz, Austria
jkh@bimgs1.kfunigraz.ac.at

Nick M. Hoekzema, Sterrekundig Instituut Utrecht
Princetonplein 5, 3584 CC Utrecht
The Netherlands
N.M.Hoekzema@fys.ruu.nl or nmh@worldonline.nl

Rajmal Jain, Udaipur Solar Observatory,
(PRL/DOS) Off bari Road,P.B. No. 198,

Dewali, Udaipur - 313001, India
rajmal@uso.ernet.in rajain@solar.stanford.edu

Edisher Kaghashvili, Abastumani Astrophysical Observatory
eddy@dtapha.kheta.ge

R. Kariyappa, Indian Institute of Astrophysics
Koramangala, Bangalore 560034, INDIA
rkari@iiap.ernet.in ; RKariyappa@solar.stanford.edu

Stephen L. Keil, National Solar Observatory,
P. O. Box 62, Sunspot, NM 88349, USA
skeil@sunspot.noao.edu

Christoph U. Keller, National Solar Observatory, P.O. Box 26732,
Tucson, AZ 85726, USA
ckeller@noao.edu

Michael Knoelker, HAO/NCAR, P.O. Box 3000
Boulder, CO 80307-3000, USA
knoe@hao.ucar.edu

Markus Koschinsky, Universitaets-Sternwarte
Geismarlandstrasse 11, D-37083 Goettingen, Germany
mkosch@uni-sw.gwdg.de

John W. Leibacher, National Solar Observatory, P.O. Box 26732,
Tucson, AZ 85726, USA
@noao.edu

Hoasheng Lin, National Solar Observatory,
P. O. Box 62, Sunspot, NM 88349, USA
lin@sunspot.noao.edu

Bruce Lites, HAO/NCAR, P.O. Box 3000
Boulder, CO 80307-3000, USA
lites@hao.ucar.edu

William C. Livingston National Solar Observatory, P.O. Box 26732,
Tucson, AZ 85726, USA
wcl@noao.edu

Mats Löfdahl, Stockholm Observatory
SE-133 36 Saltsjobaden, SWEDEN
mats@astro.su.se

Valentin Martinez Pillet Instituto de Astrofisica de Canarias La Laguna, E-38200, Tenerife (Spain)
vmp@ll.iac.es

Shibu K. Mathew, Udaipur Solar Observatory, Post Box : 198, Devali, Udaipur -313001, Rajasthan, INDIA
shibu@uso.ernet.in

Peter Nisenson, Center For Astrophysics, 60 Garden St. Cambridge, MA 02138, USA
pnisenson@cfa.harvard.edu

Rick Paxman, ERIM International, PO Box 134008, Ann Arbor, MI 48113-4008,USA
paxman@erim-int.com

Debi Prasad C, Space Science Laboratory, ES82, Marshall Space Flight Center/NASA, Huntsville, AL, 35812
debipra@magaxp1.msfc.nasa.gov

Kristof Petrovay, Instituto de Astrofisica de Canarias, La Laguna (Tenerife), E-38200 Spain
kpetro@iac.es

Richard R. Radick, Air Force Research Lab,
P. O. Box 62, Sunspot, NM 88349, USA
radick@sunspot.noao.edu

S.Paul Rajaguru, Dept. of Physics, Indian Institute of Science, Bangalore
rajguru@physics.iisc.ernet.in

Mark P. Rast, HAO/NCAR, P.O. Box 3000, Boulder, CO, USA
mrast@hao.ucar.edu

Thomas R. Rimmele, National Solar Observatory,
P. O. Box 62, Sunspot, NM 88349, USA
rimmele@sunspot.noao.edu

Robert J. Rutten, Sterrekundig Instituut, postbus 80 000, NL 3508 TA Utrecht, The Netherlands
R.J.Rutten@astro.uu.nl

Jorge Sanchez Almeida, Instituto de Astrofisica de Canarias, E-38200 La Laguna, Tenerife, Spain
jos@iac.es

Monica Sanchez Cuberes, Instituto Astrofisico de Canarias, C Via Lactea s/n, 38200 La Laguna. Tenerife, SPAIN
msanchez@ll.iac.es

Göran B. Scharmer, Royal Swedish Academy of Sciences, Stockholm Observatory, SE-133 36 Saltsjöbaden, Sweden
scharmer@astro.su.se

Rolf Schlichenmaier, Kiepenheuer-Institut fuer Sonnenphysik (KIS), Schoeneckstr.6, D-79104 Freiburg, Germany
schliche@kis.uni-freiburg.de

George W. Simon, Air Force Research Lab,
P. O. Box 62, Sunspot, NM 88349, USA
simon@sunspot.noao.edu

Luigi A. Smaldone, Dipartimento di Scienze Fisiche, Universita' di Napoli "Federico II", Mostra d'Oltremare Pad 19, 80125 NAPOLI (Italy)
smaldone@na.infn.it

Ray M. Smartt, National Solar Observatory,
P. O. Box 62, Sunspot, NM 88349, USA
smartt@sunspot.noao.edu

Sebastian Steffens, Astronomisches Institut der Universitaet Wuerzburg, Am Hubland, D-97074 Wuerzburg
steffens@astro.uni-wuerzburg.de

Oskar Steiner, Kiepenheuer-Institut fuer Sonnenphysik (KIS), Schoeneckstr.6, D-79104 Freiburg, Germany
schliche@kis.uni-freiburg.de

Louis Strous, Lockheed Martin Solar and Astrophysics Lab, O/H1-12 B/252
3251 Hanover St., Palo Alto, Ca. 94304
LStrous@solar.stanford.edu

Yoshinori Suematsu, National Astronomical Observatory, PO Box: 2-21-1 Osawa, Mitaka, Tokyo 181-8588, Japan
suematsu@solar.mtk.nao.ac.jp

John H. Thomas, University of Rochester, Dep. of Physics and Astronomy, Rochester, NY 14627
thomas@me.rochester.edu

Alan Title, Lockheed Martin Solar and Astrophysics Lab
O/H1-12 B/252 3251 Hanover St., Palo Alto, Ca. 94304
title@nice.lmsal.com, atitle@solar.stanford.edu

Alexandra Tritschler, Kiepenheuer Institut fuer Sonnenphysik
Schoeneckstr. 6, D-79104 Freiburg, Germany
ali@kis.uni-freiburg.de

Han Uitenbroek, Harvard-Smithsonian Center for Astrophysics
60 Garden Street, Cambridge, MA 02138, U.S.A.
huitenbroek@cfa.harvard.edu

Binxun Ye, Beijing Astronomical Observatory, Science Park, Datun Rd, Beijing, 100101, China PR
bxye@public.bta.net.cn

Deng Yuanyong, Beijing Astronomical Observatory, Chinese Academy of Sciences, Beijing 100080, P. R. China
dyy@sun10.bao.ac.cn

Adriaan A. van Ballegooijen, Center for Astrophysics, 60 garden street, MS 15, Cambridge MA 02138, USA
vanballe@cfa.harvard.edu

V. K. Verma, U. P. State Observatory, Naini Tal-263129, INDIA
verma@upso.ernet.in , vkv61@hotmail.com

v.d. Luehe, Oskar, Kiepenheuer Institut fuer Sonnenphysik, Schoeneckstr. 6, D-79104 Freiburg, Germany
ovdluhe@kis.uni-freiburg.de

Jack Zirker, National Solar Observatory,
P. O. Box 62, Sunspot, NM 88349, USA
zirker@sunspot.noao.edu

High Resolution Solar Physics: Theory, Observations, and Techniques
ASP Conference Series, Vol. 183, 1999
T.R. Rimmele, K. S. Balasubramaniam, R.R. Radick, eds.

The Case for High Resolution in Solar Physics

John H. Thomas

University of Rochester, Rochester, NY 14627, USA

Abstract. High-resolution observations of the Sun are required in order to answer many important questions in solar and stellar physics. We now know that several important solar phenomena must be studied with an angular resolution of 0.1 arcsecond or better. We need high resolution in the broadest sense, encompassing spatial, temporal, and spectral resolution, which together place high demands on telescope aperture and overall system efficiency, and in the case of ground-based telescopes, require post facto or real-time image correction. The scientific case for high resolution is argued here by means of some examples of important solar phenomena occurring at length scales at or below our current limit of resolution. These examples include small-scale magnetic flux tubes, the structure and dynamics of a sunspot penumbra, granular convection and the excitation of p-modes, and magnetic effects on p-modes and higher frequency acoustic waves. For each example, in addition to theoretical considerations, I discuss some recent observational successes which have led to new knowledge and which bolster the case for further effort and investment in high-resolution solar telescopes and instrumentation.

1. Introduction

The quest for high resolution in solar physics has always been driven by important scientific questions. For over a hundred years, the main driver was the solar granulation (see Bray, Loughhead & Durrant 1984). For the past twenty years or so the study of the small, intense flux tubes that constitute most of the photospheric magnetic flux have been a strong driver.

Observations of solar features with sizes below the resolution limit of 0.2 arcseconds of existing solar telescopes are needed in order to study several important physical phenomena that occur at these small scales. Several examples of these phenomena will be discussed in this paper. A new generation of ground-based solar telescopes with apertures in the range 2-4 m is required. In order for these telescopes to achieve their resolution limit during a substantial fraction of their observing time, real-time correction of the effects of atmospheric seeing by means of adaptive optics will be necessary.

The case for high-resolution observations of the Sun has been presented many times in recent years, in several contexts. Strong scientific arguments for high resolution were presented in support of the NASA Solar Optical Telescope (SOT) mission and for the Large European Solar Telescope (LEST). Neither of these projects was carried out, but not, in my view, for any lack of scientific justification.

In 1985 the AURA Board of Directors commissioned a study of the role of large-aperture, high-resolution, ground-based solar observations, motivated by scientific opportunities, concerns about current facilities at the National Solar Observatory, and suggestions from NOAO management that a large-aperture solar telescope should be considered as a long-range goal for the NSO. (So the idea of a new "flagship" advanced solar telescope for NSO, which will receive much attention at this workshop, goes back more than 13 years now.) The committee appointed by the AURA Board to carry out this study consisted of R. M. MacQueen (chair), R. G. Athay, J. L. Linsky, R. Rosner, J. H. Thomas, and A. M. Title. Our committee took the view that a very strong case could be made for a new-generation, large solar telescope on the basis of a number of scientific problems of fundamental importance in solar physics, and indeed in astrophysics generally, that could be addressed by such a facility. The MacQueen Committee report (MacQueen et al. 1985) discussed the following examples of physical processes on the Sun which require high-resolution, high-throughput observations for a new or better understanding:

1. Convection in a compressible, radiating atmosphere. To fully resolve the temperature and flow field in solar granular convection requires 0.1 arcsecond angular resolution and temporal resolution of a few seconds extending over ten minutes or more.

2. Structure of stellar atmospheres. We need to understand the structure of the magnetic field in the solar atmosphere, especially that of the small-scale, kilogauss magnetic flux tubes in the photosphere. What is the detailed structure of the solar atmosphere, and what is the meaning of mean models of the atmosphere? Observations of the thermodynamic, magnetic, and hydrodynamic structure needed to answer these questions require angular resolution of 0.1 arcseconds, spectral resolution of 10 mÅ in several spectral lines simultaneously, and temporal resolution of 100 ms.

3. Equilibria and evolution. Can we think of the solar atmosphere as being in equilibrium, or must we consider it to be a dynamical system driven by thermodynamic and magnetic instabilities? The resolution requirements are similar to those above.

4. Mass and energy transport. What are the properties of MHD waves in the solar atmosphere, especially waves in the small-scale flux tubes, and what is their role in atmospheric heating? What is the role of steady flows, such as the Evershed flow? How can we understand diffusive transport in the solar atmosphere, and the nonlocal transport of heat by particles with mean free paths longer than the local scale height?

Today these scientific problems requiring high resolution are perhaps better defined, and significant progress has been made, but they are still not solved. Some of them will be discussed at this meeting. Also, there has been significant technical progress in demonstrating our ability to achieve 0.1 arcsecond angular resolution from the ground through image reconstruction.

Let's consider next just what we mean by "high resolution," taking a broad view of the term. Then I will discuss some particular scientific problems in solar physics which demand high resolution for their solution, pointing out some recent observational successes that support the case.

2. What Do We Mean by High Resolution?

When we speak of high resolution in solar astronomy, we most often are thinking of high angular resolution and efforts to achieve images limited only by the diffraction limit of the telescope. But in the most general sense, high resolution includes:

high spatial resolution,
high temporal resolution,
high spectral resolution,

and all three of these are important to our understanding of small-scale phenomena on the Sun. Also, we should keep in mind that we need high resolution not just in broad-band images, but also in (more or less in increasing order of difficulty) narrow-band images, Doppler images, magnetograms, and vector polarimetry.

Achieving high temporal and spectral resolution simultaneously requires a high throughput of photons, which in turn requires a large telescope aperture. Indeed, the requirement for high throughput may easily be a stronger driver toward large aperture than angular resolution. The very real problem of "photon starvation" in high-resolution solar observations is illustrated by the following example.

Suppose we are trying to observe a solar feature of diameter d equal to the diffraction limit of our telescope, and we want to follow the detailed evolution of this feature in time. The diameter d of a resolution element at the diffraction limit of a telescope is inversely proportional to the telescope aperture D. The number of photons per unit time in the image of the solar feature, which scales as D^2d^2, is thus independent of the telescope aperture D. Furthermore, the time scale τ for evolutionary changes in a solar feature generally decreases with decreasing size; τ typically scales as $d/V \propto d$, where V is either a fluid velocity or a wave speed, so the number of photons collected in time τ scales as D^{-1}. In order to resolve the time behavior of our solar feature, we need a cadence time step Δt equal to, say, 0.1τ. If we intend to do speckle image reconstruction, we will need many exposures per cadence time step, say 100 images (a low estimate), so the time step between images is $\delta t = 0.01\Delta t = 0.001\tau$. The actual exposure time must be considerably less than δt to allow time for data readout. Now, imagine that our observations are being made with a narrow-band filter for high spectral resolution. We now find that we are running low on photons and have insufficient signal-to-noise ratio in our images. We need a telescope with an aperture larger than the minimum suggested by the Rayleigh criterion in order to have enough throughput of photons. It is also clear that the overall system efficiency becomes an issue and we need very efficient filters with high transmission, such as Fabry-Perot interferometers, and CCD detectors with high quantum efficiency.

Another consideration pushing us to large telescope aperture is the need to do high-resolution imaging in the near infrared. For example, 0.1 arcsecond angular resolution at a wavelength of 1.5 microns requires a telescope aperture of 3 m.

It is important to remember that the solar features and phenomena that we wish to study are three-dimensional, and that spatial resolution involves both angular resolution perpendicular to the line of sight and resolution along the line of sight. Because of foreshortening and obscuration near the limb, most small-scale structures are best seen near the center of the solar disk, so angular resolution and resolution along the line of sight may usually be thought of as horizontal and vertical (or height) resolution on the Sun. The local density scale height in the low photosphere is less than 100 km, and we expect significant height variations of solar features on this scale. We achieve height resolution through high spectral resolution, by measuring spectral properties such as line width, bisector shape, Doppler shift, and polarization as functions of depth in a spectral line.

High-resolution solar observations from the ground also require some form of correction for the degrading effects of atmospheric seeing. At a good site, the relatively simple technique of frame selection – taking a rapid sequence of exposures and selecting the ones with the smallest wavefront errors – has proved to be quite effective. This is especially true when the best frames are selected for storage automatically in real time through the use of a computational algorithm which measures, for example, the rms image contrast or the contrast a high spatial frequencies. Various post facto methods of solar image correction, such as speckle interferometry and phase diversity, have been employed with considerable success (see, for example, Harvey & Breckenridge 1973; Radick 1993; Denker 1998, and references therein); several results using these techniques will be presented at this meeting.

In order for a large-aperture, ground based solar telescope to be cost effective, however, it will require real-time image correction by means of adaptive optics. Because a point source is not available on the Sun, a wavefront sensor for solar adaptive optics has to operate with low-contrast, extended sources such as sunspots, pores, or granulation as target images. Since a sunspot or pore is not always available in the desired field of view, the granulation pattern is the only reasonable choice for the target image. Techniques for wavefront sensing currently under development or consideration include curvature sensing, phase diversity, and a correlating Shack-Hartmann method. Progress in solar adaptive optics will be reported by several groups at this meeting.

3. Some Solar Phenomena Requiring High-Resolution Observations

Now I would like to discuss a few specific examples of solar phenomena for which an understanding will require observations with an angular resolution of 0.1 arcsecond or better, as well as good spectral and temporal resolution. I also will describe some recent high-resolution observations that have improved our understanding of these phenomena and have strengthened the case for even higher resolution.

Much of the motivation for high resolution comes from the small-scale structures associated with the magnetic field in the solar photosphere, such as penumbral filaments and umbral dots in sunspots, and intense magnetic flux tubes in the quiet network. These features are beautifully illustrated in the G-band image in Fig. 1, taken at the Swedish Vacuum Solar Telescope on La Palma during a period of excellent seeing. This image, which is an excellent example of the resolution that can be achieved with the largest existing solar telescopes, shows a sunspot, a trailing pore, and many network bright points in the dark intergranular lanes. These bright points, which generally coincide with intense magnetic flux tubes, show brightness variations over length scales down to the diffraction limit of existing solar telescopes, about 0.2 arcseconds. Relating these structures in intensity to the structure of the magnetic field itself will require high-resolution polarimetry at scales of 0.1 arcsecond and smaller, placing great demands on telescope aperture, image quality, and overall system efficiency.

3.1. Intense magnetic flux tubes in the photosphere

Perhaps the strongest arguments for high resolution concern the need to resolve the small-scale, intense magnetic flux tubes that dominate the magnetic structure at the solar surface. Indeed, most of the first day of this workshop is devoted to small-scale magnetic flux tubes. We recognize that nearly all aspects of solar activity - chromospheric and coronal heating, flares, high-speed wind streams, mass ejections, radio, x-ray, and gamma ray emission - are dominated by the magnetic field, and that the physical processes responsible for these phenomena occur mostly at sub-arcsecond spatial scales

More than 90% of the photospheric magnetic flux outside of sunspots is concentrated into small, intense flux tubes with field strengths of 1-2 kG and diameters generally less than 400 km. Resolving these features is crucial to our understanding of solar activity and chromospheric and coronal heating, and indeed to similar processes on other late-type stars. The need to resolve intense magnetic flux tubes at a scale of about 0.1 arcsecond has been the main scientific driver for high-resolution solar physics in recent years.

There has already been considerable success in the effort, using post facto image restoration. Keller & von der Lühe (1992) developed a method of speckle polarimetry which uses real-time frame selection and differential speckle imaging with a polarimeter to produce diffraction-limited magnetograms. They take simultaneous broad-band and narrow-band images. The broad-band images, with their high signal-to-noise ratio, are used to determine the instantaneous optical transfer function, which is then used to correct the narrow-band polarimetric images used to construct the magnetograms.

Using this method, Keller (1992) succeeded in resolving individual flux tubes with diameters of about 200 km with the Swedish Vacuum Solar Telescope. He also showed that, while some intense magnetic flux tubes do indeed coincide with bright points in the continuum (with contrast of 30% or more), as had been suggested by observations at lower resolution, the correspondence is not one-to-one, there being flux tubes without associated bright points and vice versa.

Keller has continued his work on speckle polarimetry of magnetic flux tubes using the Zurich Imaging Polarimeter (ZIMPOL) and the Universal Birefringent Filter (UBF) at the Sacramento Peak VTT. He uses a broad-band filter and then a beam splitter in front of the UBF, directing one beam to a white-light camera and the other beam through the UBF. The ZIMPOL modulator is placed in

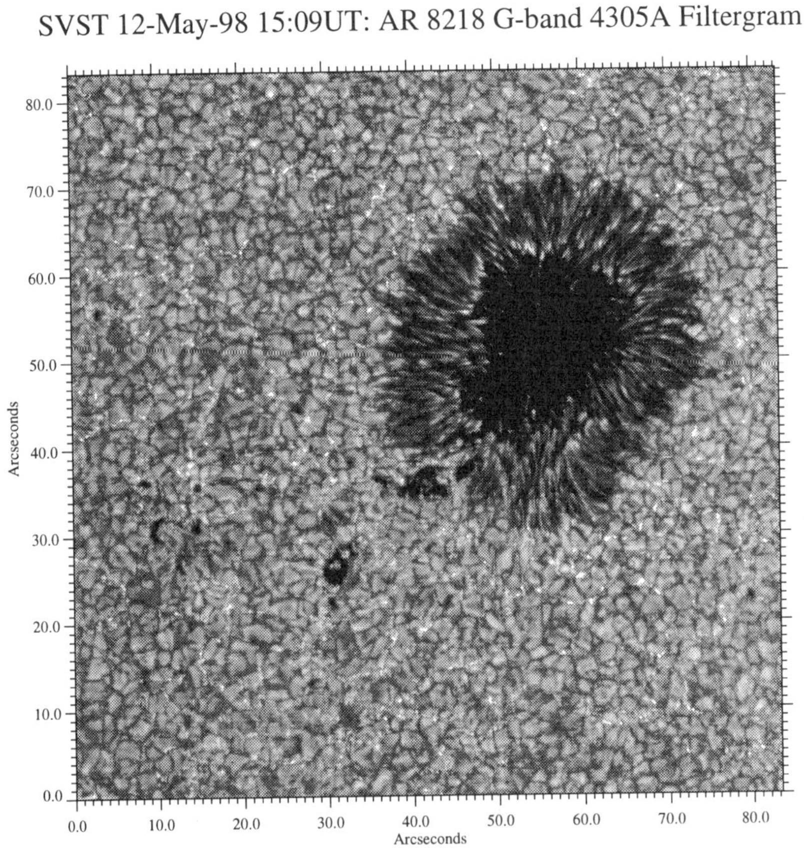

Figure 1. High resolution G-band image of a sunspot and surrounding photosphere, taken during a period of excellent seeing at the Swedish Vacuum Solar Telescope on La Palma, Canary Islands. Many network bright points, which generally coincide with intense magnetic elements, can be seen in the dark intergranular lanes. (Courtesy of T. Berger.)

front of the beam splitter, and the modulation is done at 42 kHz to freeze the seeing fluctuations. The ZIMPOL produces images in the two states of circular polarization, which are then used to construct magnetograms. Exposures are taken at a rapid rate of five per second with an exposure time of 20 ms. The 300 images recorded in a minute are then used in the speckle reconstruction.

Fig. 2 shows an example of Keller's recent results (unpublished) with this setup, for magnetic flux tubes in the network near disk center. On the left is a white-light image and on the right is the corresponding magnetogram using Ca I 610.3 nm. The pixel size in these images is 0.1 arcsecond, and the images are almost diffraction limited, the diffraction limit being 0.17 arcsecond at this wavelength. The magnetogram shows several resolved flux tubes with diameters in the range $0.2 - 0.3$ arcseconds and field strengths of order 1 kG.

Keller has now resolved more than 70 network magnetic flux tubes in his speckle magnetograms, and hence is able to do a crude statistical analysis of their properties. He has found that only about half of all the network magnetic elements are spatially resolved at the 0.2 arcsecond level achievable with his current technique. Thus, fully half of all intense flux tubes in the network have diameters smaller than about 0.2 arcseconds. Resolving these smaller flux tubes presents a strong challenge to high-resolution solar astronomy.

From a theoretical viewpoint, the diameter of a magnetic flux tube can be estimated by setting the tube's diffusive decay time equal to the turnover time of a convective turbulent eddy. Thus, the estimated size of the smallest flux tubes depends critically on the assumed value of the effective magnetic diffusivity. The tube diameter scales as the square root of the effective diffusivity. If we could determine from observations a minimum, cutoff diameter of the smallest flux tubes, this would establish that turbulent diffusion is indeed the mechanism controlling the diffusion of magnetic flux, on scales much larger than that of molecular diffusion.

It is easy to see that speckle reconstruction of sequences of magnetograms tracing the evolution of individual magnetic flux tubes places severe demands on telescope aperture and overall system (telescope, instrumentation, and detector) efficiency. Beyond this lies the even greater challenge of measuring the full vector magnetic fields of these flux tubes by means of the full set of Stokes profiles (Stokes polarimetry). These measurements, together with simultaneous measurements of Doppler velocity at several heights, should reveal the dynamics of the convective collapse of the flux tubes and perhaps also reveal motions associated with the various wave modes of the flux tubes, which may play an important role in chromospheric and coronal heating.

3.2. The structure of a sunspot penumbra and the evershed flow

Sunspots are the ideal objects for testing magnetohydrodynamic theory under astrophysical conditions. They provide us with a considerable challenge, especially in understanding the structure and dynamics of the filamentary penumbra. Our understanding of the sunspot penumbra has undergone significant changes in recent years as a result of theoretical advances and observations with improved resolution (see the review by Thomas & Weiss 1992). Whereas formerly we thought of the penumbra as a shallow, horizontal magnetic layer lying over field-free gas, we now know that the penumbra is a deep structure containing

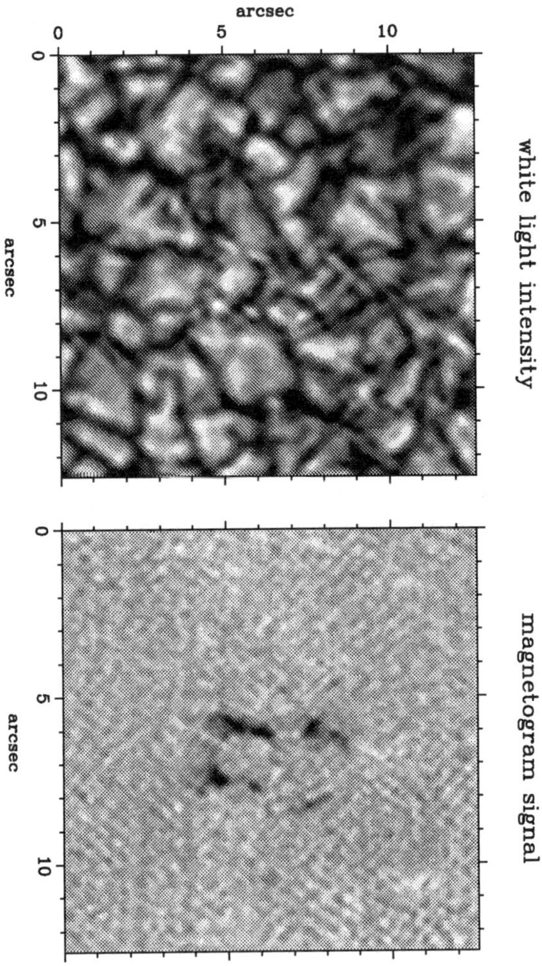

Figure 2. High-resolution white-light image and magnetogram of intense magnetic flux tubes in the network obtained by speckle polarimetry. (Courtesy of C. U. Keller.)

more than half of the magnetic flux emerging through the solar surface within a sunspot.

Resolving the filamentary structure of the penumbra has been a challenge to observers for many years. Some thirty years ago the observations of Beckers and Schröter here at Sacramento Peak showed that the magnetic field in the dark penumbral filaments is nearly horizontal, while the field in the bright filaments is significantly inclined to the horizontal (Beckers & Schröter 1969). This picture has now been well confirmed by more recent high-resolution observations (Degenhardt & Wiehr 1991; Title et al. 1993; Lites et al. 1993; Stanchfield, Lites & Thomas 1997). These observations have revealed the "fluted" structure of the penumbral magnetic field, with radial spokes of inclined magnetic field separated by regions of nearly horizontal magnetic field.

Another important recent advance in our knowledge of the penumbra is the apparent resolution of a mystery associated with the Evershed effect. The most plausible mechanism for the Evershed flow is that of siphon flows along individual magnetic flux tubes (Meyer & Schmidt 1968; Thomas 1988; Degenhardt 1989; Montesinos & Thomas 1997 and references therein). The siphon flow is driven along an arched magnetic flux tube by the drop in gas pressure between the two footpoints of the arch. In this configuration, there is a downflow of gas into the Sun at the downstream footpoint where the flux tube dives back below the solar surface. Theoretical models suggested that the downstream footpoint should in many cases be very near the outer penumbral boundary, but until recently these downflows had not been observed. The observations of Börner & Kneer (1992), Rimmele (1995), and Stanchfield, Lites, & Thomas (1997) presented increasing evidence for the downflowing extension of the siphon flows.

The most convincing evidence, however, comes from the recent observations of Westendorp Plaza et al. (1997). Their method uses an inversion technique based on response functions that give the sensitivity of the Stokes profiles to variations in physical quantities, making it possible to determine the height variation of these quantities. Thus, they were able to construct three-dimensional maps of both the vector magnetic field and the Doppler velocity in the penumbra, using data obtained with the Advanced Stokes Polarimeter (ASP) at the VTT at Sacramento Peak. They showed that much of the outward Evershed flow in the penumbra dives back down below the surface near the outer edge of the penumbra. This is an excellent example of how a combination of high angular and spectral resolution can yield important details of solar phenomena.

3.3. Granulation and the excitation of solar p-modes

There is now renewed interest in high-resolution observations of the solar granulation because of the intense magnetic flux tubes in the intergranular lanes and because of the highly localized sources of excitation of the solar p-modes, which seem to lie in the intergranular lanes.

Helioseismology, based on observations of the 5-minute p-mode oscillations, is now one of our most important tools for studying the internal structure and dynamics of the Sun. Although the nature of the resonant p-modes themselves has been well understood for some time, it is only recently that we have developed an understanding of the mechanism that excites these modes. We have progressed from the idea that the p-modes are excited stochastically by turbulent

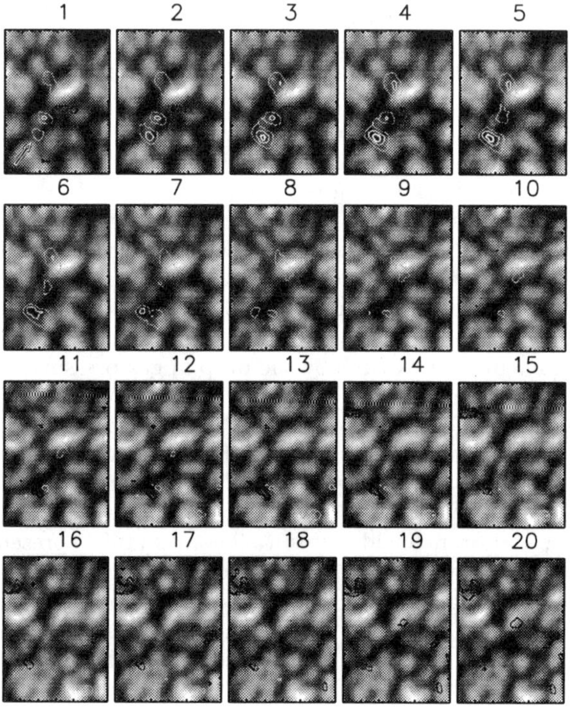

Figure 3. A time sequence of images, at 30s intervals, showing the evolution of an acoustic event occurring in a dark intergranular lane. Superimposed on these broadband images of the granulation are contours of upward (white contours) and downward (black contours) acoustic flux, with contour levels (0.6, 1.0, 1.4, -0.2, -0.5)$\times 10^8$ ergs cm^{-2} s^{-1} (From Rimmele et al. 1995, Fig. 1.)

convection just beneath the photosphere (Goldreich & Keeley 1997; Goldreich & Kumar 1988) to the idea that most of this excitation is due to random acoustic events occurring very close to the solar surface (Brown 1991; Kumar & Lu 1991), perhaps as little as 200 km deep (Goode, Gough & Kosovichev 1992).

For some time it was assumed that the convective excitation was due primarily to the deceleration of the updrafts in granules, acting as pistons to drive the oscillations. But then numerical simulations of granulation (Nordlund 1985; Rast & Toomre 1993) showed the presence of strong, narrow, supersonic downflows in the intergranular lanes. These downflows, which are much stronger than the broader, slower upflows, generate a lot of acoustic noise. It was suggested that this noise is the principal exciter of the p-modes.

These ideas prompted observational searches for the photospheric signatures of the localized acoustic events that excite the p-modes (Goode, Gough & Kosovichev 1992; Brown et al. 1992; Restaino, Stebbins & Goode 1993). The recent high-resolution observations of Rimmele et al. (1995) and Goode et al. (1998) have succeeded in identifying these acoustic events and showing that they occur preferentially in the dark intergranular lanes. The observations were made by Rimmele with the VTT at Sacramento Peak using a very narrow filter (20 mÅ passband), consisting of the Universal Birefringent Filter (UBF) coupled with a Fabry-Perot interferometer, which allowed him to scan the full line profile of the magnetically insensitive Fe I 543.4 nm line. He was thus able to measure the Doppler shift as a function of depth in the line, which corresponds to velocity as a function of altitude over the range of heights of formation of the line. These observations would not have been possible without the use of the correlation tracker.

Fig. 3 shows a typical example of the acoustic events observed by Rimmele et al. (1995). In this time sequence of images of the white-light granulations pattern, the acoustic event is shown by means of contours of upward and downward acoustic flux. The acoustic event appears in the first frame as an upward acoustic flux in a dark intergranular lane. The upward acoustic flux intensifies and reaches its peak value in about two minutes (in frame 4), then fades away over the next 5 minutes (by frame 10), and then is replaced by a downward acoustic flux at the same location which grows (frames 14-18) and then fades away(frames 15-20). Typically, at the site of an acoustic event, the continuum intensity darkens over several minutes leading up to the event, forming an intergranular lane, and then abruptly darkens further just before the acoustic event. The more powerful acoustic events are accompanied by greater darkening, suggesting that the sites of the most rapid radiative cooling at the surface produce the most acoustic noise.

Goode et al. (1998) distinguished the seismic events from motions associated with the resonant p-modes themselves by observing the phase of the velocity as a function of height: the seismic events have a consistent pattern of an upward propagating wave followed by a reflected, downward propagating wave some 4 to 5 minutes later. The p-mode oscillations, on the other hand, are standing waves with only very small phase changes with height due to dissipation.

These acoustic events that apparently excite the solar p-modes are occurring on spatial scales at and below our current limit of spatial resolution. They are obvious candidates for higher-resolution observations. We need better resolution to examine the structure and evolution of the granular convective flow and the associated acoustic events. The study of these acoustic events would benefit greatly from high-resolution observations in the infrared, in spectral lines such as Fe I 15648 which are formed very deep in the photosphere.

3.4. Magnetic effects on solar p-modes

Since the discovery of the solar 5-minute oscillations (Leighton, Noyes, & Simon 1962), it has been known that the amplitude of these oscillations is reduced in regions of strong magnetic field. More recently, it has been found that the amplitude of untrapped, higher frequency oscillations, just above the acoustic cutoff frequency, is increased in patches surrounding regions of strong magnetic field. These patches of enhanced high-frequency power are seen on a small scale in photospheric velocity (Brown et al. 1992) and on a larger scale in Ca II K intensity (Braun et al. 1992; Toner & LaBonte 1993).

Recently we have found (Stanchfield & Thomas 1998) that the suppression of the p-modes by the magnetic field takes place on very fine scales within an active region, with areas of reduced p-mode power tracing out the magnetic field on scales of a few arcseconds. Our observations were taken with the VTT at NSO/Sacramento Peak, using the ASP to obtain maps of the vector magnetic field (in Fe I 630.15 and 630.25 nm) and the UBF to obtain maps of photospheric Doppler velocity (in Fe I 557.6 nm). We computed pixel-by-pixel power spectra of velocity oscillations and then produced spatial maps of the integrated oscillatory power in two frequence bands: the p-mode band, 2.5–3.9 mHz; and the high-ν band, 5.5–7.0 mHz. (The effective acoustic cutoff frequency of the photosphere is 5.5 mHz.)

Fig. 4 shows power maps for the p-mode band and the high-ν band for a two-hour run on an active region around a small pore, along with a map of total magnetic field strength for the same region taken just before the run. Here we can see p-mode suppression on fine scales, with patches of low p-mode power generally coincident with patches of highest field strength. Indeed, many features in the magnetic field map can be matched with features in the p-mode power map. The high-ν power map also shows suppression of power coincident with areas of highest field strength, but also shows patches of enhanced power surrounding these areas of highest field strength.

An interesting aspect of these observations is the fact that, of the two patches of highest field strength, the patch to the left corresponds to a small pore visible in white light during the run, whereas the patch on the right is the site of a developing pore which did not appear dark in white light until shortly after this two-hour run. Thus, the p-mode suppression and high-ν suppression and halo signal the formation of a pore before it is detectable as a dark area in continuum intensity.

These results are still not fully understood. The highly localized, fine-scale suppression of power in the 5-minute band may reflect the influence of the magnetic field on the localized acoustic sources driving the p-modes, rather than the resonant p-modes themselves, which generally have horizontal wavelengths greater than the size of the magnetic patches. The enhanced high-ν power surrounding the strongest magnetic elements might be due to the influence of the adjacent, relatively rigid magnetized gas on the frequency distribution of power

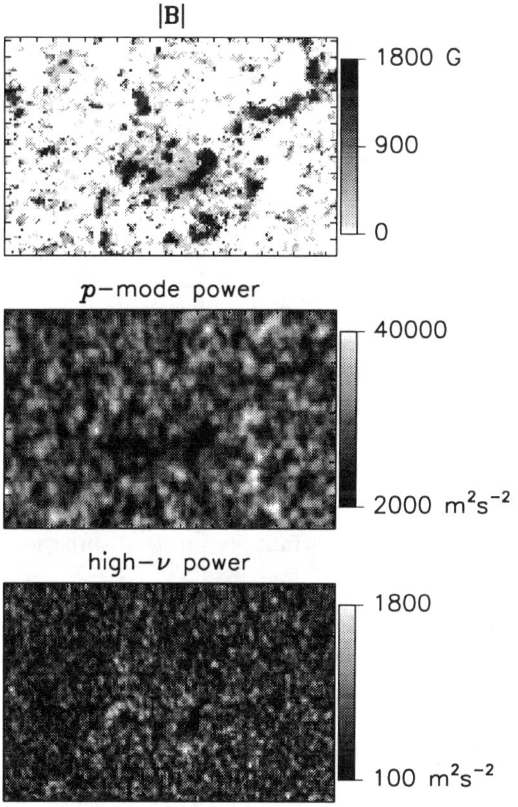

Figure 4. Spatial maps of total magnetic field strength and integrated oscillatory power in two frequency bands: the p-mode band (2.5–3.9 mHz) and the high-ν band (5.5–7.0 mHz). The magnetic field was measured shortly before the two-hour run of Doppler velocity used to produce the power maps. Note that the areas of reduced p-mode power generally coincide with areas of strong magnetic field, even on fine scales. High-ν power is also suppressed in areas of strongest magnetic field, but is enhanced in patches ("halos") adjacent to these areas. Tick marks are 5 arcseconds apart. (From Stanchfield & Thomas 1998.)

in the nearby acoustic sources. More observations with high spatial resolution are clearly needed to resolve these issues.

4. Concluding Remarks

The examples discussed above serve to demonstrate the need for high resolution in solar physics. Other equally relevant and important examples could have been chosen; for example, solar flares. Also, the examples given here have concentrated on phenomena occurring at photospheric heights; we have said little about the equally important need for high resolution in studying the structure and dynamics of the solar chromosphere and corona. Finally, I have concentrated solely on ground-based observations, which constitute the main area of emphasis in this meeting, and have said little about high-resolution observations from space. These other topics will be discussed at this meeting, and in particular we will have presentations of some of the remarkable new results on the chromosphere and corona from the TRACE mission.

I am well aware of the fact that we who are attending this Workshop hardly need convincing of the importance of high-resolution observations in solar physics. It is up to us now to convince our colleagues in other branches of astronomy and astrophysics, the funding agencies, and the general public of the importance of this work for our understanding of the Sun and for astrophysics generally. Two major studies just getting under way offer very important opportunities for us to present our case: the NAS decadal review of astronomy, and the NOAO Future Directions study. We must make the most of these opportunities, and this Workshop is an important step in that direction.

Acknowledgments. I thank Tom Berger, Christoph Keller, and Thomas Rimmele for allowing me to present some of their unpublished results and for providing Figs. 1-3.

References

Beckers, J. M., & Schröter, E. H. 1969, Solar Physics, 10, 38

Braun, D. C., Lindsey, C., Fan, Y., & Jefferies, S. M. 1992, ApJ, 392, 739

Bray, R. J., Loughhead, R. E., & Durrant, C. J. 1984, The Solar Granulation, 2nd ed. (Cambridge: Cambridge Univ. Press)

Börner, P., & Kneer, F. 1992, A&A, 259, 307

Brown, T. M. 1991, ApJ, 371, 396

Brown, T. M., Bogdan, T. J., Lites, B. W., & Thomas, J. H. 1992, ApJ, 394, L65

Degenhardt, D. 1989, A&A, 222, 297

Degenhardt, D., & Wiehr, E. 1991, A&A, 252, 821

Goldreich, P., & Keeley, D. A. 1977, ApJ, 212, 243

Goldreich, P. & Kumar, P. 1988, ApJ, 326, 462

Goode, P. R., Gough, D. O., & Kosovichev, A. 1992, ApJ, 387, 707

Goode, P. R., Strous, L. H., Rimmele, T. R., & Stebbins, R. T. 1998, ApJ, 495, L27

Harvey, J. W., & Breckenridge, J. B. 1973, ApJ, 182, L137

Keller, C. U. 1992, Nature, 359, 307

Keller, C. U., & von der Lühe, O. 1992, A&A, 261, 321

Kumar, P. and Lu, E. 1991, ApJ, 375, L35

Leighton, R. B., Noyes, R. W., & Simon, G. W. 1962, ApJ, 135, 474

Lites, B. W., Elmore, D. F., Seagraves, P., & Skumanich, A. P. 1993, ApJ, 418, 928

MacQueen, R. M., Athay, R. G., Linsky, J. L., Rosner, R., Thomas, J. H., & Title, A. M. 1985. Large aperture, ground-based solar observations. Report to the Board of Directors, AURA.

Meyer, F., & Schmidt, H. U. 1968, Zeits. Angew. Math. Mech., 48, 218

Montesinos, B. & Thomas, J. H. 1997, Nature, 390, 485

Nordlund, A 1985, Solar Physics, 100, 209

Radick, R. R. (ed.) 1993, Real Time and Post Facto Solar Image Correction (Sunspot, NM: NSO Sacramento Peak).

Rast, M. P., & Toomre, J. 1993, ApJ, 419, 224,240

Restaino, S. R., Stebbins, R. T., & Goode, P. R. 1993, ApJ, 408, L57

Rimmele, T. R. 1995, ApJ, 445, 511

Rimmele, T. R., Goode, P. R., Harold, E., & Stebbins, R. T. 1995, ApJ, 444, L119

Stanchfield, D. C. H., II, & Thomas, J. H. 1998, preprint

Stanchfield, D. C. H., II, Thomas, J. H., & Lites, B. W. 1997, ApJ, 477, 485

Thomas, J. H. 1988, ApJ, 333, 407

Thomas, J. H., & Weiss, N. O. 1992, in Sunspots: Theory and Observations, ed. J. H. Thomas & N. O. Weiss (Dordrecht: Kluwer), 3

Title, A. M., Frank, Z. A., Shine, R. A., Tarbell, T. D., Topka, K. P., Scharmer, G. B., & Schmidt, W. 1993, ApJ, 403, 780

Toner, C. G., & LaBonte, B. J. 1993. ApJ, 415, 847

Westendorp Plaza, C., del Toro Iniesta, J. C., Ruiz Cobo, B., Martínez Pillet, Lites, B. W., & Skumanich, A. 1997, Nature, 389, 47

Group Discussion

Goode : Could the excess power above the acoustic cut-off over small fields be due to mode conversion of the absorbed p-mode power? After all, the absorbed p-mode power is much greater than the excess power above the cut-off.

Thomas : Mode conversion by some nonlinear process is a possible explanation of the excess high-frequency power surrounding strong magnetic regions. Another possibility is that the presence of the strong field acts as a wall for the acoustic sources just outside the field, somehow enhancing the production of acoustic power. Theoretical models of both of these processes are needed.

Simon : Why does the penumbra end where it does and why is this penumbra-photosphere boundary so sharp? Your figures suggest that the magnetic field is approximately horizontal at this boundary. Are these local co-spatial?

Thomas : Most of the magnetic field at the outer penumbral undary is at least slightly inclined to the horizontal, rising to form the magnetic canopy outside the sunspot. In this case, the sharpness of the penumbral boundary seen in white light is due to the expanding flux tubes changing abruptly, from opaque to transparent. Some of the flux tubes dive back below the solar surface near the penumbral boundary. It is also important to note that the penumbral boundary is irregular in time, with dark extensions moving in and out radially.

Sanchez-Almeida: Isn't it too optimistic, your statement that half of the quiet-Sun magnetic concentrations are spatially resolved? In other words, does this statement refer to very strong magnetic flux concentrations?

Thomas: Yes, my statement that half the flux tubes have not yet been spatially resolved is based on Christoph Keller's study of intense flux tubes in the quiet network, and this only applied to those flux tubes.

Ayres : What about the Parker Committee?

Thomas: The Parker Committee report has not yet been made public. However, from what I have learned privately, their recommendations will be quite supportive of a new high-resolution telescope for the NSO.

High Resolution Solar Physics: Theory, Observations, and Techniques
ASP Conference Series, Vol. 183, 1999
T.R. Rimmele, K. S. Balasubramaniam, R.R. Radick, eds.

Small-scale Magnetic Flux Tubes in the Photosphere: A Simulation Perspective

Oskar Steiner

Kiepenheuer-Institut für Sonnenphysik, Schöneckstrasse 6
D-79104 Freiburg, Germany

Abstract. We propose that the observed fragmentation of magnetic elements is due to their inherent liability to the interchange instability. The convective collapse and numerical simulations of the formation of intense magnetic flux sheets, including spectral signatures of the formation process, are discussed. We suggest that the mean downflow measured in magnetic elements that are due to a fraction of an ensemble of magnetic elements which are in the formation phase at any given time. It is shown that observed Stokes V profiles of magnetic elements are actually an average of spatially strongly varying profiles with a delicate balance of positive and negative contributions to the area asymmetry.

1. Evolution and Interchange Instability of Network Flux-Tubes

Magnetic elements, when observed at disk center, are often best visible in "G-band filtergrams", which are typically obtained with a wide-band filter of 12Å FWHM bandpass centered around 4305Å. In this wavelength region, the solar spectrum shows vibration-rotation bands of the CH-molecule, which probably act to enhance the contrast of magnetic elements. First observations of the network in this wavelength window, named G-band by Fraunhofer, have been carried out by Muller (1985) at the Pic du Midi observatory in the French Pyrénées. More recently, this observational technique has been used at the Swedish vacuum tower telescope at La Palma, where time series of up to 4 h under excellent seeing conditions have been obtained (Berger *et al.*, 1995, 1998; Berger & Title, 1996). The myriad of bright grains or "crinkles" that are visible in these images are known as G-band bright points and they can be taken as good proxies for magnetic elements according to Berger (1996).

Time series of G-band filtergrams (Berger & Title, 1996) reveal a strongly dynamic evolution of magnetic elements. They are located in intergranular lanes and at vertices of granules and move along intergranular lanes with velocities up to 5 $\mathrm{km\,s^{-1}}$. Frequently, bright points split into smaller ones, while the formation of larger points by coalescence of two or more fine points is also observed. The lifetime of a single bright point, *i.e.*, the time span from its first appearance as an isolated object to its fragmentation, coalescence with other bright points, disappearance (possibly due to cancellation with opposite polarity elements), or fading, is of the order of the granular life time of 6–8 min.

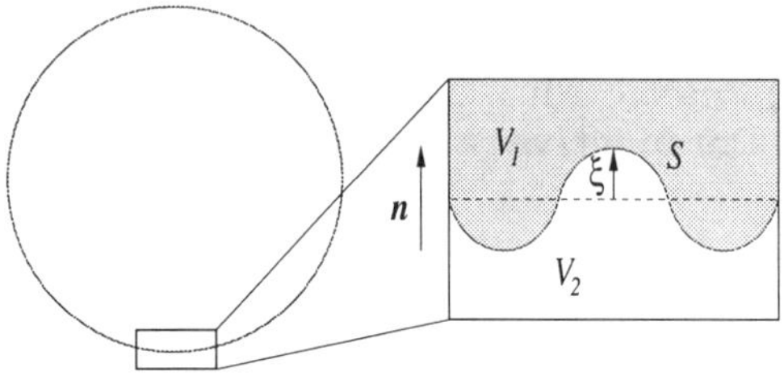

Figure 1. Cross section of a cylindrical flux tube (left) and an enlarged portion of the surface with interchange of magnetic with non-magnetic plasma ("fluting") taking place.

One ubiquitous process that seems to act over the whole range of point sizes is the fragmentation of bright points. Correspondingly, the number of points increases with decreasing diameter down to the resolution limit of the G-band images of about 0.3 arcsec. It is observed that the fragmentation of bright points is often the result of a perturbation by surrounding granules, leading to a deformation of the bright point prior to its breakup. In the following we explain that this observation suggests that the fragmentation of magnetic elements is due to their liability to the interchange instability, an inherent property of flux tubes, first noticed by Parker (1975) and Piddington (1975).
If the surface of a cylindrical straight flux tube (Fig. 1) is perturbed in such a way that the volumes occupied by the magnetic and non-magnetic plasma remain the same, then the total energy of the configuration does not change, meaning that no effort is needed to deform the flux-tube surface. In a gravitationally stratified medium, the situation is somewhat more complicated. In summary it can be said that vertically oriented axisymmetric flux tubes in a stratified atmosphere are stable against interchange instability only if they expand strongly enough with height, so that buoyancy can stabilize the tube (Meyer, Schmidt, & Weiss 1977). This is the case for sunspots or pores but *not* for tubes with a magnetic flux smaller than about 10^{20} Mx, a limit well above the typical flux of magnetic elements of 5×10^{17} Mx.

The instability can be remedied if the magnetic field is twisted, in which case the toroidal field component acts like a "surface tension" inhibiting the interchange from occurring, or if the flux tube is surrounded by a whirl flow in the height range where the interchange instability occurs as was first noticed by Schüssler (1984). Bünte, Steiner, & Pizzo (1993b) found the necessary velocity to be 2–4 $\mathrm{km\,s^{-1}}$ on the tube surface, depending on the flux-tube diameter, the field strength, and the adopted atmosphere within and outside the flux tube. It is likely that a bathtub effect develops at vertices of granules and leads to the necessary whirl flow, but it will probably be disturbed on the granular time scale of about 6 min. Observation of this whirl flow is quite difficult since spectroscopy on a scale of $0.1''$, or local correlation tracking on subarcsecond scales (Simon & Weiss, 1997) is needed.

Bünte (1993) shows that in many cases the interchange instability of flux tubes is only severe in a limited height range around $\tau_c = 1$ so that single

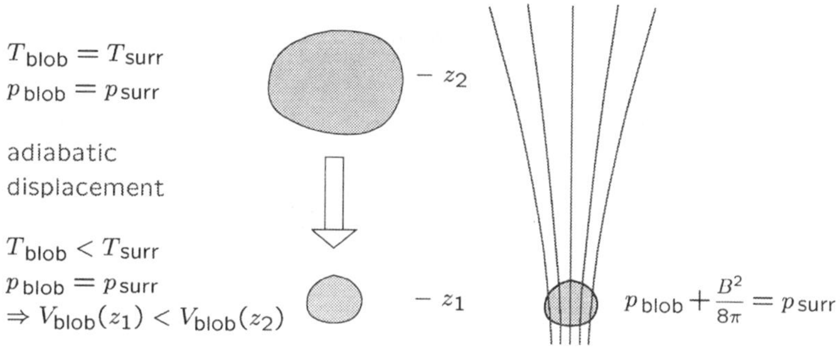

Figure 2. Superadiabatic effect. As a gas ball moves adiabatically downwards in a superadiabatically stratified atmosphere its volume shrinks (left) and the magnetic field becomes concentrated (right).

magnetic elements at the solar surface are again bundled up into a larger, stable flux tube sufficiently far below and above the surface.

2. The Formation of Photospheric Magnetic Flux Tubes

In the preceding section we have shown that magnetic flux tubes are subject to continuous fragmentation, presumably due the their inherent liability to the interchange instability – yet we know that over 90% of the magnetic flux in quiet Sun regions is concentrated into intense magnetic flux tubes with a field strength of 1 – 2 kG. Consequently, there must be an efficient process that reassembles weak field flux-tubes and/or continuous weak magnetic field into intense magnetic flux tubes. It is well known that due to the "frozen-in condition" of magnetic field in the photospheric solar plasma, the field is swept to the boundaries of supergranular cells and to the intergranular lanes by the local horizontal flow (e.g., Parker, 1963; Weiss, 1966; Galloway, Proctor, & Weiss 1977).

Subsequently, the magnetic field is further intensified by the *superadiabatic process* (Parker, 1978). Assuming that the downflowing gas inside the nascent magnetic flux tube to behave like a thermal insulator (because the magnetic field suppresses convective heat transfer) its temperature increases nearly adiabatically with depth. The important point is that the upper 3 Mm of the convection zone is significantly superadiabatic. This situation leads to a growing temperature deficit of the descending gas within the tube with respect to the surroundings, which implies a strong compression of the gas including the frozen-in magnetic field (Fig. 2). At and above the surface ($\tau_{500} = 1$), the horizontal inflow of plasma into the forming tube is suppressed by the growing field and the downflow is not replenished from the top because of the vanishingly small density there (Schüssler, 1990). Thus, the flux tube becomes almost evacuated throughout the photosphere and the field strength is limited only by the gas pressure of the surrounding atmosphere leading to a strength of up to 1800 G at the continuum optical depth unity of the surrounding atmosphere.

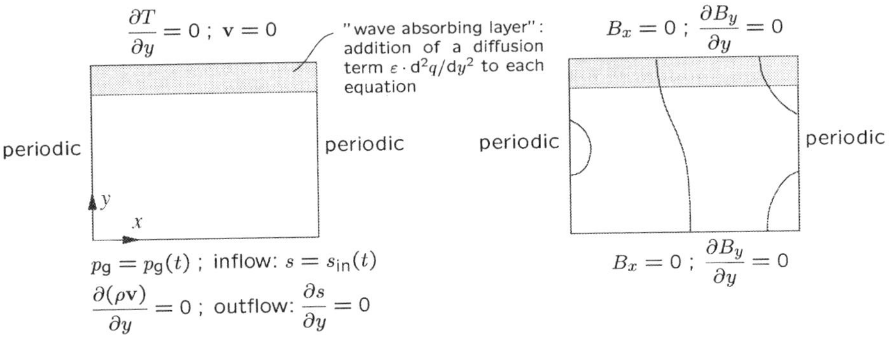

Figure 3. Boundary conditions on the MHD-equations for the simulation of magnetic flux-sheet formation. The left panel shows the conditions for the hydrodynamic variables, the right panel those for the magnetic field.

Based on a linear analysis of the vertical motion in a slender flux tube (Roberts and Webb, 1978), Spruit and Zweibel (1979) derived a critical value $\beta_{\rm crit} = p_g/(B^2_{\rm crit}/8\pi) \approx 1.8$, above which limit convective instability ensues. Webb and Roberts (1978) as well as Unno and Ando (1979) came to a similar conclusion. Spruit (1979), using a quasi static approach, computed the final state to which this instability, which he called *convective collapse*, leads to. Nonlinear numerical solutions of the same problem have been obtained by Venkatakrishnan (1983) and Hasan (1984). Hasan(1985) and Venkatakrishnan (1985) have taken non-adiabatic effects into account. A linear stability analysis including non-adiabatic effects was carried out by Webb and Roberts (1980), Venkatakrishnan (1986), and Hasan (1986). Further improvement for treating radiative exchange was done by Massaglia, Bodo, & Rossi (1989), while Takeuchi (1993, 1995) stresses the importance of the adopted boundary conditions.

All these calculations rely on the thin-flux-tube approximation, use for the initial state a constant $\beta_{\rm plasma}$ along the flux-tube axis, and neglect the perturbation of the surrounding plasma by the flux tube as well as any interaction between the surrounding convective flow and the flux tube.

In order to overcome these limitations, while at the same time keeping the number of free parameters as small as possible, one has to solve the system of the magnetohydrodynamic (MHD) equations in a computational domain that includes the atmospheric layers most critical to the driving of the convective collapse instability and wide enough to encompass a sufficiently large environment of the flux tube. Thus, at least two spatial dimensions – depth and one horizontal direction – are needed. Since radiative cooling plays a decisive role in the intensification process and since one wishes to include the photospheric part in a manner as realistically as possible (e.g., for the subsequent derivation of synthetic spectral lines), the inclusion of radiation transfer is an indispensable requirement.

A simulation of the magnetic field intensification that fulfills these requirements was carried out by Grossmann-Doerth, Schüssler, & Steiner 1998), where they used boundary conditions as summarized in Fig. 3.

The "wave absorbing layer" damps outgoing waves in order to keep reflection at the closed top to a minimum and to suppress inwards propagating waves. The gas pressure is kept constant along the bottom boundary. Its value is adjusted in time so that the total mass within the computational domain remains constant. The actual equations being solved, their finite volume representation, and the employed numerical technique are discussed in Steiner, Knolker, & Schüssler (1994).

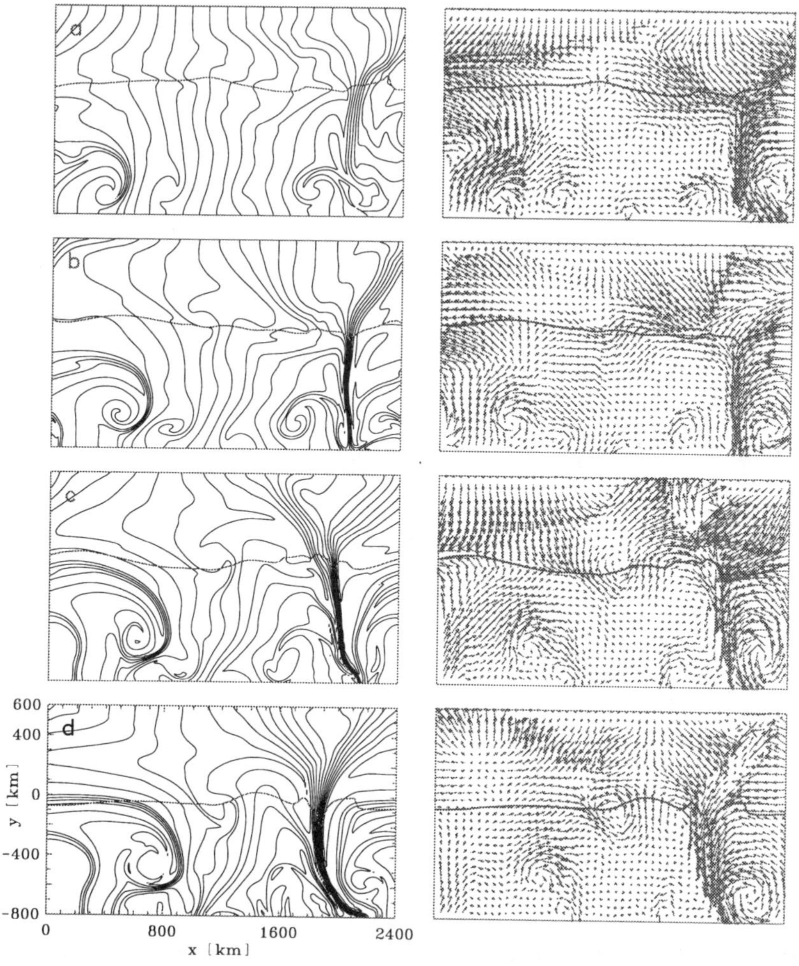

Figure 4. Formation of a magnetic flux sheet. Magnetic field lines (left) and velocity (right) at times 60, 115, 175, and 230 s after imposing an initially homogeneous magnetic field of 100 G.

The simulation is started with a plane-parallel stratification according to a standard model of the solar atmosphere. The convectively unstable stratification is perturbed by a small velocity field, which subsequently grows until the system relaxes to instationary convective motion. Then a homogeneous vertical magnetic field is imposed onto the entire computational domain and the ensuing interaction between the magnetic field and the convective flow can be followed.

Fig. 4 shows a simulation with an imposed vertical homogeneous field of a strength of 100 G, other examples are discussed in Grossmann-Doerth *et al.* (1998). The field becomes rapidly concentrated at the location of the preexisting downflow and a flux sheet with a field strength of about 2000 G at $\tau_{500} = 1$ forms within a time of about 3 min. Due to the relatively small magnetic flux density of the initial field the resulting flux-sheet width remains small (about 100 km at $\tau_{500} = 1$) and the backreaction of the field onto the convective motion is relatively feeble, at least in the initial phase. The downflow, which is further strengthened by radiative cooling of the plasma caught within the growing magnetic field, persists down to the open bottom of the computational domain and thus the flux sheet extends through the bottom boundary. Due to lack of material replenishing the downflow from the sides, the flux sheet becomes partially evacuated as can be seen from the depression of the $\tau_{500} = 1$ surface at that location of strong field concentration. This result changes drastically when the initial field is stronger (Grossmann-Doerth, Schüssler, & Steiner 1998). For example, if the same convective pattern as the one used in the simulation giving rise to Fig. 4 is superimposed with a 400 G field, a rebound of the downflow ensues followed by an upward traveling strong shock front.

3. The Spectral Signature of Flux-Sheet Formation

Fig. 5 shows a time sequence of Stokes V profiles of Fe I 525.02 nm corresponding to the simulation shown in Fig. 4. The profiles have been computed by integration of the radiation transfer equation for polarized light along many vertical lines of sight distributed over the entire computational domain, followed by an averaging over the domain width. This averaging procedure mimics a spatially unresolved observation, but since the vertical magnetic field is strongly concentrated over only the small width of the flux-sheet of about 100 km, the resulting Stokes V is essentially an average over just this small region. Note that a similar statement would *not* apply to Stokes I.
Due to the descending gas within and in the vicinity of the magnetic flux concentration during the formation process, all Stokes V profiles of the sequence shown in Fig. 5 are redshifted (with the exception of the first one at time $t = 0$). This can be best seen by examination of the zero-crossing shift, the amount by which the zero-crossing wave-length of the Stokes V profile is shifted with respect to the rest wavelength. How does this redshift of the zero-crossing wavelength of synthetic Stokes V profiles compare to observations of magnetic elements?

3.1. A brief history of the Stokes V zero-crossing shift

Early measurements of the zero-crossing shift of Stokes V profiles in plage and network regions resulted in downflow velocities of up to 2.2 $\mathrm{km\,s^{-1}}$, e.g., Giovanelli & Ramsay (1971), Harvey, Livingston, & Slaughter (1972), Stenflo (1973), Harvey (1977), Wiehr (1985), Scholiers & Wiehr (1985).

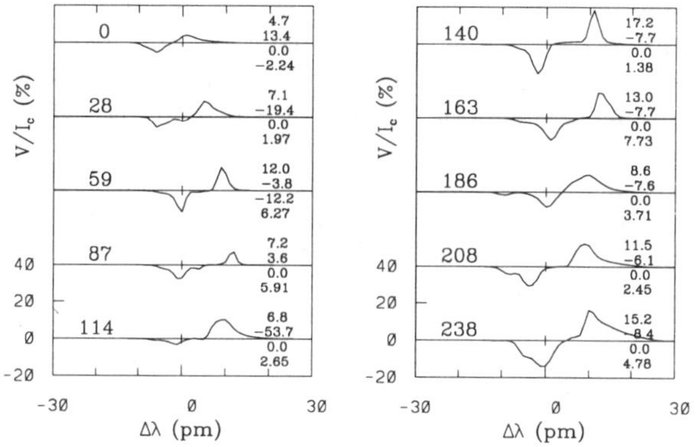

Figure 5. Temporal sequence of Stokes V of FeI 525.02 nm corresponding to the simulation of Fig. 4. In each panel are given on the left times in s, on the right (top to bottom) mean amplitude in % of I_c, amplitude and area asymmetry in %, and zero crossing shift in pm.

In 1986, Solanki and Stenflo noted that a spurious zero-crossing shift is measured as a consequence of insufficient spectral resolution. Spectral smearing of an asymmetric Stokes V profile with the blue lobe being larger than the red one (which is usually the case for plage and network magnetic elements at disk center) results in a redshifted zero-crossing wavelength. Thus, the measured zero-crossing shifts were considered to be artifacts.

Subsequently, high spectral resolution measurements, mainly carried out with the Fourier transform spectrometer that was commissioned at Kitt Peak in the late 1970s (Brault, 1979), resulted in no significant zero-crossing velocities, *e.g.*, Stenflo and Harvey (1985), Solanki (1986), Stenflo, Solanki, & Harvey (1987), Wiehr (1987), Muglach and Solanki (1992), Fleck (1991), Fleck & Deubner (1991) (see also Solanki, 1993, for a review).

"Epur si muove" – And yet it does move! Grossmann-Doerth, Keller, & Schüssler (1996) found a mean downflow velocity in network and plage magnetic elements of 0.8 $\mathrm{km\,s^{-1}}$ using the Zürich CCD-demodulator polarimeter (ZIMPOL) (Povel, Aebersold, & Stenflo 1990) at the McMath-Pierce telescope. With a very different instrument, namely with HAO's Advanced Stokes Polarimeter (ASP) (Elmore et al. 1992) at the Richard B. Dunn Telescope, Sigwarth, Balasubramaniam, & Knölker (1998) obtain a mean downflow velocity of 0.73 $\mathrm{km\,s^{-1}}$ in magnetic elements of the quiet Sun, and 0.48 $\mathrm{km\,s^{-1}}$ in active region plages. These mean velocities are an average over 24,000 magnetic elements with Stokes V signals $> 0.20\%$ of the continuum intensity in the case of Grossmann-Doerth *et al.*, and 35,000 network elements and 93,000 active region elements, respectively, with Stokes V amplitudes $> 0.15\%$ in the case of Sigwarth *et al.* If the averaging would be restricted to the weakest network elements, say with amplitudes below 1%, the zero-crossing redshift would be rather in the order of 1 $\mathrm{km\,s^{-1}}$. Evidence for a mean downflow in magnetic elements was also found by Martínez Pillet, Lites, & Skumanich (1997). Note that single weak magnetic elements show a scatter of up to ± 5 $\mathrm{km\,s^{-1}}$ around the mean value.

The difference with respect to older measurements resulting in no significant zero-crossing shift is due to the high sensitivity of the ASP and ZIMPOL polarimeters and to the better spatial resolution compared to the FTS data, which permits the sampling of magnetic elements with small magnetic flux. These elements became swamped by the elements of largest flux within a resolution patch in earlier measurements. Magnetic elements with a Stokes V signal ($\geq 2\%$) show negligible zero-crossing shifts also in these latest measurements.

Because the weakest Stokes V profiles that enter the statistics of Grossmann-Doerth *et al.* (1996) and Sigwarth *et al.* (1998) are strongly affected by noise, one might argue that noise could introduce once again a spurious zero-crossing redshift similar to the effect of insufficient spectral resolution. However, Sigwarth *et al.* show that noisy profiles with *negative* area asymmetry show a redshifted zero-crossing as well, contrary to what one would expect if this argument were true. This strongly suggests that the measured mean downflow in magnetic elements is real.

3.2. Convective collapse contribution to the zero-crossing redshift

The scatter plots of the zero-crossing shift versus Stokes V amplitude shown in the works of Grossmann-Doerth *et al.* (1996) and Sigwarth *et al.* (1998) represent magnetic elements in various stages of their evolution. Many of them may be in a quiescent phase of their life, giving rise to "normal" Stokes profiles with a small zero-crossing shift. Others correspond to magnetic elements in the stage of dissolution, coalescence with other elements, cancellation with opposite polarity field, or in the stage of formation through convective collapse. We expect the latter to produce a net zero-crossing redshift according to Fig. 5 and to contribute to an ensemble-averaged positive zero-crossing redshift.

4. Stokes V Area and Amplitude Asymmetries

Stokes V profile asymmetries provide a useful diagnostic for the gas motion within magnetic elements and in their close surroundings. Solanki & Pahlke (1988) have given a easy-to-remember formula relating the sign of the area asymmetry to the gradients of the absolute magnetic field strength and the velocity along a line of sight:

$$\frac{\mathrm{d}|B|}{\mathrm{d}\tau} \cdot \frac{\mathrm{d}v(\tau)}{\mathrm{d}\tau} = \begin{cases} < 0 & \Rightarrow \delta A > 0 \\ > 0 & \Rightarrow \delta A < 0\,, \end{cases} \tag{1}$$

where $\delta A = (A_\mathrm{b} - A_\mathrm{r})/(A_\mathrm{b} + A_\mathrm{r})$, with A_b and A_r being the area of the red and blue lobe of the Stokes V profile, respectively. A rigorous derivation of Eq. 1 can be found in Sanchez Almeida, Collados, & Del Torro (1989) and an extension to transverse fields in Sanchez Almeida (1998).
Fig. 6 shows the result when applying Eq. 1 to a line of sight along the flux-tube axis, when an accelerated downflow or upflow is present. An accelerating downflow occurs during the convective collapse while an accelerating upflow ensues in the "inverse collapse" phase that may take place after the rebound of a downflow.

These two situations yield a negative area asymmetry, which is at variance with the observed positive values of δA. However, lines of sight that pass through

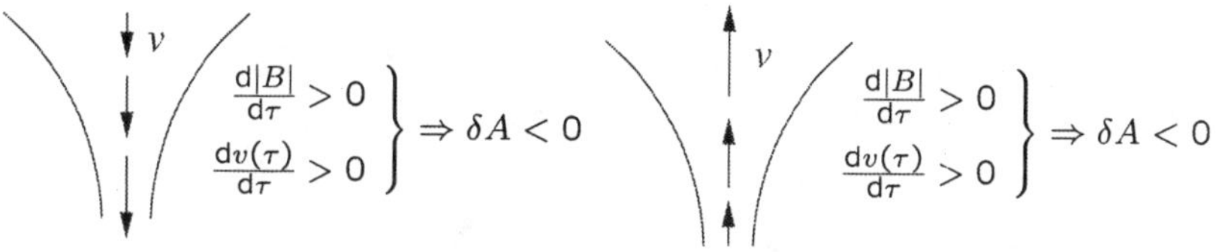

Figure 6. Sign of the Stokes V area asymmetry formed along the flux-tube axis with an accelerating downflow (left) or upflow (right).

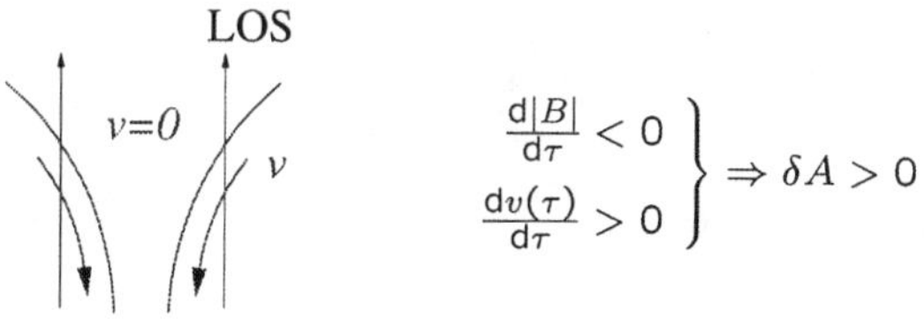

Figure 7. Sign of the Stokes V area asymmetry formed along lines of sight that pass through the expanding flux-tube boundary.

the expanding boundary of the flux tube (the onset of the magnetic canopy) as sketched in Fig. 7 contribute with a positive δA (Grossmann-Doerth, Schüssler, & Solanki 1988,1989; Bünte, Solanki, & Steiner 1993a).
This is due to the downflow in the close vicinity of the flux tube, which is driven by the radiative heat loss of the surrounding gas into the flux tube (Deinzer *et al.*, 1984). Thus, at the location where the line of sight crosses the flux-tube boundary, a strong negative field strength gradient co-locates with a strong positive velocity gradient, producing a positive area asymmetry. This positive contribution seems most of the time to outweigh the negative contribution of the tube core.

How realistic are the sketches shown in Figs. 6 and 7? We have no means to answer this question observationally, because it addresses fine-structure of magnetic elements that is far beyond the spatial resolution capabilities of present time solar telescopes, but we may attempt an answer by examination of snapshots from a simulation. Fig. 8 shows an example of a snapshot from a simulation on the interaction of granular flow with a magnetic flux sheet (Steiner *et al.*, 1998 [panel b of Fig. 1]), where the temperature, the velocity field, and representative magnetic lines of force are indicated. The horizontally running curve shows the surface of optical depth $\tau_{500} = 1$. In contrast to the simulation of Fig. 4, this one was started with a mature strong magnetic flux concentration and the bottom boundary was closed to the fluid so that the flux sheet remained intact over a simulation period of 40 min.

When computing the Stokes V profiles that emanate from vertical lines of sight distributed over the horizontal interval between 600 and 1380 km of Fig. 8, we find that positive area asymmetries result in the two regions labeled with "+", while the lines of sight in the middle region labeled with "−" contribute negative area asymmetries, exactly as expected from the simple sketches of Figs. 6 and 7, based upon Eq. 1. Half of the snapshots to which series Fig. 8 belongs show a similar behaviour with respect to δA, while in the other half the flow within the

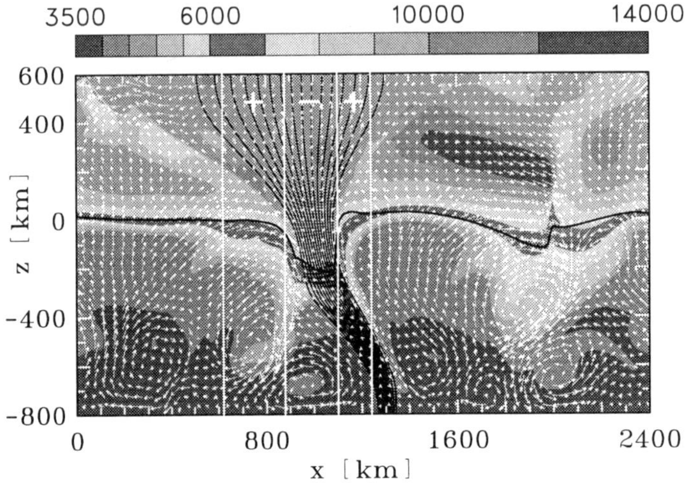

Figure 8. Simulation snapshot (see text for explanations).

flux sheet does not correspond to either case of Fig. 6 and therefore produces a positive δA too.

We conclude that the measured, spatially unresolved Stokes V area asymmetry of network elements results from a delicate balance of negative and positive contributions due to the internal structure of the elements. The simulations also suggest that the amplitude asymmetry is not subject to a similar balance and hence shows larger and almost always positive values. Bellot Rubio, Ruiz Cobo, & Collados (1997) were led to a similar conclusion by the inversion of observed Stokes V profiles based upon a multi-component model. However, the uniqueness of this inversion was put into question by Frutiger & Solanki (1998).

A statistical analysis similar to those made by Grossmann-Doerth *et al.* (1996) and Sigwarth et al. (1998) has been carried out with synthetic profiles using the above-mentioned simulation series and considering the individual snapshots to be independent single magnetic elements (Grossmann-Doerth, 1997). Mean values for δA, δa, and $v_{\rm zc}$ of 14.9%, 0.4%, and 0.74 km s^{-}1, respectively are obtained. This compares very well with the observational results of Sigwarth *et al.* (1998) for magnetic elements of quiet Sun regions (15%, 6%, and 0.73) with the exception of the area asymmetry, which is rendered too small by the simulation. One reason for this might be that the boundary of real flux tubes is very "sharp", *i.e.*, the transition from internal to external atmosphere is confined within about 10 km. This leads to strong gradients in magnetic field strength and velocity and therefore to an enhanced area asymmetry compared to the case of the simulation where the flux sheet boundary is more diffuse. Another reason one could think of is that for geometric reasons more lines of sight pass through the canopy region in the case of a flux tube compared to a sheet. However this effect is canceled by a reduced expansion rate of the tube versus the sheet for the same geometric reason.

5. Conclusions

1. The fragmentation of G-band bright points is probably due to the interchange instability of small-scale magnetic flux tubes.

2. Magnetic flux-tubes in the stage of formation produce red-shifted Stokes profiles. Consequently, they must be an important contributor to the observed mean redshift of the Stokes V zero-crossing of magnetic elements in the network and in plage regions.

3. The net Stokes V area asymmetry of a single magnetic element is the result of a delicate balance between a positive contribution by lines of sight that pass through the expanding tube surface (canopy) and a negative contribution by lines of sight through the core of the flux tube.

4. The simulation of the quiescent stage of flux tubes reproduce the observational results except of a discrepancy in the area asymmetry of Stokes V that is possibly due to the boundary layer of flux-sheets being smaller than the resolution limit of the simulation of 10–20 km.

Acknowledgments. I would like to thank M. Schüssler for critically reading the manuscript and U. Grossmann-Doerth for providing Fig. 5.

References

Bellot Rubio, L.R., Ruiz Cobo, B., & Collados, M., 1997, A&A, 478, L45
Berger, T.E., 1996, Thesis, Stanford University
Berger, T.E. & Title, A.M., 1996, ApJ, 463, 365
Berger, T.E., Löfdahl, M.G., Shine, R.S., & Title, A.M., 1998, ApJ, 495, 973
Berger, T.E., Schrijver, C.J., Shine, R.A., Tarbell, T.D. Title, A.M., & Scharmer, G., 1995, ApJ, 454, 531
Brault, J.W., 1979, Oss. Mem. Arcetri, 106, 33
Bünte, M., 1993, A&A, 276, 236
Bünte, M., Solanki, S.K. & Steiner, O., 1993a, A&A, 268, 736
Bünte, M., Steiner, O., & Pizzo, V.J., 1993b, A&A, 268, 299
Deinzer, W., Hensler, G., Schüssler, & M., Weisshaar, E.: 1984, A&A, 139, 435
Elmore, D.F., Lites, B.W., Tomczyk, S., Skumanich, A.P., *et al.*, 1992, SPIE (Society of Photo-Optical Instrumentation Engineers), 1746, 22
Fleck, B., 1991, Rev. Mod. Astron., 4, 90
Fleck ,B. & Deubner, F.-L., 1991 in Mechanisms of Chromospheric and Coronal Heating, (ed. P. Ulmschneider, E.R. Priest, & R. Rosner), Springer-Verlag, p. 19
Frutiger, C. & Solanki, S.K., 1998, A&A, 336, L65
Galloway, D.J., Proctor, M.R.E., & Weiss, N.O., 1977, Nature, 266, 686
Giovanelli, R.G. & Ramsay, J.V., 1971, in proc. IAU Symp. 43, p. 293
Grossmann-Doerth: 1997, *The Betaslab Brief*, Kiepenheuer-Institut, internal report
Grossmann-Doerth, U., Keller, C.U., & Schüssler, M., 1996, A&A, 315, 610
Grossmann-Doerth, U., Schüssler, M., & Solanki, S.K., 1988, A&A, 206, L37
Grossmann-Doerth, U., Schüssler, M., & Solanki, S.K., 1989, A&A, 221, 338

Grossmann-Doerth, U., Schüssler, M., & Steiner, O., 1998, A&A, 337, 928
Harvey, J.W., 1977, in Highlights of Astronomy, (ed. E.A. Müller), Vol 4, Part II, p. 223
Harvey, J.W., Livingston, W., & Slaughter, C., 1972, in Line formation in the presence of magnetic fields, High Altitude Observatory, NCAR, Boulder, CO, p. 227
Hasan, S.S., 1983, in proc. IAU Symp. 102, p. 73
Hasan, S.S., 1984, ApJ, 285, 851
Hasan, S.S., 1985, A&A, 143, 39
Hasan, S.S., 1986, MNRAS, 219, 357
Martínez Pillet, V., Lites, B.W., & Skumanich, A., 1997, ApJ, 474, 810
Massaglia, S., Bodo, G., & Rossi, P., 1989, A&A, 209, 399
Meyer, F., Schmidt, H.U., & Weiss, N.O., 1977, MNRAS, 179, 741
Muglach, K. & Solanki, S.K., 1992, A&A, 263, 301
Muller, R.A., 1985, Sol. Phys., 100, 237
Parker, E.N., 1963, ApJ, 138, 552
Parker, E.N., 1975, Solar Phys., 40, 291
Parker, E.N., 1978, ApJ, 221, 368
Piddington, J.H., 1975, Astrophys. Space Sci., 34, 347
Povel, H., Aebersold, H., & Stenflo, J.O., 1990, Applied Optics, 29, 1186
Roberts, B. & Webb, A.R., 1978, Sol. Phys., 56, 5
Sanchez Almeida, J., 1998, ApJ, 497, 967
Sanchez Almeida, J., Collados, M., & Del Torro Inesta, J.C., 1989, A&A, 222, 311
Scholiers, W. & Wiehr, E., 1985, Sol. Phys., 99, 349
Schüssler, M., 1984, A&A, 140, 453
Schüssler, M., 1990, in proc. IAU Symp. 138, p. 161
Sigwarth, M., Balasubramaniam, K.S., & Knölker, M., 1998, A&A, in prep.
Simon, G.W. & Weiss, N.O., 1997, ApJ, 489, 960
Solanki, S.K., 1986, A&A, 168, 311
Solanki, S.K., 1993, Space Sci. Rev., 63, 1
Solanki, S.K. & Pahlke, K.D., 1988, A&A, 201, 143
Solanki, S.K. & Stenflo, J.O., 1986, A&A, 170, 120
Spruit, H.C., 1979, Sol. Phys., 61, 363
Spruit, H.C. & Zweibel, E.G., 1979, Sol. Phys., 56, 5
Steiner, O., Grossmann-Doerth, U., Knölker, M., & Schüssler, M., 1998, ApJ495, 468
Steiner, O., Knölker, M., & Schüssler, M., 1994, in Solar Magnetism (eds. R.J. Rutten & C.J. Schrijver, NATO ASI Ser. C-433), Kluwer, Dordrecht, 181
Stenflo, J.O., 1973, Sol. Phys., 32, 41
Stenflo, J.O. & Harvey, J.W., 1985, Sol. Phys., 95, 99
Stenflo, J.O., Solanki, S.K., & Harvey, J.W., 1987, A&A, 171, 305
Takeuchi, A., 1993, Publ. Astron. Soc. Japan, 45, 811
Takeuchi, A., 1995, Publ. Astron. Soc. Japan, 47, 331
Unno, W.& Ando, H., 1979, Geophys. Astrophys. Fluid Dynamics, 12, 107
Venkatakrishnan, P., 1983, J. Astrophys. Astr., 4, 135
Venkatakrishnan, P., 1985, J. Astrophys. Astr., 6, 21
Venkatakrishnan, P., 1986, Nature, 322, 156
Webb, A.R. & Roberts, B., 1978, Sol. Phys., 59, 249
Webb, A.R. & Roberts, B., 1980, Sol. Phys., 68, 87

Weiss, N.O., 1966, Proc. Roy. Soc. Lon., A 293, 310
Wiehr, E., 1985, A&A, 149, 217
Wiehr, E., 1987, in Cool Stars, Stellar Systems, and the Sun V, Lecture Notes in Physics Vol. 291, Springer-Verlag, p. 54

High Resolution Solar Physics: Theory, Observations, and Techniques
ASP Conference Series, Vol. 183, 1999
T.R. Rimmele, K. S. Balasubramaniam, R.R. Radick, eds.

Dynamics of Magnetic Elements in the Photosphere and the Formation of Spicules

A. A. van Ballegooijen and P. Nisenson

Harvard-Smithsonian Center for Astrophysics, Cambridge, MA 02138, USA

Abstract. We consider the proper motions of photospheric magnetic elements, and the effects of these motions on flows at larger heights. We summarize the results from recent analyses of high resolution G-band data obtained at the Swedish Vacuum Solar Telescope (SVST) on La Palma. The G-band images show small bright features which are known to correspond to kilogauss magnetic fields. We measure the motions of these bright points and find that the autocorrelation time of the bright point velocity is about 100 s. From the observed continuum intensity images, we derive a model of the granulation flow velocity as function of time and position on the Sun. We use this flow model to simulate the horizontal motions of photospheric magnetic elements, assuming that the elements are passively advected by the granulation flow. We find that this passive advection model is in reasonable agreement with the observed spatial distribution of the G-band bright points. Finally, we use potential-field modeling to extrapolate the magnetic and velocity fields from the photosphere, where the flux tubes are well separated, to a height of 1500 km in the chromosphere, where the flux tubes fill the available volume due to the spreading out of the field lines. We find that strong shear flows occur near separatrix surfaces where neighboring flux tubes slide past each other in the chromosphere. We propose that spicules are formed in these separatrix layers.

1. Introduction

Observations of the Sun with high spatial resolution show network bright points and filigree (Dunn & Zirker 1973), which are small bright features located within intergranular lanes (see review by Muller 1994). The bright points are seen in the wings of strong spectral lines such as Hα and Ca II H & K, in spectral lines formed in the photosphere, and even at continuum wavelengths. The widths of these structures are less than 200 km, at the limit of resolution of ground-based solar telescopes. The bright points are associated with strong magnetic fields (e.g., Simon et al. 1988; Title et al. 1989, 1992; Keller 1992) and correspond to magnetic flux tubes with kilogauss field strength (see review by Solanki 1993). The filigree are chains of bright points which fill the intergranular lanes (von der Lühe 1994). The dynamical behaviour of bright points has been studied by a number of authors (e.g., Muller et al. 1994; Strous 1994; Berger & Title 1996;

Berger et al. 1998). These studies show that the bright points are continually pushed around by the granulation flow. The bright points continually split and merge, with a mean time between splitting events of few hundred seconds (Berger et al. 1998).

In the next section, we summarize results from recent analyses of the dynamics of photospheric magnetic elements and their interaction with the solar granulation flow, as derived from observations obtained at the Swedish Vacuum Solar Telescope (SVST) on La Palma. As part of this study, we developed a model of the chromospheric mass flows induced by photospheric flux tube motions. The model predict enhanced flow velocities near separatrix surfaces where different flux tubes slide past each other in the chromosphere. In §3 we propose that Hα spicules originate in such separatrix layers.

2. Dynamics of Magnetic Elements

We recently analyzed SVST observations of photospheric bright points and presented a model of the advection of magnetic flux tubes by the granulation flow (van Ballegooijen et al. 1998, hereafter paper I). The data were collected on 1995 October 5 between 10:57 and 12:08 UT (see Löfdahl 1996; Löfdahl et al. 1998; Berger et al. 1998), and were made with two CCD cameras, one for observing the G-band (4305 Å), the other for observing the nearby continuum (4686 Å). Both cameras were equipped with phase-diversity beam splitters, and were operated with a cadence of 23.5 s. The field of view analyzed in paper I was a 29×29 arcsec area near solar disk center containing an enhanced network region.

In the first part of our study, we measured the proper motion of G-band bright points and derived an estimate of the autocorrelation time of the bright-point velocity. Following Berger et al. (1998), we construct "magnetic" difference images ($I_{\rm magn} = I_{4305} - I_{4686}$) which suppress the solar granulation pattern and show the bright points with unprecedented clarity. Comparisons with the continuum images show that the bright points form linear structures ("filigree") which fill the intergranular lanes and sometimes surround the granules on all sides. We study the dynamics of these linear structures using an object tracking technique which assigns multiple tracer particles to each structure and allows for the fact that the filigree continually split and merge. The velocities of the tracer particles are a measure of the transverse velocities of the filigree structures. We find that these velocities have a correlation time of about 100 s, significantly less than the typical lifetime of a granule cell. This suggests that magnetic elements in intergranular lanes are buffeted by the converging flows from neighboring granules and are shaken by these flow on short time scales.

In the second part of the study, we use the observed granulation intensity patterns to derive a 2D model of the granulation flow velocity. Let $v_x(x, y, t)$ and $v_y(x, y, t)$ denote the horizontal velocity at a fixed geometrical depth of 40 km below the visible photosphere, where x and y are horizontal coordinates on the Sun, and t is the time. The granulation flow model is based on the assumption that the vertical vorticity inside the granules is small, so that the velocity can be approximated as the gradient of a flow potential, $\phi(x, y, t)$. We use a simple 2D dynamical model to link $\phi(x, y, t)$ to the horizontal temperature variations, and

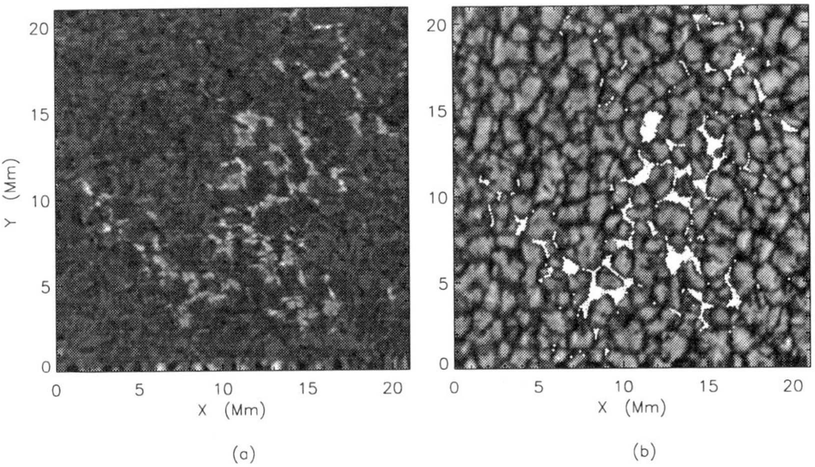

Figure 1. Magnetic patterns derived from high resolution observations at the Swedish Vacuum Solar Telecope at La Palma, on 1995 October 5, 12:08 UT. (a) Difference image obtained by subtracting simultaneous G-band and continuum images. (b) Results from a 2D simulation of the dynamics of magnetic elements, superposed on the observed granulation intensity image.

hence to the emergent intensity of the granulation, $I_{\rm conv}(x, y, t)$. We calibrate the velocity amplitude by comparing our results with the more sophisticated 3D simulations of the solar granulation by Stein & Nordlund (1994).

We use this 2D granulation flow model to simulate the motions of magnetic elements in the observed enhanced network region. Each element is represented by one or more rigid, finite-size "corks" which are passively advected by the granulation flow. The rigidity of the corks simulates the fact that the magnetic pressure in the photospheric flux tubes cannot exceed the external gas pressure. The initial positions of the corks coincide with the observed initial positions of G-band bright points and filigree structures. We simulate the motions of the corks over time, and compare their spatial distribution with the observed distribution of the G-band bright points. We find that the corks collect in irregular shaped clusters located at the vertices of the intergranular lanes. Linear structures extend from these clusters out into the lanes, and have shapes which are similar to the observed filigree. These linear extensions of the clusters are due to the narrow width of the velocity lanes (about 100 km), which prevent the magnetic elements from all collecting in one round cluster at the vertex point. Although the model does not reproduce the exact location of each and every bright point, it does reproduce the general character of the observed patterns. Fig.1 shows a comparison of observed and predicted bright point patterns for the last frame of the time series.

In the third part of our study, we use potential-field modeling to predict the magnetic and velocity fields in the chromosphere above the observed network region. The photospheric flux tubes expand with height, and above some level in the chromosphere the magnetic field fills the available volume. Since we are interested only in the magnetic structure in the photosphere and chromosphere, we assume that all flux tubes have the same magnetic polarity (sign of the vertical field). We approximate the magnetic field inside the flux tubes as a potential field from a collection of discrete sources located 160 km below the photosphere. Each source is the sum of a monopole and a (vertically oriented) dipole, and the (x, y) positions of the sources are taken from the cork simulation discussed above. To obtain tube-like structures, we trace magnetic field lines downward from a height of 1500 km in the chromosphere, i.e., well above the "merging" height of the flux tubes. We assume that the magnetic field is embedded in a massless ideal plasma; this allows us to use the ideal MHD equations to compute the velocity field at a height of 1500 km in the chromosphere (for details, see paper I).

We find that the chromospheric velocity field is extremely non-uniform: the largest velocities occur near *separatrix surfaces* where neighboring flux tubes interact. These separatrices are present wherever two neighboring photospheric flux tubes come into contact with each other in the chromosphere. The velocity enhancements are due to the presence of magnetic nulls (points where $\mathbf{B} = 0$) at locations where the magnetic fields from neighboring flux tubes cancel each other. The nulls are located in the upper photosphere or low chromosphere, and lie on the magnetic canopies between neighboring flux tubes. The magnetic field lines which pass near such a null point fan out more or less radially along the separatrix surface above the null point. As neighboring flux tubes slide past each other, the field lines sweep rapidly along the separatrix surface, with velocities that are oppositely directed on the two sides of the separatrix plane.

3. Formation of Spicules

In the potential-field model described in paper I, the inertia of the chromospheric plasma is neglected, and therefore the velocity increases without bound as we approach the separatrix surface. However, in reality the field lines near the separatrix are mass loaded, and the magnetic torques near the null point are insufficient to rapidly sweep the field lines past the separatrix plane. As a result, the field lines near the separatrix lag behind the motion of the main body of the flux tubes; there is a *sepatarix layer* in which the magnetic field becomes strongly sheared. Potential-field models cannot accurately describe the dynamics of plasma and magnetic field in this separatrix layer. Based on estimates of the magnetic torques involved, we estimate that the region of strong magnetic shear may be several hundred kilometers thick. The Lorentz forces in this layer drive strong shear flows, and likely also cause upward acceleration of plasma within the separatrix layer.

We propose that quiet-Sun Hα spicules may have their origin in these separatrix layers (see Fig.2). Observations of limb spicules show that the inclination angles of spicules have a broad angular distribution which is centered on the vertical direction (e.g., Mosher & Pope 1977; Heristchi & Mouradian 1992). This

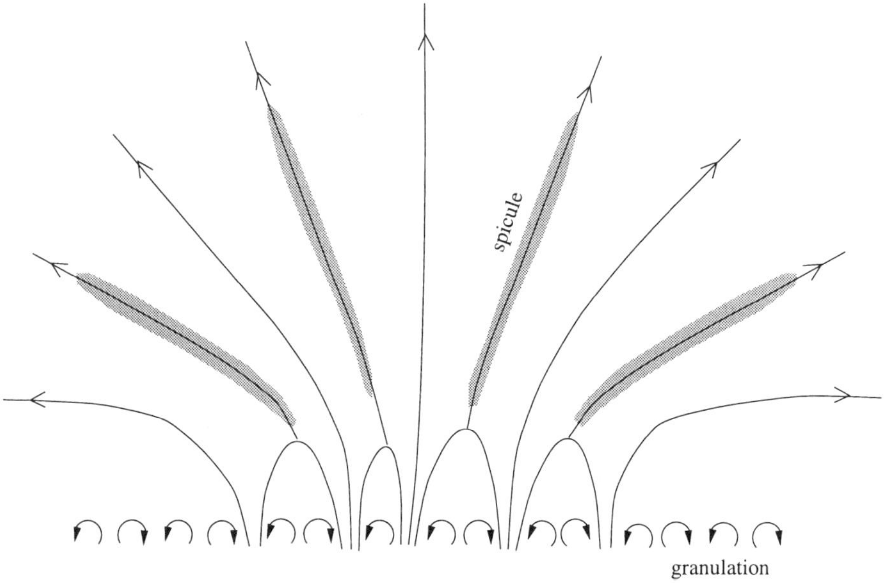

Figure 2. Model of the structure of a magnetic network element on the quiet Sun. Note that the spicules originate in the separatrix layers between neighboring flux tubes.

suggests that spicules originate from within unipolar network elements, and do not require the presence of opposite polarity fields in the photosphere. High-resolution magnetograph observations show that each supergranular network element (size $\sim$ 3000 km) consists of a number of discrete flux tubes, all with the same polarity of the vertical field. These flux tubes merge at some height in the upper photosphere or low chromosphere, hence the chromosphere above a network element likely contains several separatrix layers where the different flux tubes interact. Each flux tube is differently affected by the granulation flow, causing the tubes to move horizontally relative to each other. As discussed above, this results in shear flows and upward acceleration of plasmas in the separatrix layers in the chromosphere. We propose that this leads to the formation of spicules. A key feature of this model is that it does not require reconnection of opposite polarity magnetic fields in the photosphere. To test this model against spicule observations, more detailed MHD models of separatrix layers are needed.

4. Future Work

We are planning further observations of G-band bright points and their interaction with the solar granulation flow. The goal is to detect small-amplitude, short-period motions of magnetic elements inside intergranulation lanes, and thus put observational constraints on the amplitudes of transverse MHD waves in photospheric magnetic flux tubes. To detect transverse motions with periods less than 100 seconds and velocity amplitudes of order 1 km/s, it is necessary to measure bright point positions (relative to the granulation) once every 10 seconds with very high positional accuracy. We hope to achieve this accuracy by recording many images within each 10-second interval. This will provide us with many samples of the seeing-induced image distortions, and will hopefully allow us to better correct for these distortions. In a later phase, we will combine such measurements with Hα observations in order to test the model of spicules described in §3.

Acknowledgments. This work was supported by the National Science Foundation (grant ATM-9811523), and by the National Aeronautics and Space Administration (grant NAG5-7146).

References

Berger, T.E., & Title, A.M. 1996, ApJ, 463, 365

Berger, T.E., Löfdahl, M.G., Shine, R.A., & Title, A.M. 1998, ApJ, 495, 973

Dunn, R.B., & Zirker, J.B. 1973, Solar Phys., 33, 281

Heristchi, D., & Mouradian, Z. 1992, Solar Phys., 142, 21

Keller, C.U. 1992, Nature 359, 307

Löfdahl, M.G. 1996, Ph.D. thesis, Stockholm University

Löfdahl, M.G., Berger, T.E., Shine, R.A., & Title, A.M. 1998, ApJ, 495, 965

Mosher, J.M., & Pope, T.P. 1977, Solar Phys., 53, 375

Muller, R. 1994, in *R.J. Rutten & C.J. Schrijver (eds.), Solar Surface Magnetism* (Kluwer Acad. Publ.), p. 55

Muller, R., Roudier, Th., Vigneau, J., & Auffret, H. 1994, A&A, 283, 232

Simon, W.G., Title, A.M., Topka, K.P., Tarbell, T.D., Shine, R.A., Ferguson, S.H., Zirin, H., & the SOUP team. 1988, ApJ, 327, 964

Solanki, S. 1993, Space Science Reviews, 63, 1

Stein, R.F., & Nordlund, Å. 1994, in *D. Rabin, J. Jeffries & C. Lindsey (eds.), Infrared Solar Physics*, IAU Symp. 154 (Dordrecht: Kluwer), p. 225

Strous, L. 1994, in *R.J. Rutten & C.J. Schrijver (eds.), Solar Surface Magnetism* (Kluwer Acad. Publ.), p. 73

Title, A.M., Tarbell, T.D., Topka, K.P., Ferguson, S.H., & Shine, R.A. 1989, ApJ, 336, 475

Title, A.M., Topka, K.P., Tarbell, T.D., Schmidt, W., Balke, C., & Scharmer, G. 1992, ApJ, 393, 782

van Ballegooijen, A.A., Nisenson, P., Noyes, R.W., Löfdahl, M.G., Stein, R.F., Nordlund, Å., & Krishnakumar, V. 1998, ApJ, 509, in press

von der Lühe, O. 1994, A&A, 281, 889

High Resolution Solar Physics: Theory, Observations, and Techniques
ASP Conference Series, Vol. 183, 1999
T.R. Rimmele, K. S. Balasubramaniam, R.R. Radick, eds.

High Resolution Observations of the Dynamics of Magnetic Elements

M. Sigwarth

Kiepenheuer-Institut für Sonnenphysik, D-79104 Freiburg i.Br., Germany and National Solar Observatory, Sacramento Peak, Sunspot, NM 88349, USA

K.S. Balasubramaniam

National Solar Observatory, Sacramento Peak, Sunspot, NM 88349, USA

M. Knölker

High Altitude Observatory, Boulder, CO 80307, USA

Abstract. The HAO/NSO Advanced Stokes Polarimeter was used to obtain time series of high-resolution and low-noise Stokes-V and I spectra of small scale magnetic elements in the quiet sun. The time series provide the opportunity to investigate the evolution of individual magnetic elements and their interaction with convective motions and oscillations. It reveals that the 5-minute oscillation have a minor influence on the flux tube dynamics, whereas the magnetic elements interact strongly with convective motions. The formation of magnetic elements is found to coincide with accelerated downflows within the magnetic field exceeding the downflows in the non magnetic environment by a factor of 5 to 7. During this formation process, in general, positive amplitude and area asymmetries of the V-profiles decrease and finally become negative.
These observational results qualitatively agree with results from 2D numerical MHD simulation and support the formation of kG flux tubes by convective collapse.

1. Introduction

A major fraction of the magnetic flux seen in circular polarization outside of sunspots reside in small-scale (< 1 arcsec) strong (> 0.1 T) flux tubes, the so called *magnetic elements* (Livingston & Harvey 1969; Howard & Stenflo 1972; Frazier & Stenflo 1972; see also review by Solanki 1993). During the last 3 decades there has been much debate on the presence of a net downflow inside small scale flux tubes (Stokes-V *zero-crossing redshift*) . Recently high resolution spectro-polarimetric measurements (Grossmann-Doerth et al. 1996; Martínez Pillet et al. 1997; Sigwarth et al. 1999) revealed, that flux tubes have a highly dynamic behavior, i.e. upflows and downflows of up to several kms^{-1} are present within these magnetic elements. Averaged over a large number

of individual magnetic elements, there remains a downflow of 0.5 to 1 kms^{-1} and a positive area and amplitude asymmetry of the Stokes-V profiles. These observations are in accordance with results from 2-dimensional numerical MHD-simulations which offer a strong interaction of flux tubes with convective motions (granulation) in the photosphere leading to significant flows within flux tubes as well as at the flux-tube interface to the non-magnetic surroundings (Steiner et al. 1998).
To identify the individual processes leading to the observed flows and asymmetries (e.g. during the formation and decay of flux tubes or the stochastic interaction of flux tubes with the granulation), spatial and temporal high-resolution time series of individual magnetic elements are required. The observations presented here are an initial attempt to reach this goal.

2. Observation and data analysis

The data were obtained in July 1997 at the *Richard B. Dunn Solar Telescope* (DST) of the National Solar Observatory at Sacramento Peak, Sunspot, New Mexico. The HAO/NSO Advanced Stokes Polarimeter (ASP; Elmore et al. 1992) was used to measure simultaneously the full Stokes vectors for the FeI lines 630.15 and 630.25 nm. By spatially scanning the spectrograph slit, spatial maps for the Stokes spectra were obtained. A correlation tracker was used to reduce image motion and therefore to enhance the spatial resolution. In order to achieve a high temporal resolution only small scans of 10 steps (corresponding to 3.75 arcsec in the scanning direction) were performed. The time resolution between the consecutive scans was 1 min. Along the entrance slit of the spectrograph a spatial range of 75 arcsec was covered. Three time series of 20 min duration each (i.e. 20 consecutive scans with the ASP for each series) and high spatial resolution of 0.8 to 1 arcsec were achieved, covering an area of magnetically quiet sun at disc center.
From Stokes-I spectra the continuum intensity and Doppler-shifts of the FeI 630.25 linecore were determined. For the Stokes-V analysis we used only profiles with amplitudes exceeding 0.15% V/I_c. With an RMS noise of $< 3 \cdot 10^{-4} I_c$ for Stokes-V in the continuum (after low-pass filtering), this corresponds to a minimal signal-to-noise ratio of ≈ 5. V-profiles of unusual profile shape (e.g. profiles with significantly only one wing or symmetric profiles) were excluded. For all selected profiles the blue and red wing amplitude (a_b, a_r) and area (A_b, A_r) and the zero-crossing wavelength (λ_{zc}) were determined. The profile asymmetries expressed in terms of the relative area asymmetry $\delta A = (|A_b| - |A_r|)/(|A_b| + |A_r|)$ and amplitude asymmetry $\delta a = (|a_b| - |a_r|)/(|a_b| + |a_r|)$ were calculated. Using terrestrial O_2 lines, the Doppler-shifts of the Stokes-I linecore and Stokes-V zero-crossing were transformed into absolute calibrated velocities. The accuracy of the velocity-calibration is $\pm$ 170 ms^{-1} (see also Sigwarth et al. 1999 and Sigwarth 1999 for details on the V-analysis and the absolute velocity calibration). Spatial maps from the first time series for the continuum intensity, the Doppler velocity (from Stokes-I), the V amplitude and the V zero-crossing velocity and amplitude-asymmetry are shown in Fig. 1. Clearly visible are the evolution of the granulation (e.g. an *exploding granule* during the second half of the series) and the 5-minute oscillations.

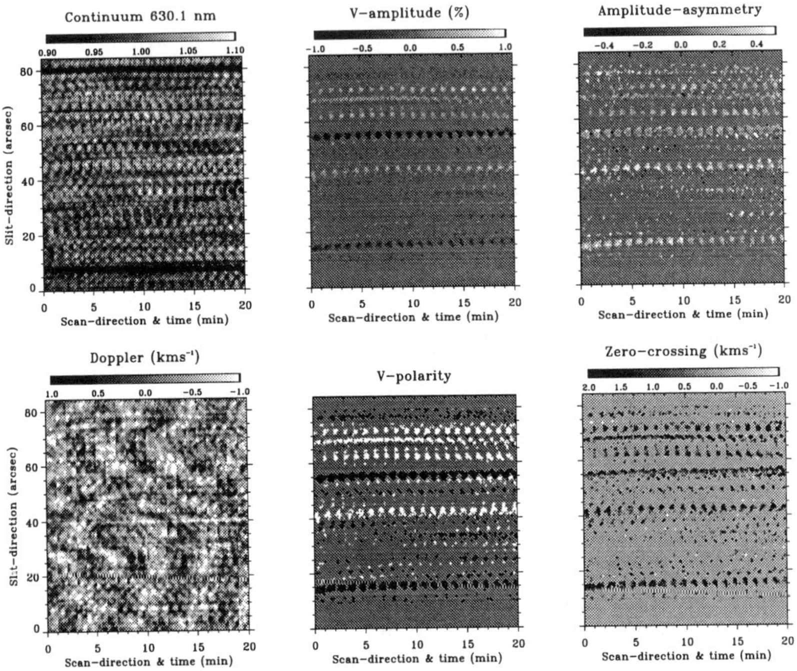

Figure 1. Spatial maps for the first time series. Each map consist of 20 consecutive spatial scans over a region of 3.75x74 arcsec^2. The upper row shows the continuum sprectroheliogramms at 630.1 nm, the maps for the Stokes-V amplitude and the V-profile amplitude-asymmetry. In the second row the Doppler-velocity from Stokes-I line core shifts, the polarity of all significant V-signals and the Stokes-V zero-crossing velocity are shown. All informations were obtained from FeI 630.25. The bars on top of the plots denote the plot scaling.

From these data, the spatial and dynamic evolution of several larger network-elements as well as small-scale magnetic structures can be investigated.

3. Results

On average over all Stokes-V profiles a significant redshift of the zero-crossing of 1.2 kms^{-1} and a positive amplitude- and area profile asymmetry are found (see Tab. 1). Corresponding to the results from a statistical analysis of an intensive set of V profiles from network, intra-network and active region magnetic fields (Sigwarth et al. 1999), we also find an increasing variation of velocities and profile-asymmetries towards smaller V amplitudes, i.e. for smaller fill fraction within the resolution element, in the data presented here. By comparing individual magnetic features of different fill fraction, this behavior can be demonstrated directly. Fig. 2 shows a detail slice from the first series, including a network element or magnetic knot (negative polarity/black scaled) and a smaller magnetic

# spectra	δa (%)	δA (%)	v_{zc} (kms^{-1})
17534	13.5	22.2	1.25

Table 1. Total number of analyzed profiles and mean values of the amplitude δa and area δA asymmetry and the zero-crossing velocity (v_{zc}) from FeI 630.25nm.

element (positive polarity/white scaled). The knot was visible over the whole observing period of $\approx$ 2h, whereas the small element disappeared during the second series (i.e. after 90 min). For the central part of the knot where the V amplitude reaches up to 2.5%, the zero-crossing velocity and amplitude asymmetry does not change substantially, whereas at the border of the knot and for the small element, where the V amplitude varies only between 0.15 and 0.5%, these values show a significant variation with time and spatial location. The mean values over the entire magnetic knot did not vary substantially with time whereas for the small element there is a significant variation. Magnetic knots are thought to be built up from a dense cluster of small flux tubes (Beckers & Schröter, Soltau 1997). Therefrom, the *dynamic behavior* can be interpreted as an averaging effect, where in the central part of the knot the measured V-profile is averaged over a larger number of individual flux tubes within the resolution element, whereas at the border or for the small element, single or only a few individual flux tubes are considered. Also the higher fill fraction, the magnetic fields might lead to a general suppression of convective and turbulent motions, leading to a generally suppressed dynamics. The 5-minute oscillations are clearly visible in the Doppler signal, whereas they have no major influence on the zero-crossing velocity. For a detailed frequency analysis longer time series are necessary. In general, the zero-crossing velocity, i.e. the motion within the magnetic field only, exceeds the Doppler velocity, i.e. the motion (both, convective and oscillatory) in the mainly non-magnetic environment, significantly.

3.1. Formation of magnetic elements

While searching for indications on how concentrated magnetic elements are formed in the photosphere, we found two examples for such events. The detail plots for these are shown in Fig. 3. In the first example (series I) a single element is formed out of several smaller patches until the middle of the time series, which then starts to become spatially more and more concentrated. At the same time the V amplitude increases by a factor of 2. The whole process took place while a pronounce intergranular region is formed at the same location as visible from the continuum and Doppler maps. Together with the increasing amplitude, the zero-crossing velocity increases steadily to an downflow of almost 4 kms^{-1} while the amplitude asymmetry slightly decreases. For better comparison the individual values and the mean values for each single scan of the considered magnetic structure is plotted in Fig. 4. Before the formation or concentration starts at about minute 10, the asymmetries and the zero-crossing velocity varies strongly while in the second half of the series there seems to be a close relation between the strongly increasing v_{zc} and the smoother evolution

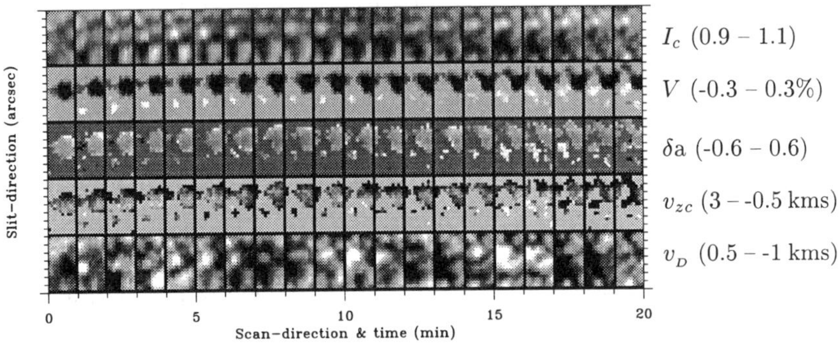

Figure 2. Detail of series I showing a magnetic knot and a small element. The first row consists of 10 consecutive spetroheliogramms in the continuum. The second row shows the Stokes-V amplitude, with negative polarity (dark scaled) for the knot and positive polarity (bright scaled) for the element. In the third row the amplitude asymmetry is displayed followed by the zero-crossing velocity (dark-downflow/bright-upflow). The last row shows the Doppler-velocity. The size of a single scan shown here correspond to 3.75x7.1 arcsec^2. The tickmarks along the slit direction correspond to 1 arcsec (in both directions). The scaling is given on the right side of each single row.

of the profile asymmetry. Again, the 5-minute oscillations seem to have almost no influence on the velocity signal from Stokes-V.

Between the first and second series there is a time gap of 70 minutes – the further evolution of the newly formed element cannot be followed. The second example (detail from the second series) shows beside a larger structure of positive polarity (white scaled) a new element of negative polarity appearing at about the middle of the series. First with a very weak V-signal which increases while the magnetic patch becomes more and more concentrated until it covers finally only one pixel which corresponds to a size of only 270x270 kms^{-1}. Again the zero-crossing velocity increase substantially from a slight upflow to a strong downflow of 2 kms^{-1} while the asymmetries become smaller and finally change to negative sign. The individual values for the parameters are also shown in Fig. 4. The newly formed element cannot be identified in the third series, which started only 2 minutes after the second series stopped. Because towards the end of the series the V amplitudes starts to decrease the element possibly starts to decay.

4. Conclusions

The analysis of the Stokes-V profile asymmetry and zero-crossing shift for individual magnetic elements with the respect to the spatial location and the temporal evolution reveal a highly dynamic character of these structures when

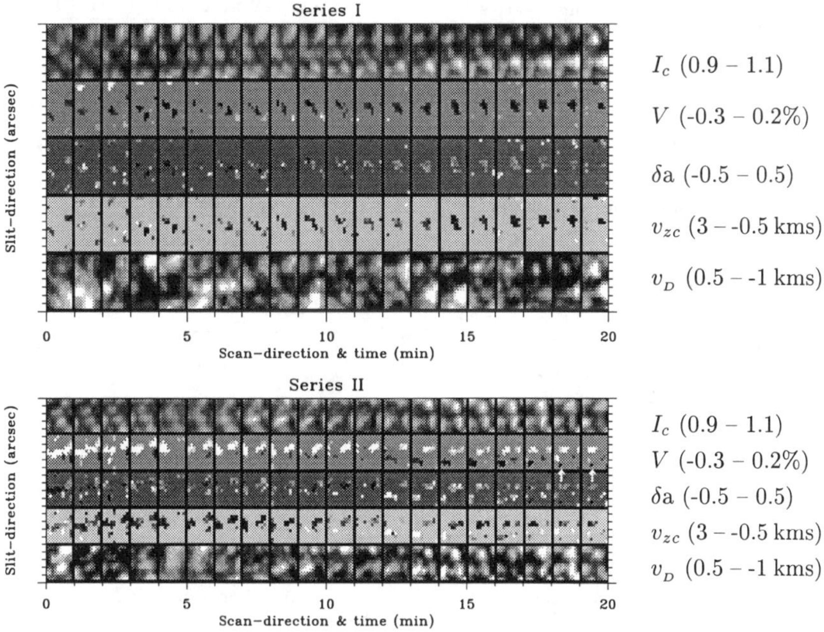

Figure 3. Formation and concentration of magnetic elements. The upper figure show a detail of series I, the lower of series II. The white arrows point towards a small magnetic element of negative polarity, which covers only a single data point. For further informations see also Fig. 2.

observed at high spatial and temporal resolution. The results confirm a prior statistical investigation, where a strong dependence of the dynamic within magnetic fields on the filling fraction and the convective motions in the non-magnetic environment was found. These results lead to the conclusion, that spatially and temporally integrated observations are not able to solve the highly dynamic behavior of individual flux tubes in the photosphere due to averaging over several individual structures and rapidly changing states.

To investigate the influence of photospheric oscillations on individual flux tubes more precisely, longer time series are necessary in order to perform a frequency analysis. So far we find that the 5-minute oscillations did not affect significantly the dynamic behavior of magnetic elements.

The two examples found for the formation of concentrated magnetic elements give strong evidence that these formation occurs due to *flux expulsion* and convective collapse as predicted since a long time from theoretical models and numerical simulations (Parker 1963; Parker 1978; Weiss 1981a; Weiss 1981a Spruit & Zweibel 1979). The fact that in both cases a primarily spatially spread out and weak V signal becomes spatially concentrated within intergranular regions gives evidence to the expulsion of the magnetic flux by convective motion. The non-stationary accelerated downflow found within the flux-tube(s) during the

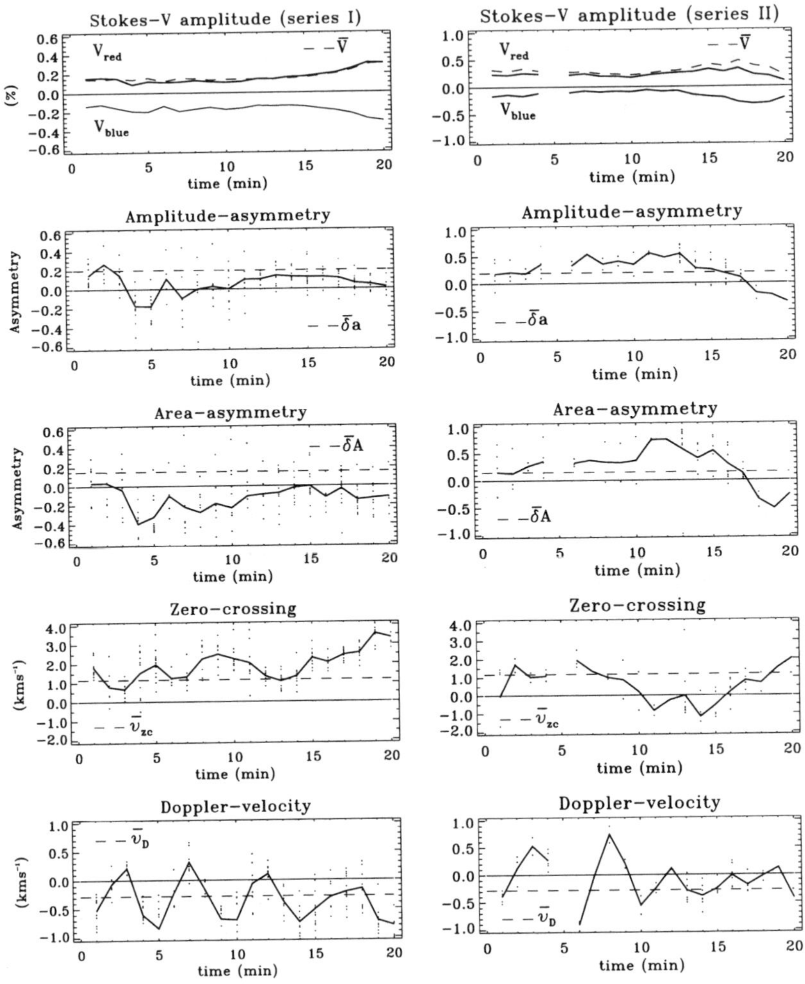

Figure 4. Temporal development of Stokes-V parameters and the Doppler-velocity during the formation and concentration of magnetic elements. The solid lines denote the mean value over the investigated magnetic element within each scan, the dots represent the value for each data point with a significant V-signal. The horizontal dashed lines denote the mean-value over all Stokes-V profiles of the observation. The negative mean Doppler velocity (blue-shifted/upflow) is due to the *convective blueshift* of the line. The left column shows the values for the example of the first series, the right for such of the second series. See also Fig. 3.

formation and concentration of the structure is in accordance with the convective collapse scenario. To confirm these, more of such events should be observed together with a precise measurements of the field strength. A higher spatial resolution will increase the magnetic filling fraction per pixel for small flux tubes and therefore allow for more precise Stoke-profile measurements and analysis. Or measurement in the IR at reasonable spatial resolution may allow for direct measurements of the Zeeman splitting.

References

Beckers J. M., Schröter E. H., SP 4, 142

Elmore D. F., Lites B. W., Tomczyk S., et al., 1992, Proc. SPIE 1746, 22

Frazier E. N, Stenflo J. O., SP 27, 330

Grossmann-Doerth U., Keller C. U., Schüssler M., A&A 315, 610

Howard R., Stenflo J. O., SP 22, 402

Livingston W. C., Harvey J. W., 1969, SP 10, 294

Martínez Pillet V., Lites B. W., Skumanich A., 1997, ApJ 474, 810

Parker E. N., 1963, ApJ 138, 552

Parker E. N., 1978, ApJ 221, 368

Sigwarth M., 1999, Thesis, Univ. Freiburg (written in German)

Sigwarth M., Balasubramaniam K. S., Knölker M., et al., 1999, submitted to A&A

Solanki S. K., 1993, Space Sci. Rev. 63, 1

Soltau D., 1997, A&A 317, 586

Spruit H. C., Zweibel E. G., 1979, SP 62, 15

Steiner O., Grossmann-Doerth U., Knölker M., et al., 1998, ApJ 495, 468

Weiss, N. O., 1981a, Journal of Fluid Mechanics 108, 247

Weiss, N. O., 1981b, Journal of Fluid Mechanics 108, 273

High Resolution Solar Physics: Theory, Observations, and Techniques
ASP Conference Series, Vol. 183, 1999
T.R. Rimmele, K. S. Balasubramaniam, R.R. Radick, eds.

Radiative Transfer On Unstructured Triangular Grids

J.H.M.J. Bruls, P. Vollmöller and M. Schüssler

Kiepenheuer-Institut für Sonnenphysik,
Schöneckstrasse 6, D-79104 Freiburg, Germany

Abstract. The basic problems involved in the design of the radiative transfer module for our new magnetohydrodynamics (MHD) code are discussed. Special attention is given to the difficulties arising from the triangular grid structure and the need to minimize the communication overhead in the code to enable it to run efficiently on distributed memory parallel computers. The computation of the radiative heating rate, required as source term in the energy equation of magnetohydrodynamics, involves several integration steps, each of which is scrutinized. In particular, the influence of the cell size, the accuracy of the angular integration of the intensity, and the accuracy of the integration of the radiative heating rate over each cell are considered. Since theoretical estimates of possible errors are in general cumbersome to obtain, simple model problems serve as a base for the accuracy estimates.

1. Introduction

As part of the development of a magnetohydrodynamics (MHD) code, initially for 2-D and eventually for 3-D, we study the problems related to the radiative transfer part of that code. The 2-D code uses triangular cells with adaptive cell refinement; initially the triangles will be regular in shape, but in a later stage the grid becomes unstructured. Since in the 2-D code all variables are implicitly translation invariant in the third dimension and since radiative transfer is essentially a 3-D process, the radiative transfer module can be the same for both codes. It therefore suffices to study the behavior and accuracy in the 2-D case.

The system of MHD equations, describing the interaction of matter and magnetic fields, will be solved by means of a Riemann solver, a finite-volume method, resulting in cell-centered values for all hydrodynamic variables. In principle these are constant on each cell and have jumps at the cell boundaries; a higher-order reconstruction method can be applied to obtain e.g. piecewise linear variation as a function of the spatial coordinates. The interaction of matter and radiation is described by the radiative heating rate, $Q_{\rm R}$, a source term in the energy equation. Here we deal with the computation of the radiative heating term, and the difficulties that arise in the specific case of a triangular grid with cell sizes that are imposed by the MHD solver. In practice, the adaptive grid refinement criteria, which are based on the gradients of temperature, T, and possibly mass density, ρ, automatically inhibit too large cell-to-cell variations of T and ρ. Since the maximum allowable cell-to-cell variations of T and ρ are still free parameters at this point, it is useful to quantify the accuracy of $Q_{\rm R}$. Even though the radiative transfer is expected to be quite sensitive to cell size effects, one may hope that the accuracy with which $Q_{\rm R}$ can be determined does not require stricter limits on the cell sizes than the ones imposed by the MHD solver in order to obtain a certain accuracy in other variables.

In order to compute $Q_{\rm R}$ we should in principle solve for all intensities $I_\nu(\vec{\Omega})$ in all directions $\vec{\Omega}$. However, since it is not feasible to solve the entire non-LTE radiative transfer and statistical equilibrium problem consistently, it does not make much sense to compute the detailed frequency dependence of I either. At best, we can divide the spectrum in a number of broad bands and compute wavelength-averaged values within each band and possibly account for scattering effects in an ad-hoc way (e.g., Nordlund 1982, Nordlund & Stein 1989, Stein & Nordlund 1998). This requires a significant amount of work that is largely related to the opacities and the frequency averaging procedure, but it does not add new aspects to the method required to solve the radiative transfer problem on a triangular grid. Therefore, in a first step we ignore the frequency dependence and scattering effects and adopt the Rosseland mean opacity, $\kappa_{\rm R}$. We keep in mind, however, that $\kappa_{\rm R}$ is specifically designed to yield the correct radiative flux in combination with the diffusion approximation, which is valid only in the optically thick parts of the atmosphere. In the optically thin parts, the Planck or intensity mean opacities would be more suitable, but there radiation plays a minor role in the energy balance anyway. In the energetically most interesting intermediate region, where the radiation takes over from the convection as the main energy transport mechanism, none of these mean opacities is really appropriate. Consequently, comparisons with frequency-dependent calculations are necessary and will be made at a later stage.

In frequency-independent form ($\kappa = \kappa_{\rm R}$) and suppressing all spatial dependences the two possible expressions for the cell-averaged radiative heating rate per unit volume are

$$Q_{\rm R}^J = \frac{4\pi}{V}\int_V \kappa\rho(J-S)\;{\rm d}V \tag{1}$$

$$\begin{aligned} Q_{\rm R}^F &= -\frac{1}{V}\int_V \nabla\cdot\vec{F}\;{\rm d}V \\ &= -\frac{1}{V}\int_{\partial V}(\vec{F}\cdot\vec{n})\;{\rm d}A. \end{aligned} \tag{2}$$

The mean intensity $J = \frac{1}{4\pi}\int I(\vec{\Omega})\;{\rm d}\vec{\Omega}$ and the radiation flux $\vec{F} = \int I(\vec{\Omega})\,\vec{\Omega}\;{\rm d}\vec{\Omega}$ both require an angular integration of the intensities.

In the next section we discuss the solution method for the radiative transfer equation, compare the merits of both definitions of $Q_{\rm R}$, and explore the accuracy of $Q_{\rm R}$ for a plane parallel quiet Sun model atmosphere (Maltby et al. 1986) with a small downward extension into the convection zone (Spruit 1977).

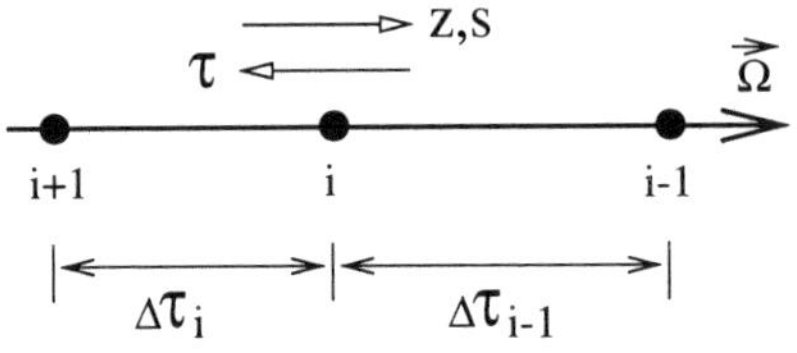

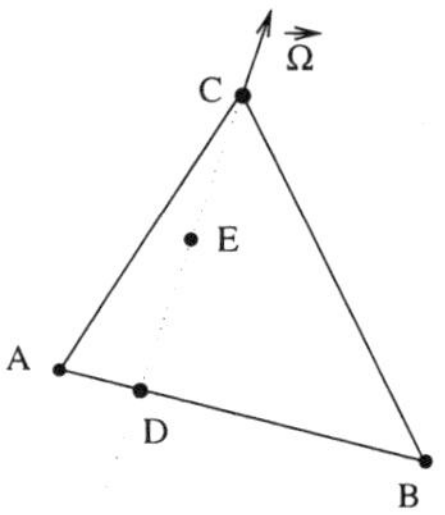

Figure 1. Definition of the geometry for the short-characteristics solution of the radiative transfer on a ray in direction $\vec{\Omega}$.

Figure 2. Radiative transfer on a single triangle. I_C can be computed once I_A and I_B are available for interpolating I_D.

2. Radiative Transfer

2.1. Numerical solution of the radiative transfer equation

On a ray in the direction $\vec{\Omega}$, the radiative transfer equation to be solved is

$$\frac{dI(\vec{\Omega})}{d\tau} = I(\vec{\Omega}) - S, \tag{3}$$

where the source function $S = \sigma T^4/\pi$ is the frequency-integrated Planck function, and the optical path length $d\tau$ corresponding to an infinitesimal interval ds along the ray is defined through $d\tau = -\kappa\rho ds$. In the present case, the source function does not contain a scattering contribution, so that the radiative transfer problem reduces to a formal solution. Given the incident intensity I_o at the inflow boundary at some optical depth τ_o, the intensity $I(\tau)$ at optical depth $\tau < \tau_o$, is given by

$$I(\tau) = I_o\, e^{-(\tau_o - \tau)} + \int_\tau^{\tau_o} S(t)\, e^{-[t-\tau]}\, dt. \tag{4}$$

The first term represents the attenuated incident intensity and the second term accounts for the radiation that is emitted between the inflow boundary and optical depth τ.

Since we are interested in minimizing the communication overhead, we require a numerical solution method that is local. The method of choice is the short-characteristics formal solver (Mihalas, Auer & Mihalas 1978, Olson & Kunasz 1987, Kunasz & Olson 1988, Kunasz & Auer 1988); it is not as sophisticated as e.g. the Feautrier solver, which automatically incorporates a scattering component in the source function at a penalty of being highly non-local. The strength of the short-characteristics method lies in its simplicity. Basically, it is the discrete form of the formal solution of the radiative transfer equation. Consider the situation shown in Fig. 1: given the incident intensity I_{i+1} at grid

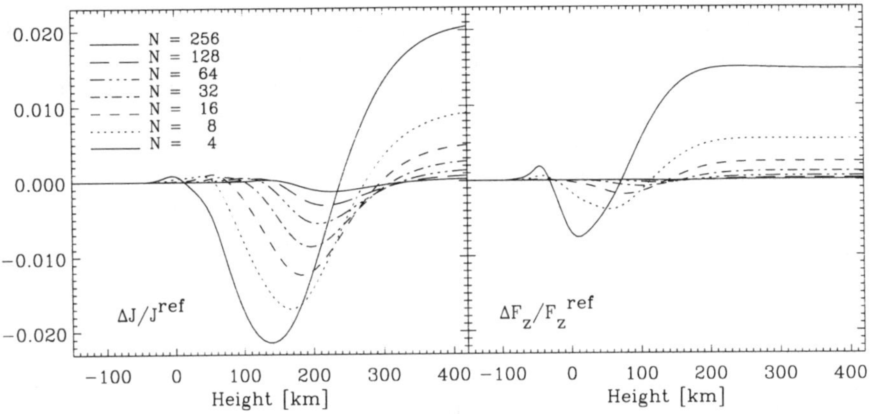

Figure 3. Comparison of the relative error of the mean intensity J (left panel) and the z-component of the flux vector for angular quadratures of different order N, computed for the 1-D quiet Sun model on a fine depth grid. The total number of directions considered is $M = N(N+2)$. All deviations ΔJ and ΔF are relative to the $N = 512$ quadrature, and the scales are the same for both panels.

point $i+1$, characterized by optical depth τ_{i+1} measured along the ray in the direction $-\vec{\Omega}$ starting from $\tau(z=\infty)=0$, the intensity I_i is given by

$$I_i = I_{i+1}\, e^{-\Delta\tau_i} + \int_{\tau_i}^{\tau_{i+1}} S(\tau)\, e^{-(\tau-\tau_i)}\, d\tau, \tag{5}$$

where $\Delta\tau_i = \tau_{i+1} - \tau_i = \int_{z_{i+1}}^{z_i} \kappa(s)\rho(s)\, ds$. Accurate numerical evaluation of the integral in Eq. (5) would in general be cumbersome, but after approximating $S(\tau)$ by a linear or quadratic function of τ it can be written analytically as

$$\Delta I = \int_{\tau_i}^{\tau_{i+1}} S(\tau)\, e^{-[\tau-\tau_i]}\, d\tau \approx \alpha_i S_{i-1} + \beta_i S_i + \gamma_i S_{i+1}. \tag{6}$$

The coefficients α_i, β_i and γ_i are functions of $\Delta\tau_i$ and $\Delta\tau_{i-1}$ only, and after approximating $\kappa(s)$ by a linear or quadratic function, those can be written analytically as well.

Initially, the short-characteristics method alone will suffice to solve the entire radiative transfer problem. Eventually, scattering may also be incorporated, in which case it will have to be complemented with a rapidly-converging iteration procedure.

2.2. Angular integration

The integration of the intensity $I(\vec{\Omega})$ over all directions $\vec{\Omega}$, i.e. 4π steradian of solid angle, is expressed as a quadrature sum over a discrete set of directions. For a given number of directions, M, the choice of the discrete directions $\vec{\Omega}_m$ and their weights w_m is subject to several criteria, the most important being that the lowest moments of the intensity, in particular the mean intensity $J = \sum w_m I_m$ and the flux $\vec{F} = 4\pi \sum w_m I_m \vec{\Omega}_m$ obtain their correct values for an isotropic radiation field. A more complete discussion of angular quadratures can be found in Carlson (1963); for our simulations we chose quadrature type A of Carlson. Figure 3 compares the errors introduced in J and in F_z, the vertical component of the flux vector, as a function of the order N of the quadrature; the total number of directions, M, is given by $M = N(N+2)$. For a given quadrature, the discretization errors in F_z are markedly smaller than those in J. Errors are typically in the percent range for the A4 quadrature with 3 directions per octant; significant improvements can only be obtained at a high computational cost.

2.3. Radiative transfer on a triangular grid

The most obvious way to obtain $Q_{\rm R}$ would be through Eq. (1), computing J directly for each cell center, but this straightforward approach requires excessive and poorly-defined interpolation. It is easier, more efficient and more accurate to compute the intensities I at the vertices and afterwards use either J or $\vec{F}$ to compute the cell-average of $Q_{\rm R}$.

First consider the radiative transfer problem within a single triangle (Fig. 2), where we want to compute $I_{\rm C}(\vec{\Omega})$. Starting from C, tracing the ray in the upwind direction, locate point D where the ray enters the cell. $I_{\rm C}$ can be computed by means of a short-characteristics integration over the interval DC provided that the incident intensity $I_{\rm D}$ is known. In general, point D does not coincide with a vertex of the grid and $I_{\rm D}$ is interpolated from $I_{\rm A}$ and $I_{\rm B}$, which must therefore be known beforehand: this requires that the intensities at the vertices are computed in the proper order. The intensities at the periodic side boundaries are computed first, by integration along rays that start at the top or bottom boundary, where the incident intensities are prescribed by the boundary conditions. Alternatively, if such long rays interfere with parallel processing, those intensities may be obtained by iteration. The radiative transfer problem for the entire computational domain then simplifies to a sequence of radiative transfer problems on single triangles.

Since the short-characteristics integration would reduce to a trivial multiplication in case of constant data on the triangles, we will assume T and ρ to be linear functions of the spatial coordinates x and z, and therefore also of the coordinate s parametrizing the interval DC; note, however, that any other functional dependence on the spatial coordinates can be treated without additional effort. Those relations provide T and ρ at any position within a triangle, and allow κ and S to be computed there. The simplest short-characteristics integration method uses $\kappa_{\rm C}$ and $\kappa_{\rm D}$ to define a linear $\kappa(s)$ relation on interval DC, which together with the relation for $\rho(s)$ yields an analytic expression for $\Delta\tau_{\rm DC}$. That in turn is used together with $S_{\rm C}$ and $S_{\rm D}$ to define a linear $S(\tau)$ to evaluate ΔI. This approach has limited applicability: given the strongly non-linear dependence of κ especially on T, a small cell-to-cell variation of T already leads to significant non-linear variation of κ, so that the optical path length $\Delta\tau_{\rm DC}$ is misrepresented. Straight application of the short-characteristics method with quadratic $S(\tau)$ is not possible, since it requires a point downwind from point

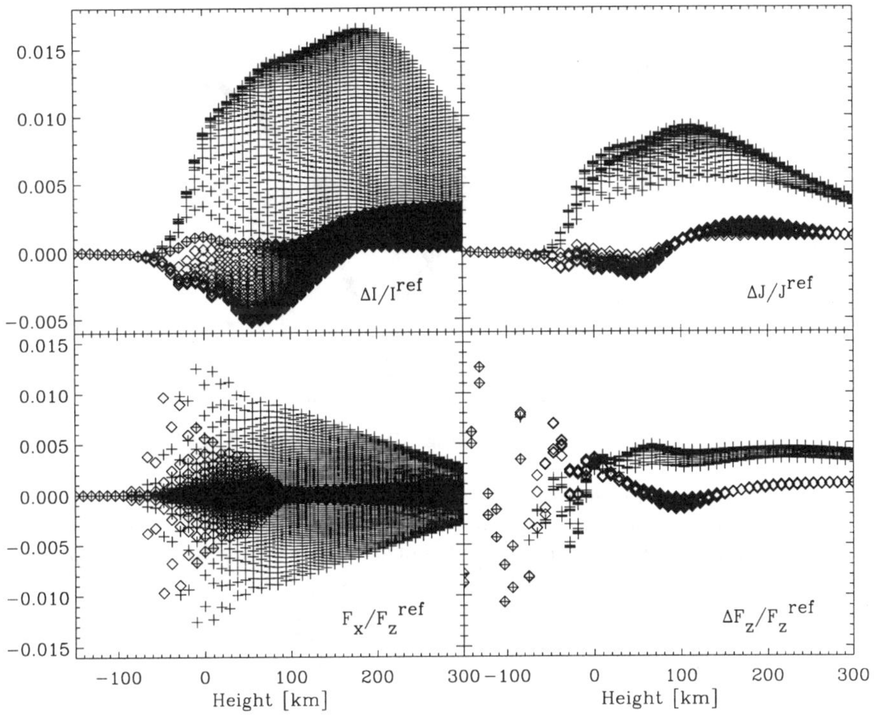

Figure 4. Comparison of errors in I, J and $\vec{F}$ computed for the plane-parallel quiet Sun model on a triangular grid with 8192 cells, extending 500 km horizontally (x-direction) and 600 km vertically (z-direction). Each cell measures about 10 km. Crosses: straight linear interpolation for the incident intensities; diamonds: $\kappa\rho$-weighted interpolation. The errors were computed w.r.t. the 1-D reference solution on a fine depth grid. Top left: intensity in the direction $\vec{\Omega} = (0.882, 0.333, 0.333)$; top right: mean intensity J; bottom right: vertical component, F_z, of the flux vector. The bottom left panel shows the ratio of the horizontal and vertical flux components, which should be zero in this case.

C in order to define the necessary quadratic functions for $\kappa(s)$ and $S(\tau)$. Such point would be located outside of the triangle and break the local character of the method. However, without significant adverse effects the method can be reformulated to use an auxiliary point E exactly halfway between D and C to define those quadratic functions. This significantly increases the maximum allowable cell-to-cell variation of T at a given accuracy level in I.

The common practice to obtain $I_{\rm D}$ from $I_{\rm A}$ and $I_{\rm B}$ by linear interpolation is an important source of error (Kunasz & Auer 1988, Auer & Paletou 1994). If one of these intensities is significantly larger than the other, e.g. because it is located inside a beam, this leads to artificial broadening of the beam. In our test computations for a plane-parallel atmosphere, this effect manifests itself

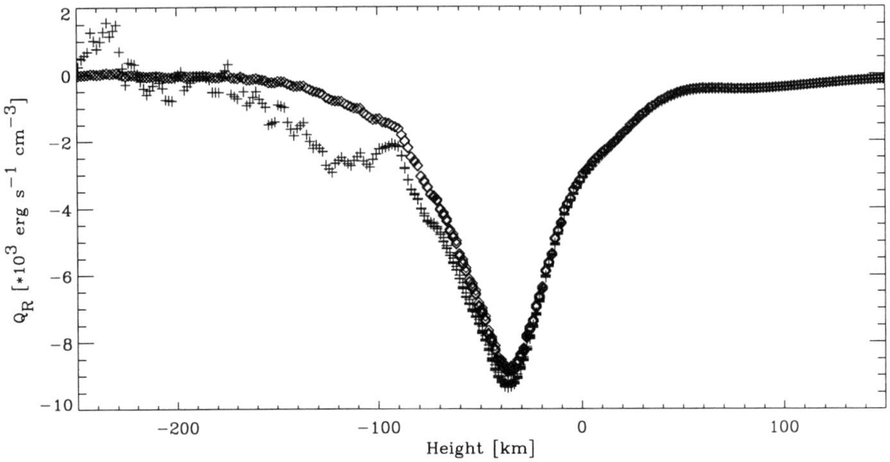

Figure 5. Comparison of Q_R^J (crosses) and Q_R^F (diamonds) for the plane-parallel quiet Sun model on 65536 cells (cell size 4 km). On the scale of this plot, the reference solution (not drawn) coincides with Q_R^F.

through an increase of the intensity (at a given height) with increasing distance from the inflow side boundary (Fig. 4, the fan of crosses in the top left panel). As expected, the effect in I is most prominent for rays that are strongly inclined and the largest relative errors occur slightly above the region with largest non-linear behavior of $\kappa(z)$. In cartesian grids, a partial fix for this problem can be obtained by using higher-order interpolation, although no interpolation method can correctly describe a true discontinuity. Since higher-order interpolation is not an option in our triangular grid, we propose to use $\kappa\rho$-weighted interpolation of I_A and I_B. This biases I_D towards the intensity at the point where the physical parameters most closely resemble those at point D. This procedure strongly reduces the errors in I (Fig. 4, diamonds in the top left panel) so that the errors in J (top right panel) become considerably smaller.

The bottom panels of Fig. 4 show the effects of the interpolation on the flux vector. The horizontal component F_x, which should be zero for a plane-parallel atmosphere, becomes non-zero. Weighted interpolation reduces F_x in most of the atmosphere, but does not do really well in the range slightly below $z = 0$. The growth of the relative error in F_z in the optically thick layers ($z < 0$) has a different origin: with increasing depth the radiation field rapidly approaches isotropy and the accuracy of F_z, which results from subtracting nearly equal intensities in opposite directions, decreases.

2.4. Radiative heating rate

At first sight, the error comparison in Fig. 4 suggests that Q_R^J, with a suitable quadrature for the cell-averaging procedure, should be more accurate than Q_R^F. However, it suffers from severe accuracy problems. Even though J is computed very accurately, the $J - S$ subtraction becomes very inaccurate in the deep atmosphere, where J approaches S, and any numerical error is amplified by the

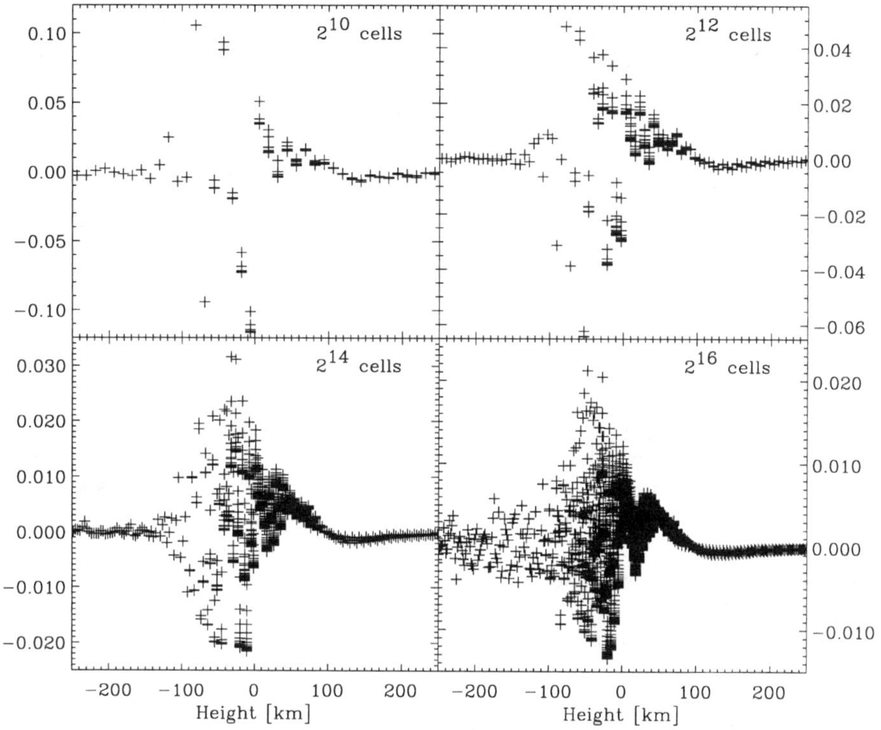

Figure 6. Comparison of Q_{R} accuracy for cell sizes of 30, 15, 8 and 4 km, corresponding to maximum temperature variations within a single cell of 15, 8, 4 and 2%, respectively. Plotted are the deviations of Q_{R} from the reference solution, expressed as a fraction of the maximum cooling rate. Atmosphere and computational box as in Fig. 4. Note the different ordinate scales for each panel.

factor $\kappa\rho$, which grows rapidly with depth. Higher up in the atmosphere, the larger error in J plays an important role. Figure 5 demonstrates that Q_{R}^{F}, with a simple 3-point quadrature for the surface integral, is more stable. It is not very sensitive to small errors in $\vec{F}$, such as the ones shown in the bottom panels of Fig. 4, because it essentially uses the spatial derivative of $\vec{F}$. The relatively large but smoothly varying value of F_x has negligible influence on Q_{R}^{F}. In the deeper atmosphere, the flux itself is small, so that the absolute error in Q_{R}^{F} is small, even if F_z is not very accurate.

Given the sensitive dependence of several procedures on the cell size, we compared the results obtained for Q_{R} for various grids. Cell sizes range from 4 to 30 km. Figure 6 shows that for the plane-parallel quiet Sun model the largest errors in Q_{R} occur in the region where the cooling rate is largest, and that the relative error there is comparable to the maximum temperature variation within a single cell, which varies almost linearly with the cell size. Only for the smallest cells we see a deviation from this linear trend, probably because the cell size no longer dominates the total error. The small error in the upper layers is deceptive; due to the small heat capacity in those layers a heating rate that is only a small fraction of the maximum rate can have marked influence. A comparison of the actual values of $Q_{\rm R}^F$ and $Q_{\rm R}^J$ with the reference solution reveals that both methods produce good results for this model, but that $Q_{\rm R}^J$ is actually smoother and slightly more accurate. Note, that in addition to the errors shown here, there is also a small hidden error due to the angular quadrature that usually works in favor of $Q_{\rm R}^F$.

3. Conclusion

We have implemented radiative transfer routines to compute the radiative heating rate, $Q_{\rm R}$, in gray approximation for a 2-D atmospheric model on a triangular grid. $Q_{\rm R}$ can be computed from the mean intensity, J, or from the radiation flux, $\vec{F}$. We studied the accuracy of the short-characteristics radiative solver on such a grid, and the accuracy of the angular integration required to compute J and $\vec{F}$. We compare the merits of the two alternatives to compute $Q_{\rm R}$ and conclude that $Q_{\rm R}^J$ has severe accuracy problems in the deep atmosphere and minor deviations higher up, whereas $Q_{\rm R}^F$ is stable and accurate throughout most of the atmosphere. As expected, in the upper atmosphere the 3-point quadrature used for computing $Q_{\rm R}^F$, in combination with a strong decrease of the interaction taking place in a single cell, leads to mild deviations from the reference solution. If anything, this will get worse in a model with more structuring, so that $Q_{\rm R}^J$ is the preferable method for the upper atmosphere.

References

Auer, L. H., & Paletou, F. 1994, Astron. Astrophys. **284**, 675

Carlson, B. G. 1963, in B. Alder & S. Fernbach (Eds.), Methods in Computational Physics, Vol. 1, p. 1

Kunasz, P. B., & Auer, L. 1988, J. Quant. Spectrosc. Rad. Transfer **39**, 67

Kunasz, P. B. & Olson, G. L. 1988, J. Quant. Spectrosc. Rad. Transfer **39**, 1

Maltby, P., Avrett, E. H., Carlsson, M., Kjeldseth-Moe, O., Kurucz, R. L. & Loeser, R. 1986, Astrophys. J. **306**, 284

Mihalas, D., Auer, L. H. & Mihalas, B. R. 1978, Astrophys. J. **220**, 1001

Nordlund, Å. 1982, Astron. Astrophys. **107**, 1

Nordlund, Å., & Stein, R. F. 1989, in R. J. Rutten & G. Severino (Eds.), Solar and Stellar Granulation, NATO ASI Series C-263, Kluwer, Dordrecht, p. 453

Olson, G. L., & Kunasz, P. B. 1987, J. Quant. Spectrosc. Rad. Transfer **38**, 325

Spruit, H. C. 1977, Magnetic flux tubes and transport of heat in the convection zone of the sun, PhD Thesis, Rijksuniversiteit Utrecht, Utrecht

Stein, R. F & Nordlund, Å. 1998, Astrophys. J. **499**, 914

High Resolution Solar Physics: Theory, Observations, and Techniques
ASP Conference Series, Vol. 183, 1999
T.R. Rimmele, K. S. Balasubramaniam, R.R. Radick, eds.

Radiative Transfer Effects and Convective Collapse: Size(flux)-Strength Distribution for the Small-Scale Solar Magnetic Flux Tubes

S.Rajaguru[1] & S.S.Hasan

Indian Institute of Astrophysics, Bangalore-560 012, India

Abstract. The effects of radiative energy exchange on the convective instability of a weak field magnetic structure, which lead to a prediction and a physical explanation of the magnetic flux dependent field strength, are examined in detail using a real model stratification for the photospheric and convection zone structure of the Sun. Adopting the generalised Eddington approximation for the radiative transfer, which is valid both in the optically thick and the thin limits, we model the lateral radiative energy exchange by the tube with the external medium with a self-consistent inclusion of vertical radiative losses.

1. Introduction

One of the important properties of the small-scale solar magnetic structures, as established by the observations (Stenflo and Harvey 1985, Zayer et al. 1989,1990, Schüssler 1991), is that while the strong field *network elements* have field strengths very weakly dependent on the flux per element, Φ, with typical values of Φ about 1×10^{18} Mx and higher, the *inner network* weak field structures have a typical value of about 500 G (Keller et al. 1994, Solanki et al. 1996, Lin 1995) for their strength and Φ about 1×10^{17} Mx or lower with a strong dependence between the two. If the convective collapse of a weak field tube is the global process responsible for the formation of the strong field tubes (Parker 1978, Webb and Roberts 1978, Spruit and Zweibel 1979, Spruit 1979, Hasan 1983, 1984,Venkatakrishnan 1985, Steiner 1996) that comprise the *network*, with strengths weakly dependent on Φ, then it should be explained why tubes with smaller fluxes, viz., the *inner network* elements, do not collapse to kG strength. Efficient radiative exchange with the surroundings by a small flux tube (Hasan 1986, Venkatakrishnan 1986) offers a natural explanation. Here, having in mind a quantitative comparison with the above mentioned observationally established properties of the solar flux tubes, we employ a semi-empirical model of the photospheric and the convection zone structure of the Sun and study in detail the effects of radiation on the convective instability and the wave motions.

2. Equations

We add to the familiar thin flux tube equations (Roberts & Webb 1978) the following non-adiabatic energy equation(see e.g., Cox 1980),

$$\frac{\partial p}{\partial t} + v\frac{\partial p}{\partial z} - \frac{\Gamma_1 p}{\rho}\left[\frac{\partial \rho}{\partial t} + v\frac{\partial \rho}{\partial z}\right] = -\frac{\chi_T}{\rho c_v T}\nabla.\mathbf{F}, \tag{1}$$

where p, ρ, and $\mathbf{F}$ denote the fluid pressure, density, and radiative energy flux, respectively, (see Cox (1980) for the definitions of other thermodynamic quantities); all the variables are evaluated on the tube axis (r=0). The radiative flux $\mathbf{F}$ is calculated in the generalised Eddington approximation following Unno and Spiegel (1967): the mean intensity J that is needed to evaluate the flux,

$$\mathbf{F}_R = -\frac{4}{3\kappa\rho}\nabla J \tag{2}$$

, is found by reducing the exact relation

$$\nabla.\mathbf{F}_R = 4\kappa\rho(S - J) \tag{3}$$

to a form appropriate for a thin tube, which reads

$$\frac{1}{3}\frac{\partial^2 J}{\partial \tau^2} + \frac{4}{3}\left(\frac{J_e - J}{\tau_a^2}\right) = J - S, \tag{4}$$

where $d\tau = \kappa\rho dz$, $\tau_a = \kappa\rho a$, S is the source function which we take it to be the Planck function, a is the tube radius, and J_e is the mean intensity in the external medium. J_e is found by solving the equation

$$\frac{1}{3}\frac{\partial^2 J_e}{\partial \tau_e^2} = J_e - S_e. \tag{5}$$

2.1. Equilibrium

The equilibrium stratification of the external medium that we use here is the one determined to match the combined $VAL-C$ (Vernazza et al. 1981) and Spruit (1977) models (see Hasan, Kneer and Kalkofen (1998) for details about the construction of this external quite Sun model). The Rosseland mean opacities are calculated by interpolation from the tables of Kurucz (1993) for the upper layers and from those of Rogers and Iglesias (1992) for the deeper regions. We assume temperature equilibrium, $T = T_e$, which implies that β, defined as $\beta = 8\pi p/B^2$, is constant with z (if the dependence of μ is neglected). The pressure and density are thus determined from

$$p = \frac{\beta}{1+\beta}p_e \tag{6}$$

$$\rho = \frac{\beta}{1+\beta}\rho_e. \tag{7}$$

The extent of the flux tube covers the atmosphere from the temperature minimum in the chromosphere to 5000km deep in the convection zone with the photospheric surface ($\tau = 1$) being assigned $z = 0$. We measure the positive z downwards.

2.2. Linear stability: the perturbation equations

With the assumption that the perturbations in the ambient medium are negligible, small perturbations inside the tube about the equilibrium described above obey the following equations in the linear limit:

$$\frac{\partial \xi}{\partial z} = \frac{B'}{B} - \frac{\rho'}{\rho} - \left[\frac{d(\ln \rho)}{dz} - \frac{d(\ln B)}{dz}\right]\xi \tag{8}$$

$$\frac{\partial^2 \xi}{\partial t^2} = -Hg\frac{\partial}{\partial z}\left(\frac{p'}{p}\right) - g\left(\frac{p'}{p} - \frac{\rho'}{\rho}\right) \tag{9}$$

$$\frac{B'}{B} = -\frac{\beta}{2}\frac{p'}{p} \tag{10}$$

$$\frac{\partial}{\partial t}\left(\frac{p'}{p}\right) - \Gamma_1\frac{\partial}{\partial t}\left(\frac{\rho'}{\rho}\right) + \left[\frac{d\ln p}{dz} - \Gamma_1\frac{d\ln \rho}{dz}\right]\frac{\partial \xi}{\partial t} = -\frac{\chi_T}{\rho c_v T}\nabla.\mathbf{F}', \tag{11}$$

where ξ denotes the vertical displacement and H is the pressure scale height. The perturbation of equation (3) can be done in a straightforward manner:

$$\nabla.\mathbf{F}_\mathbf{R}' = \frac{dF_R'}{dz} = 4\pi\kappa_a\rho(S' - J') + \nabla.\mathbf{F}_\mathbf{R}\left(\frac{\kappa'}{\kappa} + \frac{\rho'}{\rho}\right) \tag{12}$$

. The perturbation in the mean intensity J' is determined by perturbing and linearizing the transfer equation (4) which in the first order moment form is

$$\frac{d}{dz}\left(\frac{J'}{J}\right) = \frac{dlnJ}{dz}\frac{\eta'}{\eta} - \frac{dlnJ}{dz}\frac{J'}{J} + \frac{dlnJ}{dz}\frac{\mathcal{H}'}{\mathcal{H}} \tag{13}$$

$$\frac{d}{dz}\left(\frac{\mathcal{H}'}{\mathcal{H}}\right) = \left(\frac{\Delta_t}{2\epsilon H_p} - \frac{dln\mathcal{H}}{dz}\right)\frac{\eta'}{\eta} + \frac{\Delta_t}{2\epsilon H_p}\frac{a'}{a} - \frac{(1+\nabla_c)(4+3\tau_a^2)}{16\epsilon H_p}\frac{J'}{J} + \frac{1}{qH_p}\frac{T'}{T} - \frac{dln\mathcal{H}}{dz}\frac{\mathcal{H}'}{\mathcal{H}}, \tag{14}$$

where $\mathcal{H} = \mathbf{F}/4$ is the Eddington flux, $\eta = \kappa\rho$, H_p is the pressure scale-height at the bottom, and the various other quantities are as defined below.

$$\nabla_c = \frac{J}{S} - 1 \tag{15}$$

is a measure of departure from radiative equilibrium,

$$\Delta_t = \frac{J - J_e}{S} \tag{16}$$

is the ratio of the excess of the mean intensity inside the tube to the Planck function, ϵ and q are the ratios

$$\epsilon = \frac{\tau_r}{\tau_{th}} \tag{17}$$

and

$$q = \frac{\tau_N}{\tau_{th}}, \tag{18}$$

where

$$\tau_{th} = \frac{\rho c_v T H_p}{\mathcal{H}}, \tag{19}$$

is the radiative relaxation time over the length of one pressure scale-height at the bottom,

$$\tau_r = \frac{\rho c_v a^2}{K}, K = \frac{16\sigma T^3}{3\kappa\rho} \tag{20}$$

is the radiative relaxation time across the tube in the optically thick limit and

$$\tau_N = \frac{c_v T}{4\kappa S} \tag{21}$$

is the radiative relaxation time that one obtains in the optically thin limit (with Newton's law of cooling)(Spiegel, 1957), and

$$\tau_a = \kappa\rho a \tag{22}$$

is the depth dependent optical thickness of the tube. We reduce the perturbation equations to a final set of four equations for the four variables ξ, $p^{'}/p$, $J^{'}/J$ and $\mathcal{H}^{'}/\mathcal{H}$. The optically thick reduction of the set of equations corresponds to taking the limit τ tending to infinity and replacing the mean intensity by the Planck function. The optically thin case is got in the limit τ_{th} approaching zero.

2.3. Boundary conditions

We use closed mechanical boundary conditions and as thermal conditions we impose that there is no incoming radiation from above at the top boundary and that the perturbations are adiabatic at the bottom boundary.

3. Results and Discussion

The action of radiation that we bring out in this study of convective instability and wave motions of the tube is explained conveniently with the help of the graphs shown in Figs. 1 and 2 where we plot the growth rates and frequencies of the fundamental mode, which is the most unstable, respectively, as a function of the surface tube radius a_0 for various values of the plasma β.

3.1. Convective instability

The onset of the convective instability corresponds to the cusps in the curves of fig.1 where the overstable mode's frequency becomes zero and the growth rate shoots up sharply. Comparison with the results obtained with only the lateral exchange in the diffusion approximation (Venkatakrishnan 1986, Hasan 1986) or in the Newton's law of cooling reveal that these earlier treatments overestimate the degree of instability: the growth rates obtained with the present more accurate treatment of radiation with the inclusion of vertical exchange of

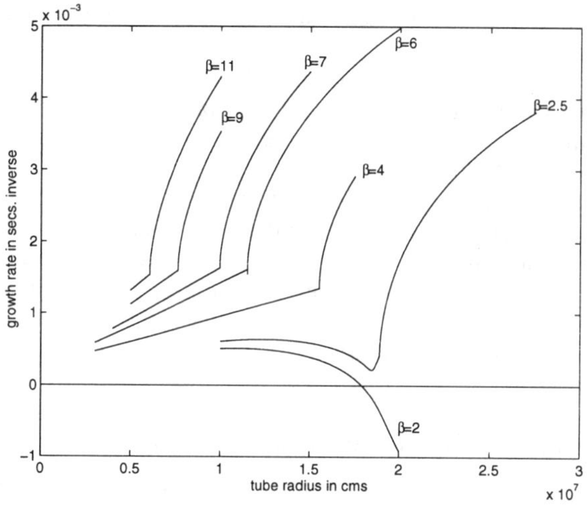

Figure 1. Growth rates of the fundamental mode as a function of the tube radius a_0 at the surface $z = 0$ $(\tau = 1)$ for various values of the plasma β.

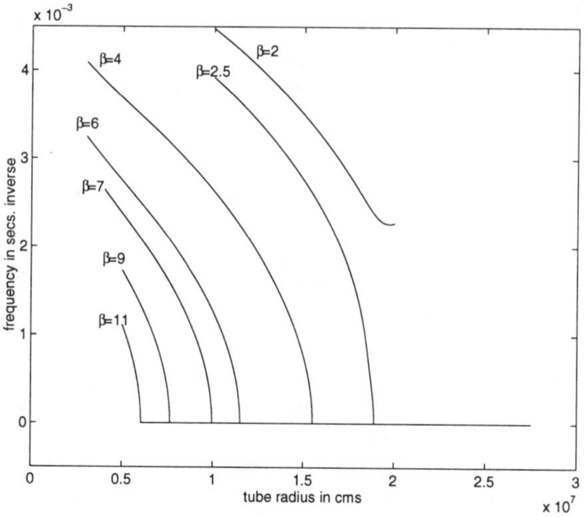

Figure 2. The frequencies of the fundamental mode as a function of the tube radius a_0 at the surface $z = 0$ $(\tau = 1)$ for various values of the plasma β.

radiation are appreciably smaller. Moreover, for a given value of β, i.e., for a tube of given strength, the onset of convective instability requires the size of the tube be greater in the present case than is required when the diffusion approximation is used.

The convective instability is completely suppressed for tubes with the plasma β smaller than 2.45 whatever be its size; this corresponds to a field strength of about 1160 Gauss at $\tau = 1$ inside the tube. This has to be compared with the value of 1350G ($\beta = 1.83$) that Spruit and Zweibel obtained in the adiabatic case. We point out here that the field strength of 1160G that we obtain here does not necessarily imply that all collapsing tubes of weaker fields will attain this unique value and become stable; this value represents a necessary strength for stability against convective collapse and thus can be considered as a minimum strength for stability. A collapsing weaker tube of sufficient size can of course attain an equilibrium collapsed state of field strength higher than this value (cf. Spruit 1979).

3.2. Size-strength distribution for the solar tubes

From the positions of the cusps in the curves of Fig. 1, i.e. from the positions that mark the onset of the convective instability, we pick up the values of the plasma β and the radius a_0; the resulting radius-field strength dependence is shown in Fig. 3, and the corresponding flux-strength distribution in Fig. 4. Comparison of this curve with those observationally produced(Solanki et al. 1996, Lin 1995) shows a remarkable agreement leading to the conclusion that indeed the convective collapse is the cause of the formation of the flux elements on the Sun's surface. Our refined, realistic, and exact treatment of the convective collapse process on the Sun reinforces the conclusions drawn from earlier simplified treatments (Venkatakrishnan 1986, Hasan 1986) and verifies the original suggestion and the physical explanation by Parker (1978) for the concentrated small scale magnetic structures on the Sun.

3.3. Overstability

The characteristics of the overstable mode are explained with the help of Fig. 1 again. The growth rates in the present Eddington approximation are lower than those obtained in the earlier treatments where the vertical losses are not taken care of and use either the diffusion approximation or the Newton's law of cooling. The differences thus demonstrate that overstability is hindered by the vertical losses

We point out that while there is no damping out of oscillations when only lateral exchange takes place and the oscillations' growth rate only asymptotically goes to zero in the limit of large radii, i.e., in the adiabatic limit, the inclusion of vertical radiative losses make the oscillations damp out for radii greater than a particular finite value which is determined by the β. Thus it is clarified that the horizontal exchange between vertically oscillating fluid elements acts to amplify the oscillations while the vertical losses always try to smooth out the fluctuations thereby introducing damping. This *radiative damping* is quite severe for the overstable mode of an intense flux tube on the Sun that it gets completely damped out for tubes of radii larger than a certain critical value.

Finally we note that a tube which can undergo convective collapse for radii greater than a critical value for a given field strength remains overstable for all smaller radii that it can take.

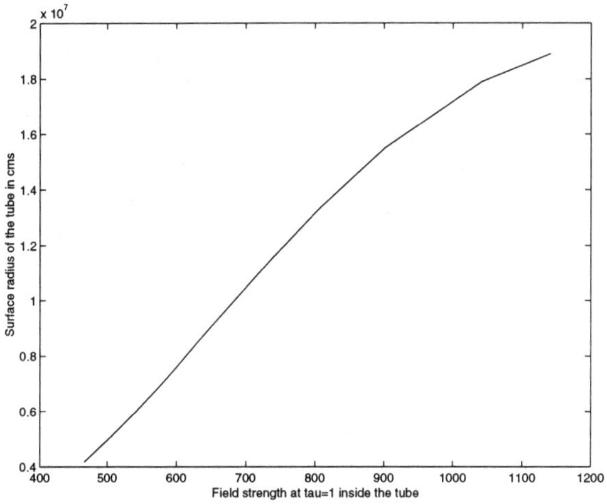

Figure 3. Size-strength relation for the stable tubes; see the text for details

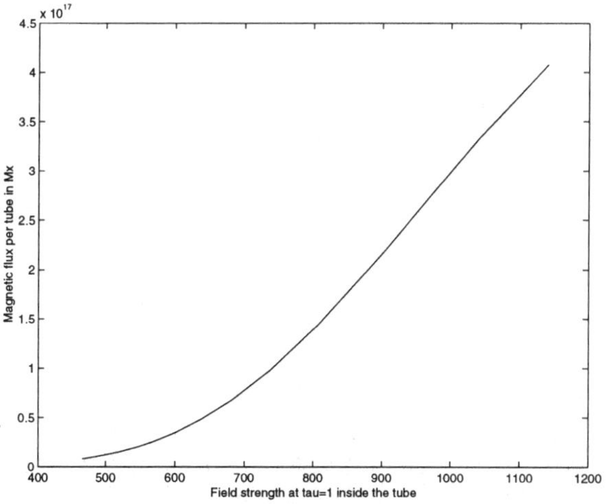

Figure 4. The magnetic flux-strength relation corresponding to the size-strength relation of the previous graph

4. Conclusions

- We have demonstrated that radiative transport has a marked effect on the size-field strength relation for solar flux tubes. Our results can be be applied to solar flux tubes more reliably in view of the more refined treatment of radiative transfer.

- We have generalized the necessary condition for the onset of convective instability in the presence of radiative energy exchange. We find that radiation has a stabilizing influence which is greater for tubes with small radius.
- Overstability of the longitudinal slow mode is shown to be dependent on the tube radius: there is a critical tube radius above which a strong field convectively stable tube's oscillations get damped as a result of vertical radiative losses.

References

Cox, J.P., 1980, Theory of Stellar Pulsation, Princeton University Press.

Hasan, S.S., 1984, ApJ, 285, 851

Hasan, S.S., 1984, A&A, 143, 39

Hasan, S.S., 1986, MNRAS, 219, 357

Hasan, S.S., Kneer, F., & Kalkofen, W. 1998, A&A, 332, 1064

Keller, C.U., et al., 1994, A&A, 286, 626

Kurucz, R.L., 1993, CD-Rom 13

Lin, H., 1995, ApJ, 446, 421

Parker, E.N., 1978, ApJ, 211, 368

Roberts, B. & Webb, A.R., 1978, Sol. Phys. 56, 5

Rogers, F.J., & Iglesias, C. 1992, ApJS, 79, 507

Schussler, M., 1991 in The Sun: A Laboratory for Astrophysics, 191-220, ed. by J.T.Schmelz and J.C.Brown

Solanki, S.K., et al, 1996, A&A, 310, L33-L36

Spiegel, E.A., 1957, ApJ, 126, 202

Spruit, H.C., 1977, Sol. Phys. 55, 3

Spruit, H.C., 1979, Sol. Phys., 61, 363

Spruit, H.C., & Zweibel, E., 1979, Sol. Phys., 62, 15

Steiner, O., Grossman-Döerth, U., Schussler, M., & Knölker, M., 1996, Sol. Phys. 164, 223

Stenflo, J.O., & Harvey, J., 1985, Sol. Phys., 95, 99

Unno, W., & Spiegel, E.A., 1966, PASJ, 18, 85

Venkatakrishnan, P., 1986, Nature, 322, 156

Vernazza, J., Avrett, E., & Loeser, R., 1981, ApJSuppl. 45, 635

Webb, A.R., & Roberts, B., 1978, Sol. Phys., 59, 249

Zayer, I., Solanki, S.K., & Stenflo, J.O., 1989, A&A, 211, 463

Zayer, I., Solanki, S.K., Stenflo, J.O.,& Keller, C.U., 1990, A&A, 239, 356

High Resolution Solar Physics: Theory, Observations, and Techniques
ASP Conference Series, Vol. 183, 1999
T.R. Rimmele, K. S. Balasubramaniam, R.R. Radick, eds.

Structure of a Facular Region From the Inversion of High Spatial and Temporal Resolution Stokes Spectra

L.R. Bellot Rubio, B. Ruiz Cobo, M. Collados

Instituto de Astrofísica de Canarias, 38200-La Laguna, Tenerife, Spain

Abstract. A recently developed inversion technique is applied to high spatial and temporal resolution Stokes profiles of a map recorded with the Advanced Stokes Polarimeter in order to investigate the spatial distribution of the properties of solar magnetic elements belonging to a facular region near disk center. The thin flux tube approximation is adopted in an attempt to describe the structure of these elements. The inversion code retrieves, in a self-consistent manner, the thermal, magnetic and dynamic parameters determining the internal and external atmospheres of the flux tubes. We find that the atmospheres resulting from individual pixels show very similar structures despite the large differences between their emergent Stokes I and V spectra. Within the facular region analyzed, there exist smooth pixel-to-pixel variations of the physical parameters which are highly suggestive of a well defined large-scale organization. No significant correlations between the magnetic field and filling factor or between the internal temperature and magnetic field are observed. According to our results, the facula seems to be composed of individual tubes sharing the same properties, the only significant difference being the filling factor. At $z = 0$ km, the average magnetic field and filling factor turn out to be 1460 G and 0.27, respectively.

1. Introduction

It is now widely accepted that flux tubes are the basic constituents of plage regions. While their general properties are well established, we know little about the full stratification of the various physical parameters describing the structure of these elements. Indeed, there exist very few investigations aimed at determining the thermal, magnetic and dynamic structure of active region faculae. Here we present preliminary results of an analysis which intends to fill this gap. More specifically, our goals are: (a) assessing whether the simple thin flux tube scenario is capable of reproducing the shape of the Stokes profiles emerging from individual pixels within a facular region, and (b) determining both the horizontal and vertical distributions of the physical parameters in the facula. Hopefully, this will help us ascertain whether or not there is a range of different magnetic elements with different properties in active region plages.

For this investigation, high temporal and spatial resolution polarization profiles are needed. This is so because it is important to try to isolate individual elements in specific evolutionary stages in order to determine their properties accurately. We also need sophisticated analysis techniques, since we want to infer the whole stratification of the various physical parameters with depth.

2. Observations and data analysis

We have analyzed the Stokes I and V profiles of the Fe I lines at 6301.5 and 6302.5 Å emerging from individual pixels of a bidimensional map of active region NOAA 7197 obtained on 18 June 1992 with the Advanced Stokes Polarimeter at the Sacramento Peak VTT. The observations were performed near disk center ($\mu = 0.97$). The spectral resolution provided by the spectrograph was 28 mÅ at 6302 Å. The integration time for each slit position was 4.3 s. The spatial resolution of the observations is about 1 arcsec. Fig.1 shows the magnetic flux image corresponding to this map as derived from the results of the Milne-Eddington inversion carried out by Martínez Pillet, Lites & Skumanich (1997). A fairly round sunspot, along with several facular regions, are seen in the magnetic flux image. Our results refer to 116 pixels contained within the box centered at $(32, 2)$ Mm. This region was selected because of its relatively large distance to the central sunspot. The profilcs wc consider have Stokes V amplitudes of at least 0.025 and Stokes Q and U amplitudes below 0.005 (in units of the continuum intensity). Further details on the data reduction, particularly on the removal of the instrumental polarization, can be found in Martínez Pillet et al. (1997). The absolute wavelength calibration of the spectra used here is described in Bellot Rubio, Ruiz Cobo & Collados (1999b).

For the analysis of the observations we have applied the LTE inversion code developed by Bellot Rubio, Ruiz Cobo & Collados (1999a). The code implements the simplest two-component scenario for unresolved magnetic elements, namely the thin flux tube model. The inversion itself proceeds by minimizing the sum of squared differences between the observed and synthetic Stokes spectra. The solution of this problem is achieved iteratively by use of a Marquardt's least-squares algorithm and the so-called Response Functions (see Ruiz Cobo & del Toro Iniesta 1994, for a definition). The code infers the whole set of free model parameters that reproduce the shape of the observed Stokes spectra. In particular, it is capable of dealing with *arbitrary* stratifications of the various physical quantities.

Thin flux tubes are axisymmetric, horizontally homogeneous structures. Vertical thin flux tubes are modeled by specifying two individual atmospheres: the internal atmosphere, representative of the axis of the tube, and the external atmosphere, representing the ambient medium in which the tube is embedded. The free parameters of the model, those to be sought during the inversion, are the temperature and LOS velocity stratifications in both atmospheres, the internal and external microturbulent velocities (assumed height-independent), the same macroturbulence in both atmospheres, and the magnetic field strength, the radius of the tube, and the external gas pressure at the base of the photosphere. An additional free parameter is the amount of stray light contamination. From these parameters, the stratification of the external and internal gas pressures, the magnetic field strength and the shape of the tube are obtained by demanding hydrostatic equilibrium, horizontal pressure balance and magnetic flux conservation.

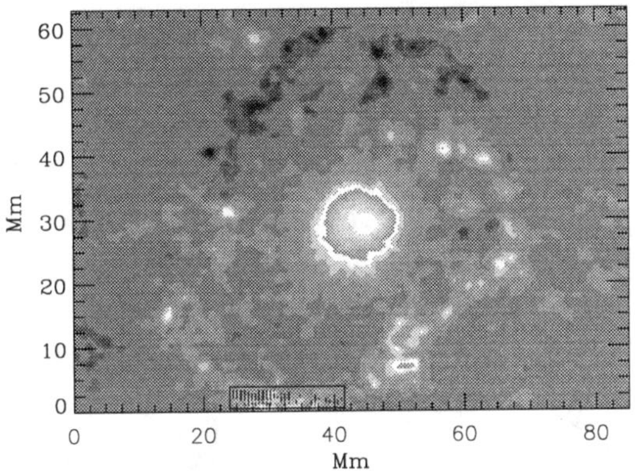

Figure 1. Magnetic flux image of the ASP map corresponding to NOAA 7197. White and black indicate emerging and returning flux, respectively. The background color indicates pixels for which no Milne-Eddington inversions are available.

3. Results

Fig.2 summarizes the results of the inversions. The best-fit synthetic profiles, which reproduce those emerging from the 116 individual pixels, are shown in the upper panels along with the residuals of a typical fit. These residuals are only slightly larger than the noise in the original spectra, which testifies to the quality of the fits. Shown in the central and bottom panels are the temperature, LOS velocity, gas pressure and magnetic field strength stratifications for the internal and external atmospheres resulting from the inversion of the 116 individual pixels.

Inspection of Fig. 2 reveals that the simple thin flux tube scenario is capable of reproducing the shape of the Stokes I and V profiles emerging from facular regions. We want to emphasize that the flux tube model has demonstrated to be able to match real observations only very recently. Moreover, we note that both the internal and external models show the same qualitative properties, despite the large differences among the various Stokes I and V profiles from which they were extracted.

As it is clear from the central right panel of Fig. 2, the inversion code infers the existence of velocity gradients in the internal atmospheres representing the axis of the tubes. Mass motions are virtually absent in the upper layers, but nonnegligible in the lower layers. At $z = -100$ km, internal velocities of ~ 2 km s^{-1} are typical. These velocity gradients have been shown to be capable of reproducing the asymmetrical shapes of the Stokes V profiles observed in plage regions (see Bellot Rubio, Ruiz Cobo, & Collado 1999b) as well as the behavior of the observed Stokes V zero-crossing wavelengths of a large number of Fe I and Fe II lines (Bellot Rubio et al. 1998).

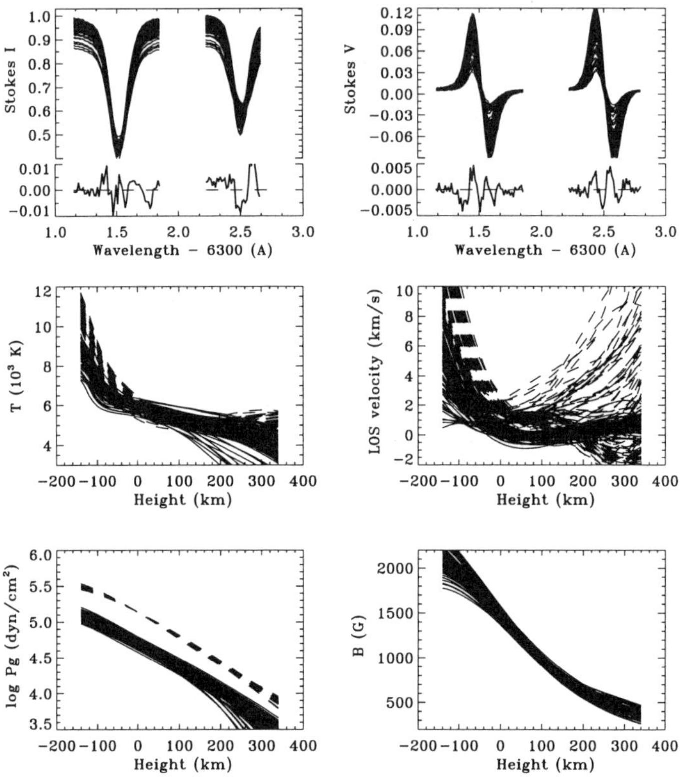

Figure 2. Top: Best-fit synthetic Stokes I and V profiles of 116 individual pixels, and residuals for a typical fit. Middle and bottom: Model atmospheres resulting from the inversions. The internal and external models are represented by the solid and dashed lines, respectively.

Fig.3 shows the distribution of magnetic field strengths and filling factors at $z = 0$ km. The filling factor is defined as the fractional area of the resolution element occupied by the flux tubes at a given height. Both distributions are quite sharp, with very few values deviating significantly from their mean values. This is an important result which suggests that different tubes within the same facular regions have similar magnetic fields and sizes or packing modes. At $z = 0$ km we find a mean magnetic field strength $\bar{B} = 1463$ G which is to be compared with the value $\bar{B} = 1375$ G derived by Martínez Pillet, Lites, & Skumanich (1997) from a much larger set of pixels belonging to the same ASP map. The difference between Martínez Pillet et al.(1997),mean magnetic field and ours is due to the different scenarios adopted. Milne-Eddington inversions provide one single magnetic field value, whereas flux tube inversions recover the full stratification of the magnetic field strength with height. Our results suggest

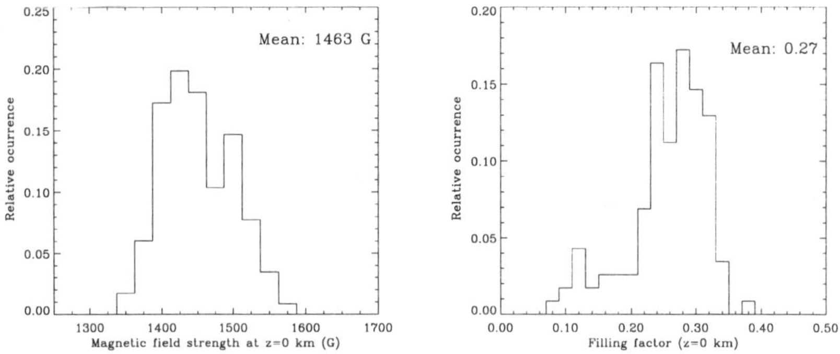

Figure 3. Histograms of magnetic field strengths (left) and filling factors (right) at $z = 0$ km

that Milne-Eddington inversions measure the magnetic field strength at $z \sim 110$ km ($\log \tau_5 \sim -1.8$).

4. Large scale distribution of the physical quantities

To investigate the spatial distribution of the physical properties of the flux tubes belonging to the facular region under consideration, we plot in Fig. 4 various selected parameters at $z = 0$ km as a function of pixel position for two adjacent rows of the ASP map.

A significant result derived from Fig. 4 is that, rather than very large differences, there exist smooth pixel-to-pixel variations of the physical parameters. This is highly suggestive of some sort of large scale distribution within the facula. Neighboring flux tubes have similar thermal, dynamic and magnetic structures. Note also that there are correlations and anticorrelations between the different parameters depicted in Fig. 4.

To study this in more detail, in Fig. 5 we present scatter plots of the internal temperature and magnetic field strength as a function of the filling factor at three different geometrical heights (-100, 0, and 100 km). Inspection of Fig. 5 (upper panels) reveals that the larger the filling factor, the cooler the internal temperature. This effect is more apparent above $z = 0$ km and is consistent with the fact that larger tubes are optically thicker (the limiting case in which the filling factor is unity would correspond to a small pore). The lower panels in Fig. 5 show that no significant correlation between magnetic field strength and filling factor exists at any height. It thus seems that the magnetic field is essentially independent of the radii and/or packing of the tubes.

In Fig. 6 we plot the values of the internal temperature and internal velocity as a function of the magnetic field strength at the same three geometrical heights. While the internal temperature does not seem to correlate with the magnetic field, the internal LOS velocity turns out to be somewhat smaller the larger the magnetic field. This behavior is more apparent in the lower layers, and could reflect the more efficient suppression of mass motions by larger magnetic fields.

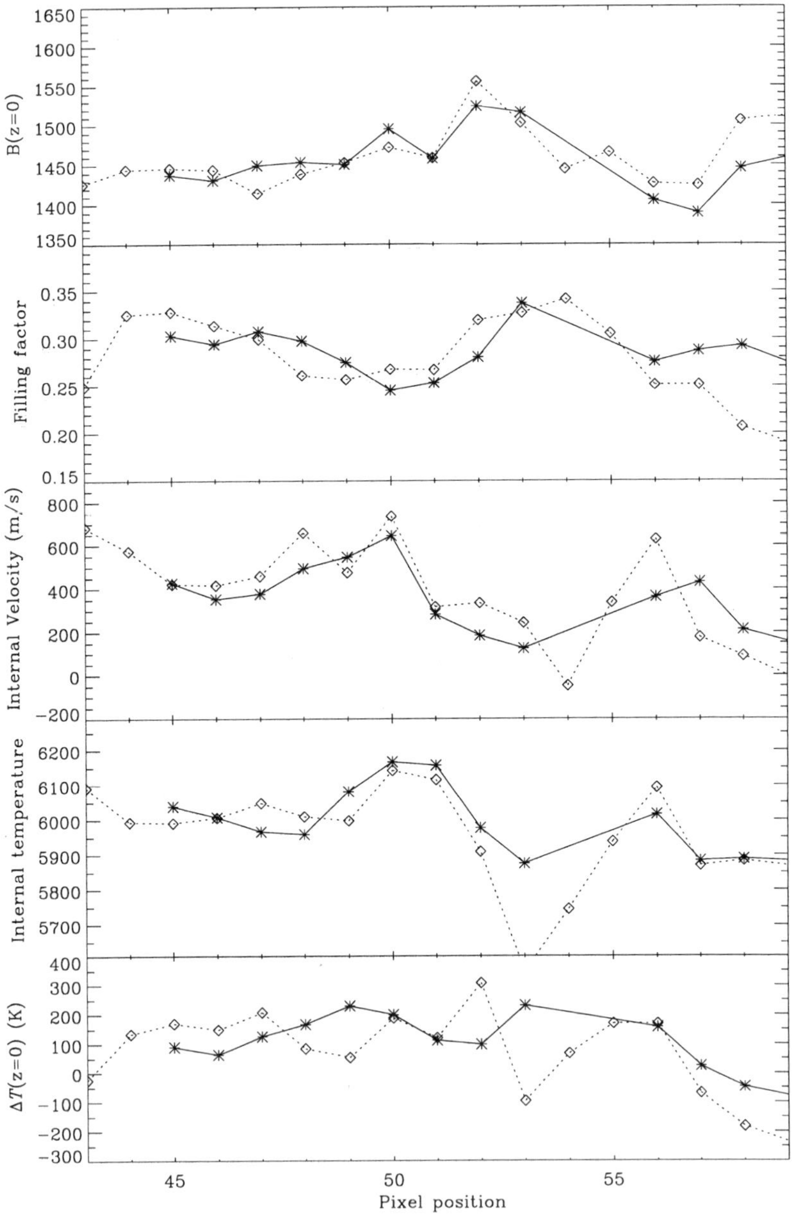

Figure 4. Magnetic field strength, filling factor, internal velocity, internal temperature and temperature excess (defined as the difference between the internal and external temperatures) at $z = 0$ km for nearby pixels belonging to two adjacent rows (asterisks and diamonds, respectively).

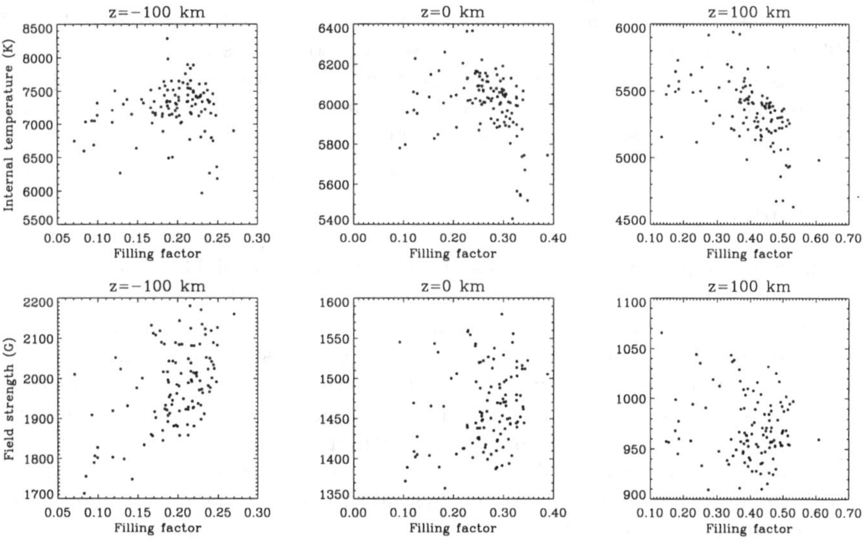

Figure 5. Internal temperature (upper panels) and magnetic field strength (lower panels) vs filling factor at three different heights: $z = -100$ km, $z = 0$ km and $z = 100$ km.

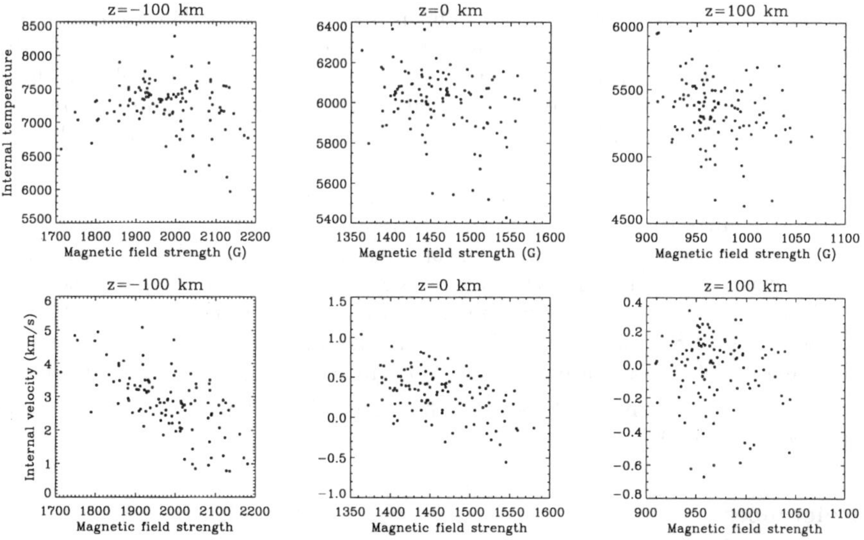

Figure 6. Internal temperature (upper panels) and internal velocity (lower panels) vs magnetic field strength at three different heights: $z = -100$ km, $z = 0$ km and $z = 100$ km.

According to our experience, these correlations are not artifacts of the inversion. Indeed, the numerical simulations by Bellot Rubio, Ruiz Cobo, & Collados (1999a) have demonstrated that the inversion code does not introduce significant cross-talks among the retrieved parameters. Additionally, it is important to emphasize that the results of the inversions are largely independent of the particular initial guess models used to start the procedure.

5. Summary

We have presented preliminary results of the inversion of 116 high spatial and temporal resolution Stokes I and V profiles of the lines Fe I 6301.5 and 6302.5 Å recorded by the Advanced Stokes Polarimeter in a plage region near disk center. The thin flux tube model, implemented in the inversion code, turns out to be capable of explaining the complete shape of the observed spectra. From the fits to the observed profiles, the thermal, dynamic and magnetic structure of the solar magnetic elements belonging to this facular region is retrieved. By allowing for arbitrary stratifications of the parameters, our analysis provides information about the depth-dependence of the physical structure of facular elements. This information is complemented by that resulting from the variations seen across the solar surface.

In our sample, we find an average magnetic field strength and filling factor at $z = 0$ km of 1463 G and 0.27, respectively. These values compare rather well with those obtained in previous analyses. The large scale organization of the facula is suggested by the absence of large pixel-to-pixel variations in the physical parameters. Our scatter plots indicate that the larger the filling factor, the cooler the internal atmosphere. This is probably due to 2D radiative transfer effects. No significant correlations between the magnetic field and the filling factor or between the internal temperature and the magnetic field are found, which suggests that individual tubes within the facula have similar properties. The only significant difference is the filling factor (i.e., the diameter of the tubes and/or the packing mode of individual elements).

The analysis of the whole facular region, involving the inversion of ~ 2000 spectra, is currently in progress. Such an analysis will allow us to generalize the results presented here concerning the spatial distribution of the physical parameters across the facula. In addition, it will be possible to compare the results of these inversions with those from the Milne-Eddington analysis of Martínez Pillet, Lites, & Skumanich (1997). Hopefully, this will serve to better interpret the outcome of Milne-Eddington inversions in active regions outside of sunspots.

Acknowledgments. This work has been funded by the Spanish DGES (project PB 95-0028).

References

Bellot Rubio L.R., Ruiz Cobo B., & Collados M. 1997, ApJ, 478, L45
Bellot Rubio L.R., Ruiz Cobo B., & Collados M. 1998, A&A, in press
Bellot Rubio L.R., Ruiz Cobo B., & Collados M. 1999a, ApJ, in press
Bellot Rubio L.R., Ruiz Cobo B., & Collados M. 1999b, ApJ, in press

Martínez Pillet V., Lites B.W., & Skumanich A. 1997, ApJ, 474, 810
Ruiz Cobo B., & del Toro Iniesta J.C. 1994, A&A, 283, 129

High Resolution Solar Physics: Theory, Observations, and Techniques
ASP Conference Series, Vol. 183, 1999
T.R. Rimmele, K. S. Balasubramaniam, R.R. Radick, eds.

Origin and Propagation of Fluctuations of Turbulent Magnetic Fields

K. Petrovay

Instituto de Astrofísica de Canarias, La Laguna (Tenerife), E-38200 Spain

Abstract. The degree of linear polarization has recently been found to show wide random variations over the solar disk. These variations are presumably at least partly due to fluctuations in the flux density of turbulent photospheric magnetic fields and associated variations in the degree of Hanle depolarization. In order to understand the origin of such large scale fluctuations of the turbulent magnetic flux density we develop a phenomenological model to calculate the spatial Fourier spectrum of the fluctuations of turbulent magnetic fields in the solar photosphere and convective zone. It is found that if the model parameters are fitted to turbulence closure models and numerical experiments the characteristic scale of the fluctuations is by about an order of magnitude larger than the turbulence scale (the scale of the granulation), owing to the more effective quenching of small-scale fluctuations by turbulent diffusion.

1. Introduction

In a highly conductive and strongly turbulent plasma like the solar photosphere random small-scale deviations of the flow field from mirror symmetry are expected to incessantly generate small magnetic flux loops of random orientation. This is a miniature analogue of the large-scale dynamo process occurring in a globally non-mirror-symmetric flow, for which reason it is known as small-scale dynamo action (Leorat, Pouquet, & Frisch 1981, Kida, Yanase, & Mizushima 1991, Petrovay & Szakály 1993). Owing to the random orientation of the loops, no net large-scale field will arise, but a non-zero mean magnetic energy density B^2 and mean unsigned flux density $|B|$ results. Most of this turbulent flux resides in magnetic structures with scales much smaller than the characteristic scale of the turbulent velocity field (which is the granular scale of ~ 1000 km in the solar case). Traditional Zeeman magnetography is "blind" to these fields as the net circular polarization of the mixed polarity small-scale field cancels out in a resolution element.

It has long been recognized that the Hanle effect (magnetic depolarization of linearly polarized radiation) may offer a way to detect the turbulent fields (Stenflo 1982). The observational study of turbulent fields has, however, been hampered by the shortage of observational data, by the lack of a reliable radiative transfer theory for polarized radiation in magnetic fields and by the fact that besides turbulent fields, the Hanle effect is also due to resolved magnetic elements (network, ephemeral active regions) and to the overlying canopy fields (especially for lines formed higher in the atmosphere). Nevertheless, in recent years important advances have been made in all these areas (Faurobert-Scholl 1993, Faurobert-Scholl et al. 1994, Bianda, Solanki, & Stenflo 1998, Bianda, Stenflo, & Solanki 1998, Landi Degl'Innocenti 1996, Trujillo Bueno & Landi Degl'Innocenti 1997, Landi Degl'Innocenti 1998).

On of the most intriguing recent discoveries is the finding of Stenflo, Keller, & Gandorfer (1998) that the degree of linear polarization Q/I shows large amplitude random variations over the solar disk. While, as mentioned above, canopy fields, network elements and ephemeral active regions may also contribute to the observed Hanle depolarization, it is still likely that at least part of this observed spatial variation is due to the presence of similar fluctuations in the flux density $|B|$ of the turbulent photospheric magnetic field.

The presence of fluctuations should not come as a surprise from a theoretical point of view. After all, in the turbulent solar photosphere any physical quantity should show fluctuations, and even variations of an amplitude comparable to the mean value are commonplace. What is more surprising is the *spatial scale* of the fluctuations. While the observations have a resolution of about 1" along the slit, the dominant variations seem to occur on the much larger scale of $\sim 10^4$ km. Variations on the granular scale are of much smaller amplitude. (See e.g. Fig. 3 in Stenflo, Keller, & Gandorfer 1998.) The real theoretical challenge is therefore to understand how the dominant scale of fluctations of turbulent flux density can be so much larger than the turbulence scale?

One popular explanation for the existence of large-scale photospheric structures (such as supergranulation) is that they are the "imprints" of processes going on deeper down in the convective zone where the characteristic scales (determined by the pressure scale height H) are larger. Alternatively, it is of course also possible that the large scales are due to some local photospheric process like an inverse cascade. In order to resolve this problem, we need a model for the generation and transport of turbulent magnetic flux that takes into account both the generation and saturation processes constituting the small-scale dynamo and the turbulent transport of $|B|$ throughout the underlying convective zone. In the following such a model will be presented.

2. The Spectrum of Fluctuations: a Linear Model

The evolution equation for the unsigned flux density $|B|$ of the turbulent field should read something like

$$\partial_t|B| = \underbrace{\nabla(\beta\nabla|B|)}_{\text{turbulent diffusion}} + \underbrace{|B|/\tau_+}_{\substack{\text{linear generation} \\ \text{(small-scale dynamo)}}} - \underbrace{(|B|/B_0)^q\,|B|/\tau_-}_{\substack{\text{nonlinear saturation} \\ q>0}} \qquad (1)$$

As the transport of the highly intermittent magnetic field in a turbulent plasma is due to advection of flux tubes irrespective of their polarity, the transport terms are expected to be identical to those for the signed field B. For simplicity, here we consider an isotropic turbulent diffusion with diffusivity $\beta = lv/3$ (first term on the r.h.s.), l, and v being the correlation length and r.m.s. amplitude of the turbulent velocity field. The linear generation term corresponding to a small scale dynamo would lead to an exponential growth of a homogeneous $|B|$ field to infinity, were it not for a higher order term leading to the saturation of $|B|$ at a finite value B_0, induced by the curvature force (last term on r.h.s.). Numerical simulations and closure calculations (Kida, Yanase, & Mizushima 1991, Nordlund et al. 1992, Durney, De Young, & Roxburgh 1993, De Young 1980) show that B_0 is about an order of magnitude lower than the equipartition flux density: $B_0 \sim B_{\rm eq}/10$.

As the coefficients $1/\tau_+$ and $1/\tau_-$ are functionals of the turbulent velocity field they are expected to show considerable fluctuations around their mean values. Hence, the flux density $|B|$ determined by the equilibrium of generation and nonlinear saturation processes will also fluctuate around its mean value B_0. It is plausible to assume

$$\overline{1/\tau_-} = \overline{1/\tau_+} = 1/\tau \equiv v/l. \tag{2}$$

Furthermore, introducing the notation $|B|/B_0 = 1 + b$, it greatly simplifies the treatment if one assumes $b \ll 1$. (In Section 3 we will consider the problem to what extent this simplification affects the results.) This allows us to linearize equation (1):

$$\partial_t b = \nabla(\beta \nabla b) - qb/\tau + (1/\tau_+ - 1/\tau_-)' \tag{3}$$

Now the second term on the r.h.s. describes the net mean restoring effect of generation and saturation terms, tending to reduce the fluctuation b, while the last term is the *fluctuation generation term* or *forcing term* arising owing to the turbulent fluctuations in the coefficients.

We further introduce the (physically plausible) assumption that the Fourier spectrum of this forcing term is dominated by those horizontally isotropic modes whose vertical phase is such that at the depth where the pressure scale height equals the inverse of their horizontal wavenumber k they are maximal:

$$(1/\tau_+ - 1/\tau_-)' = \frac{1}{\tau} \sum_k \hat{b}_f(k, z, t) \cos[\ln P - \ln P_0(k)] \exp[i(kx + ky)] \tag{4}$$

(Note that throughout this paper we take the logarithmic pressure $\ln P$ as independent variable in the vertical direction, the depth z being just a shorthand notation for a function $z(\ln P)$, determined by a convection zone model. x and y are the horizontal coordinates.) For the solution of equation (3) we take a similar *Ansatz*:

$$b = \sum_k \hat{b}(k, z, t) \cos[\ln P - \ln P_0(k)] \exp[i(kx + ky)] \tag{5}$$

Let us consider one mode only. (Note that in this case $\hat{b}$ can always be considered real as the initial phase can be chosen freely by a displacement of the time axis.) We substitute (4) and (5) into (3), simplify, and take the time average ($\overline{d_t b} = 0$ in a stationary state). To simplify the notation, from this point onwards we omit the hats, and denote simply $b(k, z) = \overline{b(k, z, t)}$ and $b_f(k, z) = \overline{b_f(k, z, t)}$. With this we arrive at

$$d_z(\beta \, d_z b) + (2k^2 \beta + q/\tau) \, b = b_f/\tau \tag{6}$$

This equation determines the Fourier amplitude $b(z,k)$ of each mode of horizontal wavenumber k in the spectrum of turbulent magnetic field fluctuations. Vertical diffusion is now separated in the first term; horizontal diffusion and the restoring force constitute the second and third term on the l.h.s.

The r.h.s. is the Fourier amplitude of the forcing. b_f is the fluctuation amplitude produced in time τ by the forcing if other terms were not present. As the forcing is due to the action of the fluctuating flow field on the existing magnetic field, it is plausible to assume $b_f \propto B_k/B_0$, where B_k is the corresponding Fourier amplitude in the spectrum of $|B|$. This spectrum may be determined from closure calculations, numerical simulations and observations. Its simplest representation is by two power laws joining in a peak at $k \sim 10/l$. Thus, we represent the r.h.s. by

$$b_f(z,k)/\tau(z,k) = A\,10^{2/3-p}\,(k/k0)^p, \qquad k_0 \sim 1/H(z), \tag{7}$$

where $p = p_1$ for $k < 10k_0$ and p_2 otherwise. (Note that in fact a two-peaked spectrum might be more realistic as the spectrum of τ is peaked at a lower wavenumber of $k \sim 1/l$; our purpose here however is just to present a simple example calculation.) On the basis of the observed properties of photospheric magnetic fields and motion, $p_1 \sim 1.5$ seems to be a realistic choice. The value of p_2 will be found to be irrelevant to the solution below.

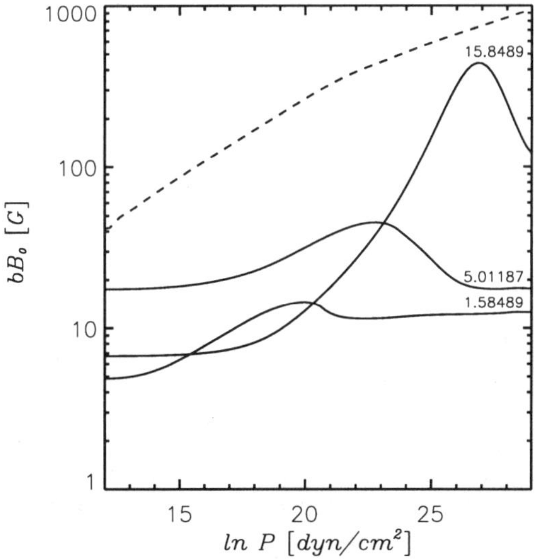

Figure 1. Vertical profiles of some Fourier modes in the spectrum of fluctuations of the turbulent magnetic field. The modes are labelled by their value of $1/k$ in megameter units. Dashed: B_0

For the solution of equation (6) the parameters β and τ are interpolated from a convective zone model. q is evaluated by fitting a solution of equation (1), with the diffusive term and the perturbations of the coefficients neglected, to profiles of $|B|^2(t)$ resulting from the closure model of De Young (1980); this yields $q \simeq 0.1$. The boundary conditions may be chosen as closed ($F \equiv \beta\, d_z b = 0$) or open ($F = vb$); experimenting has shown that this does no exert a very strong influence on the resulting mode profiles. For the numerical solution of equation (6) we take $\ln P$ as independent variable and use a relaxational method.

The vertical profiles of some Fourier modes in the spectrum of fluctuations of the turbulent magnetic field are shown in Fig. 1. As expected, the modes with larger horizontal scales have their maxima in deeper layers.

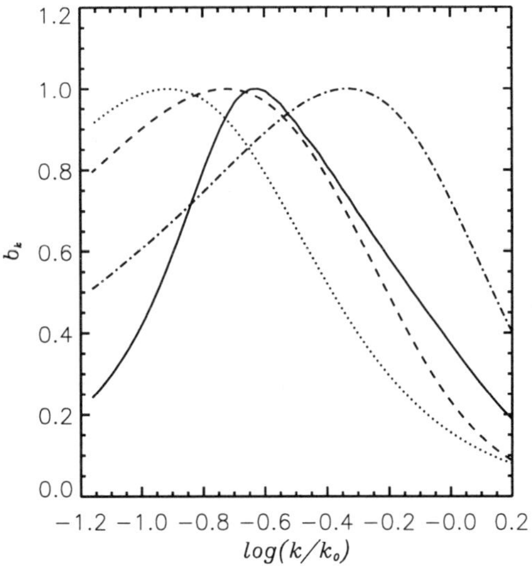

Figure 2. Fourier spectra of the fluctuations of the turbulent magnetic flux density as functions of the horizontal wavenumber. Solid: $p_1 = 1.5$, $q = 0.1$; dashed: $p_1 = 0.5$, $q = 0.1$; dash-dotted: $p_1 = 1.5$, $q = 1.0$; dotted: $p_1 = 0.5$, $q = 0.1$ with vertical transport switched off

Computing a large number of modes and plotting their amplitudes near the surface against k yields the Fourier spectrum of the fluctuations in the turbulent magnetic field (Fig. 2). Experimenting with the parameters in equation (6) we find that the choice of p_2 is irrelevant for the spectrum, A and $B_0/B_{\rm eq}$ determine its amplitude, while p_1 and q determine the shape of the spectrum. A striking feature of the spectra in Fig. 2 is that their maxima fall to significantly lower wavenumbers than $k_0 = 1/l$, i.e. the characteristic scale of the fluctuations in turbulent magnetic flux density is much larger than the typical scale of turbulent motions. The physical background of this phenomenon is that the high wavenumber components are more efficiently suppressed by diffusion ($2k^2\beta\, b$ term in eq. (6)). The wavenumber of the maximum increases with q (dashed vs. dash-dotted curves), as a stronger nonlinear saturation reduces the role of the diffusive terms in the equation. On the other hand, the spectral amplitude at even lower wavenumbers depends strongly on the spectral index

p_1 of the forcing (dashed vs. solid curves) —clearly, the shallower the forcing spectrum, the more energy is input at the larger scales.

3. The Effect of Nonlinearity

The results presented in the previous section were computed under the assumption that the fluctuations of $|B|$ have a small relative amplitude. This assumption is rather dubious in the light of the large Fourier amplitudes found in the model (Fig. 1). In order to have an idea about the extent to which nonlinear effects may modify the linear results, in this section we present an alternative model that calculates the fluctuating field without the assumption of linearity, at the cost of a strong simplification of the geometry: only $k = 0$ modes are considered. This obviously implies that spatial spectra or correlation lengths cannot be studied; instead, we will compare the *temporal* autocorrelation of the fluctuations in the linear and nonlinear cases.

For $k = 0$ modes $d_x = d_y = 0$ so we write equation (1) in the form

$$\partial_t |B| = d_z(\beta\, d_z |B|) + |B|/\tau\, [1 - (|B|/B_0)^q] + B_f/\tau \tag{8}$$

where the last term corresponds to fluctuation forcing by to the turbulent fluctuations of the velocity field. We model this term as a Gaussian stationary random process with correlation time $\tau(z)$, vertical correlation "length" 1 (in $\ln P$ units), and a mean displacement of $0.3B_0$ over τ .

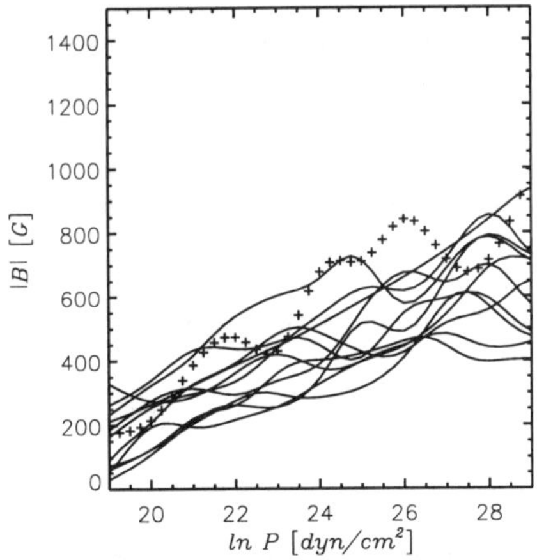

Figure 3. Snapshots of the solution of equation (7) with $q = 1$ at different instants of time. Crosses: initial profile.

Equation (8) is then integrated numerically starting from an arbitrary perturbed initial state. An example solution is presented in Fig. 3. Fig. 4 shows the autocorrelation of the fluctuating flux density as a function of the time shift for different cases. It is apparent (dotted vs. dashed curves) that the correlation time is primarily determined by the value of the q nonlinearity parameter, lower q values corresponding to longer correlation times. This result is a close analogue to the findings of Section 2 with respect to the spatial correlations. On the other hand, switching off the diffusive term in equation (8) (dash-dotted vs. dashed curves) has no significant effect, showing that diffusive quenching of small spatial scales is *not* the mechanism responsable for the extended correlations here. Replacing equation (8) with its linear equivalent (solid vs. dashed curves) does not lead to great modifications in the results. This may reassure us to some extent of the reliability of the findings of Section 2 above.

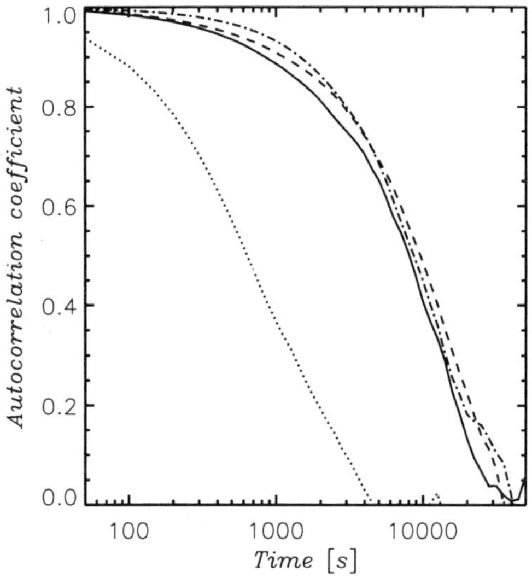

Figure 4. Autocorrelation of the fluctuating flux density as a function of the time difference at a fixed point in the convective zone in the nonlinear model with $q = 0.1$. Dashed: reference model; dotted: $q = 1$; dash-dotted: reference model with diffusive term switched off; solid: linear equivalent

4. Conclusion

The results of Section 2 now enable us to answer the question posed at the end of the introduction. Turbulent transport processes indeed result in a dominant scale for the fluctuations of the turbulent magnetic flux density that is an order of magnitude larger than the turbulence scale. The physical mechanism behind this phenomenon is that small-scale fluctuations are more efficiently damped by diffusion, shifting the spectral peak to lower wavenumbers. The low value of the q nonlinearity parameter suggested by closure calculations and simulations also favors larger dominant scales.

On the other hand, the possibility that deeper structures are "imprinted" on the photospheric pattern can apparently be discarded. Switching off the first term (vertical diffusion) in equation (6) does not lead to any reduction of the dominant scales (dotted vs. dashed curves in Fig. 2; dash-dotted vs. dashed in Fig. 4).

All this shows that the observations of strongly varying Hanle depolarization over the solar disk may be understood from a theoretical point of view even if all the depolarization is attributed to turbulent fields. Further advances on both the observational and theoretical side may make more detailed comparisons between the models and the observations possible, thereby offering the prospect of a direct observational diagnostics of the properties of the turbulent dynamo.

Acknowledgments. This work was funded by the DGES grant no. 95-0028.

References

Bianda, M., Solanki, S. K., & Stenflo, J. O. 1998, A&A, 331, 760

Bianda, M., Stenflo, J. O., & Solanki, S. K. 1998, A&A in press

De Young, D. S. 1980, ApJ, 241, 81

Durney, B. R., De Young, D. S., & Roxburgh, I. W. 1993, Solar Phys., 145, 207

Faurobert-Scholl, M. 1993, A&A, 268, 765

Faurobert-Scholl, M., Feautrier, N., Machefert, F., Petrovay, K., & Spielfiedel, A. 1994, A&A, 298, 289

Kida, S., Yanase, S., & Mizushima, J. 1991, Phys. Fluids A, 3, 457

Landi Degl'Innocenti, E. 1996, Solar Phys., 164, 21

Landi Degl'Innocenti, E. 1998, Nature, 392, 256

Leorat, J., Pouquet, A., & Frisch, U. 1981, J. Fluid Mech., 104, 419

Nordlund, Å., Brandenburg, A., Jennings, R. L., Rieutord, M., Ruokolainen, J., Stein, R. F., & Tuominen, I. 1992, ApJ, 392, 647

Petrovay, K., & Szakály, G. 1993, A&A, 274, 543

Stenflo, J. O. 1982, Solar Phys., 80, 209

Stenflo, J. O., Keller, C. U., & Gandorfer, A. 1998, A&A, 329, 319

Trujillo Bueno, J., & Landi Degl'Innocenti, E. 1997, ApJ, 482, L183

Group Discussion

Sanchez-Almeida : The angular resolution of the Hanle effect measurements is rather poor. Couldn't it be that the large-scale variation that you mentioned has to do with this limitation?

Petrovay : No. The angular resolution is 1", and the angle of the variation in question is 10-20".

High Resolution Solar Physics: Theory, Observations, and Techniques
ASP Conference Series, Vol. 183, 1999
T.R. Rimmele, K. S. Balasubramaniam, R.R. Radick, eds.

Axisymmetric Sunspot Oscillations

T. J. Bogdan

High Altitude Observatory, National Center for Atmospheric Research [1]
P.O. Box 3000, Boulder CO, USA

Abstract. A discussion of some elementary considerations encountered in the study of the oscillations of axisymmetric sunspot equilibria is presented. An expansion method is proposed for converting the relevant oscillation equations to a form that is better suited to numerical computation. In so doing, it is demonstrated that thin flux tube and flat canopy approximations are capable of providing physical insight onto the morphology of the oscillations in the deep submerged root of the sunspot and the far-flung outlying penumbral canopy.

1. Introduction

As the study of the solar acoustic oscillations has revealed much about the hidden structure of the solar interior, one may well anticipate that analogous efforts applied to the magneto-atmospheric oscillations of sunspots have the potential to diagnose the subsurface structure of these, the largest of the solar surface magnetic flux concentrations. Although observational efforts to detect and categorize the form and substance of these oscillations have been hampered by the practical limitations of scattered light and insufficient spatial and temporal resolution, a coherent global picture of sunspot oscillations is beginning to take shape (e.g., Lites 1992).

On the theoretical side, different, although equally perplexing, practical difficulties have also been encountered. They stem from the intrinsic anisotropic nature of the Maxwell stress tensor, and the continual variation of the inclination of the magnetic field lines with height *and* radius in the sunspot magneto-atmosphere (see Fig.1). Consequently the magnetic and thermal stratifications of the sunspot interior continually change their relative orientation between the umbra and penumbra, and this leads to the non-separability of the governing oscillation equations. Oscillation modeling activities have for the most part focused on delineating the properties of uniformly magnetized cylinders and slabs, where a precise orientation between the magnetic and thermal stratification is in force throughout the entire equilibrium (e.g., Thomas 1983). The purpose of this contribution is to outline the initial steps toward more sophisticated model calculations.

---

[1] The National Center for Atmospheric Research is sponsored by the National Science Foundation

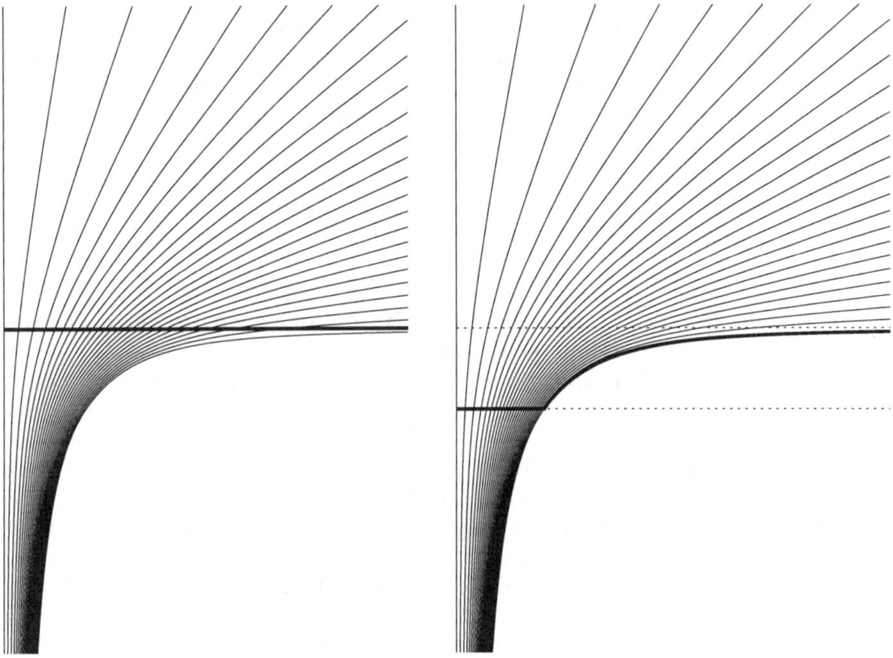

Figure 1. Meridional slices through two axisymmetric potential sunspot models. The thin solid lines show selected magnetic lines of force, while the thick solid line indicates the location of the plasma-vacuum transition. The density stratification is described by equation (1) with some constant value of $\epsilon < 1$. In the left panel, the Wilson depression W is zero, while in the right panel it is equal to the vertical separation of the two horizontal dotted lines.

2. A Current-Sheet Sunspot Model

The next step in complexity beyond the uniformly magnetized cylinder and the magnetized slab models previously employed to study sunspot oscillations is illustrated schematically by Fig. 1. The atmosphere both outside *and* inside the axisymmetric sunspot equilibrium is described by the *same* isentropic polytrope, with a specified polytropic index, $m = 1/(\gamma - 1)$, and a constant temperature gradient. In order that the sunspot contain a nontrivial magnetic flux, it is necessary to reduce the gas pressure within the sunspot relative to that of the surrounding quiet Sun. Within the framework established above, this can be accomplished in two distinct ways: by sliding the internal atmosphere downward with respect to the external atmosphere by some distance W–i.e., the Wilson depression, and by taking the internal pressure to be a constant fraction ϵ of the external pressure. In terms of the fluid density, ρ, these two operation are summarized by,

$$\rho(\mathbf{x}) = \begin{cases} \rho_\odot \left(1 - \frac{z}{mH_\odot}\right)^m & \mathbf{x} \in \text{quiet Sun} \\ \epsilon\rho_\odot \left(1 - \frac{z+W}{mH_\odot}\right)^m & \mathbf{x} \in \text{Sunspot} \end{cases} \qquad (1)$$

where $\rho_\odot$ and $H_\odot$ are the prescribed density, and density scale height, at the level $z = 0$ in the surrounding quiet Sun.

Since the internal sunspot atmosphere is hydrostatic over its entire cross-section, the Lorentz force must vanish within the spot. This in turn forces the magnetic field to be force-free. Because α (i.e., $\nabla \times \mathbf{B} = \alpha \mathbf{B}$) is strictly constant along a given magnetic line of force, and all magnetic field lines must thread through the plasma-vacuum interface at the top of the internal polytrope (see Fig. 1), it follows that the sunspot magnetic field must in fact be *potential*, with $\alpha \equiv 0$. The potential field is surrounded by a current sheet, across which the magnetic field is discontinuous. The limiting value of $B^2/8\pi$ attained as the current sheet is approached from the inside of the spot is simply the discrepancy in gas pressure between the internal and external polytropic atmospheres, consistent with equation (1).

It should be noted that this current-sheet sunspot constitutes a very poor model of a real sunspot, and was already abandoned in the 1980's from the sunspot modeling repertoire (Skumanich 1992). On the other hand, this simplified configuration is particularly well-suited for exploring the properties of sunspot oscillations in that it encompasses a diverging magnetic field geometry, without the unnecessary complications of distributed volume currents and buoyant restoring forces.

3. The Oscillation Equations

In cylindrical coordinates (ϖ, z), the linearized adiabatic equations for axisymmetric magneto-atmospheric oscillations are,

$$\ddot{\xi}_\varpi = -\frac{\partial}{\partial \varpi}\frac{p'}{\rho} + \frac{B_z}{4\pi\rho}\left[\frac{\partial^2}{\partial z^2} + \frac{\partial}{\partial \varpi}\frac{1}{\varpi}\frac{\partial}{\partial \varpi}\varpi\right](B\hat{\mathbf{n}} \cdot \boldsymbol{\xi}) \qquad (1)$$

$$\ddot{\xi}_z = -\frac{\partial}{\partial z}\frac{p'}{\rho} - \frac{B_\varpi}{4\pi\rho}\left[\frac{\partial^2}{\partial z^2} + \frac{\partial}{\partial \varpi}\frac{1}{\varpi}\frac{\partial}{\partial \varpi}\varpi\right](B\hat{\mathbf{n}}\cdot\boldsymbol{\xi}) \tag{2}$$

$$p' = \rho\Big[g_\odot \xi_z - c^2(z)\,\nabla\cdot\boldsymbol{\xi}\Big] \tag{3}$$

$$B\hat{\mathbf{n}}\cdot\boldsymbol{\xi} = B_z\xi_\varpi - B_\varpi\xi_z \tag{4}$$

when specialized to the current-sheet geometry depicted in Fig. 1. In these equations $\boldsymbol{\xi}$ is the Lagrangian displacement of a fluid parcel from its equilibrium position, p' is the pressure fluctuation, $c(z)$ is the adiabatic sound speed, $\mathbf{B}(\varpi, z)$ is the potential magnetic field, $\hat{\mathbf{n}}$ is the outward-directed unit vector normal to $\mathbf{B}$, and the constant gravitational acceleration is $g_\odot$.

Although equations (2) and (3) are rather compact and symmetrical in their outward appearance, the presence of $\mathbf{B}(\varpi, z)$ on either side of the Laplacian operator is responsible for the non-separable character of the problem. One possible avenue for solving these equations is through an expansion procedure based on prescribed radial planforms F_n and G_n,

$$\xi_\varpi(\varpi, z, t) = \sum_{n=0}^{\infty} G_n(k\varpi)\,u_n(z,t) \tag{5}$$

$$\xi_z(\varpi, z, t) = \sum_{n=0}^{\infty} F_n(k\varpi)\,w_n(z,t) \tag{6}$$

$$G_n(x) \equiv \frac{(-1)^n}{n!}\left(\frac{x}{2}\right)^n J_{n+1}(x) \qquad F_n(x) \equiv \frac{(-1)^n}{n!}\left(\frac{x}{2}\right)^n J_n(x) \tag{7}$$

where $J_n(x)$ is the familiar Bessel function, and the u_n and w_n functions are to be solved for using equations (2)–(5). The constant k plays the role of the horizontal wavenumber, but strictly speaking the oscillation eigenfunctions cannot be characterized by a single horizontal wavenumber.

If this *ansatz* is substituted into the oscillation equations (2) and (3), and the coefficients of each distinct planform are individually equated to zero, there results a coupled set of *separable* partial differential equations (in z and t) for the unknown functions u_n and w_n (Bogdan 1999). For the trivial special case of a *uniform* vertical magnetic field ($B_\varpi = 0$, $B_z =$ constant), this system of coupled equations can be solved by inspection. In this case, $u_n = w_n = 0$ for $n \geq 1$, while u_0 and w_0 are the solutions of the uniformly magnetized cylinder equations originally written down by Ferraro & Plumpton (1958). Under more general circumstances, a practical computational algorithm involves setting $u_n = w_n = 0$ for $n \geq N$, and solving the finite set of equations for the remaining u's and w's. One must then sequentially increase the truncation value N until the computed oscillations are no longer sensitive to further incremental changes in N.

4. The Thin Flux Tube Limit

The expansion method described in the previous section can be carried out with many different choices for the radial planforms, subject only to a certain minor limitation—i.e., that the right-sides of equations (2) and (3) must map exclusively onto G and F planforms, respectively. The specific choice described by equation (8) is motivated by the particularly simple and elegant manner in which the expansion *ansatz* can be resummed to yield the thin flux tube approximation. This remarkable transformation is described in the present section, although it should be borne in mind that equation (8) may not, in fact, be the optimal choice of radial planforms for other purposes or applications.

If a $e^{i\omega t}$ time-dependence is assumed and then factored out of the u's and w's, a set of coupled ordinary differential equations (in z) results, with coefficients that depend upon the two (quasi-)separation parameters k and ω. To obtain the thin flux tube approximation from this infinite system of coupled ODE's, one takes the counterintuitive measure of imposing one additional equation for the u's and w's to satisfy. At a subsequent step in the analysis of sunspot oscillations it is necessary to match the internal magneto-atmospheric oscillations of the spot to the exciting p-modes of the surrounding quiet Sun. This is effected by enforcing the continuity of the normal displacement of the horn-shaped current-sheet, and the continuity of the normal stress across this surface. The component of the stress normal to the magnetic lines of force is given by the expression,

$$\Pi' \equiv p' - g_\odot \rho (\hat{\mathbf{n}} \cdot \hat{\mathbf{e}}_z)(\hat{\mathbf{n}} \cdot \boldsymbol{\xi}) + \frac{1}{4\pi} \mathbf{B} \cdot \mathbf{B}' \tag{8}$$

where $\mathbf{B}'$ denotes the fluctuations in the magnetic field induced by the oscillations. Now although Π' needs be continuous only across the current sheet surrounding the sunspot, it may be defined throughout the interior of the spot (where its continuity is in fact guaranteed by the governing oscillation equations). If the expansion *ansatz* of the previous section is used to evaluate the right-side of equation (9), then Π' is found to have an expansion given solely in terms of the F planforms. The coefficient multiplying the first $F_0(k\varpi)$ term in this expansion depends only on u_0 and w_0 and is given by the right-side of the following equation,

$$\Pi'_{\text{exciting p-mode}}(z) = \rho \Big\{ g_\odot w_0 - c^2 \Big(\frac{\partial w_0}{\partial z} + k u_0 \Big) - a^2 \Big[k u_0 + \frac{1}{b} \frac{db}{dz} w_0 \Big] \Big\} \tag{9}$$

where $a(z)$, $b(z)$, and $c(z)$ are the Alfvén speed, the magnetic field strength and the adiabatic sound speed *all* evaluated on the sunspot axis. In anticipation of the matching between the internal magneto-atmospheric oscillations and the surrounding p-modes, it is reasonable to associate the free function that appears on the left-side of this equation with the normal stress fluctuations produced by the external p-modes having an oscillation frequency ω.

If the u's and w's are forced to satisfy equation (10) in addition to equations (2) and (3), then it proves possible to solve the augmented infinite+1 system of ODE's sequentially! The equation which expresses the fact that the coefficients of the $F_0(k\varpi)$ planforms that appear on each side of equation (3) must be equal is

$$\omega^2 w_0 = \frac{\partial}{\partial z} \Big[g_\odot w_0 - c^2 \Big(\frac{\partial w_0}{\partial z} + k u_0 \Big) \Big] \tag{10}$$

so taken together, equations (10) and (11) provide a closed inhomogeneous second-order ODE for w_0. This pair of equations is in fact *precisely* the thin flux tube equations for the sausage mode oscillations originally written down by Defouw (1976). Equating the coefficients of $G_0(k\varpi)$ in equation (2) and $F_1(k\varpi)$ in equation (3) in turn produces a closed, inhomogeneous, second-order ODE that determines w_1 and u_1 in terms of $\Pi'_{\text{exciting p-mode}}$, u_0 and w_0. Continuing in this fashion, it proves possible to compute each successive pair of w_n and u_n in terms of the previously obtained solutions.

Perhaps the most intriguing aspect of this computational windfall concerns the lack of a role for the pseudo-horizontal wavenumber, k. In the sausage mode equations (10) and (11), it is easy to verify that the solution $w_0(z)$ is independent of the value assigned to k, as is the product $ku_0(z)$. This lack of dependence upon the choice adopted for k in fact persists in a manner of speaking to all higher levels in the expansion scheme. If one adopts the following representation for the $w_n(z)$ sequence (an analogous statement can be readily conceived for the u_n's),

$$w_n(z) = \sum_{p=0}^{n}(-1)^{n+p}\frac{n!}{p!(n-p)!}\, k^{-2p}\, W_p(z) \tag{11}$$

$$w_0(z) \equiv W_0(z) \tag{12}$$

then it transpires that every $W_n(z)$ is independent of the choice of k, although clearly the w_n in fact do care about the value of k, but in a restricted sense. It is worth reiterating that since the w_n's can be computed to any order n by the sequential process outlined above, the same property applies to the W_n's.

If the right-side of equation (12) is now substituted into equation (7), and the two dummy summation variables are shifted and exchanged, one obtains (suppressing the common factor $e^{i\omega t}$),

$$\xi_z(\varpi, z) = \sum_{n=0}^{\infty} F_n(k\varpi)\, w_n(z) \tag{13}$$

$$= \sum_{n=0}^{\infty}\sum_{m=0}^{\infty}(-1)^m\frac{(m+n)!}{m!n!}k^{-2n}F_{m+n}(k\varpi)W_n(z) \tag{14}$$

$$= \sum_{n=0}^{\infty}\frac{1}{[n!]^2}\Big(-\frac{\varpi^2}{4}\Big)^n W_n(z) \tag{15}$$

where the inner summation over m has been carried out using a formula provided by Magnus et al. (1966). Equation (16) shows that the pseudo-horizontal wavenumber is in fact entirely absent from the resummed result and completes the demonstration that the thin flux tube approximation is contained within the expansion proposed in §3 for the radial planforms given by equation (8). It is worth pointing out that the thin flux tube expansion embodied by equation (16) is precisely that suggested by Ferriz-Mas & Schüssler (1989) based on an entirely independent line of argument.

5. The Flat Canopy Limit

The recovery of the thin flux tube approximation from the expansion method described in §3, and the attendant advantages which that offers for treating the oscillations in the thin root-like structure of the current-sheet sunspot shown in Fig. 1, suggests that one seek an analogous simplification in the thin flat canopy brim that lies just below the vacuum-plasma interface at large radial distances from the spot axis. This endeavor presents a number of interesting complications that are not encountered in the thin flux tube analysis. To appreciate their physical significance, it proves useful to express the Lagrangian displacements in terms of a pair of potentials Φ and Υ according to the following prescription,

$$B^2\,\xi_\varpi = B_\varpi \mathbf{B}\cdot\nabla\Phi + B_z\Upsilon \tag{16}$$

$$B^2\,\xi_z = B_z \mathbf{B}\cdot\nabla\Phi - B_\varpi\Upsilon \tag{17}$$

$$p' = -\rho\ddot{\Phi}\,. \tag{18}$$

Notice that Υ is effectively the fluid displacement transverse to the equilibrium magnetic lines of force, while Φ is essentially the gas pressure fluctuation. In terms of these potentials, the oscillation equations read,

$$\ddot{\Upsilon} = \left\{B_z\frac{\partial}{\partial\varpi} - B_\varpi\frac{\partial}{\partial z}\right\}\ddot{\Phi} + a^2(\varpi,z)\left[\frac{\partial^2}{\partial z^2} + \frac{\partial}{\partial\varpi}\frac{1}{\varpi}\frac{\partial}{\partial\varpi}\varpi\right]\Upsilon \tag{19}$$

$$\ddot{\Phi} = -\frac{g_\odot}{B^2}\Big(B_z\mathbf{B}\cdot\nabla\Phi - B_\varpi\Upsilon\Big) + c^2(z)\;\nabla\cdot\boldsymbol{\xi} \tag{20}$$

where a is again the Alfvén speed.

A careful examination of this pair of equations is particularly useful for illustrating the essential physical properties which impact the construction of the asymptotic structure of the oscillations in the canopy (present in the left, but not the right, panel of Fig. 1). Along the thick solid line the magnetic field strength is finite but the density vanishes, so the Alfvén speed is infinite. Accordingly, the Υ function must to leading order satisfy Laplace's equation with a small residual correction left to balance the remaining terms in equation (20). Indeed along the entire vacuum-plasma interface plane threaded by the ambient magnetic field the global relationship between Υ and its gradient is set by the requirement that the perturbed surface magnetic field must match smoothly onto a potential vacuum magnetic field. This elliptic character of the problem (brought about by the neglect of the displacement current in Maxwell's equations) shows that the transverse motions of the outlying canopy are dependent upon fluid motions near the sunspot axis and vice versa. The same physics is obliged to apply to real sunspot oscillations to the extent that the Alfvén transit time from the spot center to the outer penumbral canopy can be regarded as being short compared to the oscillation period.

In sharp contrast, the behavior of the gas pressure fluctuations in the canopy has a more local and highly-oscillatory component. Whereas the Alfvén speed diverges as the vacuum-plasma interface is approached from below, the sound speed becomes arbitrarily small. As the variations in the specific volume of the plasma should be finite at this interface (i.e., $\nabla\cdot\boldsymbol{\xi}$ does not diverge), it follows that the second term on the right-side of equation (21) is negligible and can be omitted near the underside of this interface. The remaining terms then provide a local relationship (as opposed to the global one Laplace's equation imposes on Υ) between Φ and its gradient, which is inhomogeneous owing to the vertical displacement of the plasma against gravity contributed by the Υ potential. This freedom to add an arbitrary amount of Φ subject only to the constraint that it bears a particular ratio to its local gradient leads to the presence of a second independent mode of oscillation that complements the elliptic Υ counterpart inferred from equation (20). These independent Φ and Υ oscillations can be associated with the slow and fast magneto-atmospheric waves in a low-beta plasma.

However, the singular behavior of the slow mode is now immediately clear. The further one looks out into the canopy in the left panel of Fig. 1, the smaller is the ratio of B_z to B_{ϖ}. If B_z is set precisely to zero then the freedom to specify the slow-mode amplitude is lost and Φ is slaved to Υ. To retain this necessary degree of freedom the equations have no choice but to increase the size of the Φ gradients to compensate for the radial falloff of B_z. This implies that despite the thin geometrical appearance of the canopy, in terms of gas pressure fluctuations and field-aligned motions, one anticipates rapid spatial variations from magnetic field line to magnetic field line across this layer. And thus as far as the slow magneto-atmospheric wave motions are concerned, the canopy depicted in the left panel of Fig. 1 remains "thick" at large radial distances from the spot. This behavior bears a close resemblance to phase-mixing and resonance absorption, and in the present circumstances it most likely represents the direct manifestation of these concepts in the adopted magnetic field geometry. These predictions again apply to real sunspot oscillations in so far as penumbral magnetic fields are highly-inclined but not *precisely* horizontal, and the relevant acoustic transit times can be regarded as being long compared to the oscillation period.

I thank Andy Skumanich for his critical comments on an earlier draft of this article.

References

Bogdan, T. J. 1999, ApJ, in press

Defouw, R. J. 1976, ApJ, 209, 266

Ferraro, V. C. A. & Plumpton, C. 1958, ApJ, 127, 459

Ferriz-Mas, A. & Schüssler, M. 1989, Geophys. Astrophys. Fluid Dyn., 48, 217

Lites, B. W. 1992, in Sunspots: Theory and Observations, (eds. J. H. Thomas & N. O. Weiss), Kluwer, Dordrecht, p. 261

Magnus, W., Oberhettinger, F. & Soni, R. P. 1966, Formulas and Theorems for the Special Functions of Mathematical Physics, Springer-Verlag, New York, p. 129

Skumanich, A. 1992, in Surface Inhomogeneities in Late-Type Stars (eds. P. B. Byrne & D. J. Mullan), Springer-Verlag, Berlin, p. 94

Thomas, J. H. 1983, Ann. Rev. Fluid Mech., 15, 321

Group Discussion

Rutten : You indicated that the trail of power at 5mHz produces enhanced 3mHz power higher up. Does this enhancement extend to higher frequencies?

Bogdan : The enhancement of the field-aligned motions predicted by the calculation of the Eigenfunctions sets in once the frequency of the oscillations exceeds the effective acoustic cut-off frequency of the umbral atmosphere, about 5.2 mHz. Beyond this frequency the field-aligned velocity scales inversely with the square-root of the ambient density, irrespective of the further increase of the oscillation frequency.

Steiner : Tom, can you say something about the magnetic field strength oscillation in your simulation? The amplitude of this oscillation has recently been measured, using MDI data by Ruedi et al., and A. Norton(LA), seeing in the order of 10G.

Bogdan : For a uniformly magnetized cylinder with field strength of 3000G, one expects photosheric field fluctuations on the order of 0.5-3G for a field-aligned doppler velocities of 40m/sec at oscillation periods of 5 min.

High Resolution Solar Physics: Theory, Observations, and Techniques
ASP Conference Series, Vol. 183, 1999
T.R. Rimmele, K. S. Balasubramaniam, R.R. Radick, eds.

The Spectrum of Intensity Fluctuations Across Penumbral Filaments

J. Sánchez Almeida & J. A. Bonet
Instituto de Astrofísica de Canarias, E-38200 La Laguna, Tenerife, Spain

Abstract. We estimate the typical Fourier spectrum corresponding to the spatial fluctuations of intensity in penumbrae. Our high angular resolution sunspot images show spectra whose amplitudes follow the Modulation Transfer Function of the observation. In other words, the real penumbral spectra across filaments seem to be flat up to the cut-off frequency (equivalent to $0\overset{\prime\prime}{.}28$). Present high angular resolution observations do not resolve the penumbral structure.

1. Motivation of the Work

Sánchez Almeida (1998) argues that the real penumbral filaments have to be just a few kilometers wide. Such a small-scale random distribution is needed to balance the observed horizontal and vertical magnetic field gradients, as required by the conservation of magnetic flux. In simpler words, if the filaments were as wide as it is commonly accepted (a few hundred kilometers), then the vertical magnetic field gradients (derived from line asymmetries) and the horizontal gradients (deduced from pixel-to-pixel variations) are not consistent with each other.

If the scenario sketched above does indeed represent typical penumbrae, the fluctuations of brightness associated with the penumbral magnetic structure should be random. Obviously, the random variations only show up whenever the separation between points is larger than the spatial smearing produced by radiative transfer, say, larger than the mean-free-path of the photospheric photons (MFP). Since the MFPs are still smaller than the resolution of the best photospheric observations, the fluctuations in good penumbral images have to be random. Random functions have flat spectra and, consequently, the observed penumbral spectra should to be flat up to the cut-off frequency. The work presented to the workshop was aimed at testing this clear-cut prediction.

2. Estimate of the Fourier Spectra

Fourier transforms were estimated using a series of high angular resolution sunspot images gathered near the disk center with the Swedish Vacuum Solar Tower on La Palma (Spain). The observations, corresponding to the green continuum, are described in detail by Sobotka et al. (1995). We selected the intensity variations along various rings centered on the umbral core. These signals were 1D Fourier transformed to estimate the amplitude spectra. Representative spectra are shown in Figure 1. Although the observed amplitudes fall off towards high frequencies, this damping is just that to be expected for an originally flat spectrum. The response of the optical system to a flat spectrum (i.e., the so-called MTF) is overplotted in the figure for comparison. The flatness of the amplitude spectra holds right up to the cut-off frequency of our observations (~0.″28).

We want to stress that the conclusion of the work, namely, the flatness of a typical penumbral spectrum, does not depend on which sunspot image we choose among the good ones in our series. It also holds for various independent estimates of the MTF, and if the Fourier transforms are computed in 2D with a subsequent selection of the frequencies across filaments.

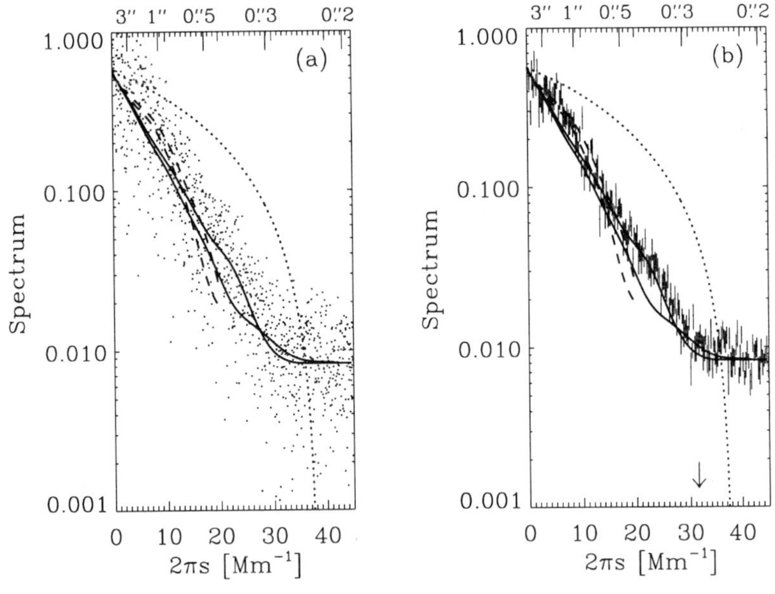

Figure 1. Amplitude spectra of the spatial fluctuations of intensity across penumbral filaments. (a) The dots represent independent determinations whereas the solid and dashed lines estimate the MTF which characterizes the smearing produced by our optical system (including seeing). The dotted line shows the MTF of an ideal telescope. The symbol s stands for the spatial frequency; the corresponding scale, s^{-1}, is shown on top of the figure. (b) Same as (a) but bars corresponding to the mean amplitude ± its standard deviation substitute the individual measurements. As the mean observed spectrum mimics the MTF, the real penumbral spectra seem to be flat. The cut-off frequency, pointed out by an arrow, corresponds to 0.″28.

3. Conclusions

The spectrum of intensity fluctuations associated with the penumbral filaments is flat up to $0\farcs28$ (Figure 1). This observational result is consistent with a penumbra made out of irregular km-wide magnetic fibrils, as claimed by Sánchez Almeida (1998). It also points out that the penumbral structure is not yet resolved.

4. Further Details

Refer to Sánchez Almeida & Bonet (1998) for details of the work summarized here.

References

Sánchez Almeida, J. 1998, ApJ, 497, 967

Sánchez Almeida, J., & Bonet, J. A. 1998, ApJ, 505, 1010

Sobotka, M., Bonet, J. A., Vázquez, M., & Hanslmeier, A. 1995, ApJ, 447, L133

Group Discussion

von der Luhe : 1) How does the diffraction-limited MTF compare with the ones you designed? 2) Doesn't the observed dynamics of penumbrae-fine structure indicate more coherence of the intensity patterns?

Sanchez-Almeida : 1) They are quite different. You can see both of them in the figure containing the observed spectrum. 2) The penumbral irregularities inferred in this work occur across field lines. Along field lines (i.e., along the radial direction in a round sunspot), the filaments should keep their identities over much larger scales. Actually, the motions observed in the penumbrae tend to occur along field lines.

Keller : I have been doing speckle reconstructions of penumbrae with the McMath-Pierce telescope approaching 0.1 resolution. As far as I remember, the power spectra are relatively flat out to that resolution, which is consistent with your hypothesis.

Sanchez-Almeida : I am glad to hear this.

Beckers : Many years ago, Jack Harvey used a dual aperture on the McMath-Pierce telescope(15 degrees cm baseline) to check for the existence of very small-scale structure. He observed fringes in spot umbrae but not in the photosphere. That experiment should be carried out in penumbrae at short wavelengths (eg. 330mm.). It should give you the power at scales of about 30km.

Sanchez-Almeida : I am glad to hear this. I did not know it. there are works showing power in the penumbral spectra up to a very small scales. For example, Stachnik et al.(1983, ApJ, 271, L37) find power up to 0.11. Repeating Harvey's experiment sounds very interesting to me. Radiative transfer smearing should already be important at 30km scales, therefore I would expect a damping of the observed power at these scales.

Chae : 1) I am impressed by the result confirming that there should be very small of 1-10km in penumbral filaments, which is very similar to the value reported to be a typical size of transition region filamentary structures inferred from UV experiments. 2) Is there any physical reason that solar magnetic fields should have filamentary structures of 1-10km?

Sanchez-Almeida : 1) Filamentary structures with such small size seem to be part of all magnetic structures including prominences, the the corona, etc. 2) I do not know why they have to be 10km instead of 30 or 0.1km. However, a back-of-the-envelope estimate of the smallest scales expected in photospheric magnetic field concentrations point out km-size scales(see, e.g., Sanchez-Almeida, 1998, in the 3D structure of solar active regions, eds. Alissandrakis & Schmieder, ASP Conf. Ser., in press).

Simon : I cannot understand how random superposition of dozens or hundreds of tiny filaments can produce relatively stable filaments which often last for several hours and contain organized structures such as penumbral grains.

Sanchez-Almeida : See the reply to O. von der Luhe.

High Resolution Solar Physics: Theory, Observations, and Techniques
ASP Conference Series, Vol. 183, 1999
T.R. Rimmele, K. S. Balasubramaniam, R.R. Radick, eds.

A Model for Penumbral Phenomena

R. Schlichenmaier

Kiepenheuer-Institut für Sonnenphysik, Schöneckstr. 6,
D-79104 Freiburg, Germany

Abstract. We have investigated the dynamical evolution of a thin magnetic flux tube that is embedded in a sunspot penumbra (Schlichenmaier et al. 1998a,b). Initially, the tube is positioned along the magnetopause. During the evolution, it rises upward and migrates inward toward the umbra. Our simulation suggests that many observed penumbral phenomena can be explained as the consequence of a rising magnetic flux tube within the penumbra. In particular, we can address the following features: 1) Existence and migration of penumbral grains 2) the Evershed flow 3) asymmetric Stokes-V profiles 4) the surplus brightness of the penumbra relative to the umbra. However, we need to stress that we have only simulated the evolution of a single tube that is embedded in a constant background model, i.e. the back-reaction of the tube on the background is not included.

1. Introduction

Sunspots have lifetimes that typically exceed several days. Since the dynamical time scale is much shorter, e.g. the travel time of an Alfvén wave across the sunspot roughly amounts to 1 hour, sunspots are to some extend static phenomena. Hence it is reasonable to assume magnetohydrostatic equilibrium and to describe sunspots as a monolithic magnetic flux tube, that fans out with height. For a discussion of physical sunspot models, see Jahn (1997). A sophisticated sunspot model was presented by Jahn & Schmidt (1994). Their self-consistent model in magnetohydrostatic equilibrium includes a deep penumbra and hence describes three stratifications, i.e., the umbra, the penumbra, and the quiet sun. These stratifications are separated by two current sheets which generate the global magnetic field. Their model is constrained by the observed sizes of the penumbra and the umbra and by the radiative fluxes that emerge from the umbra, penumbra, and quiet sun. Even though the magnetic field of the model is two-dimensional, thermodynamic variables are homogeneous in each stratification and only depend on depth. Most compelling about this sunspot model is the fact that it is capable of producing the observed radial dependence of the *mean* magnetic field in the photosphere.

However, we know from observations at high angular resolution ($< 0.5''$) that especially the penumbra is highly inhomogeneous and highly dynamical on small spatial scales:

1. In the continuum, *bright and dark filaments* as well as *penumbral grains* are visible. With photographs at hand that had an angular resolution of 0.3″, Muller (1973b) realized that bright filaments consist of chains of penumbral grains. He described a penumbral grain as a feature that has an elongated comet-like photometric contour (Muller 1973a), its tail always pointing outwards, i.e. away from sunspot center. Penumbral grains are observed to migrate mainly inwards (Muller 1973a). In addition, Denker (1998) and Sobotka, Brandt, & Simon (1998) find bright features that migrate outward in the outer parts of the penumbra. Denker describes them as granules that squeeze through from the quiet sun and which are subsequently pushed outward.

   Recently, Speckle imaging and phase diversity methods have been used in order to achieve close to diffraction limited spatial resolution. It has been found that in some cases the intensity of individual penumbral grains exceeds the intensity of the brightest granules (Sütterlin & Wiehr 1998, Tritschler 1999).

2. Photospheric absorption lines show line shifts and line asymmetries within the penumbra. This is known as the *Evershed effect* (Evershed 1909). Assuming that these line shifts are caused by a flow, one finds that the flow is almost horizontal and directed outwards (e.g. Beckers & Schröter 1968). At high spatial resolution the Evershed effect is correlated with dark penumbral regions (Wiehr & Degenhardt 1994). The observed line asymmetries are such that the line wing is more shifted than the line core (e.g. Degenhardt 1993). The line asymmetry increases with formation height of the spectral line, whereas the line shift decreases with formation height (Bumba 1960). For a strong line (formation height $h \geq 500$ km), Balthasar, Schmidt, & Wiehr(1997) find no line core shift and a shift of the line wing by typically 1.5 km s^{-1}. Simultaneously, a weak line ($h \approx 90$ km) shows a line core shift of 2.5 km s^{-1} and no significant line asymmetry.

3. Another peculiarity of penumbral line profiles is the fact that Stokes-V line profiles of magnetically sensitive absorption lines are asymmetric, and moreover, the kind and degree of the asymmetry depends on the spatial location in the sunspot (e.g. Sanchez Almeida & Lites 1992, and references therein). It can be shown on theoretical grounds, that Stokes-V profiles have to be antisymmetric in the absence of velocity fields. In order to obtain *asymmetric Stokes-V profiles*, velocity gradients are needed. Additionally, Sanchez Almeida & Lites (1992) have shown that in order to reproduce the degree of asymmetry of observed Stokes-V profiles an inhomogeneous magnetic field is needed to enhance the asymmetries that are caused by velocity gradients. In a later work, Sanchez Almeida et al. (1996) demonstrate that the observed line asymmetries can be generated naturally by assuming a micro-structured magnetic atmosphere, in which structures of the magnetic and velocity fields have scales that are much smaller than the mean free path of photons, i.e., scales of a few kilometers. Alternatively, asymmetric Stokes-V profiles can be reproduced qualitatively by assuming a horizontal flow channel that is embedded in a highly inclined magnetic field (Solanki & Montavon 1993).

So far, no self-consistent model has been presented that incorporates these small-scale penumbral phenomena. A natural first step towards an understanding of penumbral peculiarities is to investigate the dynamics of a single thin magnetic flux tube that is embedded in a sunspot model that describes the average thermodynamic and magnetic properties of a sunspot. We have performed such a numerical simulation (Schlichenmaier et al. 1998b), which we shortly summarize in Sect. 2. Based on our results, we present new physical insights on the small-scale dynamics in the penumbra. Our model is capable to qualitatively reproduce the above mentioned peculiarities of the penumbra. In Sect. 3 we will elaborate on the link between our model and the observations.

2. Model

Our model is described in detail in Schlichenmaier et al. (1998b). Here, we only list its essential features, and shortly describe the tube's evolution in a more qualitative way. A video sequence of the evolution can be viewed via the world wide web: `http://www.kis.uni-freiburg.de/~schliche/video.html`.

The three dimensional problem of ideal MHD is reduced to the description of a one dimensional magnetic flux tube that evolves in a two dimensional background by taking advantage of the thin-flux-tube approximation and by assuming an axi-symmetric sunspot model. As a background model, we have prescribed a tripartite sunspot model from Jahn & Schmidt (1994). Within the thin-flux-tube approximation, the lateral pressure balance of the total (gas + magnetic) pressure of the tube with its surroundings is achieved instantaneously and the magnetic flux is conserved along the tube. Special features of our code are: 1) We allow for radiative heat exchange with the surroundings, using a simplified description in which the entropy change is proportional to the temperature difference and the inverse radiative relaxation time. 2) We include partial ionization of Hydrogen and Helium. 3) The tube is embedded in a two dimensional magnetic environment, which implies a force term in the perpendicular component of the equation of motion, namely the gradient of the background magnetic pressure.

Initially, the tube lies along the magnetopause in magnetohydrostatic equilibrium. The initial equilibrium goes unstable due to radiative heat exchange. The tube being at penumbral temperature gets heated by the hotter plasma of the adjacent quiet sun. Thereby the tube expands, gets less dense and becomes buoyant. Buoyancy forces overcome restoring magnetic forces and the tube starts to rise. We note that it is essential that the magnetopause of the background model exceeds a critical inclination angle. Only then, buoyancy forces can overcome the restoring magnetic forces.

During the subsequent evolution the tube rises through the sub-photospheric penumbra. There, the stratification is convectively unstable. Above the photosphere the tube ceases to rise, since there the stratification is convectively stable and, more important, the tube's surplus heat is lost efficiently by radiation whereby the tube gets compressed and more dense. During the rise, the footpoint of the tube, i.e. the intersection of the tube with the photosphere at $\tau = 2/3$, migrates inwards.

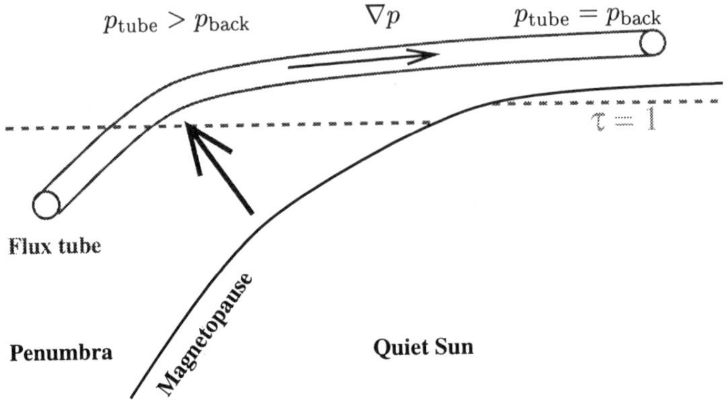

Figure 1. Build-up of surplus gas pressure as the plasma inside the tube rises. The resulting gas pressure gradient accelerates the plasma.

As one may expect, the initial hydrostatic equilibrium along the tube gets perturbed as the tube evolves, and a flow along the tube develops. As the plasma inside the tube rises, the magnetic field strength of the tube decreases more rapidly than the background magnetic field strength. Due to the condition of total (gas + magnetic) pressure equilibrium between the tube and its surroundings, a surplus gas pressure builds up in the plasma that rises within the tube. The resulting pressure gradient accelerates a flow along the tube that points upwards beneath the photosphere and outwards towards the quiet sun above the photosphere (see Fig. 1).

Fig. 2 shows the photospheric part of the tube after 1.5 hours. At this stage the flow exerts a centrifugal force at the turning point in the photosphere, which prevents the tube from sinking. In fact, the tube runs into a quasi-stationary equilibrium, in which the centrifugal force is balanced by the magnetic tension that tries to drag the tube down. As a result, the footpoint stops migrating inward. Here, buoyancy forces are negligible, since the sub-photospheric part of the tube is close to vertical.

3. Link Between Model and Observation

The consistent upward flow along the tube that bends horizontally above the photosphere is the most interesting feature of our model, since it brings up hot plasma that causes bright elongated features within the penumbra. At the same time, after it has cooled by radiation, the flow can account for the Evershed effect in the dark penumbral filaments, and possibly for the asymmetric Stokes-V profiles.

3.1. Penumbral grains: existence and migration

At the snapshot after 1.5 hours the status of the tube is as follows: At the footpoint the tube is very hot. As discussed above, the magnetic field strength is substantially reduced at the footpoint, and hence the tube's radius is enhanced.

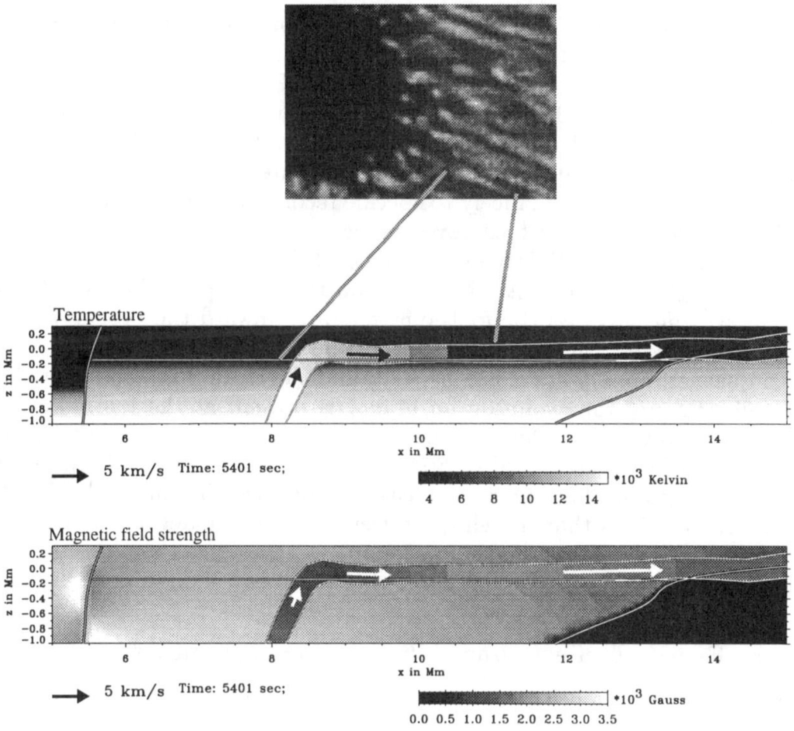

Figure 2. Snapshot after 1.5 hours. From top to bottom: Observed penumbral grain (Courtesy: H. Schleicher), temperature, magnetic field strength. Arrows in tube indicate the flow velocity.

As the plasma flow cools by radiation, the internal energy and the tube's gas pressure decrease, implying an increasing magnetic pressure and a decreasing radius of the tube. Thus, from the footpoint outwards, the brightness and the radius of the tube decrease. Such a tube has a comet-like shape and can be identified with an observed *penumbral grain* as it is indicated in Fig. 2. In the same image one can also see that penumbral grains have different lengths. We surmise two reasons for why bright features may have different lengths: 1) The equilibrium height of the horizontal part of the flux tube may vary due to inhomogeneities in the penumbral atmosphere. Hot plasma that reaches higher layers cools more rapidly due to expansion of the tube induced by the decreasing background pressure and due to a cooler ambient medium, whereas tubes at lower layers expand less, and there the ambient medium is hotter. Lower lying tubes then produce a longer bright tail than tubes that lie higher. 2) Flux tubes that are highly inclined at the footpoint show a foreshortening effect that makes them shorter, and more horizontal flux tubes appear longer. Also, consistent with the previous point, more inclined flux tubes may reach higher up before they reach their equilibrium. From these arguments we would expect that penumbral grains that are observed close to the umbra are not as long as grains further out. This expectation relies on the assumption that flux tubes get more vertical as they approach the umbra.

At the footpoint, the tube is as hot as 13000 K (see Fig. 2). Investigating the cooling behavior of such a hot tube in a cool atmosphere (T = 4800 K) by solving the radiative transfer problem (Schlichenmaier,Bruls, & Schussler 1998c), we have demonstrated that the effective temperature of such a tube does not exceed 6500 K. This result fits nicely to speckle reconstructed temperature maps (Sütterlin & Wiehr 1998), which reveal effective temperatures of penumbral grains that in some cases slightly exceed the 6400 K level.

In our model, the footpoint migrates inward with proper velocities of more than 1 km s^{-1}, initially. As the footpoint migrates inward the proper velocity decreases, reaching 0.1 km s^{-1} after some 1.5 hours. At later stages, the tube reaches a quasi-stationary state (see Sect. 2) and the footpoint stops migrating. Since at all stages of the evolution an upflow is present at the footpoint, this migrating footpoint can be identified with a *migrating penumbral grain.* Tracking the motion of individual grains, Sobotka et al. (1998) find that the proper velocity of many grains does indeed decrease during their lifetime. Additionally, they measure velocities that match quite well with our model results (see also Muller 1973a). Note that we make no attempt to reproduce the dependence of the velocity on position, since we are limited to the study of one single flux tube.

3.2. The Evershed effect: line shift and line asymmetry

In our model, the height of the horizontal part of the tube is determined by the balance between two forces: The gradient of the background magnetic pressure, $\propto \nabla B^2_{\rm back}$, pushes the tube vertically upwards and thereby, due to the convectively stable stratification, increases the density difference between the tube and its surroundings. At some point, the resulting anti-buoyancy force balances the gradient of the background magnetic pressure. In our model, the tube is elevated above the continuum somewhat less than 100 km where it has a diameter of some 50 km. However, from the above argument it is obvious that the height of elevation depends on the details of the magnetic properties of the background model.

Fig. 3 shows a sketch of a flow channel with an optical thickness of 0.1 in the continuum at 500 nm. It spans the range from 50 km to 100 km above the background continuum. Line contributions that stem from that height are Doppler shifted. Fig. 3 also depicts the line wing and line core contribution functions of a typical photospheric absorption line with a formation height of some 200 km. From this sketch we expect that mainly the line wing is shifted by a low lying flow channel, which leads to an asymmetric line profile that has no distinct shift of the line core. In contrast, a fairly symmetric line profile is expected for absorption lines that have a formation height of less than 100 km, because for these lines the contribution functions of the line wing and of the line core are affected by the flow channel. Additionally, the contribution functions of weaker lines have smaller widths, and hence a larger fraction of the contribution functions are affected by the flow channel. Thus, one would expect a larger shift for weaker lines. These expectations are consistent with the observed dependence of line shifts and line asymmetries on the formation height as mentioned in Sect. 1.

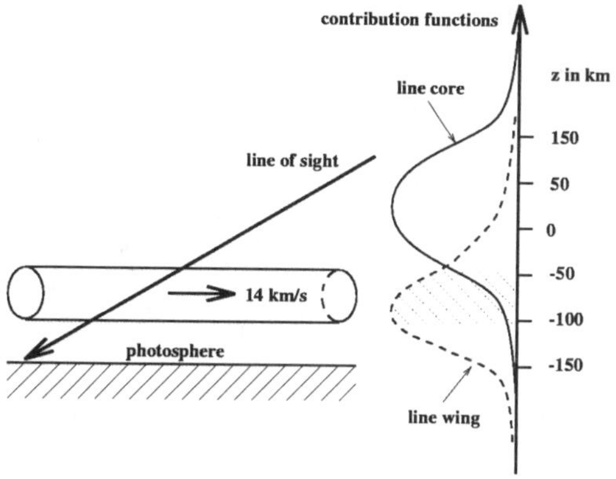

Figure 3. Line asymmetries. A thin flow channel only affects a part of the contributing layer of typical absorption line.

The observed Evershed flow velocity typically ranges between 1 km s^{-1} and 3 km s^{-1}. The measured value of the velocity crucially depends on spatial resolution and on how the observer interprets shifted and asymmetric line profiles in terms of velocity. Assuming that the Evershed effect is not spatially resolved and that the line asymmetry is caused by a line satellite, one can deduce a flow velocity ≥ 5 km s^{-1} (Wiehr 1995). Additionally, we want to point out yet another effect that may lead to even higher velocities: Contribution functions have a width of typically more than 100 km in depth. Hence, a flow channel of 50 km in diameter only partially affects the line profile. That means that even if such a thin flow channel were horizontally resolved, it is not resolvable in depth. The observed line profile always has unshifted contributions from layers where there is no flow present. Hence, the observed line asymmetries and line shifts might be caused by flow channels in which the velocity is substantially higher than 5 km s^{-1}. Even a flow velocity as high as 14 km s^{-1} as it is predicted by our model is consistent with observations.

Finally, we want to mention that our model suggests the presence of a plasma flow in bright features with velocities up to 6 km s^{-1}. Since this part of the tube is hot and optically thick, it is appropriate to use hot absorption lines that have a high excitation potential such as the CI 5380 line. Indeed, Doppler shifts of the CI line are observed within the penumbra (Rimmele 1995, Stanchfield et al. 1997) indicating that an outflow is present in bright features. Obviously, this hot flow cannot be observed using cool FeI lines.

3.3. Asymmetric Stokes-V line profiles

Solanki & Montavon (1993) have found that a horizontal flow channel can, at least qualitatively, reproduce the observed Stokes-V profile within the penumbra. Now, we have a quantitative model at hand and it will be interesting to see whether synthetic line profiles based on our model can reproduce the observations. However, these calculations have yet to be performed.

3.4. Surplus brightness of the penumbra relative to the umbra

We suggest that the observed mean surplus brightness of the penumbra relative to the umbra is caused by upflows along magnetic flux tubes. These upflows of hot sub-photospheric plasma cause penumbral grains and bright filaments which brighten up the photospheric penumbra.

As already mentioned, we have shown (Schlichenmaier et al. 1998c) that no matter how hot a flux tube is, its effective temperature within a penumbral atmosphere does not exceed $T_{\rm eff} = 6500$ K. Assuming a two-component model for the penumbra in which the bright component has $T_{\rm eff} = 6500$ K and the dark component has $T_{\rm eff} = 4000$ K, one can estimate that 40% of the penumbra needs to be covered with bright filaments if the mean penumbral flux is 75% of the quiet sun flux ($T_{\rm eff} = 5777$ K). Taking into account that hot tubes heat up some fraction of the adjacent cooler background, this number compares quite well with observations. For example, Denker (1998) finds that 46% of the penumbra is covered with bright features, and thus confirms Muller (1973b) who finds 43% of the penumbra being covered with bright features.

Acknowledgments. I am grateful to M. Stix for proof-reading the article.

References

Balthasar,H., Schmidt,W., & Wiehr,E. 1997, Solar Phys., 171, 331

Beckers,J.M., Schröter,E.H. 1968, Solar Phys., 10, 384

Bumba,V. 1960, Izvestia Krymsk. Astrofiz. Observ., 23, 253

Degenhardt,D. 1993, A&A, 277, 235

Denker,C. 1998, Solar Phys., 180, 81

Evershed 1909, MNRAS, 69, 454

Jahn,K. 1997, in First Advances in Solar Physics Euroconference: Advances in the Physics of Sunspots, A.S.P. Conf.Ser. Vol. 118, (eds. B. Schmieder et al.), 122

Jahn,K., & Schmidt,H.U. 1994, A&A, 290, 295

Muller,R. 1973a, Solar Phys., 29, 55

Muller,R. 1973b, Solar Phys., 32, 409

Rimmele,T.R. 1995, A&A, 298, 260

Sanchez Almeida,J., Lites,B.W. 1992, ApJ, 398, 359

Sanchez Almeida,J., Landi degl'Innocenti,E., Martinez Pillet,V., & Lites,B.W. 1996, ApJ, 466, 537

Schlichenmaier,R., Jahn,K., & Schmidt,H.U. 1998a, ApJ, 493, L121

Schlichenmaier,R., Jahn,K., & Schmidt,H.U. 1998b, A&A, 337, 897

Schlichenmaier,R., Bruls,J.H.M.J., & Schüssler,M. 1998c, in preparation

Sobotka,M., Brandt,P., Simon,G.W. 1998, in preperation for A&A

Solanki,S.K., Montavon,C.A.P. 1993, A&A, 275, 283

Stanchfield II, D.C.H., Thomas,J.H., Lites,B.W. 1997, ApJ, 477, 485

Sütterlin,P., & Wiehr,E. 1998, A&A, 336, 367

Tritschler,A. 1999, this volume
Wiehr,E., 1995, A&A, 298, L17
Wiehr,E., & Degenhardt,D. 1994, A&A, 287, 625

High Resolution Solar Physics: Theory, Observations, and Techniques
ASP Conference Series, Vol. 183, 1999
T.R. Rimmele, K. S. Balasubramaniam, R.R. Radick, eds.

New Results of 2D-Spectroscopy of Sunspots

H. Balthasar

Astrophysikalisches Institut Potsdam,
Telegrafenberg, D-14473 Potsdam, Germany

Abstract. Two-dimensional spectra of sunspots were obtained at the German Vacuum Tower telescope (VTT) on Tenerife using the 2D-spectrometer of the Göttingen observatory. The velocity field inside and outside the spot is investigated in order to check whether the Evershed-effect is restricted to the penumbra or whether there is a continuation in the moat of the spot. Information about the magnetic field is derived from spectral line profiles. The larger spots show temporal variations of the magnetic field at the umbra-penumbra transition and small spots in the central part.

1. Introduction

It is established now that 5–minute oscillations of Doppler velocities are present in sunspots, but their amplitude is reduced (e.g. Balthasar, Küveler & Wiehr 1987; Braun, Duvall & LaBonte 1987). The question whether the magnetic field participates in the oscillations or not, is discussed. Mogilevskij, Obridko & Shelting (1973) reported periods of 5 and 3 minutes, as well as Gurman & House (1981). Their results were doubted by Bachmann (1983) who found seeing influences in his data. Thomas, Cram & Nye (1984) could not detect any significant peak in their power spectra of the magnetic field. Recently, Efremov & Parfinenko (1996) detected 5 and 2 minutes, and shorter periods. Horn, Staude & Landgraf (1997) found 5 and 3 minute oscillations of the magnetic field strength. This result is confirmed by Rüedi et al. (1998) from data taken with the Michelson Doppler Imager (MDI) onboard the SOHO-spacecraft. In their measurements, the magnetic oscillations in sunspots are restricted to rather small locations, most of them close to the umbra-penumbra boundary. They occur at different places for 5 and for 3 minute oscillations. Lites et al. (1998) also investigated a sunspot for magnetic field variations, but they could only detect an upper limit of 4 G.

2. Observations

The preceding spot of group NOAA 8076 was observed on August 28, 1997 at the Vacuum Tower Telescope (VTT) on Tenerife (Series X6). A second, but much shorter series was recorded on August 22, 1997 centred on a small decaying single spot (NOAA 8073).

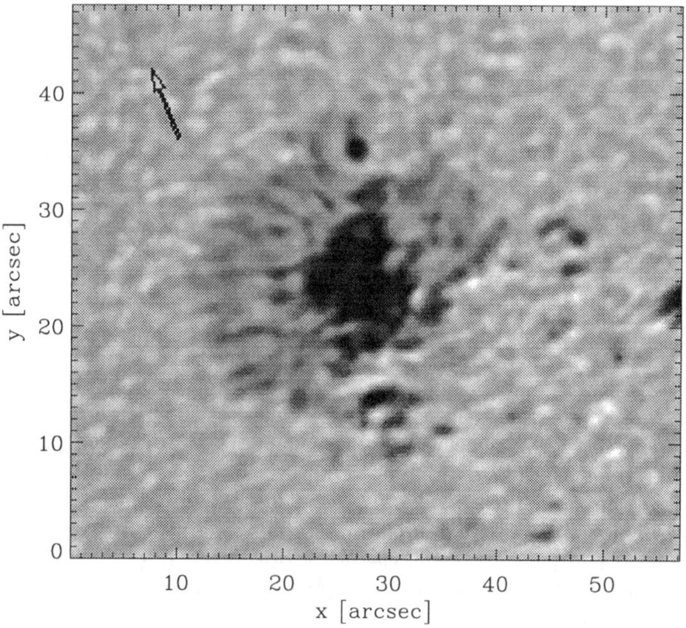

Figure 1. Selected white light image of the leading spot of NOAA 8076. The arrow points toward disk centre. The local contrast has been enhanced by an unsharp masking technique.

Profiles of the magnetically sensitive line Fe I 684.3 nm (Zeeman-triplet, g_{eff}=2.5) were obtained with the Göttingen 2D-spectrometer described by Bendlin, Volkmer & Kneer (1992). The central component of this spectrometer is a tunable Fabry-Perot interferometer. The final field of view was 57 × 48 arc sec^2. The pixel–to–pixel distance corresponds to 0.2 arc sec, but due to seeing influences, the effective spatial resolution was about 1.0 arc sec. After the single frame exposure of 60 ms, the FPI was tuned to the next wavelength point. It took about 15 s to cover all 60 points across the line profile. The spectral resolution was 2.6 pm. More detailed parameters are given in Table 1. Image motions and distortions due to seeing were corrected by image post processing.

Due to the rather high excitation potential of 4.64 eV the line is slightly reduced in the sunspot, and therefore NLTE-effects can be neglected.

The reduction procedure was the same as that used by Balthasar et al. (1996), which is based on descriptions of Bendlin and Volkmer (1995).

The magnetic field strength and the line position were derived from a comparison of the observed line profiles with calculated ones as described by Balthasar & Schmidt (1993). Although this method cannot replace polarization measurements, it yields a reliable magnetic field strength for sunspots. The advantage compared to Horn, Staude & Landgraf (1997) is the larger field of view because no beam splitter is needed.

Table 1. The observations. Given are the total observing length T, the temporal (Δt) and spectral ($\Delta\lambda$) spacings, the distance from disk centre ϑ, the heliographic latitude Θ and the central meridian distance cmd.

spot NOAA	date	start UT	T	Δt [s]	$\Delta\lambda$ [pm]	ϑ [°]	Θ [°]	cmd [°]
8073	Aug. 22	8:21	41^m	49.2	1.29	61	20N	66W
8076	Aug. 28	8:47	$1^h\ 54^m$	53.5	1.29	41	30N	35E

3. Results

Fig. 1 shows an image of the observed spot NOAA 8076, obtained through a broad filter centered on the observed line. On the disturbed side of the main spot, also bright features occur.

From the line shifts necessarily determined for the magnetic fits, maps of the temporal averages of the Doppler velocities were obtained (Fig. 2). The centre side penumbra exhibits blueshifts compared to the spatial mean value. On the other side of the penumbra, redshifts are dominating. Velocities vary between -1810 ms^{-1} and +1600 ms^{-1}, in agreement with the wellknown Evershed effect. Slight blueshifts occur in the area of a close perturbing small spot on the limb-side of the main spot. Some pores at distance to the main spot are connected with blueshifts, although redshifts were expected from former investigations, e.g. Balthasar et al. (1998) and references therein.

The temporal averaged magnetic field strength is shown in Fig. 3. It reaches a maximum value of 2270 G in the umbra. No significant field strength can be detected in the quiet sun preceding the spot (upper left in Fig.3). Values around 1000 G occur on the following side, slightly enhanced in the small spots and pores. It is remarkable that the dark patch enclosed in the centreside penumbra (we call it the penumbral pore) is unconspicuous in the mean magnetic field. For each spatial point, power spectra were determined from the individual measurements of the magnetic field. Power maps for four different period ranges are shown in Fig. 4 and Fig. 5. The power pattern are rather similar for all frequency ranges, thus the power cannot be regarded as indicator of harmonic oscillations. The highest values occur in the penumbral pore and in the umbra of the perturbing small spot on the other side. In the umbra of the main spot enhanced power occurs along a double ring close to the umbra-penumbra boundary and along the umbral waist. Comparable low power is detected in the inner penumbra.

Power values for periods below 200 s are small everywhere except for the transition from the umbra to the penumbra in the main spot and the central parts of some small spots and pores. Inside the main spot, the enhanced power is concentrated in a ring. There is no clear drop of the power at these locations before the Nyquist period at 107 s is reached. This behaviour is comparable to the velocity variations in small spots investigated by Balthasar et al. (1998).

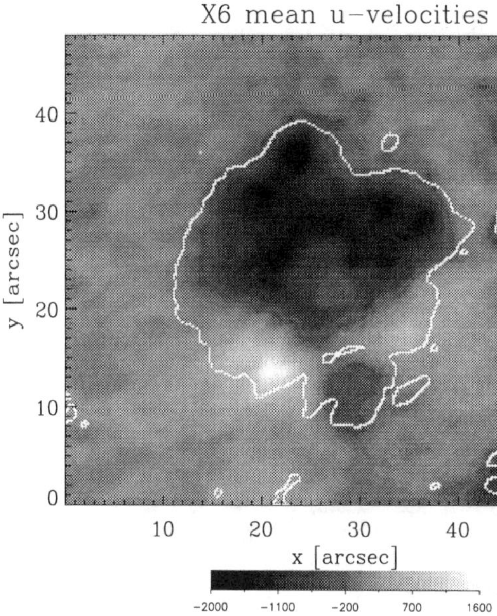

Figure 2. Temporally averaged Doppler velocity map of spot NOAO 8076. The white lines indicate the contours of the spot taken from the temporally averaged image.

Since the umbra-penumbra boundary is a location of strong intensity gradients, and the measurements of the magnetic field strength in the umbra might be influenced by the umbral blend of the line, power spectra of the white light images are calculated too. These maps look similar as those from the magnetic field strength, but really coinciding locations of high power are only the penumbral pore and the umbra of the disturbing spot. Local intensity power maxima are also found close to the umbra-penumbra boundary, but the exact positions of the intensity and the magnetic field power maxima are different. A coherence analysis indicates that there is no relation between the intensity and magnetic field variations. The coherence values of the single points vary strongly forming a pepper-and-salt pattern without any relation to the shape of the spot.

The small decaying spot NOAA 8073 also shows enhanced power of the magnetic field variations for all frequency ranges at the umbra-penumbra boundary. In this case the power is somewhat stronger on the limb side of the boundary. Again, there is no good coherence with intensity variations.

Velocity power maps show outside the spot the 5-minute oscillations, while there is almost no power for the 3 and 2-minute range. The highest values for all period ranges are found in the umbra of the main spot or of the perturbing spot on its limbside. This result was not expected for the 5-minute range, but it is in agreement with the results of Horn, Staude & Landgraf (1997). The velocity power is more concentrated to the central umbra of the main spot than

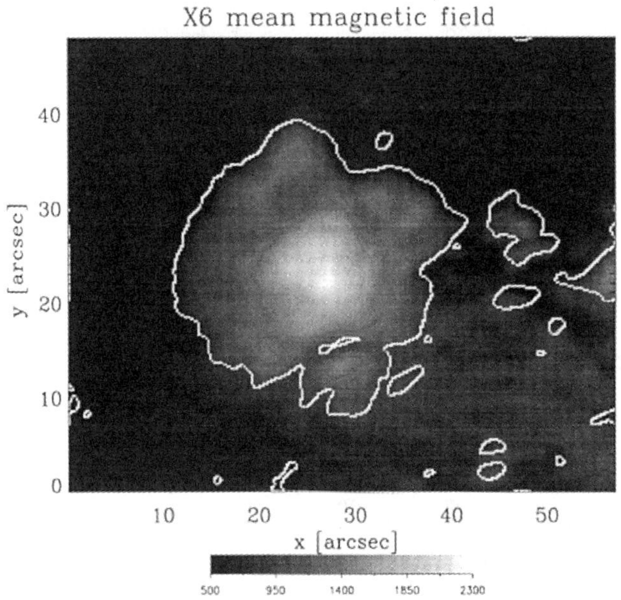

Figure 3. Temporal averaged magnetic field strength of the spot. Insignificant values below 500 G are suppressed. The white lines indicate the contours of the spot.

magnetic field or intensity power. Low power values occur in the penumbra, in agreement with former results of Balthasar (1990). The coherence analysis for the relation between the velocities and the magnetic field yields a similar result as for the relation between magnetic field in intensities.

4. Discussion

The restricted location of the temporal variations of the magnetic field cannot be explained by instrumental effects. If they would be produced by different seeing affecting the measured gradients, the variations of the whitelight images should show coinciding power - this is not the case - and the author expects a high coherence between magnetic field and white light variations for that case. Therefore the magnetic field variations are regarded to be real.

The physical origin of the variations remains open. They cannot be regarded as a superposition of a certain number of harmonic oscillations since power is enhanced of the whole available frequency range.

Preferred is the boundary between umbra and penumbra. This might indicate that the "peripathopause" could have physical background. Further, the small scale of some power peaks is hard to understand assuming a monolithic sunspot model.

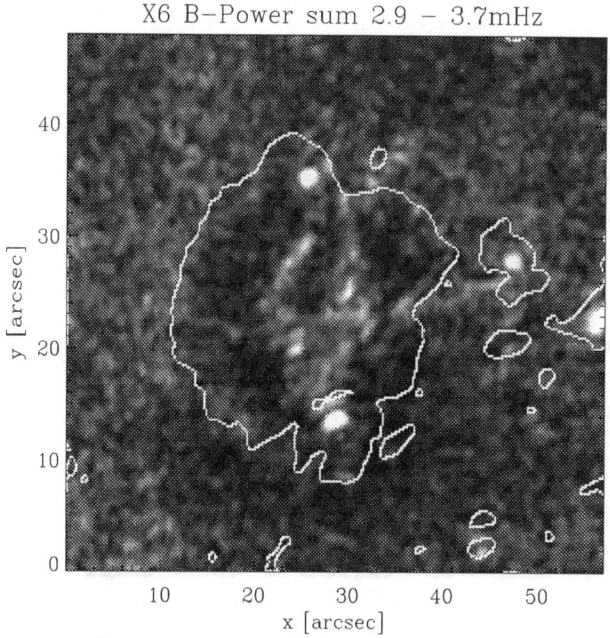

Figure 4. Power map of the magnetic field strength in the 5–minute range (2.9 - 3.7 mHz). High power values are truncated to make small differences visible.

Already during the observation of NOAO 8076, the area of the sunspot group was rather bright in the H_α-image. Very soon after the observation, a H_α-flare in the observed sunspot group is reported. Flares are discussed to excite Doppler-oscillations, thus the question arises whether flares are also related to oscillations of the magnetic field. In most of the cases with detected oscillations of the magnetic field, a flare occurred within 12 hours after the observations but none before. So it seems not very likely that flares are the reason for magnetic oscillations, but magnetic oscillations might deliver energy for a flare.

5. Acknowledgements

The author is indepted to Drs. C. Bendlin, F. Kneer and R. Volkmer for the possibility to use the Göttingen 2D-spectrometer. The 2D-spectrometer was supported by the Deutsche Forschungsgemeinschaft.

The VTT on Tenerife is operated by the Kiepenheuer-Institut für Sonnenphysik (Germany) in the Spanish Observatorio del Teide of the Instituto de Astrofísica de Canarias.

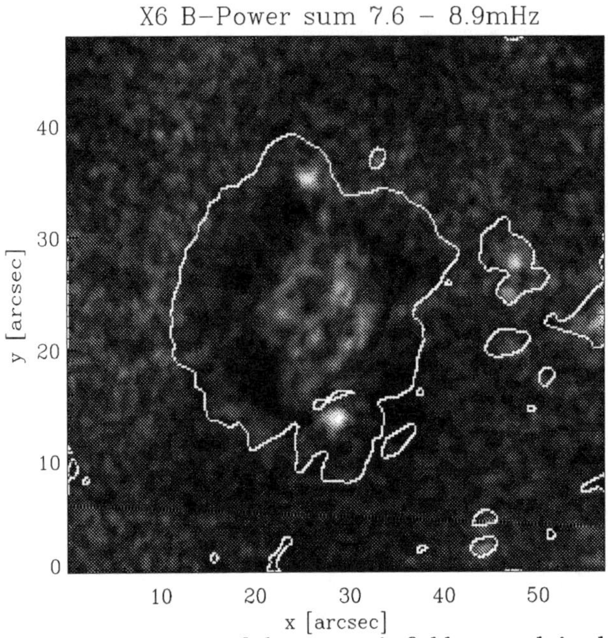

Figure 5. Power map of the magnetic field strength in the 2–minute range (7.4 - 8.9 mHz). High power values are truncated.

References

Bachmann, G. 1983, Phys. Solariterr. Potsdam No. 20, 29

Balthasar, H. 1990, Solar Phys., 125, 31

Balthasar, H. & Schmidt, W. 1993, A&A, 279, 243

Balthasar, H., Küveler, G. & Wiehr, E. 1987, Solar Phys., 112, 37

Balthasar, H., Martínez Pillet, V., Schleicher, H. & Wöhl, H. 1998, Solar Phys., (in press)

Bendlin, C. & Volkmer, R. 1995, A&AS, 112, 371

Bendlin, C., Volkmer, R., & Kneer, F. 1992, A&AS, 257, 817

Braun, D. C., Duvall, T. L. Jr. & LaBonte, B.J. 1987, ApJ, 319, L27

Efremov, V. I. & Parfinenko, L. D. 1996, Astron. Rep., 40, 89

Gurman, J. B. & House, L.L. 1981, Sol. Phys., 71, 5

Horn, T., Staude, J. & Landgraf, V. 1997, Sol. Phys., 172, 69

Lites, B. W., Thomas, J. H., Bogdan, T. J. & Cally, P. S. 1998, ApJ, 497, 464

Mogilevskij, E. I., Obridko, V. N. & Shelting, B. D. 1973, Radiofizika, 16, 1357

Rüedi, I., Solanki, S. K., Stenflo, J. O., Tarbell, T. & Scherrer, P. H. 1998, A&A, 335, L97

Thomas, J. H., Cram, L. E. & Nye, A. H. 1984, ApJ, 285, 368

Group Discussion

Keller : The measured variations in field strength might be explained by the variation in height of the line formation that occurs when the temperature varies in a region with a considerable field-strength gradient such as a sunspot. So, an observed variation in field strength might not be due to an intrinsic variation in field strength.

Balthasar : In that case, I would expect similar variations in the white light images, but in my data there is no relation between magnetic field and intensity variations at the umbral-penumbral boundary. However, at the present state, the physical origin of the observed variation is still unclear.

Sanchez-Almeida : Since you are just observing Stokes I, it may well be that part of what you identify as magnetic field strength variations are actually variations of other physical parameters (e.g., temperature, turbulence, etc.) Correct?

Balthasar : I cannot exclude such crosstalks, but I would expect similar variations regarding the crosstalking parameters. (See answer to C. Keller).

Jain: Whether the dark features in the penumbra show any significant distribution in oscillations or other properties as mentioned by you, from that the central umbra g the spot q.

Balthasar: The power signal is higher in other dark features like the penumbral or the pores outside the main spot, but there are no significant differences in the frequency distribution.

High Resolution Solar Physics: Theory, Observations, and Techniques
ASP Conference Series, Vol. 183, 1999
T.R. Rimmele, K. S. Balasubramaniam, R.R. Radick, eds.

Thermal Structure of a Sunspot: An Application of Phase Diversity

A. Tritschler and W. Schmidt
Kiepenheuer-Institut für Sonnenphysik, Schöneckstr. 6, D-79104 Freiburg, Germany

M. Knölker
High Altitude Observatory, National Center for Atmospheric Research[1], *3450 Mitchell Lane, Boulder, Colorado 80307-3000*

Abstract. We investigate the thermal structure of a sunspot in terms of brightness temperatures. The blurring caused by the combined system atmosphere/instrument in the observed sequence is minimized by applying a phase-diversity image reconstruction method.

1. Introduction

There is abundant empirical data about sunspot brightness and temperatures and also a number of semi-empirical models. These models reasonably describe the mean sunspot atmosphere (Maltby 1992), putting explicit constraints on physical processes which determine the atmospheric structure of a sunspot. Existing sunspot models (Parker 1979; Knobloch & Weiss 1984; Choudhouri 1986; Pizzo 1986; Degenhardt & Lites 1993) can neither solve the still open issue of energy deficit in umbrae, nor give a physical explanation of the small-scale dynamics of the umbral brightenings or for the filamentary pattern of the penumbra. Some residual convective energy transport, even in a sunspot umbra, is necessary to account for its observed brightness, but up to now no significant mass motion has been detected in umbrae or in umbral dots (Beckers 1977; Lites et al. 1991; Schmidt & Balthasar 1994). Hence it remains open, if and how umbral dots provide the necessary energy to the umbra. The situation changes drastically in the penumbra, where strong mass motion (Evershed flow) correlated with the familiar radially oriented filamentary pattern dominates the dynamics. Based on the interchange convection model of Jahn & Schmidt (1994), Schlichenmaier, Jahn, & Schmidt (1998) describe the penumbral fine structure as the manifestation of buoyant magnetic flux tubes driving the Evershed flow.

Observations of sunspot fine structure suffer from the degrading influence of seeing. Nowadays, availability of post-facto image reconstruction tools enable a spatial resolution up to the diffraction limit of the telescope. However, these techniques have to prove their photometric accuracy in order to allow for conclusions on the temperatures at the smallest scales observed.

2. Observations and Data Analysis

The analyzed data set covers a ~90 min sequence of a well developed sunspot (NOAA 7871) observed out of disc center at μ=0.79; it shows a considerable amount of penumbral and umbral fine structure: umbral dots, penumbral grains and a light bridge. The data were taken on May 19, 1995, with the German Vacuum Tower Telescope (VTT), Tenerife.

In order to observe simultaneously in three different continuum bands of the solar spectrum, the observations also employed the Multichannel Filter System of the VTT (Kentischer 1995). A circular field stop was used in the primary focus to limit the effective field of view to 36 arcsec in diameter. Frames were recorded in the continua at 401.9/569.1/709.1 nm with a full width at half maximum (FWHM) bandpass of 1/2/0.3 nm using three large format (1024×1024, 19 μ pixel size) CCD cameras. Spectroscopic inspection of a solar flux atlas for the quiet sun shows that neither the green nor the red continuum is severely affected by absorption lines. For the blue continuum effects of line blocking cannot be excluded. The spatial scale of the telescope in the focal plane is 4.6 arcsec mm^{-1}; the resulting image scale is ~ 0.082 arcsec $pixel^{-1}$. The exposure time was 38 msec and the cycle time was about 15 secs. Due to the rather long exposure times, we cannot assume the atmosphere to be frozen in. Therefore we do not expect to reach the diffraction limit in the reconstructed images.

Since seeing was not very uniform and continuously degrading during the observation, we selected the best 54 frames out of the original 360 frames for further analysis. The apparent spatial resolution in the best frames is about 0.5 arcsec. A defocused image was taken additionally in one of the wavelength channels (λ 569.1 nm, phase-diversity channel). The use of a beam splitter device, which is fixed in front of the CCD camera, allowed to image both the focused and the defocused image onto one CCD chip and thus guarantees the strict simultaneity between the focused and the defocused frame.

The analysis included gain correction and subtraction of dark current. Rotations and lateral shifts due to image motion are compensated for by co-aligning the individual frames using cross-correlation methods. This reduces the spatial area available for the reconstruction and the subsequent temperature analysis to ~25×25 arcsec.

3. Image Processing

The phase-diversity algorithm we used works with a single image pair to create one restored scene; it is based on the linear approach of Löfdahl & Scharmer (1994). The principles of the conventional phase-diversity technique are described by Gonsalves (1982) and Löfdahl & Scharmer (1994). A detailed review of all existing phase-diversity techniques and further references are given by Löfdahl (1996). In our application the wavefront aberration function was parametrized using 15 Zernike polynomials. Inversions were done on subfields of size 128×128 pixel (~10×10 arcsec), smoothly masked off with a modified Hanning window (apodization of 12.5 %). For the full field restoration, the whole FOV (300×300 pixel) is divided in 25 subfields with ~50 % overlap. Noise reduction is accounted for by using a modified optimum filter in each iteration step (see Löfdahl & Scharmer 1994 for a detailed description).

3.1. Wavefront scaling

In principle the information about the wavefront aberrations (given as a result of the phase-diversity algorithm) can be subsequently used to restore the other wavelength channels via a deconvolution process. In order to account for the wavelength dependence of the induced aberrations the phase aberration function is considered to be composed of two different parts corresponding to time-dependent seeing ϕ_S and to instrumentation ϕ_I, which is assumed to be constant in time during our 90 min observation. Two wavelength dependent parameters C_I and C_S are introduced to scale the phase function with λ:

$$\phi(\lambda, t) = C_I(\lambda)\phi_I(\lambda_{PD}) + C_S(\lambda)\phi_S(\lambda_{PD}, t),$$

where the paramters are given by

$$C_I(\lambda) = \frac{\lambda_{PD}}{\lambda} \quad \text{and} \quad C_S(\lambda) = \frac{\lambda_{PD}}{\lambda}\frac{(n-1)}{(n_{PD}-1)}.$$

The index PD refers to the PD channel (569.1 nm) and n denotes the refractive index of air. The parameter C_I derives from the fact that the telescope uses reflecting imaging optics and the pupil size seen in the light of the three wavelengths differs and wavefront curvature corresponds to a different number of waves at different wavelengths. For the derivation of C_S we assume a plane wavefront passing the earth's atmosphere. Considering the atmosphere as a single layer, small corrections apply to the wavefront, to account for the wavelength dependence of the refractive index of air. For simplicity, the dispersion of air at sea level taken from Allen (1955) was used.

Finally, the corresponding function $\phi(\lambda, t)$ and hence the optical transfer function used for the deconvolution were recalculated. In order to determine the fixed aberrations ϕ_I of the combined system telescope/filter device the Zernike coefficients derived from the selected subfield pairs focused/defocused were averaged in time thus following the method described by Acton et al. (1996). The corresponding time dependent seeing part ϕ_S is simply the difference between the result of the phase-diversity algorithm and the fixed term ϕ_I.

4. Brightness Temperatures

The data set is calibrated by means of solar limb-darkening in order to account for the influence of the instrument on the different wavelength channels and the deviation from $cos\theta$=1. Therefore we used the absolute intensities given by Neckel & Labs (1984) and the limb-darkening functions of Neckel (1996). Further, we calculated the continuum intensities and their center-to-limb variation with a numerical code (Großmann-Doerth, Larsson, & Solanki 1988; Großmann-Doerth 1994) solving for the radiation transfer equation using the plane parallel quiet sun stratification of Holweger & Müller (1974) The result is shown in Fig. 1 together with the emprically derived limb darkening. The curves are in good

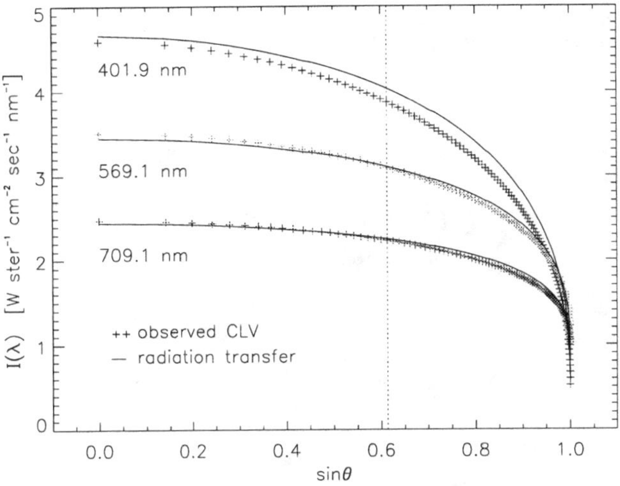

Figure 1. Observed center-to-limb variation of the absolute intensity compared to the result of a numerical calculation solving the radiation transfer equation. The dotted line indicates the sunspot's position on the disc during the observation.

agreement except for a slight deviation in the blue continuum increasing towards the solar limb, thus underestimating the true continuum due to effects of line blocking.

In the next step temperatures are derived by simply converting the intensities into brightness temperatures via the black body assumption. This gives photospheric temperatures of 6190/5945/5820 K for the empirical and 6240/5950/5650 K for the numerical calculated intensities at μ=0.79. For the further analysis and derivation of temperature maps the empirical values are used to calibrate the data set.

5. Results and Discussion

A typical temperature map is shown in Fig. 2 for the green continuum at 569.1 nm. For the in time and space averaged darkest parts of the umbra we find temperatures of 4860/4740/4540 K for the blue, green and red continuum respectively; absolute minimum values are reached at 4640/4470/4150 K.

With regard to the rather high umbral temperatures the question arises, whether remaining scattered light can account for that. Therefore we compared our intensity ratios umbra/photosphere with that given by Maltby et al. (1986). On average our ratios are ~10 % higher than those given by Maltby. Such an error would lead to an overestimate of 330-470 K in the darkest umbral parts. However, the values given by Maltby refer to rather large sunspots and not to small sunspots, like the one observed here. Sütterlin (1991) measured stray light of the VTT together with his observations of pores. Therefore we take his values of 5/4/3.5 % as an approximation and yield an overestimate

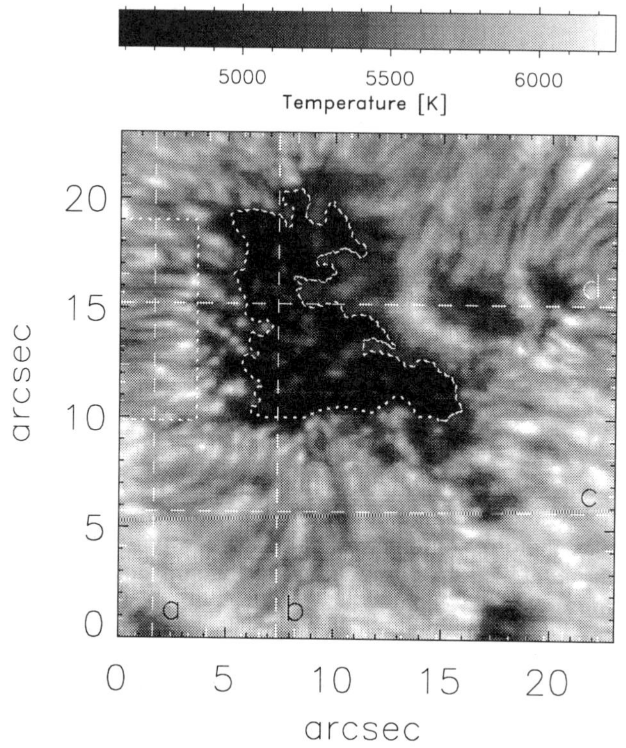

Figure 2. Temperature map derived from the phase-diversity reconstructed image in the green continuum at 569.1 nm. The dotted/dashed lines indicate the umbral and penumbral regions used for the histograms/the spatial variation of temperature in Fig. 3.

of ~110-200 K for the darkest parts of the umbra. Taken the correction for scattered light into account these values are still significantly higher than those for most semi-empirical umbral core models, such as Kollatschny et al. (1980) or the cool component of Collados et al. (1994). These models yield continuum intensities wich correspond to brightness temperatures of 4120/3940/3870 K and 3970/3780/3710 K respectively. Collados et al. (1994) however emphasize, that small sunspots genuinely show higher temperatures than large spots. Their hot umbral model yields temperatures of 5010/4710/4590 K, which is in agreement with our results.

None of the umbral dots reaches photospheric temperatures; typically they exceed the umbral minimum by 600-1200 K, depending on their location inside the umbra; umbral dot temperatures are about 5450/5280/5350 K and increase up to 5750/5580/5680 K towards the boundary umbra-penumbra. There has been an extensive debate about the true temperatures of umbral dots starting from Beckers & Schröter (1968). More recents work established that centrally located umbral dots are typically cooler than those closer to the umbra-penumbra

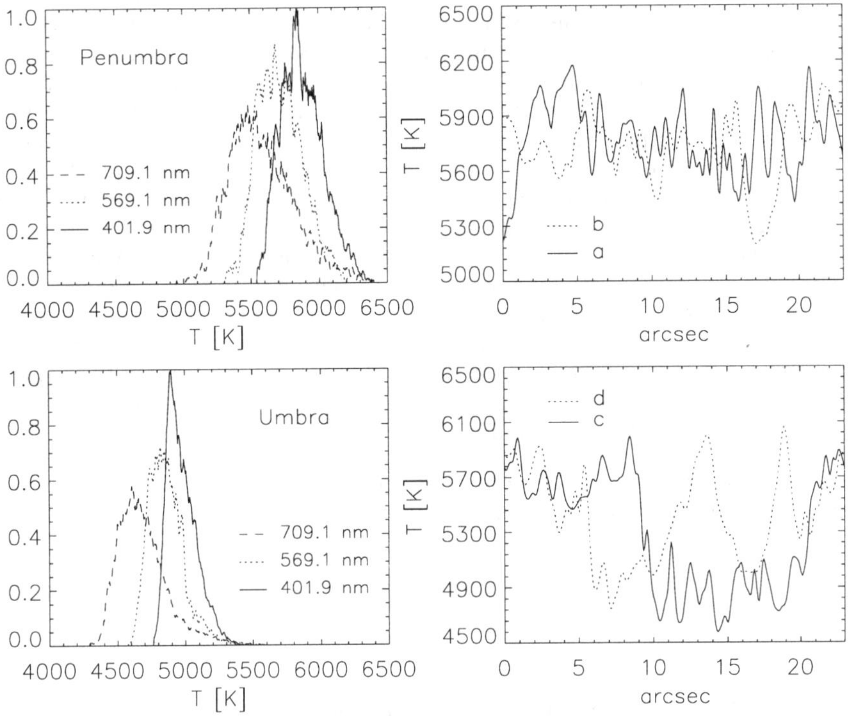

Figure 3. Left panel: histograms for the brightness temperatures derived from the umbral and penumbral regions indicated in Fig. 2. Right panel: spatial variation of temperature across the sunspot for the positions indicated in Fig. 2 in the green continuum at 569.1 nm.

boundary (Koutchmy & Adjabshirzadeh 1981; Großmann-Doerth, Schmidt, & Schröter 1986; Tritschler & Schmidt 1997; Sütterlin & Wiehr 1998).

The spatially averaged penumbra shows temperatures around 5840/5660/5550 K slightly varying in time. Spatial temperature fluctuations within the penumbra are of the order 400 K. Dark penumbral filaments thus reveal similar temperatures than those of the umbral dots. Exceptionally high temperatures are reached within some of the comet-like like shaped bright filaments, showing peak values of in the range 6250-6700/6030-6320/6050-6500 K; typically they exceed the photosphere by 200-400 K, thus being hotter than even the brightest granules (6290/6030/5950 K) in our sample. This supports the idea of Schlichenmaier, Jahn, & Schmidt (1998) of the bright heads being the footpoints of flux tubes (at τ=1), much hotter than the surrounding penumbra and even hotter than the quiet photosphere. The lightbridge exhibits high temperatures too: peak values of 6520/6250/6310 K are reached. The photospheric temperature is typically exceeded by 150-300 K.

The histograms, normalized to the maximum of the blue curve, show a shift towards lower temperatures from the blue to the red continuum as expected. However, we also note that the curves differ in their width, increasing from the blue to the red continuum. Especially the red curves exhibit not only a broader distribution but also a high-temperature tail extending farther than the green curves. This is most clearly seen in small structures such as umbral dots and penumbral grains where the red temperatures exceed the green temperatures. We interpret this as an overcorrection in the red continuum leading to brightness temperatures which are too low in dark structures and too high in bright structures. The temperature distribution in the blue continuum is relatively narrow. An essential requirement of the post-facto image restoration is the strict simultaneity of the data in all three wavelength bands. We find that the intensity ratio umbra/photosphere in its variation in time is the same for the green and the red continuum but not strictly for the blue one. This indicates that the three cameras did not take the images at the same time. Hence, the blue data set was only corrected for the fixed abberrations, leading to a general undercorrection of the structures. This is reflected in the narrow temperature distribution of the blue continuum.

Acknowledgments. We would like to thank Tim Brown and Oskar von der Lühe for many useful comments.

References

Acton, D.S., Soltau, D., & Schmidt, W., 1996, A&A 309, 661

Allen, C.W., 1955, in: Astrophysical Quanteties, Athlone Press, 115

Beckers, J.M., 1977, ApJ 213, 900

Beckers, J.M., & Schröter, E.H., 1968, Solar Phys. 4, 303

Choudhouri, A.R., 1986, ApJ 302, 809

Collados, M., Martínez Pillet, V., Ruiz Cobo, B., del Toro Iniesta, J.C., & Vázquez, M., 1994, A&A 291, 622

Degenhardt, D., & Lites, B.W., 1993, ApJ 404, 383

Gonsalves, A.R., 1982, Opt. Eng. 21, 829

Großmann-Doerth, Schmidt, W., & Schröter, E.H., 1986, A&A 156, 347

Großmann-Doerth, U., Larsson, B., & Solanki, S.K., 1988, A&A 204, 266

Großmann-Doerth, U., 1994, A&A 285, 1012

Holweger, H., & Müller, E.A., 1974, Solar Phys. 39, 19

Jahn, K., & Schmidt, H.U., 1994, A&A 290, 295

Kentischer, T.J., 1995, A&ASS 109, 553

Knobloch, E., & Weiss, N.O., 1984, MNRAS 207, 203

Kollatschny, W., Wiehr, E., Stellmacher, G., & Falipou, M.A., A&A, 86, 245

Koutchmy, S., & Adjabshirzadeh, A., 1981, A&A 99, 111

Lites, B.W., Bida, T.A., Johannesson, A., & Scharmer, G.B., 1991, ApJ 373, 683

Löfdahl, M.G., & Scharmer, G.B., 1994, A&ASS 107, 243

Löfdahl, M.G., 1996, Dissertation, Stockholm

Maltby, P., Avrett, E.H., Carlsson, M., Kjeldseth-Moe, O., Kurucz, R.L., & Loeser, R., 1986, ApJ 306, 284

Maltby, 1992, in: Sunspots: Theory and Observations, eds. J.H. Thomas and N.O. Weiss, Kluwer, 103

Neckel, H., Labs, D., 1984, Solar Phys. 90, 205

Neckel, H, 1996, Solar Phys. 167, 9

Parker, E.N., 1979, ApJ 234,333

Pizzo, V.J., 1986, ApJ 302, 785

Schlichenmaier, R., Jahn, K., & Schmidt, H.U., 1998, A&A 337,897

Schmidt, W., & Balthasar, H., 1994, A&A 283, 241

Sütterlin, P., 1991, Diplomarbeit, Freiburg

Sütterlin, P., & Wiehr, E., 1998, A&A 336,367

Tritschler, A., & Schmidt, W., 1997, A&A 321, 643

High Resolution Solar Physics: Theory, Observations, and Techniques
ASP Conference Series, Vol. 183, 1999
T.R. Rimmele, K. S. Balasubramaniam, R.R. Radick, eds.

Lifetimes and Motions of Penumbral Grains — Preliminary Results

M. Sobotka

Astronomical Institute, CZ-25165 Ondřejov, Czech Republic

P. N. Brandt

Kiepenheuer-Institut für Sonnenphysik,
D-79104 Freiburg, Germany

G. W. Simon

Air Force Research Laboratory, National Solar Observatory,
Sunspot, NM 88349, USA

Abstract. The properties of penumbral grains (PGs) in a medium size sunspot are studied from a 4.5 h observation series observed on 1993 June 5 at the Swedish Vacuum Solar Telescope (SVST), La Palma. The application of an image segmentation procedure yields a set of 605 penumbral grains which are investigated by a feature tracking algorithm.

Approximately two-thirds of the PGs move toward the umbra and one-third toward the photosphere, whereas only 5 PGs move tangentially. For the inward moving PGs we find an average lifetime of 30 minutes, for the outward moving ones of 16 minutes. Plots of the average lifetime and travel speed versus distance from the umbra are presented. There appears to be a dividing line in the penumbra, such that most PGs outside this line move toward the photosphere, and those inside move toward the umbra.

In summary, we find much shorter lifetimes than earlier authors, and no pronounced dependence of lifetime and travel speed on position in the penumbra. We discuss possible reasons for the disagreement in results, and make some comments on the differences between human and computer selection and tracking of features.

1. Introduction

Penumbral grains (PGs) are small (typically less than one arcsecond in width) elongated bright features visible in the penumbrae of sunspots. They have not been studied by many authors, because successful observations require an extended period of excellent seeing. Indeed, this is one of only a half-dozen PG observations reported in 35 years.

Neither of the first two observations (Danielson 1961, Schröter 1962), both from high-altitude balloon flights, yielded detailed results on PGs. Danielson did not actually investigate PGs but rather studied "bright penumbral filaments" and concluded that their lifetimes are about five times that of granules. Schröter's observations were too short to determine lifetimes, but he did give an estimate of PG velocities. Later Muller (1973) and Tönjes & Wöhl (1982) (hereinafter T&W) made longer and more detailed measurements from ground-based observations, and determined both velocities and lifetimes of PGs as a function of position within the penumbra. Zirin & Wang (1989) found that "both bright and dark structures in the penumbra move inward with average velocity 0.5 $\mathrm{km\,s^{-1}}$." What is most interesting and provocative in the earlier observations is the striking disagreement among the authors, especially regarding the magnitude and direction of PG motions. To illustrate this, we quote from Schröter's abstract:

> "Bright knots as seen in the penumbral filaments show changes in structure as well as a systematic motion (1–2 $\mathrm{km\,s^{-1}}$) away from the center of the spot. These 'proper motions' of the knots represent the Evershed-effect, seen here in integrated light for the first time."

whereas Muller (1973) says about PGs that:

> "They move toward the umbra of the spot. Their horizontal velocity is zero at the border penumbra-photosphere and maximum at the umbral border (0.5 $\mathrm{km\,s^{-1}}$).... The lifetime of the grains is a function of their place of origin within the penumbra: It is maximum and of the order of 3 h or more for those forming in the middle part of the penumbra, and 50 and 40 minutes respectively for the points formed in the inner and outer part of the penumbra."

That is, Schröter finds very large *outward* flows, primarily near the penumbra-photosphere (PEPH) boundary, and no appreciable motion near the penumbra-umbra (PEUM) boundary. But Muller finds only *inward* motions, with maximum velocity at the PEUM boundary, where Schröter finds none. Muller observes an essentially linear decrease of the inflow speed from umbra to photosphere. T&W also find an inward flow, which has peak magnitude in the center of the penumbra, decreasing toward both the umbra and the photosphere. Muller (1973) finds a maximum speed of 500 $\mathrm{m\,s^{-1}}$ whereas T&W find a maximum of only 330 $\mathrm{m\,s^{-1}}$. These are roughly one-fourth the magnitude reported by Schröter.

Since we had an excellent high-resolution time series of a sunspot, and had recently developed computer algorithms for the analysis of solar digital image data, we decided to study PGs anew, in hopes of resolving the contradictory earlier results.

2. Observations and Analysis

The data were obtained with a fast real-time frame selection system and a quad-cell image stabilization at the SVST on 1993 June 5, in white light at 468$\pm$5 nm. Image scale on the CCD array was 0.125 arcsecond per pixel. From a total of 760 frames, we picked the 330 best frames with a mean time lag of 44.5 s. The chosen field-of-view, picked from a much larger original image, was 288 x 288 pixels, or

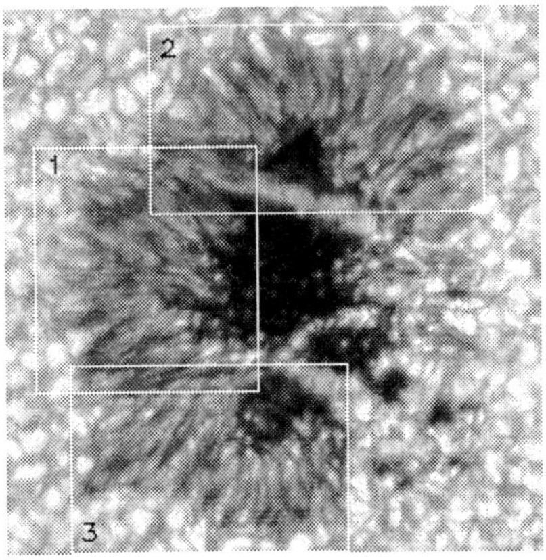

Figure 1. Sunspot in NOAA group 7519, observed 1993 June 5, with three subfields marked to show regions analyzed. The field-of-view is 36 x 36 arcseconds.

36 x 36 arcseconds. It was centered on the main spot of group NOAA 7519 which was near disk center at the time of observation. After correcting for dark current, flat field, rotation, transparency, exposure variations, and telescope MTF, and application of destretching and subsonic filtering, we selected three subfields of the sunspot (Fig. 1) that showed regular penumbrae. For each subfield we applied an image segmentation algorithm (similar to the one used by Sobotka et al. 1997) to select PGs, and then used an identification and tracking algorithm to follow each PG. To be considered a PG, each object had to:

(a) be larger than 1 pixel on each frame
(b) overlap in at least 1 pixel on subsequent frames
(c) have a maximum interruption of identification of 2 frames
(d) have a lifetime of at least 3 minutes
(e) have a mean area over its lifetime of at least 9 pixels

This procedure yielded a set of 605 PGs whose lifetimes and proper motions are investigated.

3. Results

3.1. Motions of Penumbral Grains

Of the 605 PGs, 418 (roughly 2/3) move inward toward the umbra. We label them INW. Almost 1/3 (182 PGs) move outward toward the photosphere. We call them OUT. Only 5 PGs (less than 1%) move in a tangential direction. In this subsection we consider only a restricted sample of 230 INW and 46 OUT PGs with lifetime $\tau > 15$ minutes which passed a visual consistency check following selection by the computer algorithms. We imposed the $\tau > 15$ minutes condition to obtain reliable values of velocities.

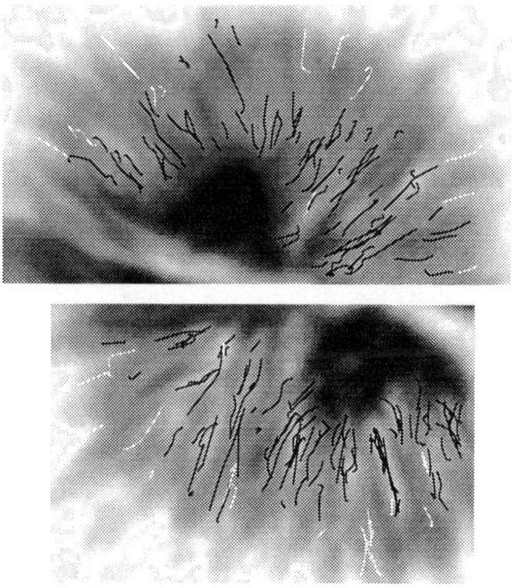

Figure 2. Trajectories of INW (black) and OUT (white) PGs, for subfield 2 (upper) and subfield 3 (lower) of Fig. 1.

There appears to be a dividing line (DL) in the penumbra, approximately 0.7 of the distance from umbra to photosphere. Outside the DL most PGs are of type OUT; inside most are INW. This phenomenon is illustrated in Fig. 2, in which we show trajectories of the INW and OUT PGs as black and white lines, respectively, for subfields 2 and 3 of Fig. 1. Most trajectories range between 0.25 arcsecond and 4 arcseconds, in a penumbra approximately 8 arcseconds wide.

For PGs with lifetimes greater than 15 minutes, we show in Fig. 3 a histogram of the time-averaged PG travel speeds. Speeds for INW PGs (solid line) range from 0.0 to 1.9 $\mathrm{km\,s^{-1}}$ with a maximum around 0.3–0.4 $\mathrm{km\,s^{-1}}$ and a median of 0.4 $\mathrm{km\,s^{-1}}$. Travel speeds of the OUT PGs (dashed line) range from 0.0 to 1.2 $\mathrm{km\,s^{-1}}$. Note that the histogram is rather flat with median value 0.6 $\mathrm{km\,s^{-1}}$.

As function of the PG's position in the penumbra, the time-averaged travel speeds are displayed in Fig. 4. For the INW PGs (shown as $\times$) speeds range from 0.3 to 0.4 $\mathrm{km\,s^{-1}}$ at the PEUM boundary (position 0.0 on the abscissa) and may exhibit a small increase toward the photosphere (position 1.0). Travel speeds of the OUT PGs (shown as squares) are 0.2 $\mathrm{km\,s^{-1}}$ at the inner penumbra, but increase suddenly to 0.7 $\mathrm{km\,s^{-1}}$ beyond position 0.6. Note, however, that the numbers of OUT PGs at positions less than 0.7, and of INW PGs at positions greater than 0.7, are very small, as can be seen in Fig. 2.

3.2. Lifetimes of Penumbral Grains

We show in Fig. 5 a histogram of PG lifetimes. The INW and OUT PGs exhibit basically different lifetimes: 33% of the INW PGs (solid line) live less than 15 minutes; only 11% live more than 1 h. They have a mean lifetime of 29.8 minutes, with a median of 22.5 minutes. In the case of OUT PGs (dashed line) 60% live less than 15 minutes; none lives more than 1 h. These have a mean lifetime of only 15.6 minutes with a median of 11.9 minutes.

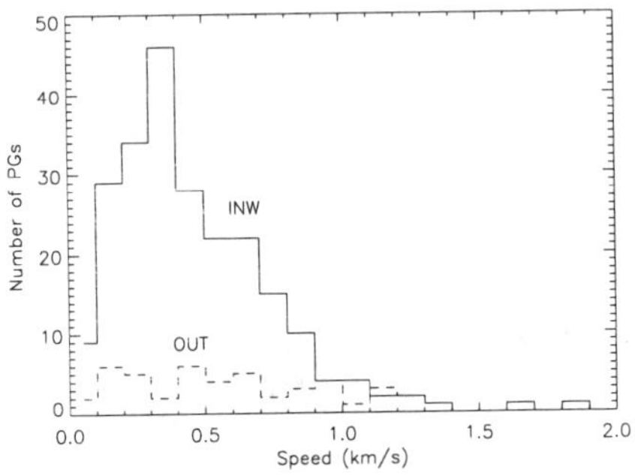

Figure 3. Histogram of average speeds of PGs. Solid line (INW), dashed line (OUT).

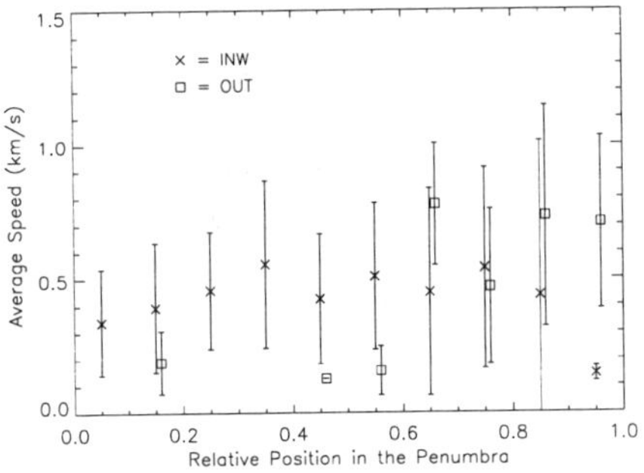

Figure 4. Average speed of PGs as function of their mean position in the penumbra. INW PGs are represented by ×, OUT PGs by squares. Position is 0.0 at umbra, 1.0 at photosphere. The bars indicate a scatter of $\pm\, 1\sigma$.

The average lifetime of PGs with $\tau > 15$ minutes as function of position in the penumbra is shown in Fig. 6. Mean lifetime of the INW PGs (shown as ×) is 35–50 minutes for those located at positions 0.0–0.7 in the penumbra, and then drops to 25 minutes near the photosphere (positions greater than 0.7). Mean lifetime of the OUT PGs (shown as squares) is 25 minutes for the whole range.

In summary, while we agree with Muller (1973) that the average lifetime of INW PGs in the inner part of the umbra is larger than for those near the PEPH boundary, we do not see the pronounced dependence of lifetime on penumbral position reported by Muller (1973) and T&W. More significantly, the lifetimes we measure are only about one-fourth as large as those of the earlier measurements.

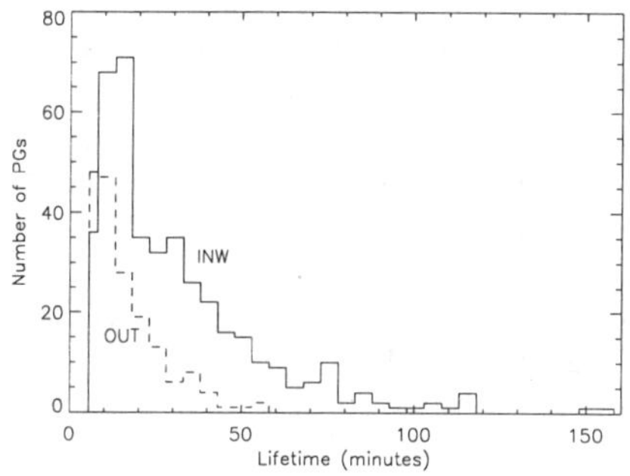

Figure 5. Histogram of PG lifetimes. Notation as in Fig. 3.

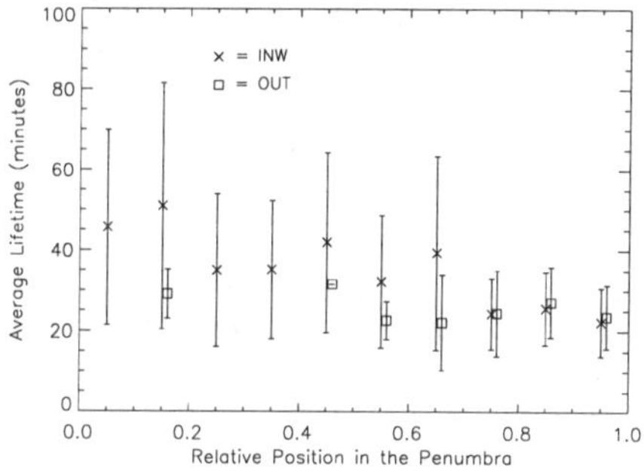

Figure 6. Lifetime of PGs as function of their mean position in the penumbra. Notation as in Fig. 4.

4. Discussion

The average PG speeds we measure are in reasonably good agreement with those of Muller (1973) and T&W, but the character is very different. Whereas they find almost exclusively INW motions, we find both INW and OUT strongly dependent on position in the umbra as described above. In addition, all three of these observations show speeds far smaller than those of Schröter. Muller (1998) reported to us that when he showed his (1973) data to Schröter in 1972 (pre-publication) Schröter was convinced by Muller's results, and was concerned that there might have been errors in his (1962) measurement reference system. Thus, for example, if there had been a systematic velocity bias in the outward radial direction, then the corrected velocity pattern would have shown smaller outflows in the outer part of the penumbra, and larger inflows in the inner part, in much better agreement with both our and Muller's (1973) results.

Recalling in more detail the difference between our velocity observations and those of Muller (1973), note that we found a DL between INW and OUT PG motions, while Muller states that all motions are INW. However, careful examination of his Fig. 3 shows that he actually found a few OUT PGs at penumbral positions 0.25–0.80, and more OUT than INW PGs in the outermost penumbra (positions 0.8–1.0). Thus the qualitative difference in our results may not be as large as it first appears.

Local correlation tracking (LCT) has been used by several authors in recent years to study motions in sunspots. Because of LCT's inherently poor spatial resolution (typically 1–2 arcseconds; i.e., each LCT "pixel" contains 64–256 of our image pixels), and its inability to distinguish between motions of bright or dark features, it is not clear that LCT of a sunspot penumbra is tracking PGs solely, or other penumbral features in addition. With these caveats in mind, we note that LCT results by other authors, using different sunspot data, have confirmed our DL observation and the dominance of OUT motions in the outer penumbra, and of INW motions in the inner penumbra. For example, Denker (1998) finds that "a line of positive divergence divides the penumbra", and also notes that "velocities at the outer boundary of the penumbra are somewhat larger than in the inner parts of the penumbra", in agreement with our Fig. 4.

Part of the discrepancy between our work and Muller's (1973) may simply be the physical properties of the particular sunspots studied. Our spot was medium-size, Muller's very large. There is no a priori reason to expect all spots to have similar velocity fields. The character of the flow may vary significantly as function of size, age, evolutionary stage (young, old, growing, decaying), and location (unipolar, leading, following, simple or complex group), etc. Thus it is possible that Muller (1973) and we are both completely correct despite our disparate results. To us, this appears to be the most reasonable explanation of the differences in our results.

On the other hand, the difference may, at least in part, lie in the methods of analysis and/or the quality of the observational data. Muller (1998) believes that "investigations based on visual identifications are superior to investigations based on computer identifications, because they are more 'intelligent'; one can see the evolution of grains in more detail, understand better what happens and take some decisions; it is less sensitive to threshold effects, because the eyes can detect small contrasts." He acknowledges that his process "is tedious and time-consuming and might not be objective enough in the decisions." This last comment by Muller is the main reason that we prefer computer identification of features. Even though the algorithms may occasionally make mistakes, they are completely objective in their decision-making. Also they are not subject to two biases inherent in visual identification; namely, the tendency of the eye-brain interaction to selectively prefer brighter (higher contrast) and larger features. The computer is ideally suited to performing tedious tasks; thus it selects far more features for analysis than the human analyst is physically or psychologically prepared to do.

With regard to the shorter lifetimes that we measure, Muller (1998) concludes "that the difference in lifetime we find in our respective investigations is mainly due to the different time interval between frames and to different ways we identify the penumbral grains. One reason may be, I think, the intensity threshold level, because I consider that a grain is still there, even if its contrast is low; if the contrast increases again, I consider it is the same grain, but you may find it is a new one." We disagree with Muller's interpretation. First, we believe that the computer is more capable of discerning small contrast differences than the eye. Second, we have attempted to account for temporary disappearances of PGs by permitting gaps of up to two times steps (89 s) between disappearance and re-appearance. As mentioned earlier, we are concerned that visual bias preferentially selects larger and higher-contrast PGs. It is obvious that such PGs will have longer lifetimes than the average PG; this effect may contribute significantly to the observed differences in our results.

We are developing a new image segmentation algorithm that will take into account the variations of image quality due to seeing. After we implement this procedure, we will update the results reported here.

Finally, what is a penumbral grain? Since it is bright, it must represent localized heating within a penumbral filament. Since we see the PGs "move", does this brightening indicate mass motion of hot material? Perhaps these bright features are the loci of the intersections with the photosphere of hot, rising, flux tubes as suggested by Schlichenmaier et al. (1998). If so, these numerical simulations must be refined to account for both the INW and OUT observed motions; at present the models only predict INW flows.

Acknowledgments. We especially thank Richard Muller for very helpful discussions, and his permission to quote his unpublished comments. We thank G. Scharmer for the opportunity to observe at the SVST, W. Wang and R. Kever for observational support, and M. Shand of DEC (Paris) for financial and technical assistance with the data acquisition system. We gratefully acknowledge funding from the Air Force Office of Scientific Research for salary (GWS) and travel support (PB and MS). The National Solar Observatory also provided travel funds for PNB and MS. The work of MS was accomplished under the Key Project K1003601 of the Academy of Sciences of the Czech Republic. The SVST is operated by the Swedish Royal Academy of Sciences at the Spanish Observatorio del Roque de los Muchachos.

References

Danielson, R. E. 1961, Astrophys. J., 134, 275

Denker, C. 1998, Solar Phys., 180, 81

Muller, R. 1973, Solar Phys., 29, 55

Muller, R. 1998, private communication

Schlichenmaier, R., Jahn, K., & Schmidt, H. U. 1998, A&A, 337, 897

Schröter, E. H. 1962, Z. Astrophys., 56, 183

Sobotka, M., Brandt, P., & Simon, G. 1997, A&A, 328, 682

Tönjes, K. & Wöhl, H. 1982, Solar Phys., 75, 63

Zirin, H. & Wang, H. 1989, Solar Phys., 119, 245

High Resolution Solar Physics: Theory, Observations, and Techniques
ASP Conference Series, Vol. 183, 1999
T.R. Rimmele, K. S. Balasubramaniam, R.R. Radick, eds.

Properties of Sunspots and Pores

C. Denker, T. J. Spirock, P. R. Goode, and H. Wang

Big Bear Solar Observatory/New Jersey Institute of Technology
40386 North Shore Lane, Big Bear City, CA 92314, U.S.A.

Abstract. In this paper, we show observations of sunspots and pores to illustrate various aspects of the high resolution observing program at the Big Bear Solar Observatory (BBSO). We address the question as to how a sunspot develops a penumbra. This process seems to occur on time scales of a few tens of minutes. Even pores with a size of a few arc seconds show fine-structures. Bright points with a diameter of a few tenth of an arc second are located inside pores and resemble umbral dots. Pores often exhibit cloud-like, fuzzy borders which may be the prenatal stages of a penumbra. Therefore, diffraction limited observations of high temporal resolution are necessary to study the temporal evolution of sunspots and pores, the onset of the Evershed flow, and the accompanying changes in the magnetic field configuration. We present speckle masking observations taken in white-light and G-band, infrared observations obtained with frame selection at 1.56 μm, and magnetograms obtained with the new digital magnetograph system. Finally, we will report on our progress in reducing the dome seeing at BBSO.

1. Introduction

Speckle masking imaging can provide almost diffraction limited observations of solar fine structures. A detailed study of sunspots and pores was recently published by Denker (1998). First high spatial resolution observations from BBSO by means of speckle interferometry were reported in Denker & Wang (1998) who studied the flare producing mechanism in an active region with a small δ-spot. Utilizing the expertise of the New Jersey Institute of Technology in near infrared imaging technology, near infrared observations are now a primary field of research at BBSO (Wang et al., 1998b). Finally, we will report on the progress in improving BBSO's magnetograph system. A description of the video magnetograph system was given by Varsik (1995) and first results from the new digital magnetograph system were presented by Wang et al. (1998a).

2. Observations

A 256 $\times$ 254 pixels, 12-bit CCD camera manufactured by Dalsa, Inc. was used to record images at approximately 90 frames s^{-1} with an exposure time of approximately 10 ms. A 320 $\times$ 240 pixels, 12-bit, room temperature, near infrared InGaAs CCD camera manufactured by Sensors Unlimited, Inc. was used to

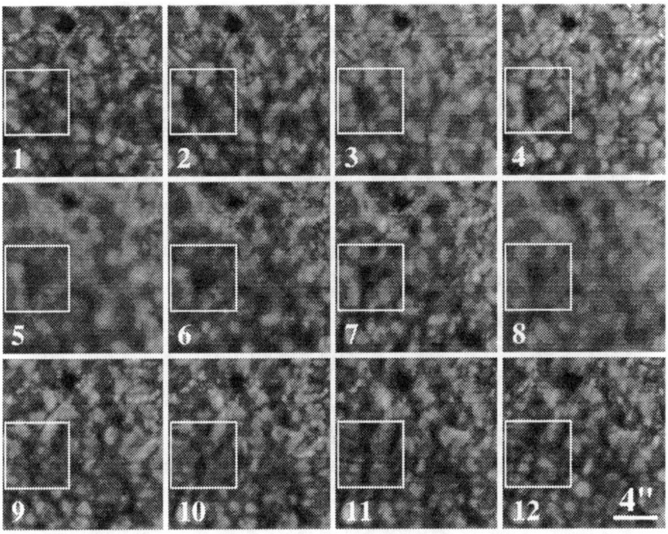

Figure 1. 12-minute time series of speckle reconstructions obtained in the G-band at 430.1 nm showing bright points and pores observed on August 14th, 1997.

record images at 30 frames s^{-1} at 1.56 μm $\pm$ 2.5 nm. The quantum efficiency is greater than 70% from 1.0 μm to 1.7 μm and the exposure time was 127 μs. The image scale at the primary focus of the 65 cm reflector is approximately $6''$ mm^{-1} which corresponds to a field of view of $76.8'' \times 57.6''$ for the near infrared CCD camera and to $24.6'' \times 24.4''$ for the visible light CCD camera. The diffraction limit of the 65 cm reflector is $\alpha_{430} = 1.22\lambda/D = 0.17''$ at 430 nm, $\alpha_{520} = 0.20''$ at 520 nm, and $\alpha_{1565} = 0.61''$ in the near infrared.

2.1. G-Band speckle reconstructions

G-band images are commonly used to follow the evolution of small-scale magnetic elements. Here, we present observations of a cluster of pores within the decaying active region NOAA 8071. We observed the pores in the G-band at 430.1 $\pm$ 0.2 nm with the Dalsa CCD camera from 16:55 to 17:19 UT on August 14th, 1997. We took 24 sequences of 200 short-exposure images and reconstructed the images with the speckle masking technique. We measured a Fried-parameter of $r_0 = 8.44 \pm 0.55$ cm, where the standard deviation reflects the temporal variation of the seeing. A detailed description of the speckle interferometric setup and further references are given in Denker & Wang (1998).

The peak brightness of G-band bright points is typically around 1.2 times the average intensity of the granulation. Extreme cases exceed 1.4 times this intensity. The size of the bright points ranges from around $1''$ down to the diffraction limit of the telescope. Some G-band brightenings are elongated and move along the intergranular lanes. This motion seems to be dominated by the

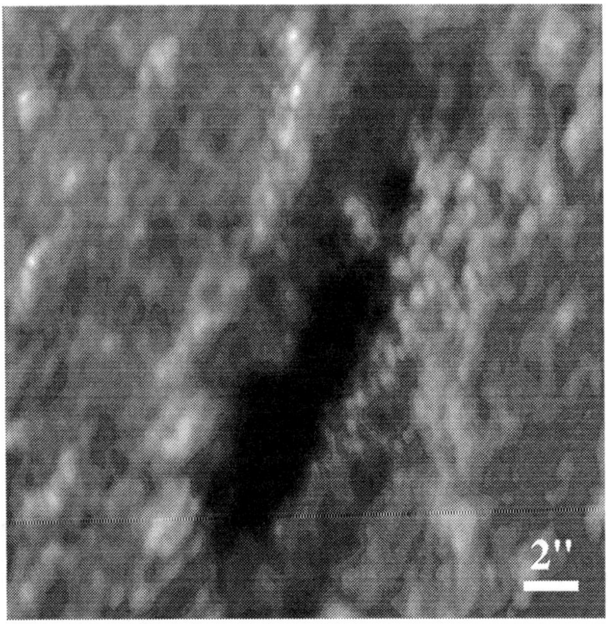

Figure 2. White-light speckle reconstruction of a sunspot observed near the solar limb on July 11th, 1998.

motion of surrounding granules, especially large, rapidly expanding granules. In the course of this motion along the intergranular lanes, some elongated brightenings begin to fragment and disappear in minutes. G-band bright points share many of the characteristics mentioned above with filigree and moustaches (e.g., Denker, 1997).

The flow pattern of granules seems also to play a crucial role in the formation and decay of pores. Fig.1 shows the evolution of G-band bright points and pores during a 12-minute interval. In one example, a large pore (white boxes) tries to pierce through the granulation. First, we find highly fragmented granules at the location of formation. Next, a pore of $4''$ diameter emerges, and the pore becomes darker and more uniform. Finally, due to the intrusion of granules the pore eventually collapses and disappears. The whole evolution of this 'hindered' pore took place in less than 10 minutes and it is worthwhile to note that high spatial *and* temporal resolution observations are essential to study this type of highly dynamic phenomenon.

2.2. White-light speckle reconstructions

Another interesting phenomenon associated with small scale magnetic fields are facular points around sunspots near the solar limb. Fig.2 shows a sunspot (NOAA 8263) and surrounding faculae observed at 520±3 nm from 17:03 to 18:03 UT on July 11th, 1998. The solar limb is just outside the field of view depicted in Fig.2 at the lower right corner. We took 60 sequences of 200 short exposure images. The Fried-parameter during the observations was $r_0 = 9.19 \pm 0.85$ cm.

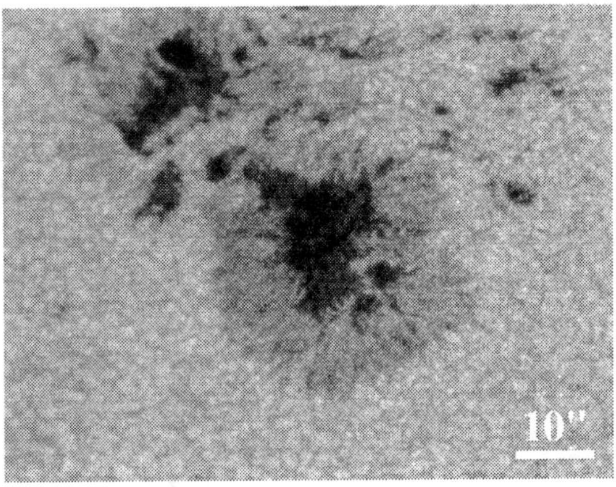

Figure 3. Sunspot observed in the near infrared on August 8th, 1998.

Many bright points are located at the inner part of the limb-side penumbra. Especially, the small light-bridge attracts many of these brightenings. In a time lapse movie, we can follow individual bright points traversing the light-bridge towards the limb-side penumbra. Bright points near the disk-side penumbra are predominantly located along the outer periphery of the penumbra. Only in isolated cases, we find brightenings traveling along penumbral filaments toward the umbra. These observations are quite different from high spatial resolution observations of sunspots near disk center where numerous penumbral grains move along the penumbral filaments towards the umbra (e.g., Denker, 1998). However, the lack of moving penumbral grains might be explained by geometrical foreshortening and the Wilson effect.

2.3. Near Infrared Observations

The near infrared CCD camera from Sensors Unlimited is the very first large format, 12-bit, InGaAs camera ever built. In August 1998, we performed a series of observations to test the performance of the CCD camera at a frame rate of 30 frames s^{-1}. This high frame rate is essential for a future near infrared magnetograph system (Spirock, Goode, & Wang 1998) where one has to scan either one spatial or the spectral dimension to obtain the full spectral information for a two-dimensional field of view. From comparison with Fried-parameters measured at 520 nm and a few measurements at 1.56 μm, we expect the Fried-parameter to be larger than $r_0 = 30$ cm at 1.56 μm for most of the observing days at BBSO.

Fig.3 shows part of active region NOAA 8293 observed with the InGaAs CCD camera in the near infrared at 1.56 μm on August 8th, 1998. The umbrae of the sunspots exhibit a multitude of umbral dots. The lower contrast in the infrared makes it more difficult to identify individual umbral dots above a surrounding background. However, we can still detect umbral dots with sizes down to the diffraction limit of the telescope. The granular pattern surrounding the

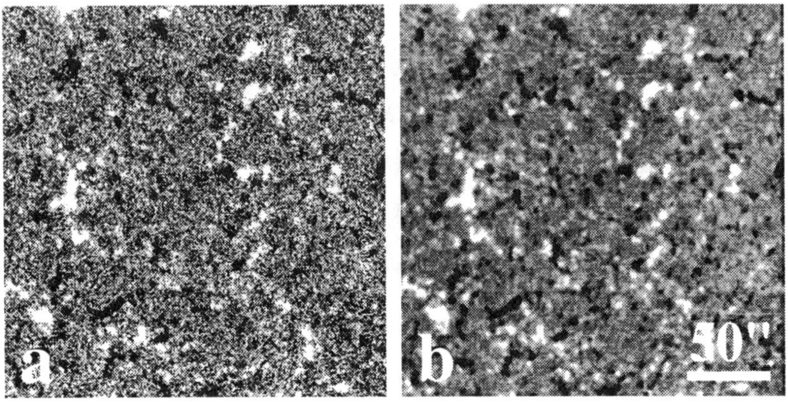

Figure 4. Magnetograms of the quiet Sun near disk center observed on July 4th, 1998. (a) 100-frame integration and (b) 16×100-frame integration after image alignment with sub-pixel accuracy.

sunspots is also clearly discernible and the granular contrast is approximately 1%. These sunspot observations represent the best near infrared observations at BBSO so far (cf., Wang et al., 1998b).

2.4. Digital magnetograph system at BBSO

The advantage of a digital magnetograph system is the accessibility of individual frames so that we can apply various post-processing techniques. In Wang et al. (1998a), we demonstrated that frame selection, image alignment, and speckle interferometric techniques can significantly improve the spatial resolution and sensitivity of magnetograms. The spatial resolution of the magnetograms was better than $1''$.

Figure 4 shows two magnetograms of the quiet Sun observed near disk center on July 4th, 1998. Since the memory of the Dalsa system was limited, we derived a magnetogram by subtracting 100 frames of left- and right-handed circular polarization. The resulting magnetogram is shown in Figure 4(a). We aligned 16 of these magnetograms with sub-pixel accuracy and averaged them to improve the signal-to-noise ratio by a factor of four. Intranetwork magnetic elements are clearly discernible in Fig.4(b). The gray-scales for both magnetograms were chosen so that the degree of polarization is displayed between $\pm 10^{-3}$. Small scale magnetic fields can be detected even below $\pm 10^{-4}$. The calibration of the system is such that 1% polarization corresponds to a flux density of 115 Gauss (Varsik, 1995), i.e., the detection threshold is approximately 1 Gauss. In time-lapse movies, we can follow the motion of individual magnetic elements and we observe flux cancellation and emergence even close to the detection threshold.

In the near future, we will improve the software so that routine observations with long integrations, on-line frame-selection, and image alignment are possible. With the implementation of a correlation tracker at the 65 cm vacuum reflector,

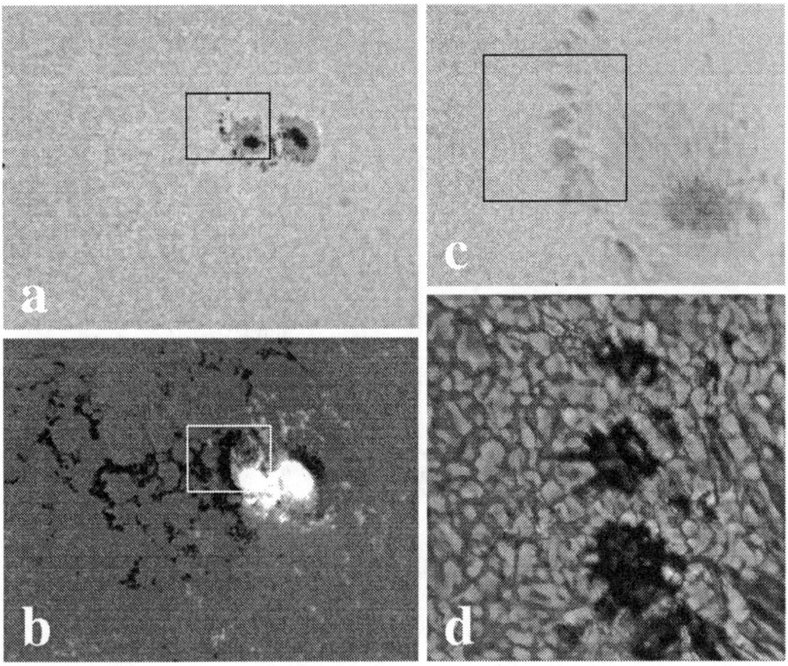

Figure 5. NOAA 8299 observed on August 16th, 1998. (a) Filtergram obtained in the line wing of the Ca I line at 610.3 nm and (b) corresponding videomagnetogram obtained with the 25 cm refractor. (c) Near infrared image of the left major sunspot filled with umbral dots. (d) White-light speckle reconstruction showing fine structures of pores in the upper left region of the sunspot group.

on-line image alignment will become superfluous so that we can improve the temporal resolution of the magnetograms. In a next generation magnetograph system, we will upgrade to a larger detector and use a tunable Fabry-Pérot system as primary filter. This will allow us to scan across the spectral line and to measure the true strength of magnetic fields.

3. Combined Observations of NOAA 8299

One unique characteristic of the telescope setup at BBSO is that the 65 cm vacuum reflector, the 25 cm vacuum refractor, and two full disk telescopes share a single equatorial mount. Each of the high resolution telescopes has three optical benches and the 65 cm telescope feeds a spectrograph at the Coudé-focus. A computer-controlled mirror system directs the light to the optical benches and the spectrograph so that seven experiments can run nearly simultaneously in addition to the synoptic full disk observations.

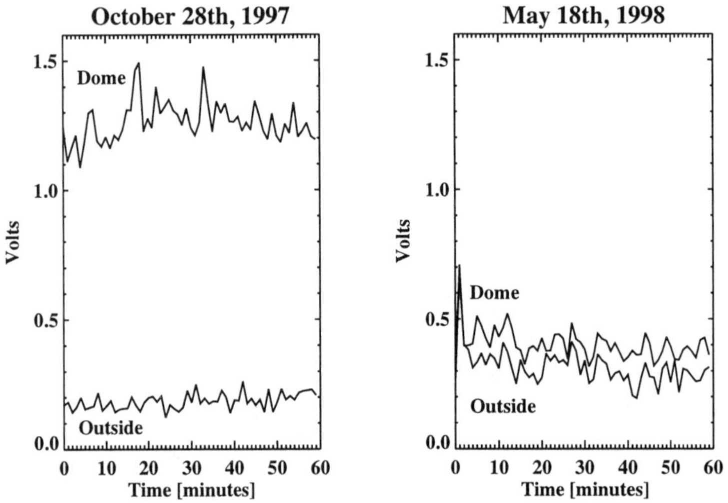

Figure 6. Comparison of scintillometer records of the dome and outside seeing before improving the dome seeing (October 28th, 1997) and afterwards (May 18th, 1998).

Our goal is to combine the observations described in the previous section. The data set shown in Fig.5 reflects the current status of the high resolution observing program at BBSO. We observed active region NOAA 8299 from 18:47 to 20:41 UT on August 16th, 1998. Fig.5(a) and 5(b) show a typical filtergram and magnetogram of NOAA 8299 obtained with the videomagnetograph system (Varsik, 1995) at the 25 cm refractor. NOAA 8299 consists of two major spots with the same polarity embedded in several smaller spots of opposite polarity. A close-up of the left major sunspot obtained in the near infrared at 1.56 μm is depicted in Fig.5(c). Again, the umbra of this sunspot is filled with umbral dots. Finally, Fig.5(d) illustrates the spatial resolution that can be achieved with the speckle masking technique.

During the almost 2-hour time-series, we took 99 sequences of 100 frames with the Dalsa CCD camera separated by 1 minute in time. The Fried-parameter during the observations amounts to $r_0 = 10.08 \pm 1.08$ cm indicating excellent seeing conditions. The white-light speckle reconstruction depicted in Fig.5(d) shows at almost diffraction limited resolution three pores, several magnetic knots, and the tips of some penumbral filaments that can be identified in Fig.5(c). Again, fragmentation of granules and concentration of filigree precede the formation of small pores and magnetic knots. A time-lapse movie shows that the intergranular lanes in the lower right corner are often co-aligned with the direction of the penumbral filaments. Interestingly, the pores exhibit various fine structures resembling umbral dots. Many of these dots originate from the fragmentation of cloud-like structures at the border of the pores. The dots move around inside the pore where they either merge with other dots or fade away.

4. Improving the Dome Seeing at BBSO

During the last year, we devoted our energies on improving the dome seeing at BBSO. The electronics and computer equipment were removed from the observing floor to reduce the heat load, and the observing floor was thermally insulated from the rest of the building. In addition, the interior of the dome was coated with TiO_2 paint to reduce local turbulence and a fan sends a gentle air current through the dome to insure that there are no large temperature gradients with the ambient air. As Fig.6 shows, all the passive measures added up so that subsequent measurements from solar scintillometry (Beckers et al., 1997) show now a clear correlation of the dome and outside seeing. All these vigorous efforts to improve the seeing break new ground for high spatial resolution observations at BBSO.

Acknowledgments. We would like to thank the staff of the Big Bear Solar Observatory for their support during our observations, especially Haifan Wang who developed the software for the near infrared CCD system. This work was supported by NSF under grant ATM-9714796, by NASA under grant NAG5-4919, and by ONR under grant N00014-9711037.

References

Beckers, J. M., Leon, E., Mason, J., Wilkins, L., 1997, Solar Phys. 176, 23

Denker, C., 1998, Solar Phys., 180, 81

Denker, C., Wang, H., 1998, Astrophys. J., 502, 493

Denker, C., 1997, Astron. Astrophys., 323, 599

Spirock, T. J., Goode, P. R., Wang, H., 1998, AGU Meeting, Boston, SH52H-01

Varsik, J., 1995, Solar Phys., 161, 207

Wang, H., Denker, C., Spirock, T., Goode, P. R., Yang, S., Marquette, W., Varsik, J., Fear, R. J., Nenow, J., Dingley, D. D., 1998a, Solar Phys., in press

Wang, H., Spirock, T., Goode, P. R., Lee, C. Y., Zirin, H., Kosonocky, W., 1998b, Astrophys. J., 495, 957

Group Discussion

Rutten: How much time do the Speckle reconstructions take?

Denker: One hour on a 266 MHz Pentium PC.

High Resolution Solar Physics: Theory, Observations, and Techniques
ASP Conference Series, Vol. 183, 1999
T.R. Rimmele, K. S. Balasubramaniam, R.R. Radick, eds.

Simultaneous High Resolution Spectroscopy of the Photosphere and Chromosphere

K. S. Balasubramaniam[1] and M. Bianda [2]

National Solar Observatory/Sacramento Peak and AFRL/VSBS
National Optical Astronomy Observatories[3] Sunspot, NM 88349, USA

Abstract. We present simultaneous high resolution spectroscopy of the photosphere and chromosphere using the HAO/NSO Advanced Stokes Polarimeter and the NSO Horizontal Spectrograph at the VTT.

The Ca II K spectral line serves as a diagnostic for plasma properties, activity levels and magnetic influence of the chromosphere in the Sun and stars. Traditionally, the K line exhibits a double reversal with well identified spectral features, K1, K2 and K3. We present observations to show that the K line exhibits multiple spectral reversals in addition to the well known features. These multiple reversals are seen only at high spatial and spectral resolution, in solar active regions. The locations of these multiple reversals are cospatial with local Hα brightenings that appear at foot-points of dark active region Hα fibrils (or active region filaments). The magnetic field structure at the photosphere at the spatial locations where these multiple reversals are observed, are either bipoles (small opposite polarity magnetic nodules) or neutral lines. The multiple reverals, when seen under high spatial resolution possibly represent reconnective process on the smallest scales.

1. Introduction

The cores of the resonance spectral line of doubly ionized calcium H (3968 Å) and K (3934 Å) exhibit properties and structure of the thermal chromosphere, much of which is associated with the structure of the photospheric magnetic field, in some way. Leighton (1959) showed a remarkable correspondence of the chromospheric calcium facular field with the photospheric magnetic field. Ca II H and K spectral lines show three well defined components (Linsky & Avrett 1970). A wide K1 absorption component. A central K2 emission or absorption component superposed on the K1 component. A smaller K3 absorption superposed on the K2 emission.

Linsky (1970), from observations in plages showed double reversals in Ca II H and K lines, and additionally observed anomalous line profiles where the red H2 and K2 features are brighter than the K2 features. He presented model calculations to understand various line formation ratios and the double reversal. Cram and Damé (1983) present a range of variations in the quiet-Sun Ca II H line profiles. while Fang et al. (1992), report characteristic K2 emission and red asymmetry in the impulsive phase of solar flares. Intensity of Ca II H and

K intensity profiles are asymmetric in sunspots, displays double peaks in plages (Martinez Pillet et al. 1990).

Dere, Bartoe & Bruckner, G. E. (1989) presented observations of transition spectral lines of C IV, 1548 and 1550 Åto show intense broadening when observed under high resolution of 1" resulting in velocities of 100–150 km/s, spread over a spatial area of roughly 2000 km and lasting for time periods of 50-150 seconds. These broad spectral lines were present all over the sun, and were interpreted as microflares.

Observations of Capella's transition region (Linsky et al. 1995) using UV resonance lines show that the resonance lines have very broad wings and are responsible for most of the heating flux, due to active regions. They interpret these broadening as the stellar analogue of the transition region explosive events (TREEs), due to the strong acceleration of the plasma away from the heating site. They in turn interpret these broadenings as a result of microflaring and reconnection.

In this paper we show that the Ca II K spectra show multiple reversals beyond the well known K1, K2 and K3 components. We also show that, in some cases, the K2 broadening is significant, spread over 3-4 Å.

2. Instrument and Observations

The observations were performed using the Richard B. Dunn Solar Telescope (RBDST; formerly the VTT) at National Solar Observatory (HAO) at Sacramento Peak. The High Altitude Observatory/NSO Advanced Stokes Polarimeter (ASP) (Lites et al. 1991; and Elmore et al. 1992) was used for full Stokes vector polarimetry in spectral lines Fe I 6301.5 and Fe I 6302.5 Å.

Using the same spectrograph as the ASP, simultaneous Ca II K spectra were obtained using an RCA 504 camera with a spectral sampling of 0.018 Å/pixel and 7.3 Å wide spectra centered about the Ca II K spectral line at about 3934Å. For each slit position, simultaneous reflection slit-jaw images were obtained using the Universal Birefringent Filter (0.25Å) imaging in 13 wavelengths of the Hα spectral line.

The observations were conducted on 1995 August 5, on Active Region 7896 (N2.2 E2.5; Fig. 3a). The observed field-of-view (FOV) was 120 $\times$ 80 arcseconds with slit-length of 80 arcseconds, slit-width $\sim$ 0.6 arcseconds, 0.75 arcseconds/step. It took 23 minutes to map active region FOV. 5 such maps of AR emergence were sequentially recorded.

3. Data Analysis

The ASP polarized spectral data were processed using the ASP software that includes dark current and gain corrections, data normalization and correction for instrumental polarization and the derivation of vector magnetic fields at all spatial locations where the total polarization content exceeded 3%.

The calcium spectra were corrected for dark current and flat-fielding effects. The reflection slit jaw images were also corrected for dark current and flat-fielding effects. We then developed algorithms to seek systematic intensities and wavelength positions of K1, K2 and K3 components, of Ca II K. Several hundred Ca II K spectra failed to fit the well known categorization of K1, K2 and K3 components. We then automated algorithms to detect wavelength and positions of K line reversals and categorization of the reversals.

4. Findings and Discussion

Figs. 1–2 show examples of multiple reversals occurring in the Ca II K line in addition to the well known K1, K2 and K3 components.

Fig. 3 shows the active region.

5. Observed Properties of Ca II K Multiple Reversals

- Additional spectral line absorption features in the Ca II K line core are often found at locations at or near plages. When observed at high resolution, the Ca II K spectral line structure is more complex than can be described by the well known K1, K2 and K3 components (Figs. 1–2)
- In many of the profiles, there is only one additional absorption component, in addition to K3 (Fig. 1). However, there are several cases where the presence of more than one additional absorption component was observed (Fig. 2a, 2d)
- These additional spectral line absorption features in the core of the Ca II K line are similar in character to the K3 absorption.
- In the majority of cases, the additional absorption component is superposed on the red wing wing of the K2 emission or absorption. However several have been observed within the central K3 absorption feature as well as in the K2 blue wing
- When adjacent spatial positions are compared, an additional absorption feature may gradually strengthen to become a K3 feature, while a previously well identified K3 fades. This make it appear as if the K3 component shifted to a new wavelength position over a small distance (Fig. 1d)
- The locations where the Ca II K spectral line has multiple absorption component (i) are spatially coincident with locations where the Hα core plages are bright, (ii) skirt and surround the brightest of the Ca II K core plages and (iii) are outside the boundaries of strong magnetic fields (Fig. 3).
- The local active region Hα loops terminate at the spatial locations where the multiple spectral reversals are seen (Fig. 3).
- The multiple reversals occur over small areas where the photospheric LOS magnetic field is identified with magnetic neutral lines, or small nodules of bipolar magnetic flux (Fig. 3).
- The spectral separation of the additional absorption component from the mean K spectral line center (over the entire region), $\Delta\lambda$, and the resulting velocity varies between -16 to 45 km s^{-1}.

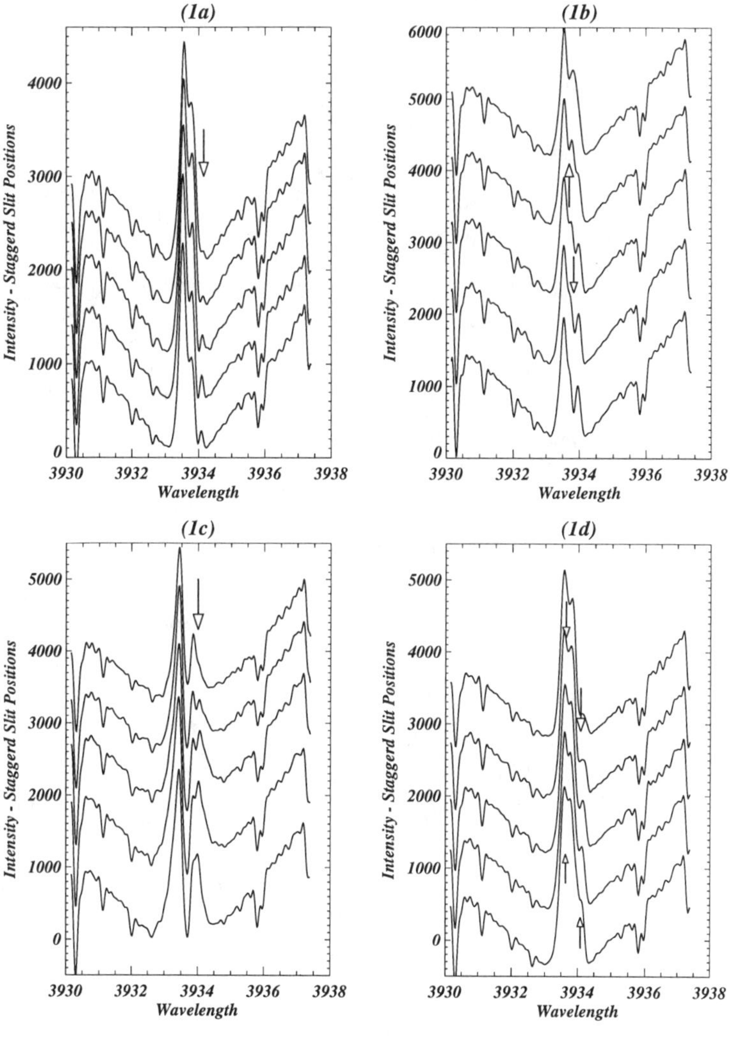

Figure 1. Plots of the Ca II K spectral variation in different spatial locations of the active region where there are multiple reversals. In each panel, the intensity is shown across adjacent spatial positions along the slit . The intensity units are arbitrary units. In each panel the spectral line are staggered in intensity to show the spatial variation. Arrow marks point to spectral positions where the multiple reversals are seen. See text for details.

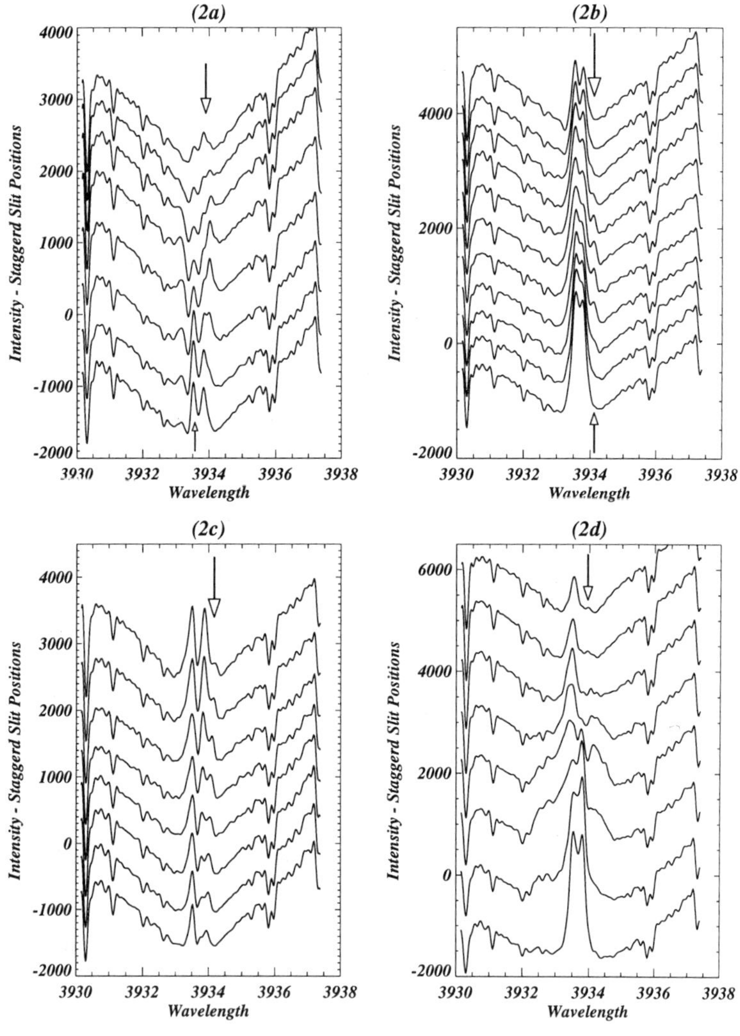

Figure 2. Plots of the Ca II K spectral variation in different spatial locations of the active region where there are multiple reversals. See caption for Fig. 1 and text.

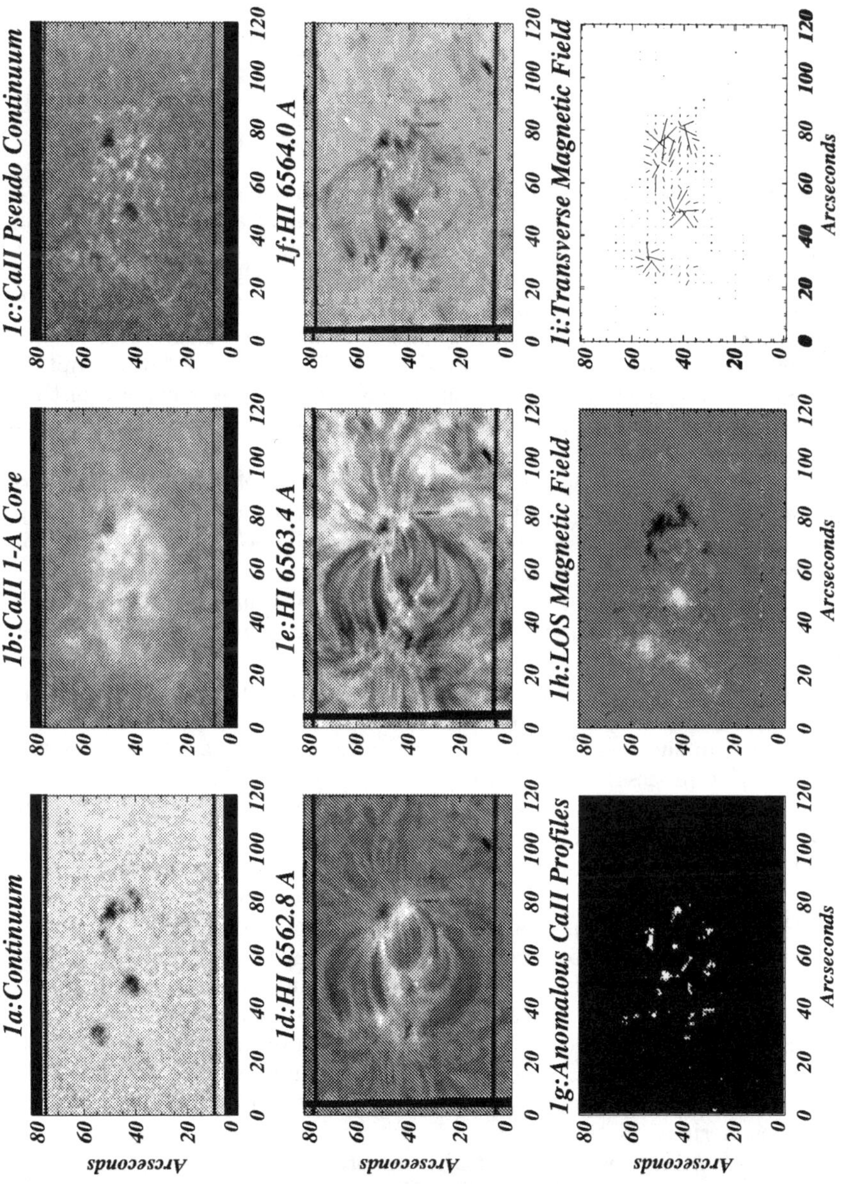

Figure 3. Filtergrams of bipolar active region (NOAA 7896) near disc center (N2.2 E 2.5) in various spectral bandwidths. (3a) True continuum (3b) logarithmically scaled Ca II K 1-Å core index. (3c) Pseudo-continuum in Ca II K, 1.8Å away from the core. (3d–f): Hα images (3g) Locations where Ca II K profile shows multiple reversals (3h) LOS magnetic field, scaled between ± 1800 G. (3i) Strength of the transverse magnetic field shows as tick marks.

- At several of these spatial locations, the K2 emission component of the spectral line is over 1.5 Å wide (separation between the K2V and K2R components) (Fig. 2d).

6. Conclusions

The presence of these red or blue-shifted additional components, with velocities ranging from -16 to 45 km s^{-1}, suggests a localized flow in the upper-photosphere/low-chromosphere boundary (between the optical depth where K3 is formed and the upper photosphere). The existence of multiple reversals (or absorptions) in the Ca II K line at certain locations in an active region could be evidence for the presence of magnetosonic waves or magnetic reconnection, continuously occurring at small spatial scales. These localized down/upflow flow jets might be triggered by the acoustic waves and/or reconnection of the magnetic field due to small-scale (2–3 arcseconds bipolar) structures and highly non-linear physical conditions.

The multiple absorption components could serve as a diagnostic tool to probe sheared magnetic fields under NLTE conditions, in the chromosphere. The multiple reversals are seen only under reasonable high spatial and spectral resolution, with short exposure times.

These observations raise a number of related questions. How would the results of Skumanich, Smythe & Frazier 1975, and Schrijver et al. 1989, relating the Ca II K-Å index to the magnetic field strength be affected by these high resolution observations of multiple reversals? Are multiple reversals influenced by the geometry of the magnetic field vector? Is the presence of these multiple absorption in the Ca II K line a precursor signature of small scale solar flares/ reconnective process?

Can similar multiple reversals be seen in other resonance lines? We suspect that the answer is yes. The observations by Martinez Pillet et al. (1990) appear to show a faint indication of multiple reversals in plage, as seen in the Ca II H line. However, there is no discussion of these reversals in the text.

What is the influence of the chromospheric oscillations on these unusual multiple reversals? These observations are not well suited for studying the influence of the oscillations due to a lack of time resolution. A separate observational project is necessary. It takes about 23 minutes to scan the active region. We have 5 such scans during good seeing. A cursory examination of adjacent 23-minute scans shows that some of the multiple reversals persist, but not with the same amplitudes. Perhaps the amplitudes of the multiple reversals are modulated by chromospheric oscillations.

The multiple reversal spectral lines presented here also show significant broadening of the whole line profile. Contrasted with the observations of Dere et al (1989), they are present only near the brightest of plages, whereas the observations of Dere et al. (1989) show that they are present everywhere on the Sun. While observations and analysis of Capella's transition region by Linsky et al. (1995) show that these line profiles are derived from the whole star, we noticed that the multiple reversals vanish when averaged over spatial resolution of a few arc seconds.

Solanki, Steiner & Uitenbroeck (1991) show detailed model calculations of Ca II K line profile in turbulent magnetic fields in a 2-D atmosphere. Model calculations for the Ca II K2V bright points for reproducing temporal variations in the spectral line by Carlsson & Stein (1992) have been successful in incorporating acoustic waves. Perhaps combinations of such model calculations in a multi-component atmosphere with propagating jets can reproduce these spectral line profiles.

Acknowledgments. We are grateful to David Elmore, Bruce Lites, Dick Mann, Eric Stratton and Steve Hegwer for the ASP observations. Jack Zirker, Steve Keil and J. Linksy provided valuable comments and criticism of this work. I gratefully acknowledge support for this work under the AFOSR Project 2311.

References

Cram, L. E., and Damé, L., 1983, ApJ, 272, 355.

Carlsson, M. & Stein, R. F. 1992, ApJL 397, L59.

Dere, K. P., Bartoe, J. -D. F., Bruckner, G. E. 1989, Solar Physics, 123, 41.

Elmore, D.F., Lites, B. W., Tomczyk, S., Streander, K. V., Skumanich, A. P., Dunn, R. B., Schneke, J. A., Leach, T. W., Chamberlin, C. W., Hull, H. K., & Lacey, L. B., 1992, Proc. SPIE 1746, 22.

Fang, C., Hiei, E., Yin, S. & Gan, W., 1992, PASJ, 44, 63.

Linksy, J. L. & Avrett, E. H., 1970, Publ. Aston. Soc. Pacific, 82, 169.

Linsky J. L., 1970, Solar Physics 11, 355.

Linsky, J. L., Wood, B. E., Judge, P., Brown, A., Andrulis, C., & Ayres, T. 1995, ApJ 442, 381.

Lites, B.W., Elmore, G., Murphy, G., Skumanich, A., Tomczyk, S. & Dunn, R.B. 1991, in Solar Polarimetry. Ed. L. November. Proc. 11th NSO/SP Summer Workshop, p 3.

Martinez Pillet, V., Garcia Lopex, R. J., del Toro Iniesta, J. C., Rebolo, R., Vazquez, M., Beckman, J. E., & Char, S ApJL, 1990, 361, L81

Schrijver, C. J., Cote, J., Zwann, C. & Saar, S. H., 1989, ApJ, 337, 964.

Skumanich, A., Smythe, C., Frazier, E. N., 1975. ApJ, 200, 747.

Solanki, S. K., Steiner, O. & Uitenbroeck, H 1991, A&A 250, 234.

Group Discussion

Beckers : In the CAII K-spectra which you showed there were some which were very broad. I was struck by the appearance of the wings of the broad lines of photosheric-like absorption lines. Do you have any idea how that should be interpreted?

Bala : No, not yet.

High Resolution Solar Physics: Theory, Observations, and Techniques
ASP Conference Series, Vol. 183, 1999
T.R. Rimmele, K. S. Balasubramaniam, R.R. Radick, eds.

High Resolution: A Retrospective

Richard B. Dunn
National Solar Observatory, National Optical Astronomy Observatories
Sunspot NM, 88349

1. Early Days

In 1953, when I came to Sac Peak "high resolution" as a field of interest did not exist! The telephone was high tech. Slide rules were the norm and computers were a dream. High resolution was limited to observations of granulation that could be recorded with short exposures that froze the smearing caused by the Earth's atmosphere (Miller, Rosch and others). Pic du Midi dominated these observations. Rosch took short bursts of exposures on film that were late frame selected. He assembled a montage of selected frames to show a time series. Later on there was great interest in the role of exploding granules.

At that time, I was interested in prominences and spicules , the later named by Walter Orr Roberts from his observations of this phenomena at Climax. At Sac Peak I took movies of spicules that were shown to the AAS by Menzel. I also surveyed the entire limb daily. Sara Lee Lippincote measured the statistics of the spicules on these surveys. The interest in the spicule observations was high, but since they occupied only a tiny fraction of the chromosphere their inclusion in the chromospheric model was thought to be unnecessary. Their questionable significance prompted Grant Athay, who was fresh from the reduction of the flash spectrum taken by Jack Evans during the eclipse at Khartoum Egypt, to give his now famous "solar dermatology" analogy, which has branded the high resolution field forever! I did not hear Grant's lecture. Did anyone here hear it? No. I understand that it went something like this:

What shall we study? What is that at the end of my arm? Lets call it a hand. What does it do? It picks up things. How does it do that? With fingers. How are they moved? With muscles and tendons attached to the bones of the fingers. The whole hand is covered with skin. How interesting. Lets turn our hand over and what do we see? Little hairs! Surely they must be useful. Lets study those!

Grant's comment was apropos. It was not until many years later when studies of the magnetic field showed that the field was quantified into tiny ropes of magnetic flux that the need for high resolution became apparent. Nowadays many solar proposals start with the justification for high resolution that "Many fundamental processes occur on the Sun at very small scale." The proposal also might point out that the surface of the Sun may be regarded as a plasma laboratory that lets one study the physics of the magnetic field and its relationship to the convective and radiation field. Such observations require high resolution, especially if one wants to quantify the measurements.

In the 60's and 70's high resolution still hadn't become a popular field of study. Now look at this workshop! You've come a long way Baby! But there is still a long way to go! So much for the retrospective! What can we do? What can be improved?

2. Go to space.

High resolution observations should be done in space where one is free from the Earth's atmosphere!

- Uniform data
- Uniform instrumental profile over the entire field of view.
- No Adaptive Optics required
- No weather problems
- Highly visible project ($$$)
- But one still needs a good telescope and a large one (2-4 m)! (The IR has become very popular now and it requires a 3-4 m telescope... maybe the IR belongs on the ground?)

I believe that a long-term goal (10-20 year?) of the ground-based high resolution program should be to justify and obtain a 2-4 meter aperture solar telescope in a satellite.

From our experience with the 1-m Solar Optical Telescope (SOT) in the late 70's and with the current budget and with the priority list of flights at NASA there is no chance of obtaining a large solar space telescope. This situation could possibly change as we build a strong, widely-supported, ground-based high resolution program that produces increasingly useful science.

High resolution on the ground will be clobbered when a large solar satellite telescope is successful. Most of our clever techniques developed for ground observations: frame selection, adaptive optics, the VTT/SP and the AST will be obsolete for high resolution. At this time, however, they are absolutely necessary to sell a large solar space telescope.

The ground-based observations should produce a lot of good science that can stand on its own and which will spark the need to go into space.

3. Telescope Resolution

Your telescope is never as good as you think it is. Nowadays wavefront measurements can tell you how good it is.

As an example, how good is the SP/VTT? What is its Strehl Ratio? Improving the resolution of the VTT has been a 25-year battle and it is still going on! Early images of granulation at full 76-cm aperture looked a little soft... They were better when the telescope was diaphragmed to 50 cm. (My filigree pictures were taken at 50 cm). The main problem is that the edge of the window heats up during the day introducing spherical aberration and a change in focus.

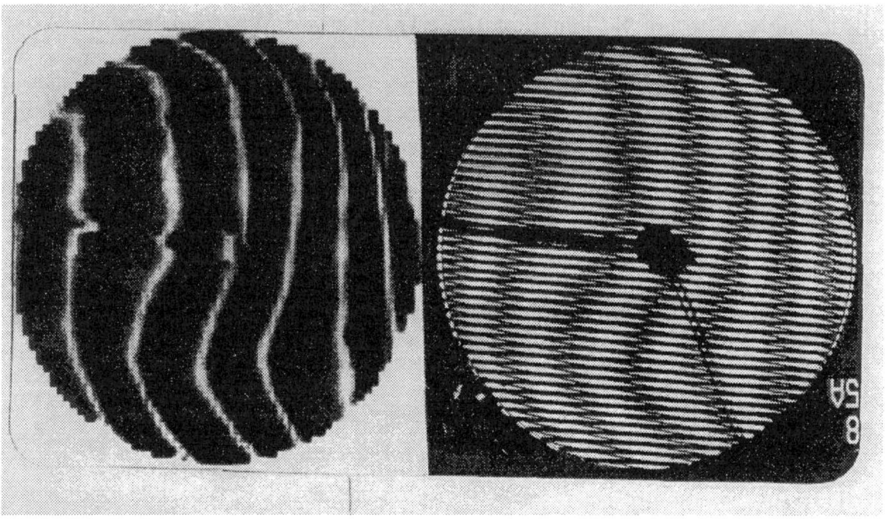

Figure 1. Interferograms of the window in the NSO/SP VTT. Two techniques for removing the effect on the interferogram of the 42-fringe wedge in the window are compared.

We tried a large variety of passive ways to reduce this effect on the edge of the window, but had to resort to an active cooling system that worked with varying degrees of success depending on the cell configuration. We have had four windows in the turret, each with a different cell design. The current active cooling system, which was designed and built by L. Wilkins, works quite well, but is not perfect.

We have an interferometer setup for studying the degradation of the wavefront caused by the window. We autocollimate the telescope using the window as a flat mirror. Two images are formed in the focal plane, one from each side of the window. The two are combined to form an interferogram when an image of the window is formed with a lens. There is a small wedge between the two surfaces of the window that causes a 42-fringe tilt in the interferogram that must be removed in order to view the two dimensional fringe pattern. Two schemes to remove the wedge are shown in Fig.1. The left hand interferogram was made by Radick using a technique pioneered by the Roddiers (Roddier & Roddier 1987). It requires an interferogram with a large number of fringes like those produced by the small wedge in our window. The interferogram is scanned and processed by a computer. The resultant wavefront can be manipulated to remove the wedge. The right hand image is a quick look scheme made by superimposing a grid with approximately 42 lines in it. The resulting Moiré pattern is approximately the same as the interferogram made with the computer. Fig.1 is a typical interferogram of the window with the cooling system operating. One can see significant departures from straight fringes.

The heating of the edge of the window continues to be the main factor in spoiling the image. We are unable to identify any additional problems, for example, from the bending of the window due to either the vacuum load or to non uniformity in the O-ring support for the window. I have tested the primary and found it to be very good over its full, 1.5 m aperture. We have recently had one of the turret flats re-figured.

According to Rimmele's Shack-Hartmann measurements that use a sunspot as a target, in the morning the Strehl Ratio of the VTT is 0.5 to 0.7. Images of bright stars taken in the early morning when the window is supposedly in thermal equilibrium show a spike on the image that changes to astigmatism later in the day as the window cooling becomes active. This variability suggests that the circulation of the liquid in the cooling system could be improved. In any case it would not take much of an improvement in the image to lay to rest the need for active optics.

Our current effort is to verify that the window is the only culprit degradating our the image. In addition to recording the interferogram of the window, we will record stellar images, and, with a star as a target, use a Shack-Hartmann to determine the overall wavefront. As a target for the Shack-Hartmann, stellar images have an advantage over the granulation in that the stellar image always shows a centroid no matter how poor the seeing. Images of granulation, on the other hand, just become washed out as the seeing degrades. Bright stars can be observed during the day with the VTT. The 100-year war goes on.....

4. Improving Frame Selection

The best high resolution pictures on the ground have been taken with the frame selection technique. One takes a burst of pictures and then selects the sharpest frame. To overcome the blurring caused by the atmosphere the exposures must be less than 10ms. Selecting the best frame in a burst is a cumbersome process when using film, but with the advent of the CCDs the frame selection technique has been automated and improved. The best frame judgment in the burst can be made by computer in real-time and only one frame in each burst need be saved. Furthermore, the quantum efficiency of the CCDs is higher than that of film so the exposures can be shortened. Frame selection has continued to be the most successful technique for obtaining high resolution pictures of scenes that can be made with short exposures (20ms). The technique has been brought to a high degree of sophistication by Scharmer, Rimmele, and others. One of the improvements has been in the registration of the scenes. In the photographic era of frame selection, residual guiding errors required registering each picture in a series by eye, a process that very few observers took the time to do. Nowadays a correlation tracker (Rimmele et al. 1991) is used to smooth out the guiding.

What can be improved in the frame selection technique? If we know the instrumental profile, we can process the frame selected scene to improve its resolution. In the past the instrumental profile has been estimated by measuring the size of the smallest feature in the scene. A better way is to measure the wavefront when the scene is exposed and then apply the corresponding point-spread function to correct the scene (Rimmele). Because the isoplanatic patch is only a few arc seconds in diameter this instrumental profile may not be correct for the entire scene. It would be better to have an instrumental profile for each isoplanatic patch. I offer one scheme that would produce an array of

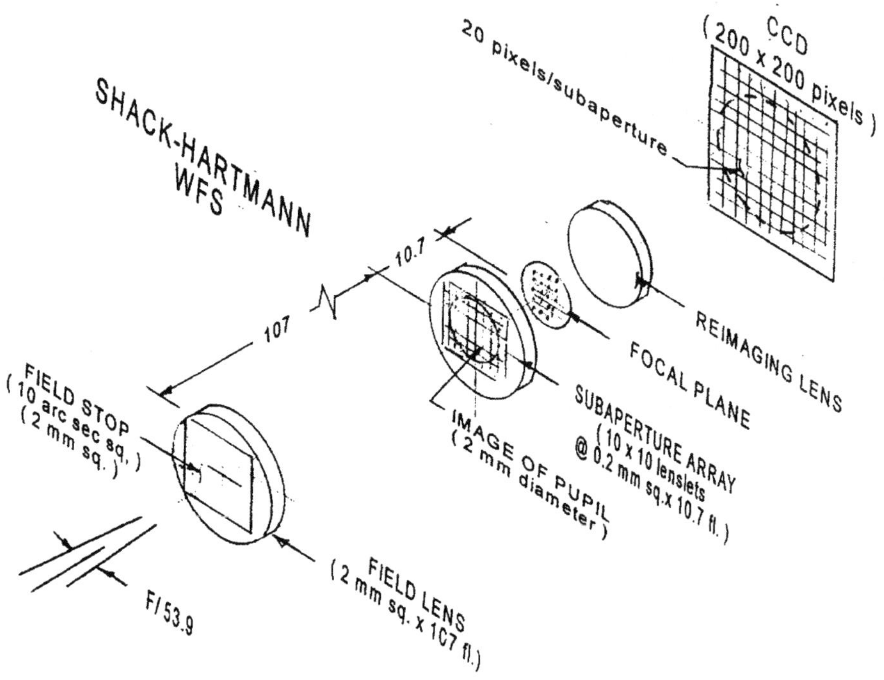

Figure 2. Shack-Hartmann WFS.

instrumental profiles over the entire scene. I call it the Areal Shack-Hartmann Wave Front Sensor, and describe it as follows: The normal configuration of a Shack-Hartmann WFS is shown in Figure 2. The focal length of the field lens is chosen so that the image of the pupil just fits 10 lenslets. A 10-arc sec field at the field lens is the same size as the 10 lenslets (subapertures). The focal length of the lenslets is chosen so that the array of 10 images of the fields made by the subapertures is the same size as the field lens and image of the pupil. Furthermore, the images of the fields made by the subapertures do not overlap each other in the focal plane of the lenslet array. A magnifier lens matches the images in the focal plane to the CCD. In our example that uses granulation as a target we have allocated 20 x 20 pixels for each subaperture requiring 200 x 200 pixels on the CCD.

The Areal Shack-Hartmann is shown in Fig.3. One hundred of the Shack-Hartmann WFSs shown in Figure 2 may be stacked side-by-side to form a 10 x 10 areal array covering 100 x 100 arc sec. The single field lens of the simple Shack-Hartmann is replaced by a 10 x 10 lenslet array. The subaperture array is 10 times larger or 100 x 100, and is a catalog item. The CCD is also 10 x larger requiring 2000 x 2000 pixels.

The frame selection technique could be extended to narrow-band wavelength observations if the throughput of the filters could be improved so that

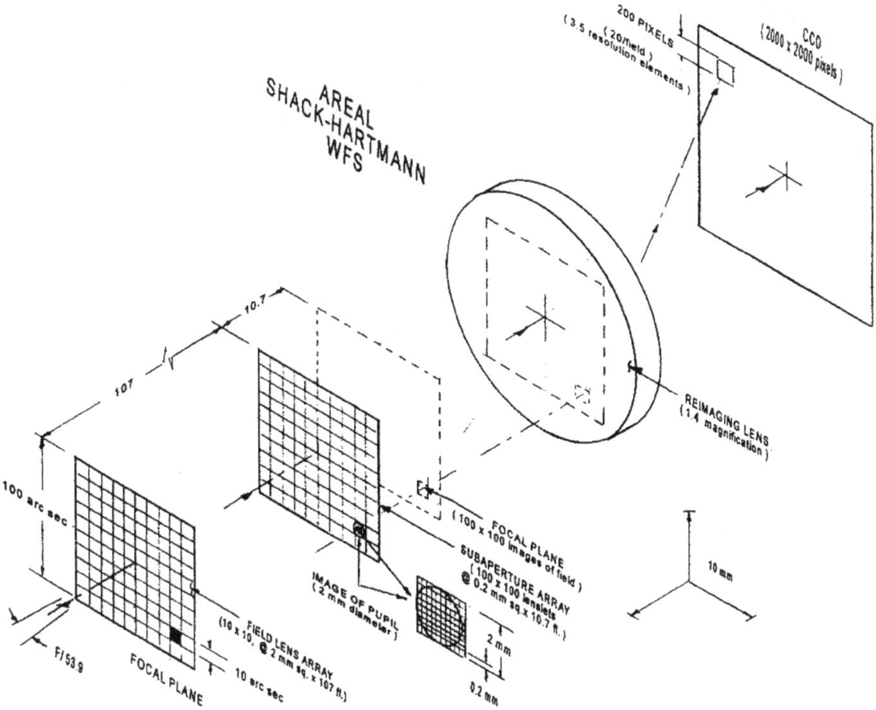

Figure 3. Areal Shack-Hartmann WFS.

the exposures could be reduced to less than 20 ms. Have we done everything we can to shorten the exposure. Coatings? Fabry-Perots? Simpler optical systems? etc?

5. Adaptive Optics

We can do science now with frame selection for scenes that can be recorded with short exposures. Adaptive Optics (AO) will allow longer exposures to be made, extending the observations to spectra and to narrow-band filters. The NSO/SP observing run with the Lockheed system showed that good uniform time series can be made with AO, however, correction of the aberrations by AO will not be perfect and the best resolution may still be obtained by frame selection. Simultaneous observations with the Areal Shack-Hartmann could further reduce the residual aberrations.

6. Conclusion

In conclusion we should improve our techniques, identify the best time series, (put them on the web?), process the data, quantify the results, encourage the theoreticians to produce models and give us feedback. We must sell the science and build community support. One way of doing this is to have workshops like this!

References

Rimmele, T., von der Lühe, O., Wiborg, P.H., Widener, A.L., Dunn, R.B., Spence, G. 1991, Solar feature correlation tracker, SPIE San Diego, July 1991

Roddier, C., Roddier, F., 1987, Interferogram analysis using Fourier transform techniques, Applied Optics, vol 26, No. 9, May 1, 1987

High Resolution Solar Physics: Theory, Observations, and Techniques
ASP Conference Series, Vol. 183, 1999
T.R. Rimmele, K. S. Balasubramaniam, R.R. Radick, eds.

The Dutch Open Telescope: History, Status, Prospects

Robert J. Rutten

Sterrekundig Instituut, Utrecht, The Netherlands

Abstract. After many years of persistent telescope design and telescope construction, R.H. Hammerschlag has installed his Dutch Open Telescope (DOT) on La Palma. I briefly review its history and design. The future of optical solar physics at Utrecht hinges on a recently-funded three-year DOT science validation period. The initial aim is to obtain high-resolution image sequences in the G band, CaIIK and H-α as proxy-magnetometry in support of SOHO and TRACE.

1. Introduction

Fig. 1 shows the cover of the proceedings of the 1980 Sacramento Peak summer workshop. It addressed essentially the same topic as the present one, but with the difference that it was organized by Dick Dunn rather then dedicated to him. He used the emptiness of the telescope platform of R.H. Hammerschlag's open tower to symbolize the need to define future solar telescope concepts and projects. At that time the open tower was ready but yet awaiting its open telescope. Now, eighteen years later, the tower and the telescope are finally joined on La Palma into what is now called the Dutch Open Telescope (DOT; Fig. 2). The DOT story is obviously a long one but it is given only briefly below. I also describe what we hope to use the DOT for the coming years and what you may want to do with it.

2. DOT principle

Fig. 2 demonstrates the open nature of the DOT. The principle of both the tower and the telescope is to minimize obstruction to the local airflow. The aim is to optimize the seeing conditions near the telescope and to minimize internal seeing within the telescope through wind flushing. There is no building or dome, only a fold-away clamshell canopy.

The airflow through the telescope must be sufficiently strong to achieve sufficient telescope flushing. At the Roque de los Muchachos the best daytime seeing tends to occur when the trade wind blows upslope from Northern directions in considerable strength, at wind speeds of 5–10 m/s. The good seeing may then last all day, making the site intrinsically different from US mountain sites (cf. Beckers & Rutten 1998) and permitting the collection of long time sequences at high resolution. The prime example is the one taken by Simon et al. (1994) at the SVST which has rms granulation contrast around and over 8% during

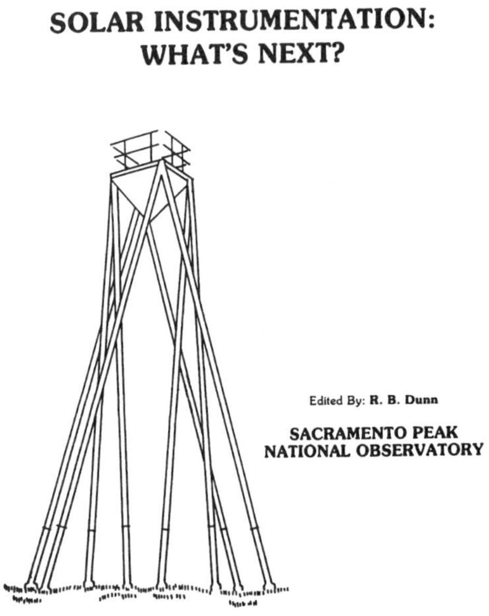

Figure 1. The cover of the 1980 Sac Peak workshop proceedings edited by Dunn (1981), derived from a photograph of Hammerschlag's open tower (then located near the Westerbork radio telescope).

eleven hours (Fig. 1 of Hoekzema, Brandt & Rutten 1998). The DOT principle aims to exploit these particular La Palma conditions.

The optical scheme is very simple. The parabolic primary mirror (Fig. 2) projects a prime-focus image on a water-cooled reflective field stop which transmits only a small part of the solar image and takes away most of the heat. Behind it, beam splitters, magnification optics and narrowband filters may be used to project images on CCD cameras mounted directly on the main axis behind the prime focus, or besides the incoming beam using 90-degree beam folding, or behind the primary mirror by deflecting the secondary beam around the prime focus stop and sending it through the central hole in the mirror. At present, the slender pipe seen in Fig. 2 contains only relay optics and a simple video camera.

The optical scheme does not possess axial symmetry because Hammerschlag prefered asymmetrical support of the prime focus structure for mechanical rigidity. However, the telescope polarization is yet favorable because it is time-independent, the DOT having a parallactic mount with hour angle and declination drives. There is no diurnal image rotation either.

Its open structure makes the DOT differ from all other solar telescopes currently in use. Almost all rely on internal evacuation to avoid internal seeing from solar heating of the telescope structure, in particular around their foci where the

Figure 2. The Dutch Open Telescope on La Palma, December 1997. Altitude about 2300 m. North is to the right, the SVST to the back of the photographer, the Carlsberg transit telescope just below the DOT, the Residencia way down past the black fire mark covering part of the HEGRA array. The Atlantic spreads to the far horizon.

solar heat is concentrated[1]. The use of vacuum windows sets a diameter limit at about one meter, also when a lens rather than a plane window is used as in the SVST (Scharmer et al. 1985) and its planned successor (the NSST, see Scharmer , Owner-Peterson & Korhanen 1999). It is therefore clear, also from the concept studies for CLEAR (Beckers 1999), that the DOT principle is of interest to the design of future solar telescopes with multi-meter aperture.

The DOT aperture, defined by its D = 45 cm, f = 200 cm primary and nearly equal to the 48 cm SVST aperture, is yet well below the 1 m limit above which the open principle may be the key strategy, but as can be seen in Fig. 2, the DOT mechanical structure accepts a larger primary, up to 100 cm without major modification.

3. DOT history

The DOT tale obviously centers on Rob Hammerschlag. He designed and built the DOT during the past decades with admirable tenacity, keeping the project going amidst storms of criticism that mostly concerned progress speed.

However, the story starts with Kees Zwaan who came up with the original idea of trying out an open telescope. This happened in the early seventies when De Jager, Zwaan and others at Utrecht were part of the extensive JOSO efforts to find the best site in Europe for ground-based solar observing and to put a large European telescope there. The latter task was later split off from JOSO to become the ill-fated LEST, a long story by itself laid down in tall stacks of orange books in many a solar physicist's office. In those years, Zwaan led a sequence of site testing campaigns that concentrated on sea level island sites off the southern coast of Portugal. He even joined a navy expedition to the Islas Selvagem, a few austere rocks in the Portuguese Atlantic. Testing these sites with small Questar-class telescopes led to the desire to have a 40 cm telescope that might be transported from site to site. The JOSO philosophy then, later jettisoned by the LEST Foundation when it simply declared La Palma to be the best site worldwide, was to nail down the number of LEST candidate sites to just a few and then use such medium-size telescopes to select and "verify" the final one through solar imaging with an aperture exceeding the Fried diameter. At the time, Rob Hammerschlag had been hired as an optics engineer at Utrecht to support a totally different project (stellar interferometry using heterodyne down-conversion, part of a non-successful and long-defunct Utrecht enterprise called laboratory astrophysics). Zwaan's enthusiasm inspired him to try his hand at a transportable open test telescope for LEST.

LEST is dead now but the DOT is finally coming alive. Over the years, Hammerschlag tended to promise completion within two years; these two years became two decades eventually, and even now the telescope is not yet ready for

[1]The NSO/Kitt Peak Pierce-McMath telescope is not evacuated, but can not be considered a high resolution optical telescope in terms of angular resolution (though it performs better in the infrared and its FTS provides superior spectral resolution). Also, as Beckers and Scharmer have pointed out during this meeting and elsewhere, some vacuum telescopes including the German VTT on Tenerife are partially open in the sense that their heliostats are outside the vacuum, are outside in the open air, and are wind flushed (and wind shaken).

science. Clearly, the project did not progress as originally anticipated. In the meantime, Kees Zwaan was retired (forcefully, there is no right to work beyond 65 in Holland) five years ago (cf. Rutten & Schrijver 1994), making me DOT project scientist by default. I might not have chosen to assume that role had I been asked two decades ago but nevertheless, I profess great admiration for Hammerschlag's efforts, tenacity and product. Even if the DOT has been overly slow in coming, certainly much slower than promised, in the end it is yet as novel and innovative a telescope as it was at its conception.

In addition, it is fair to note that other solar telescope projects have taken as long (THEMIS) or have not succeeded at all (LEST; SOT/HRSO/OSL). The most notable exception is of course Scharmer's SVST which came on-line remarkably fast and gained an outstanding track record that makes the successor NSST the most promising solar telescope project on the drawing board. In addition, Scharmer plays a key role in the DOT project by accommodating the DOT team most hospitably in the SVST building from which the DOT is operated, and by furnishing much material and political help as well as highly appreciated advice.

Hammerschlag is not a Scharmer. Göran Scharmer seems to set out on a new tack every few years, pioneering a revolutionary technique with utter simplicity and utter quality at the same time. First it was radiative transfer, then a solar refractor, then image selection and wavefront restoration; this year the NSST. Rob Hammerschlag reminds me much more of Dick Dunn even though Dick speaks faster even than I do. The similarity is that over the years, Dick Dunn has through thick and thin[2] kept going at his great telescope. The DST[3] materialized considerably faster than the DOT (be it at considerably higher cost), but in the thirty years since the initial DST completion Dick Dunn has persevered in improving it further with Hammerschlaggian tenacity. Clearly, the DST is Dick Dunn's life's work just as the DOT is Rob Hammerschlag's.

Why did the DOT take so long? First of all, it has largely been a one-man project. In the seventies, Utrecht Observatory was still rather rich (although we didn't appreciate that at the time) and had mechanical and electronics workshops large enough to start a project such as this. Even so, Hammerschlag cornered the interest and cooperation of the yet larger Central Workshop of Delft Technical University, his alma mater, where both the tower and the telescope were eventually assembled at minimal cost. During the years, steady erosion of the Utrecht astronomy funding took away both our workshops, in addition to almost all other support personnel as well as over half the staff. This demise accelerated when the newly named Sterrekundig Instituut Utrecht moved into the out-of-town physics laboratory while bequeathing Sterrewacht Sonnenborgh to amateur astronomy. It so lost its relative independence from the physics administration — which regarded the DOT project with large misgivings given its history. The manpower and the money allocated to the DOT went down to a subcritical drizzle until Hammerschlag and Zwaan managed to acquire a generous external grant from the Dutch Foundation for Technical Sciences STW. It

---

[2]What's in a name — thick thin is dik dun in Dutch, tlusty hubený in Czech.

[3]I propose that the unwieldy name "NSO/Sacramento Peak Richard B. Dunn Solar Telescope" be abbreviated in actual practice to Dunn Solar Telescope or DST.

funded the completion of the DOT and its installation on La Palma — not to provide a new toy to solar astrophysicists but out of interest in the innovative mechanical technology implemented in the DOT.

4. DOT structure

The mechanical specifications that Hammerschlag wrote down as design requirements at the start of his project constitute a second reason why the project took so long. An open telescope that relies on flushing by 5–10 m/s winds to perform near its diffraction limit should not shake or vibrate while being buffeted by such strong winds. The requirement of 0.05 arcsec pointing stability made Hammerschlag evolve from an optics engineer into a mechanical engineer; it also took a large amount of development and construction time. Many of Hammerschlag's solutions are new to astronomy; it is fitting that Dick Dunn concluded his solar telescope review in the celebratory 100th volume of *Solar Physics* with the following judgment:

> *The Utrecht telescope, designed by Hammerschlag, will also be useful for Stokes, but its aperture is somewhat small; however, even partial success of its many innovations will open up exciting new design possibilities for high resolution solar telescopes.* Dunn (1985)

These mechanical innovations should be written up and explained by Hammerschlag himself, not by a non-expert outsider like me; let me therefore just mention a few.

The principle of the 15 m high tower that lifts the telescope above the layer of ground-heated turbulence is to use triangles everywhere, as is done in all DOT support structures. The four large triangles made up by the eight steel tubes that constitute the tower legs keep the platform pairwise from tilting. It might pivot around the axis defined by the two triangles in the plane of the wind direction, but then the other two triangles inhibit such rotation and permit motion of the platform parallel to itself alone. The sun being a source at infinity makes this OK.

The massive drive wheels supplying the hour angle and declination rotation (the hour angle one alone weighs 2300 kg) are driven through self-aligning gear trains that achieve 1:75.000 reduction. The gear wheels are not fixed to any axis but have freedom of motion, enabling them to engage each other along the whole length of their teeth. This alignment inhibits the pivoting freedom that results from the single-point contact normally occurring in gear engagement.

The DOT gear trains are driven in preloaded backlash-free push-pull configuration by twin sets of brushless servo motors. They have 3–4 kW capacity but use only a few tens of Watts in normal operation in order to keep the dissipation low, three orders of magnitude below that of the oil bearings in the nearby William Herschel Telescope. The latter needed cooling plants far from the telescope to clean up the dome seeing; the DOT drive dissipation is so low that the problem does not occur.

The hour angle wheel rotates under a dust-tight cover in order to avoid problems with the frequent occurrence of Sahara dust deposition at La Palma. The declination wheel moves together with a similar cover. The driving of both wheels is done with considerable imbalance, respectively 2300 kg m and 1200 kg m, so that large variable wind loads (or ice loads) are accommodated.

The clam-shell canopy combines large strength with the possibility of being closed (or opened, less likely) in gale conditions. The strength comes from the combination of heavy steel ribs and two-dimensional curvature of the polyester fabric between the ribs, inhibiting flapping. The teflon-like fabric coating resists ice deposition, an important characteristic at the Roque de los Muchachos where below-zero winter fogs often cause heavy ice growth on metal surfaces.

The platform has an open-mesh floor, covered only in winter when the DOT will not be manned (Hammerschlag and I fulfilling considerable teaching loads). It is reached by elevator or ladders in an independent stairwell structure that may carry large ice loads and withstand large wind loads by itself (the combination is the worst and has toppled over large cranes on both the La Palma and Tenerife observatory sites).

5. DOT status

Hammerschlag and his team (in particular P.W. Hoogendoorn and F.C.M. Bettonvil, assisted by numerous young trainees from technical schools around Utrecht) installed the DOT at the International Roque de los Muchachos Observatory at a site close to SVST during 1996 and 1997. The DOT first light ceremony was held on October 31, 1997 (upper part of Fig. 3). Soon after, first images of excellent quality were obtained with a simple video camera through a modem link to the SVST. One is shown in the lower part of Fig. 3 and others may be inspected at http://www.astro.uu.nl/~rutten/dot. The simple initial equipment so served to establish the DOT's capability for high-resolution observation.

However, the first light and the first images also marked the end of the funding by the Dutch technology foundation STW as well as the final closure of our mechanical workshop. It has taken nearly a year to obtain funding to resume working on and with the DOT. Various fortunate circumstances have contributed to finally getting green light again last September. The first was the selection of the European Solar Magnetometry Network as a TMR programme by the European Commission. It provides salaries and travel allowances for three-year postdocs at eight European solar physics groups, one going to the DOT. Details are given at http://www.astro.uu.nl/~rutten/tmr. Second was the selection of a DOT science program for funding by the Dutch Science Foundation NWO from a grant program outside the regular astronomy allocation. The quality of the first DOT images had much to do with the the favorable tone of the referee reports. The third was the selection of Dutch astronomy in general as best of all Dutch science in a new NWO grant program aiming at supporting top quality university research. The four Dutch university institutes combined into what is called the Netherlands Graduate School for Astronomy (NOVA), is led by T. de Zeeuw and E.P.J. van den Heuvel, and has cornered a large allocation of "new" money of which a significant part is specifically intended to support new instrumentation. NOVA has allocated a small but politically instrumental grant to the DOT. Last but not least, the present administrations of the Faculty of Physics and Astronomy and the Sterrekundig Instituut of Utrecht University have decided

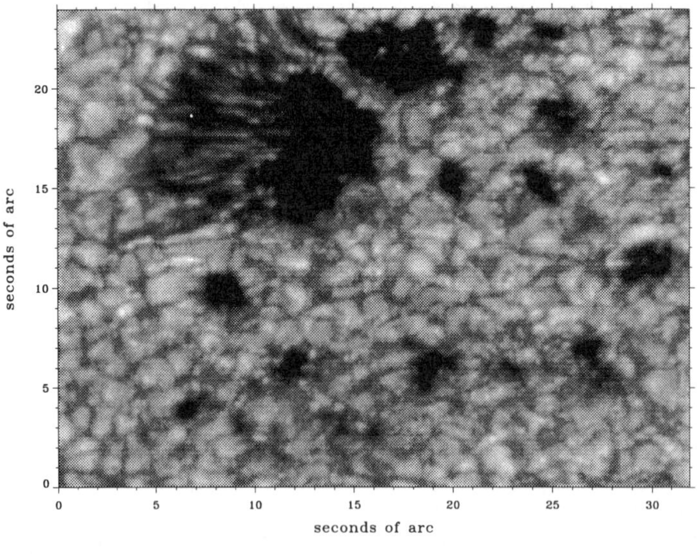

Figure 3. Top: DOT first light ceremony held on October 31, 1998, performed by His Royal Highness Willem-Alexander, Prince of Orange, and His Excellency Manuel Hermoso Rojas, President of the Canary Islands. Since it was clouded with rain coming on they (and many other dignitaries) were shown a canned movie of video images taken a few days earlier. Göran Scharmer, Oskar von der Lühe and Valentin Martínez Pillet represented the solar physics community. Bottom: DOT image taken on December 5, 1997 through a wide-band filter centered on λ = 546 nm. More photographs and images at http://www.astro.uu.nl/~rutten/dot.

that the DOT deserves a chance to prove its scientific worth. Together, these five funding agencies support a three-year "DOT science validation" period to do so.

The program for which this three-year funding is allocated consists of simultaneous imaging in three wavelength bands: the molecular CH band around $\lambda = 430.5$ nm (G band), the Ca II K line and Hα. The first two require simple interference filters; for Hα we have the tunable Zeiss filter on loan that was previously used by V. Gaizauskas at ORSO. This three-channel imaging will provide "proxy magnetometry", respectively of the deep photosphere (G band), the low chromosphere (Ca II K) and the high chromosphere (Hα). The techniques will closely follow the examples set by Scharmer and his team at the SVST in collaboration with the Lockheed-Martin group. The observations will concentrate on joint programs supporting SOHO and TRACE campaigns. Obvious research topics are the mapping of solar magnetic structures at various heights, tracing their evolution, and measuring their dynamical behavior. The DOT technical team (Hammerschlag, Bettonvil, Hoogendoorn) returned to La Palma during October 1998 to test appropriate optics designs. We hope to initialize scientific data collection by the summer of 1999.

6. DOT prospects

A dozen Utrecht-educated solar physicists is presently active outside The Netherlands, but at Utrecht itself Hammerschlag and I are the only staff members devoted full-time to the field (apart from our obligatory teaching). Hoyng, Kuijpers and Kuperus maintain only partial interest. Thus, solar physics survives at Utrecht but meagerly compared with the days of Minnaert and De Jager. None of the solar physics chairs had a solar physics successor (De Jager, Houtgast, Fokker, Zwaan) or will have one (Kuperus). There is no solar physics at the other Dutch university astronomy departments (Amsterdam, Groningen and Leiden); the Sun is not mentioned in NOVA's charter[4].

In this context, the three-year DOT science validation program and funding represent an important new impetus to optical solar physics at Utrecht. The program scale is small compared with what is desirable for running a full-fledged facility, but it does give us opportunity to hire some graduate students and postdocs which we otherwise wouldn't have had. It is clear that the future of Utrecht solar physics rides on it.

One thing that we are *not* funded to do is to experiment with the DOT as a step on the way to larger solar telescopes. It is of obvious interest to use the DOT as a testbed facility for trying out new techniques — thermal mirror control, CLEAR-like shrouds, wavefront monitoring, prime mirror adaptivity, etc. The DOT science validation program is explicitly limited to the three-channel imaging for which the funding is allocated. Therefore, if you have interest in open-principle performance experiments, that is where *you* come in. Such co-

---

[4]Fortunately, the plasma physics theory group led by Prof. Hans Goedbloed at the Dutch plasma physics laboratory Rijnhuizen (a non-university institute in Nieuwegein nearby Utrecht) has turned from tokamak theory to coronal theory. Perhaps this is a telltale sign that solar physics is too difficult for astronomers and requires takeover by regular physicists?

operation is very welcome.

Acknowledgments. The DOT was designed by R.H. Hammerschlag and built by the Sterrekundig Instituut Utrecht, the Physics Instrumentation Group of Utrecht University, and the Central Workshop of Delft Technical University. The DOT completion and installation at La Palma were funded by the Stichting Technische Wetenschappen (STW) of the Netherlands Organization for Scientific Research (NWO), and proceeded under the hospitable umbrella of the Instituto de Astrofísica de Canarias. The three-year science validation phase is funded by the Netherlands Organization for Scientific Research (NWO), the Netherlands Graduate School for Astronomy (NOVA), Utrecht University, the Sterrekundig Instituut Utrecht, and the European Commission via a TMR grant to the European Solar Magnetometry Network. The DOT team (R.H. Hammerschlag, F.C.M. Bettonvil, P.W. Hoogendoorn and E.B.J. van der Zalm) enjoys splendid hospitality of the Royal Swedish Academy of Sciences at the SVST with much support from Paco Armas, Göran Hosinsky, Rolf Kever and Göran Scharmer.

Beckers J. M., 1999, this volume
Beckers J. M., & Rutten R. J., 1998, in C. M. Tunon (ed.), Site Properties of the Canarian Observatories, in press
Dunn R. B. (ed.), 1981, Solar Instrumentation: What's Next?, Procs. 1980 Sacramento Peak Nat. Obs. Conf., Sacramento Peak Nat. Obs., Sunspot
Dunn R. B., 1985, Solar Phys. 100, 1
Hoekzema N. M., Brandt P. N., & Rutten R. J., 1998, A&A 333, 322
Rutten R. J., Schrijver C. J. (eds.), 1994, Solar Surface Magnetism, NATO ASI Series C 433, Kluwer, Dordrecht
Scharmer G., Brown D., Petterson L., & Rehn J., 1985, Appl. Opt. 24, 2558
Scharmer G. B., Owner-Peterson M., & Korhanen T., 1999, this volume
Simon G. W., Brandt P. N., November L. J., Scharmer G. B., & Shine R. A., 1994, in R. J. Rutten, C. J. Schrijver (eds.), Solar Surface Magnetism, NATO ASI Series C 433, Kluwer, Dordrecht, p. 261

High Resolution Solar Physics: Theory, Observations, and Techniques
ASP Conference Series, Vol. 183, 1999
T.R. Rimmele, K. S. Balasubramaniam, R.R. Radick, eds.

The New Swedish Solar Telescope

G.B. Scharmer

Royal Swedish Academy of Sciences
Stockholm Observatory, SE–133 36 Saltsjöbaden, Sweden

M. Owner-Petersen

Lund Observatory, Box 43, SE–221 00 Lund, Sweden

T. Korhonen

Tuorla Observatory, FIN-215 00 Piikkiö, Finland

A. Title

Lockheed Martin Solar and Astrophysics Laboratory, B/252, O/L9-41
3251 Hanover Street, Palo Alto Ca. 94304, U.S.A.

Abstract. We describe the new Swedish solar telescope which will replace the existing solar telescope (SVST) in La Palma. The new telescope has a diameter of 97 cm or twice that of the SVST and a turret design similar to that of the R.B. Dunn Telescope and the SVST. The optics consist of a singlet lens used as vacuum window and a choice of secondary optical systems. The first of these enables narrow-band imaging and polarimetry with a minimum of optical surfaces. The second optical system uses a field mirror to re-image the pupil on a 24 cm corrector which provides a perfectly achromatic image, corrected also for atmospheric dispersion.

The two high-resolution beams are used with field lenses, acting as exit vacuum windows, re-imaging the entrance pupil on an adaptive mirror which also folds the beam horizontally above the optical tables. The adaptive mirror is a bimorph modal mirror capable of correcting 20–36 Zernike aberrations. This allows near-diffraction-limited imaging a reasonable fraction of the observing time on La Palma. The adaptive optics system for the new solar telescope is under development and a prototype is undergoing tests with the SVST.

The optical quality of the turret mirrors are specified to 20 nm PV surface quality over any 300 mm diameter subaperture and 50 nm PV over the entire aperture. The specifications are similar for the 97 cm singlet objective. The high requirements for optical quality is in order to ensure accurate wavefront sensing without cross-talk from high-order telescope aberrations and to allow near-diffraction limited performance with adaptive optics.

The new telescope is expected to become operational during the first half of the year 2001. Although mainly intended for solar work, it will also be used for studies of the planets.

1. Introduction

The design of the Swedish Vacuum Solar Telescope, SVST, was based on the firm conviction, that by using a small number of optical components of excellent optical quality, it would be possible to construct a telescope which performed better than existing even larger solar telescopes (Scharmer et al. 1985). This concept was validated in December 1985, when the SVST saw first light. The SVST has produced diffraction-limited images since the first day of its operation and other solar telescopes world-wide have not been able to match that image quality consistently, in spite of significant efforts. By using real-time frame selection in combination with wavefront sensing and post-processing techniques, this telescope can produce long uninterrupted time sequences of dynamic phenomena in the solar atmosphere at a resolution which is strictly limited by the diameter of the telescope, which is 47.5 cm.

In this paper, we outline plans for a new solar telescope, to replace the existing SVST. The goal of this telescope is imaging at a resolution close to its diffraction limit of 0.1 arcsec in the blue and polarimetry and spectroscopy at approximately 0.2 arcsec resolution. To achieve this goal, a simple but excellent optical system including adaptive optics is proposed.

2. Overview of Design

The site of the SVST on La Palma is the best known site for solar telescopes in the world. By installing the new telescope on the existing tower, by keeping development and design costs down and by re-using some of the instrumentation developed for the SVST, it is possible to build an excellent meter-class telescope within a budget of 2 M$.

The SVST uses an evacuated turret, similar to that of the R.B. Dunn Telescope at the Sacramento Peak Observatory (Dunn 1964). The main advantage of this pioneering design is that the telescope needs no dome which could seriously degrade image quality by heating or cooling air in the vicinity of the telescope. Also, all mirrors are within vacuum, excluding the possibility of mirror seeing. The turret of the SVST has been on the Swedish tower for 12 years and has been exposed to the most severe weather conditions, including more than 50 m/s winds, heavy rains, complete ice coverage and dust storms. Yet it shows no signs of damage, corrosion or wear. The turret mirrors which are in the vacuum system are cleaner after ten years than a mirror exposed even to the favourable conditions in our optics labs during a few weeks. We have seen no evidence of image movement due to vibrations from either the turret or the tower in reasonable wind speeds. The mirrors in the turret are always used at fixed angles of incidence, which makes calibration and modeling of telescope polarization feasible. Significant progress at the Instituto de Astrofisica de Canarias has been

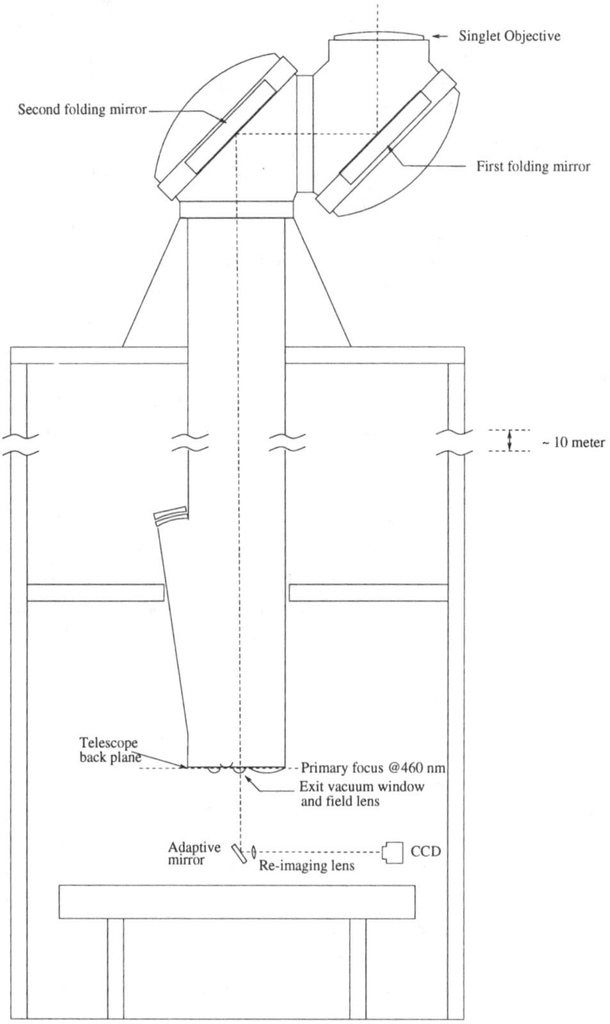

Figure 1. Schematic optics for high-resolution, narrow-band imaging with a a conventional flat adaptive mirror, used also as folding mirror. The combined vacuum window and field lens re-image the singlet objective on a small adaptive mirror. Note the simplicity of the optical system and the small number of optical surfaces.

made in such modeling of the SVST telescope polarization during the last two years (Pillet et al. 1999, these proceedings). The turret concept has proven itself at the SVST and this concept will be used also for the NSST.

We have considered modifying this turret design, such that the mirrors are used to deflect the beam by 60–70 degrees instead of 90 degrees as for the SVST. The advantages would be lower telescope polarization and reduced cost and weight for the mirrors. The disadvantage is that the turret becomes significantly less compact, leading to greater area exposed to sunlight and wind, which could degrade performance. It was finally agreed, that the advantages of modifying the turret design were less important than the possibility that e.g the telescope loses its pointing stability in moderate wind speeds. We have therefore adopted a design for the NSST which is similar to that of the SVST.

The SVST uses an achromatic doublet in the turret to form an image in the optics lab. The advantage of a doublet objective is that it gives a shallow focus curve over some reasonable wavelength range, which allows broad-band observations. However for several important applications, such as narrow-band imaging in H-alpha and Ca K, or polarimetry using narrowband filters or the spectrograph, a shallow focus curve is not needed. A doublet of 1 meter diameter is difficult to manufacture with high optical quality and would be diffraction limited with the NSST over only 1/4 of the wavelength range compared to the doublet of the SVST. For example, a doublet of 1 meter aperture would prohibit spectroscopy involving the broad wings of the Ca H or K lines with a single focus position. We therefore decided to use a singlet lens as primary image forming element and a corrector, described below, for achromatic imaging or spectroscopy. The primary optical system of the NSST consists of only one singlet objective, acting also as vacuum window, and two flat folding mirrors.

3. Singlet Objective

The choice of glass for the singlet objecive and other transmitting lenses in the optics was finally fused silica. The main argument for this is its low coefficient of thermal expansion, which gives small stresses from temperature gradients. Stresses in the glass lead to birefringence which e.g. introduces spatial cross-talk from Stokes I to linear polarization. Compensating for such effects will need deconvolution. Polarimetry of small-scale structures will require accurate determination of both the ordinary wavefront errors and the spatial variation of the orientation and magnitude of the retardation across the pupil. We believe that this will be very difficult if the retardation varies with the temperature gradient in the singlet objective. For example, inserting large polarizers in front of the singlet objective will change the illumination and thus temperature (gradient) of the objective, making such calibrations of telescope polarization inaccurate. By using fused silica instead of BK-7, we reduce this temperature sensitivity by approximately a factor 15. It should also be pointed out that birefringence leads to a reduction in the Strehl ratio which cannot be compensated by adaptive optics. We have estimated that e.g. a 2-degree temperature difference between the edge and the center of of a lens made of BK-7 would reduce the Strehl ratio to 0.8, thus imposing a fundamental limitation on image quality.

The singlet objective has a diameter of 1070 mm, a clear aperture of 970 mm, a center thickness of 80 mm and an edge thickness of 65 mm. It has a focal length of 20.8 m at a wavelength of 460 nm, is corrected for coma and has a small aspherical correction applied to its first surface. Finite element analysis

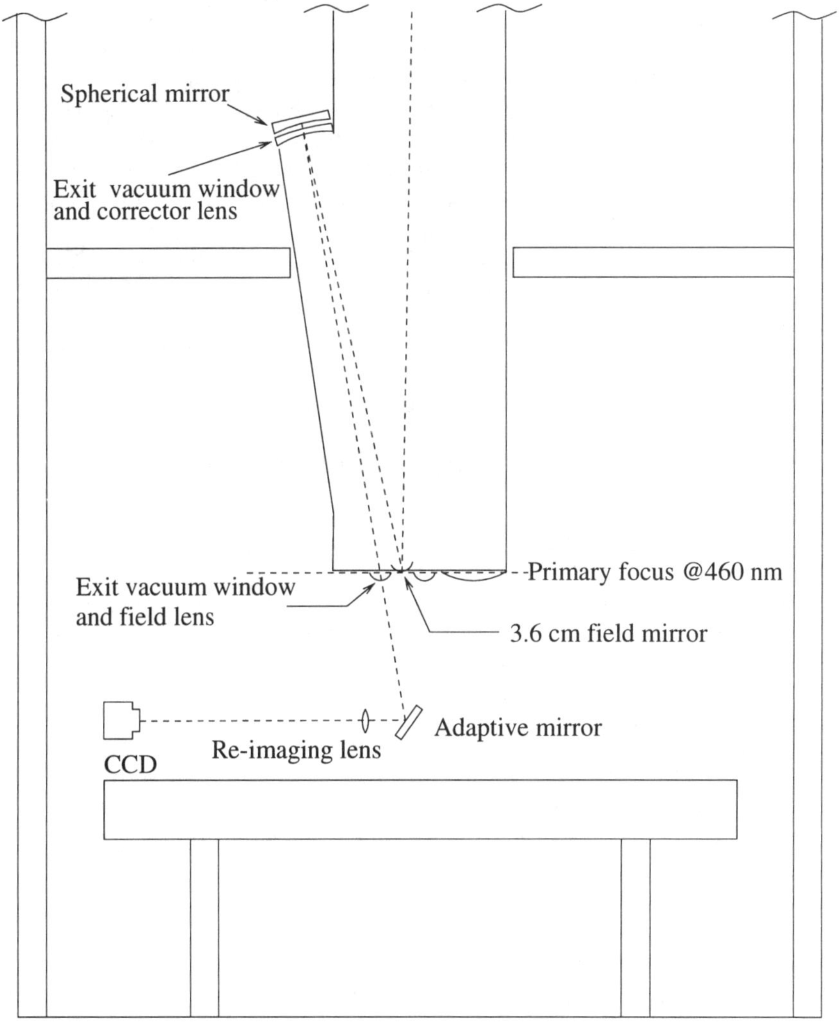

Figure 2. Lower part of the tower and optical system, showing the schematic optics of the Schupmann corrector. The beam is reflected upwards and away from the optical axis using a field mirror close to the focal plane. Located 5.25 meters up and with a diameter of 24 cm, is a corrector, consisting of a negative singlet lens and a curved adaptive mirror. This optics corrects for chromatic aberrations in the singlet. By tilting the field mirror slightly, atmospheric dispersion can be corrected.

has been used to calculate the stresses from pressure, gravity and temperature gradients and also to calculate the distortions of the surfaces. To evaluate the effects of the deformations of the surfaces of the lens, the co-ordinates of the node points are fitted to sixth-order polynomials. The corresponding coefficients are inserted in a ZEMAX ray-trace program to evaluate optical performance. The analysis shows that the pressure adds fourth- and sixth-order terms to both surfaces which are large but that the resulting spherical aberration from the vacuum load is negligible when combining the effects of the two surfaces.

In order to minimize temperature gradients in the singlet objective, it will be mounted such that its edges are exposed to air and with a shield to prevent the cell from being heated by direct sunlight, in a way similar to SVST. We believe that the spherical aberration from temperature gradients will be small enough to be corrected by the adaptive mirror. Should this not be the case, we will install a system to control the temperature of the holder of the singlet objective.

We have verified that the birefringence from stresses lead to a negligible (3%) loss of Strehl ratio. The maximum tensile stress is 4.0 MPa which is below the recommended design stress of 6.8 MPa for fused silica.

4. Secondary Optical systems

The diffraction limited field of the singlet objective is nearly 1/2 degree or 18 cm at the primary focal plane. At this field radius, the only significant aberration is a small amount of astigmatism which is easily corrected by an adaptive optics system. It is unlikely that adaptive optics will correct atmospheric aberrations over much more than 1 arcmin field. The NSST therefore uses four exit ports close to the optical axis of the singlet, each with its own secondary optical system. This is possible by restricting the field-of view for three of these secondary optical systems to approximately 2 arcmin. This allows the secondary optical systems to be used in parallell and even simultaneously depending on the choosen target on the Sun. Switching between the different secondary optical systems is done simply by changing the pointing of the telescope slightly.

4.1. Narrow-band re-imaging system

Narrow–band imaging in e.g H-alpha, Ca K or other strong lines as well as polarimetry using either filters or a spectrograph involves pass-bands of less than 0.1 nm and is possible without any chromatic correction of the singlet objective. This allows a very simple secondary optical system which is shown schematically in Fig. 1. The singlet lens and the two turret mirrors form the primary focal plane which at a wavelength of 460 nm is located at the bottom of the vacuum system. A singlet lens, acting as vacuum window re-images the 97 cm singlet objective on the adaptive mirror which also folds the beam horizonthally. A re-imaging lens produces the final image at the desired image scale.

4.2. The Schupmann system

The adaptability of the optics of the NSST is high-lighted by the secondary optical system shown in Fig. 2. This system uses a field mirror, also located close to the focal plane, to deflect the beam upwards and away from the optical axis of the telescope. This beam is maintained within the vacuum system because it is comparatively long, 5.25 meters, and because it re-images the one-meter singlet objective on a nearly 24 cm large corrector. This corrector consists of a lens and a mirror. The effect of the lens is to cancel out the chromatic aberration (variation of focus with wavelength) of the singlet objective, the effect of the mirror is to create a perfectly achromatic image at the secondary focus. This concept was proposed by Ludwig Schupmann 100 years ago.

The idea of using a singlet lens in combination with a Schupmann corrector for solar telescopes has been proposed earlier by Baker (1954), Rush and Schnable (1964) and also by Dunn as an early concept for LEST. It was later considered by Roesch (1983, private communication) for the Pic du Midi refractor, but was rejected because of strong off-axis aberrations. The Schupmann design has been implemented on several telescopes during the last 40 years with mixed results. Of these can be mentioned the two coronagraphs at Climax and at the Sacramento Peak observatories, both of which had serious optical problems. These designs used correctors with a single component, consisting of a meniscus lens silvered on its backside. For solar telescopes, a corrector consisting of a meniscus lens followed by a Zerodur mirror appears a better choice which also adds one more degree of freedom in the optimization. Ray-tracing calculations do not indicate any particular problems in aligning the corrector lens with respect to the mirror, however, it is clear that the image of the singlet needs to be accurately centered on the corrector in order to not give rise to dispersion. We believe that an arrangement as proposed for the NSST, with the Schupmann system at fixed orientation in the stable environment of the tower, with a corrector consisting of a separate lens and Zerodur mirror and combined with a wavefront sensor is a more promising concept than those of earlier designs.

Initially, we planned to replace the curved mirror with a curved adaptive mirror, but finally it was decided that using a flat adaptive mirror acting also as folding mirror, below the vacuum system was a less expensive and simpler solution. Because a folding mirror was needed anyhow, using this mirror as an adaptive mirror does not increase the number of optical surfaces. The cell of the adaptive mirror will be mounted with two piezos to allow tip-tilt correction with large stroke.

In order to give diffraction-limited image quality, either the corrector lens or the mirror needs approximately six waves PV correction for spherical aberration. The excellent performance of the telescope after correction by the Schupmann system is shown in Table 1. The design was optimized using Zemax with tilt-angles for the field mirror and the Schupmann mirror to allow straightforward implementation of baffling to reduce straylight. Early designs showed excellent NIR performance but somewhat poor performance at 400 nm. By adjusting the curvatures of the singlet objective such that its focal plane agrees with the location of the field mirror at 460 nm, a better balance between NIR and blue performance is obtained. The present design allows diffraction limited performance (Strehl ratio 0.96 or higher) over the whole 400–900 nm wavelength range at a single focus position.

An important feature of the Schupmann system is its ability to correct also for atmospheric dispersion. The Schupmann lens cancels the chromatic variations of the singlet objective only when the image of the singlet is centered

Table 1. Design performance after correction by Schupmann system. The table gives spot diameters, wavefront errors and Strehl ratios at a single focal plane for all wavelengths in the range 400 to 900 nm.

λ nm	spot RMS microns	geo radius microns	Airy radius microns	pv wf waves	RMS wf waves	Strehl
400	2.32	4.24	9.9	0.18	0.032	0.96
460	1.85	3.24	11.3	0.12	0.023	0.98
550	2.29	4.76	13.6	0.14	0.024	0.98
650	2.93	5.51	16.0	0.14	0.027	0.97
900	4.73	7.54	22.4	0.14	0.031	0.96
All	3.02	7.59				

Table 2. Atmospheric dispersion at 14.9 degrees elevation and after compensation by tilt of the Schupmann field mirror. The PV dispersion is reduced by approximately a factor 25.

λ nm	Atmospheric dispersion arcsec	Residual dispersion arcsec
400	3.43	0.08
500	1.03	−0.06
600	−0.23	−0.09
700	−0.98	−0.05
800	−1.46	0.01
900	−1.79	0.11

on the corrector lens. Displacing the the image of the singlet either by tilting the field mirror or by translating the Schupmann lens in the plane perpendicular to the optical axis, causes dispersion in the image plane. The amount of dispersion corresponds to 180 micron, or 1.8 arcsec, separation between the 400 and 900 nm wavelengths per mm decenter. Table 2 shows the atmospheric dispersion at 14.9 deg elevation for an altitude of 2000 m as calculated by Martin (1997) for wavelengths in the range 400–900 nm and the residual dispersion after optimum compensation with the Schupmann system. The Schupmann corrector allows the effects of atmosperic dispersion to be reduced by approximately a factor 25.

The simplest way of correcting the atmospheric dispersion is by tilting the field mirror, but this moves the pupil image which is undesirable for the following adaptive mirror and wavefront sensor. We have therefore choosen to install a fixed pupil stop corresponding to a diameter of the singlet of 950 mm. When compensating atmospheric dispersion at elevations below approximately 15 degrees, this causes the clear aperture of the singlet objective to be displaced slightly outside the pupil stop. The NSST will have a broadband filter which can be switched electrically between a blue and red wavelength without moving optical components. This allows the dispersion to be measured accurately, using either the correlation tracker, the adaptive optics wavefront sensor or any other CCD, and tilting the Schupmann field mirror to compensate the dispersion.

The Schupmann design allows an unobscured pupil and perfect chromatic correction by using an off-axis system and is as a result of this limited to good correction over a fairly limited field. Ray-trace calculations indicate diffraction limited performance over an approximately 3 arcmin field. In recognition of the constraints in the useful field imposed by adaptive wavefront correction, we have restricted the field further to 2 arcmin by reducing the diameter of the field mirror to 36 mm. This reduces the heat load on the Schupmann corrector and should also lead to reduction of stray-light. By locating the field mirror at the focal plane of the singlet at 460 nm, there will be vignetting of part of this field at the 900 nm wavelengths, but this has been considered an acceptable trade-off in order to enhance the performance at blue wavelengths. The unvignetted field at the 900 nm wavelength is 70 arcsec.

The Schupmann design, by requiring a large singlet lens, is of no interest to future larger solar telescopes but appears attractive for an evacuated meter-class solar telescope with adaptive optics.

5. Optical Design, Quality and Testing

It is sometimes argued that telescopes equipped with adaptive optics require only low optical quality. We believe that this is not correct. Adaptive optics systems use wavefront sensors which have a limited spatial resolution at the pupil. Telescope aberrations at spatial frequencies higher than resolved by the wavefront sensor will, because of aliasing, be interpreted as low-order aberrations by the wavefront sensor. An adaptive optics system operating in closed loop will thus introduce fixed low-order aberrations in addition to the high-order aberrations which cannot be corrected by the adaptive mirror.

Whereas we feel confident that a low-order adaptive optics system (15–36 Zernike correction) can efficiently compensate seeing in good–excellent seeing conditions ($r_0 > 12$ cm), we feel that the efficient operation of high-order AO systems to correct poorer seeing is presently not realistic.

In order not to be forced to use a high-order adaptive optics system for the NSST and in order that the accumulated telescope aberrations, including those also of the secondary optical systems and adaptive mirror, shall be negligibly small compared to those of the atmosphere in very good seeing after correction by a low-order adaptive optics system, high optical quality is needed. Accordingly, the specifications for the two turret mirrors are 20 nm surface quality PV over any subaperture with 300 mm diameter and 50 nm PV over the entire surface. The wavefront quality for the singlet objective is similar. In order to achieve such high quality for the singlet lens, high homogeneity in the refractive index is required. The fused silica blank manufactured by Corning has refractive index variations within $\pm 1.5 \times 10^{-6}$. These variations are not small enough to ensure required wavefront quality and local polishing will be required.

Optimization of the Schupmann system shows that the second surface of the corrector lens can be flat without degrading performance. This makes testing during polishing easier.

The optics will be polished by Opteon Oy in Finland. During polishing, low frequency deviations from flatness for the flat turret mirrors will be tested with the pentaprism method (Korhonen 1994). High-frequency errors will be tested using interference fringes against a reference flat. The lens will be tested both in autocollimation using the flat mirrors and by using the lens surfaces as a concave mirror. A single elements null lens is used in these tests.

The absence of plane-parallell surfaces in the optical system should eliminate the possibility of fringes.

6. Adaptive Optics

The goal of the NSST is diffraction limited imaging and near-diffraction limited spectroscopy and polarimetry in excellent seeing, corresponding to $r_0 > 20$ cm. Measurements made during the LEST site testing campaign (Brandt et al 1989), as well as experience with the SVST, indicate that such seeing conditions should occur at least 5% of the time. Simulations indicate that near diffraction-limited imaging, corresponding a wavefront rms of less than $\lambda/10$, should be achievable under such conditions with an AO system correcting the first 10 Zernike modes. In order to allow for a 50% efficiency of the AO system, at least 20 Zernike modes need to be corrected.

We are presently developing an adaptive optics system for the SVST based on a 19 electrode bimorph mirror from Laplacian Optics, capable of perfect correction of the first 15 Zernike coefficients. Recent improvements in the manufacturing process should allow this mirror to be flattened to within 1/10 wave PV, therefore allowing accurate wavefront sensing with little aliasing from high-order aberrations in the adaptive mirror. For the NSST, we expect to use an AO mirror correcting up to 36 Zernikes.

The optical arrangement is shown in Fig. 1. In order to minimize the number of optical surfaces, a lens, acting as exit vacuum window, is used as field lens. The AO mirror is located approximately 700 mm below the field lens and is used at 45 degree angle of incidence, thus folding the beam horizonthally above the optical tables. The bimorph mirror has a stretched geometry to match the elliptic pupil image. Simulations made for the CFHT by Graves (1998, private communication) show that this type of arrangement allows excellent performance of the AO system. The diameter of the adaptive mirror will be as small as possible to allow adequate stroke with a 36 mode correction on a one-meter telescope in order to minimize the risk of telescope seeing. The distance between the field lens and AO mirror will be matched to that pupil diameter.

The wavefront sensor for these adaptive optics systems is likely to be a Shack-Hartmann wavefront sensor, matched to the geometry of the adaptive mirror. We are also investigating the possibility of using phase-diversity as a real-time wavefront sensor (Scharmer 1999, these proceedings). The computer system which will be used for the wavefront sensor and adaptive mirror is described by Shand et al. (1999, these proceedings).

For future developments, it seems more important to increase the angle over which the wavefront can be corrected rather than extending the number of corrected Zernike modes. Experience with the AO system developed for the

SVST and the first year of operation of the NSST will be crucial in making priorities for further AO developments.

7. Instrumentation

The main instruments for the NSST will be the following: imaging CCD:s, including three 2000 × 2000 Kodak CCD:s, an H-alpha filter, a new Michelson Solar Polarimeter (MSP, designed by Lockheed-Martin and funded by NASA) and a short Littrow spectrograph. In addition, we will use broad-band G-band and Ca K filters.

The MSP system uses a Michelson-Lyot filter, will allow 0.14 arcsecond resolution and obtain high signal-to-noise, vector magnetograms of the solar photosphere over a 85 arcsecond field-of-view on a 10 second cadence.

For H-alpha, plans are to re-build the engineering model of the SOUP filter and to use this filter exclusively for H-alpha. This is done in order to avoid using a single tunable filter (and CCD) for too many wavelengths, which would give unacceptably long cycling times.

The Littrow spectrograph will be optimized for modest dispersion and high spatial resolution at a single wavelength. Plans are also for a spectrograph which will allow spectra to be recorded at more than one wavelength simultaneously.

For financial reasons, the instrumentation of the NSST will be restricted but significantly more powerful than that of the SVST.

Acknowledgments. Construction of the NSST is funded by the following private foundations in Sweden: Knut och Alice Wallenbergs Stiftelse, Marianne och Marcus Wallenbergs Stiftelse, Stiftelsen Marcus och Amalia Wallenbergs Minnesfond and the Royal Swedish Academy of Sciences.

The Swedish Vacuum Solar Telescope is operated by the Royal Swedish Academy of Sciences within the Observatorio del Roque de los Muchachos of the Instituto de Astrofisica de Canarias on the island of La Palma, Spain.

Darrel Torgerson is thanked for ray-tracing the Schupmann system and providing valuable information on alignment tolerances.

Mats Löfdahl is thanked for valuable help in preparing this manuscript.

We finally want to thank all the observers from the international solar community which made the SVST a successful scientific instrument and thereby contributed to making the NSST possible.

References

Baker, J. G., 1954, A.J. Vol. 59, p.74

Brandt, P., Erasmus, D. A., Kusoffsky, U., Righini, S., Rodriguez, A., & Engvold, O., 1989, LEST Foundation Techn. Rpt 38

Dunn, R. B., 1964, Appl. Opt. Vol. 3, No. 12, p. 1353

Korhonen T, K., Lappalainen, T., Sillanpaa, A. K., 1994, in Fabrication and Testing of Optics and Large Optics, Proc. SPIE 1994, (ed. V. J. Doherty), p. 225

Martin, G., 1997, http://www.eso.org/lasilla/Telescopes/NEWNTT/emmi/refr.html

Pillet, V. M., Collados, M., Sanchez Almeida, J., Gonzalez, V., Cruz-Lopez, A., Manescau, A., Joven, E., Paez, E., Diaz, J. J., Feeney, O., Sanchez, V., Scharmer, G. B., & Soltau, D. 1999, this volume.

Rush, J. H., & Schnable, G. H., 1964, Appl. Opt. Vol. 3, No. 12, p. 1347

Scharmer, G. B., Brown, D. S., Pettersson, L., & Rehn, J., 1985, Appl. Opt. Vol. 24, No. 16, p. 2558

Scharmer G. B. 1999, this volume.

Shand, M., Scharmer, G. B., & Wei, W. 1999, this volume.

High Resolution Solar Physics: Theory, Observations, and Techniques
ASP Conference Series, Vol. 183, 1999
T.R. Rimmele, K. S. Balasubramaniam, R.R. Radick, eds.

The Advanced Solar Telescope: I. Science Goals

C. U. Keller

National Solar Observatory, Tucson, AZ 85719, USA

Abstract. Many of the limitations of current ground-based observations of the sun are due to the restricted diameter of existing solar telescopes. The limited diameter leads to limited spatial resolution due to diffraction and limited photon flux. Here I present a few of the science goals requiring a large-aperture, ground-based telescope with adaptive optics: the interaction of convection and magnetic fields as a basic astrophysical process, observations of motions of magnetic flux tubes to understand the heating of chromospheres and coronae in late-type stars, and observations of the weak, mixed-polarity turbulent magnetic field to understand its origin and interaction with strong fields.

1. Introduction

Many solar observations are limited by the restricted diameter of current solar telescopes. The limitations are the spatial resolution and the photon flux. While the largest solar telescope, the McMath-Pierce main telescope, has a diameter of 1.52 m, most existing ground-based telescopes have diameters around 50-90 cm. The diameter of existing and planned balloon-borne and space-based solar telescopes is no larger. The following contribution concentrates on the science goals of a much larger, Advanced Solar Telescope with a diameter around 3 m.

The photon mean-free path and the pressure scale height in the lowest visible layer, the photosphere, are about 70 km, or $0''.1$ at disk center. This scale can barely be resolved with even the largest solar telescope. Magnetic flux concentrations are expected to occur on even smaller scales, well below 100 km. Therefore, to spatially resolve the fundamental scales in the solar atmosphere, a telescope aperture of more than 2 m is required. Since the near-infrared spectrum around 1.5 μm has many advantages, particularly for magnetic field studies, an aperture of at least 3 m is needed to reach a diffraction limit of $0''.1$ at 1.5 μm.

However, the diffraction limit is not the only driver for larger apertures. At the diffraction limit, there are about 3×10^8 photons/Å/s in the visible part of the spectrum. It is important to realize that this number is independent of the telescope diameter, as well as the distance of a star (as long as it is spatially resolved). To make things even worse, solar structures change faster on smaller scales: with observed velocities of up to 7 km/s, one can collect photons only during about 10 s to avoid spatial smearing due to evolution on scales of $0''.1$. Therefore, the total number of available photons collected with diffraction-limited spatial resolution decreases with increasing aperture. To obtain the necessary signal-to-noise ratio at a given spatial resolution, the required telescope diameter is often larger than what is required by diffraction. A future large solar telescope will therefore often not be used at its diffraction limit.

The following sections are based on a talk given to the NRC/NAS Task Group on Ground-Based Solar Research (Parker committee) in the summer of 1997. To reduce the duplication of other contributions in these proceedings and due to my limited knowledge, I will refrain from discussing science aspects related to observations in the thermal infrared and coronal observations (see for instance contributions by Ayres and by Kaghashvili and Smartt in the present proceedings).

2. High spatial resolution

2.1. Overview

Even the most common small-scale structure in the solar photosphere, the granulation, is not fully understood. Bright rims of granules (see Fig. 1 for an example) might be due to shocks or strong thermal up-flows (e.g. de Boer et al. 1992). While there is considerable observational evidence that the convective motions close to the solar surface generate the global p-mode oscillations, the exact mechanism is still under discussion (see contribution by Goode et al.).

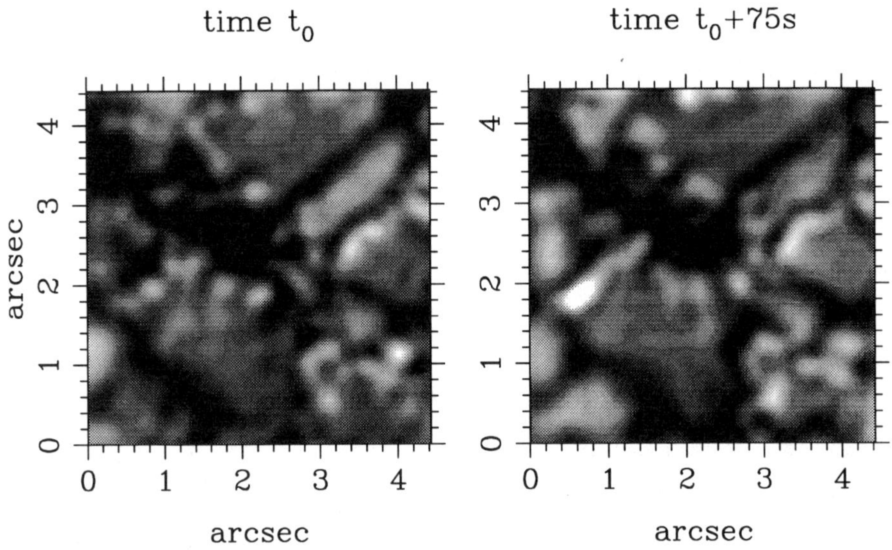

Figure 1. Phase-diverse speckle reconstruction of the evolution of granulation close to disk center observed with the 76-cm R.B.Dunn Solar Telescope and the Zurich Imaging Polarimeter (ZIMPOL) I at 656 nm. Notice the bright granular edge in the right panel. The data processing is explained in the contribution by Seldin et al.

Sunspots contain a wealth of small-scale structure. It is believed that magneto-convection is responsible for umbral dots and penumbral filaments, the

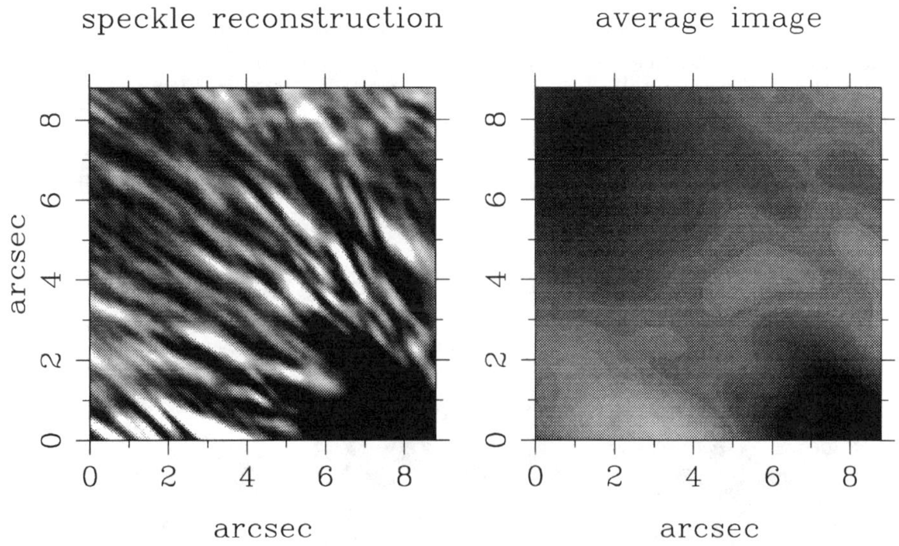

Figure 2. Speckle reconstruction and average image of part of a sunspot penumbra. 200 frames with 1.75 ms exposure time each were recorded at the 1.5-m McMath-Pierce telescope with ZIMPOL I within 30 s. The left panel shows the speckle reconstruction, while the right panel shows the average of these 200 frames after compensation of image motion. Both images are displayed at the same contrast. The smallest structures have scales of $0\farcs1$.

most prominent fine-scale structures. Figure 2 shows part of a penumbra at the highest spatial resolution ever obtained ($0\farcs11$). There are indirect observations suggesting that the fundamental scale of penumbral filaments might be much smaller than $0\farcs1$ (see contribution by Sanchez Almeida and Bonet). However, white-light snapshots (such as Fig. 2) and time sequences of these small-scale structures are not sufficient. Spectroscopic data at scales of $0\farcs1$ and below are needed for a comprehensive understanding of phenomena such as the Evershed effect (c.f. contribution by Schlichenmaier et al.).

2.2. Waves and oscillations in flux tubes

Magnetic elements are found in plages and the network (see Fig. 3). A flux tubes is a theoretical model that explains the observed properties of concentrated magnetic fields with kG field strength that are sometimes associated with brightening in the wing of strong spectral lines, in the G band, or even in the continuum. Indeed, magnetic elements, and therefore flux tubes, are the basic building blocks of the magnetic field in the network and in plages. They are the most likely channels for transporting energy to the upper atmosphere and are therefore linked to the heating mechanism working in the upper chromosphere and the corona. Furthermore, they affect convection, global irradiance and oscillations, and most likely the solar cycle in general because they contribute to the general flux loss mechanism which is important for any dynamo theory.

While theoretical models are able to resolve these structures and explore their interior dynamics (see contribution by Steiner), observations barely resolve these features. Adaptive optics and post-facto reconstruction techniques are therefore crucial components of future studies of magnetic elements. In particular, we need to understand their structure and dynamics, their connection to higher layers, their brightness, interaction with convective motions, and their internal oscillations and waves and how they relate to the heating in the upper chromosphere and the corona.

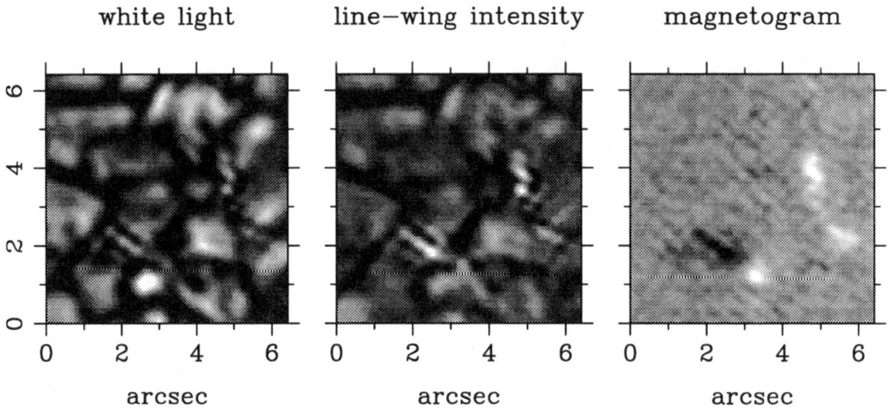

Figure 3. Speckle polarimetry reconstruction of magnetic elements in the network close to disk center. 300 frames with 20 ms exposure time were recorded at the 76-cm R. B. Dunn Solar Telescope with the ZIMPOL I within 60 s.

Current state Until now waves in flux tubes have not been detected directly, although there is indirect evidence for the existence of waves with considerable amplitudes from spatially unresolved measurements of the shape of polarized spectral lines (e.g. Solanki 1989). The major reason for the absence of direct evidence is the lack of spatial and temporal resolution of current solar telescopes due to their limited diameter. While speckle polarimetry is able to resolve the larger magnetic elements at a spatial resolution approaching $0\farcs2$ (see Fig. 3), there is no spectroscopy of individual flux tubes because of a lack of spatial resolution and photons.

Future direction Magnetic elements have sizes of about $0\farcs3$ and smaller. Typical time-scales for waves are on the order of a minute. The most likely wave mode to carry energy into the higher layers of the solar atmosphere is the kink mode (e.g. Choudhuri et al. 1993), which could be observed as a change in inclination angle of the magnetic field vector. We therefore need to obtain vector-field measurements with a resolution of at least $0\farcs1$ to spatially resolve the flux tube, a time cadence of about 10 seconds to accurately sample the temporal variation, a total length of more than 30 minutes to distinguish waves at different frequencies, and a field of view of about $10''$by $10''$to study the convective motions that might drive the waves in the flux tubes. Hence the polarization measurements need to reach an accuracy of about 10^{-4} to obtain sufficiently sensitive measurements of magnetic field angles. A rapidly scanning spectrograph or a tunable filter system is needed to obtain simultaneous full line profiles in several spectral lines. This requires a large telescope with adaptive optics.

3. High photon flux

To most night-time astronomers it is not evident that solar observations are often photon starved. As an example, let us consider vector polarimetry with a spectrograph. At the diffraction limit of a telescope, with 20 mÅ spectral pixels, a signal-to-noise ratio (SNR) of 2000, an optimistic 10% overall system efficiency, and 5 spatial or spectral positions (depending on whether a spectrograph or a tunable filter is used), we need to collect photons during 5 minutes in the visible part of the spectrum. We expect considerable evolution within 30 s or less at spatial scales of 0.″1. Hence, instead of a 1-m telescope with a diffraction limit of 0.″1, a 3-m telescope is needed to achieve the required SNR within 30 s at 0.″1 spatial resolution.

While vector-polarimetry is the prime example for an observation that requires a high photon flux at high spatial and spectral resolution, there are also applications where an even higher SNR of 10^5 to 10^6 at moderate spatial resolution is required. Measurements of very weak magnetic fields using the Zeeman effect and the study of scattering polarization close to the solar limb are examples of such observations.

The linear polarization due to scattering is observed at levels of 10^{-3} to 10^{-5} (e.g. Stenflo and Keller 1996). Such observations allow us to determine element abundances such as in the Li I lines at 670.7 nm (see Fig. 4) or isotope ratios such as in the Ba II 455.4 nm line. Moreover, these observations have provoked intensive discussions in radiative transfer physics (for instance the strange profiles of the scattering polarization in the Na I D_1 and D_2 lines, see Landi Degl'Innocenti 1998 for more details). Most importantly, the scattering polarization is reduced by the Hanle effect, which allows us to measure the field strength of unresolved, mixed-polarity magnetic fields. While measurements using the Zeeman effect in the visible and near-infrared are sensitive to field strengths down to about 100 G, the Hanle effect is sensitive in the 0.01 to 100 G range, complementing the Zeeman effect measurements. All these scattering polarization observations currently require exposure times of minutes to hours at rather moderate spatial resolution of several seconds of arc using a 1.5-m telescope (Stenflo and Keller 1997).

3.1. Mixed-polarity background field

Current state Between the network, the intranetwork (or internetwork) fields have a field strength of about 0.5 kG or less (e.g. Lin 1995). There are indirect hints of an ubiquitous mixed-polarity field background that covers the whole solar surface and has a field strength of about 4 to 40 G (e.g. Stenflo et al. 1998). First direct evidence for such an ubiquitous field have been presented by Lin and Rimmele in these proceedings. Figure 5 also shows these fields very clearly. The total flux in all these fields is at least comparable to all the flux in strong field form (magnetic elements and sunspots). However, the flux emergence rate is at

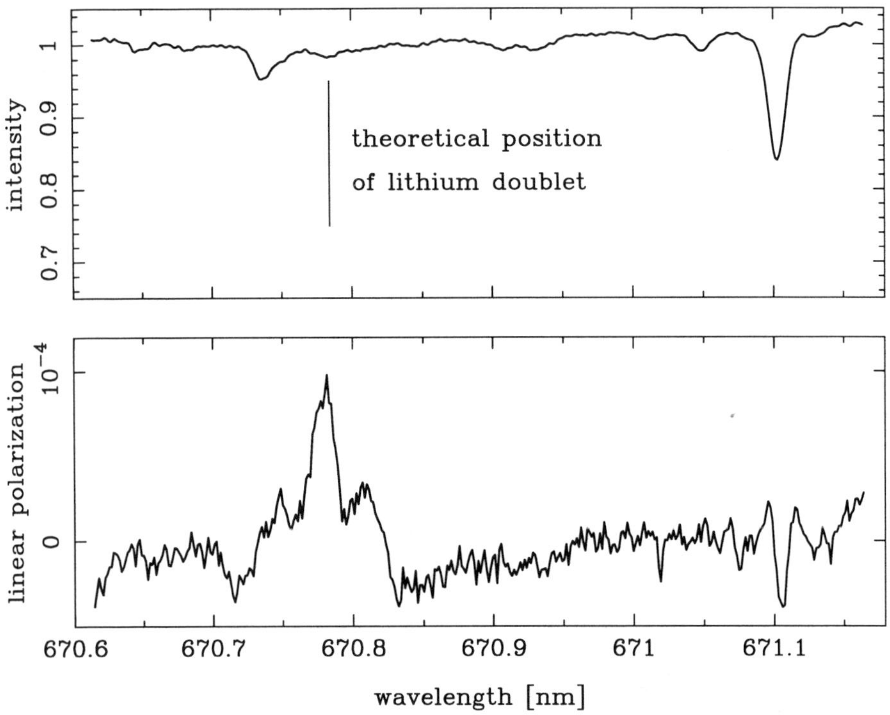

Figure 4. The solar photospheric lithium abundance is of general astrophysical interest since it is roughly 100 times less than what is expected from simple theories of star formation. The lithium doublet at 670.9 nm is very hard to measure in the solar spectrum. The weak line seen in intensity is due to CN. However, these are resonance lines and therefore exhibit linear polarization close to the solar limb due to resonant scattering.

least 10^4 times that of the rate of active regions (Zirin 1987), which makes these fields the dominant component of magnetic flux emergence and disappearance processes.

Future direction We need very high precision polarimetric measurements at the 10^{-5} level at the characteristic spatial and temporal scales of convection. By combining Zeeman and Hanle effect measurements, we can cover the 0.01 to 4000 G range in field strength and study the origin and dynamics of weak fields and their interaction with the strong fields. This requires a large aperture telescope because the current measurements hardly achieve the required spatial and temporal resolutions.

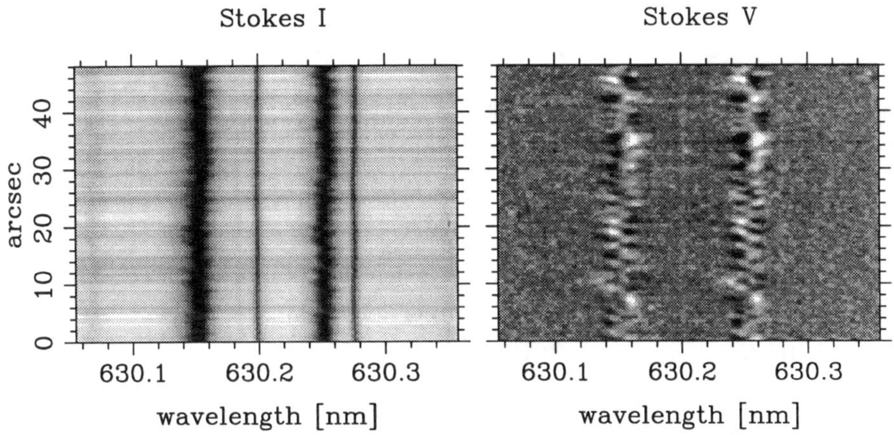

Figure 5. Stokes I and Stokes V spectra of a quiet region in the solar photosphere. The broad spectral lines are due to FeI 630.15 nm and 630.25 nm. Notice their wiggly shape due to convective motions in the solar photosphere. Black and white in the Stokes V image correspond to ± 0.003 of the average continuum intensity.

4. Outlook

The science presented above requires a large-aperture (about 3 m in diameter) telescope with adaptive optics. Other science goals not addressed here such as access to the thermal infrared and observations of the corona require an all-reflective telescope and low-scattering optics, respectively.

While this paper is mostly aimed at solar physicists, any large-aperture solar telescope must be justified to a wider, astrophysical community. In general, the sun is the premier proving ground for many theories of fundamental astrophysical importance. Some of these theories include:

- stellar and galactic dynamos
- magnetohydrodynamics
- generation of hot atmospheres in stars and accretion disks

To achieve these goals, we need higher spatial resolution and more photons than offered by current solar telescopes.

Acknowledgments. I am grateful to Rick Paxman and John Seldin for their collaboration on phase-diverse speckle reconstructions of photospheric time sequences, to Jan Stenflo and Achim Gandorfer for collaborating on the scattering polarization measurements, and to Pietro Bernasconi, Urs Egger, Peter Povel, and Peter Steiner for their help with the direct observations of the turbulent background field. The National Solar Observatory is one of the National Optical Astronomy Observatories, which are operated by the Association of Universities for Research in Astronomy, Inc. (AURA) under cooperative agreement with the National Science Foundation.

References

Choudhuri A.R., Auffret H., Priest E.R. 1993, Sol. Phys. 143, 49
de Boer C.R., Kneer F., Nesis A. 1992, A&A257, L4
Landi Degl'Innocenti E. 1998, Nature 392, 256
Lin H. 1995, ApJ446, 421L
Solanki S.K. 1989, A&A224, 225
Stenflo J.O., Keller C.U. 1996, Nature 382, 588
Stenflo J.O., Keller C.U. 1997, A&A321, 927
Stenflo J.O., Keller C.U., Gandorfer A. 1998, A&A329, 319
Zirin H. 1987, Sol. Phys. 110, 101

Group Discussion

Scharmer: My main objection is the idea of building a single large (3m) telescope to work from the visible to 12 microns. I would advocate separate visible (1.5m) and NTR-IR telescopes ($\approx$ 7m). I question the photon starvation problem as a main driver for building very large telescopes in the visible.

Keller: If a telescope has no entrance window, the thermal infrared is accessible without any additional efforts. At a certain aperture diameter windows become impractical. It is therefore likely that any new, large-aperture telescope has no entrance window, and therefore provides access to the thermal infrared. At the 1.5m McMath-Pierce telescope, we are often photon-starved. Another new 1.5m telescope will therefore not solve that problem. For the SOLIS vector-spectromagnetograph with 1" pixels, we require a 50cm aperture even when using extremely efficient optics and detectors. To do vector-polarimetry at much smaller scales, we need a much larger aperture (of about 3m diameter).

High Resolution Solar Physics: Theory, Observations, and Techniques
ASP Conference Series, Vol. 183, 1999
T.R. Rimmele, K. S. Balasubramaniam, R.R. Radick, eds.

The Advanced Solar Telescope. II Technical Aspects

J.M. Beckers

National Solar Observatory/NOAO[1]
Sunspot, NM 88349/Tucson, AZ 85726, USA

Abstract. In this paper I review the technical issues associated with the minimum requirements for an Advanced Solar Telescope (AST) as agreed upon by the scientific staff of the National Solar Observatory

1. Introduction

In order to achieve the scientific objectives discussed in the accompanying paper by Keller (1999; these proceedings), NSO's scientific staff identified the following minimum technical requirements for the Advanced Solar Telescope (AST):

- Aperture: 300 cm
- Wavelength Range: 300 nm - 30 μm
- Polarization Capability: 10^{-4} accuracy; 10^{-5} precision (photon flux allowing)
- Scattered Light: Coronagraphic, provided cost impact $\leq 10\%$
- Siting: Best seeing & sunshine site; Secondary considerations: sky brightness, IR, economy, environmental, staff needs & recruitment, etc.
- Adaptive Optics: At first light for 1.6 μm for median, 50 percentile seeing and 0.6 μm for 5 percentile seeing. Design to allow use at 0.6 μm at median seeing in the future.
- Field-of-View: Diameter $\geq$ 3 arcmin with $\leq$ 0.1 arcsec FWHM optical image quality (unless limited by the telescope Airy disk)
- Exit F-Ratio: Matched to instrumentation needs (f/30?)
- Instrumentation: Initial complement should be able to verify performance

In this paper I examine some of the technical consequences of these requirements. It is based largely on a feasibility study of a large, open, low-scattered light telescope (Beckers 1994a; Beckers et al. 1998; Siegmund 1998)

2. Telescope Aperture

A 3 – 4 meter aperture solar telescope collects about 10 KWatts of solar energy. Of this 10% is absorbed by the primary mirror, causing it to heat up by about 5K, the rest ends up in prime focus and possibly other parts of the telescope enclosure. The heating of the primary may cause it to distort unless constructed out of low thermal expansion materials like ULE or Zerodur . In vacuum solar telescopes that is the only concern since the other heating induced problem, internal telescope seeing, is eliminated. However, for a number of reasons, vacuum telescopes are not an option for the AST. No vacuum windows are available which simultaneously satisfy the requirements for diameter, optical quality, wide wavelength transmission and polarization neutrality. The requirements given in Section 1 therefore forces one into using "open telescopes". That brings with it the requirement of seeing control inside the telescope. Telescope seeing should not exceed the (best) seeing associated with the site. The latter is poorly known for solar observing sites, but median seeing values as good as 0.5 arcsec appear possible (Beckers & Rutten 1998; Beckers 1998), with best seeing of 0.25 arcsec happening occasionally. An upper limit to telescope seeing of 0.2 arcsec appears therefore a reasonable requirement.

2.1. Primary mirror seeing

Seeing generated by the primary mirror heating is probably the worst contributor to open solar telescope seeing. The current rule-of-thumb for nighttime telescopes is that under no air-flushing conditions 1 degree K of excess temperature results in 0.4 arcsec seeing. That implies 2 arcsec seeing for an excess temperature of 5K, or a requirement to maintain the mirror to within 0.5 K excess temperature with respect to the neighboring air. The tolerance for temperatures lower than the air temperature is larger, around -1.5 K (maybe more). The most obvious way to control mirror seeing is therefore the control of mirror temperature. Analysis done for the CLEAR concept (Beckers 1998) shows that cooling a 75 mm thick mirror at its back indeed achieves the required temperature control of $+0.5K/-1.5K$. Such a relatively thin mirror requires, of course, active support. The complexity of that support is similar to that of the 20 - 25 cm thick 8 meter class meniscus mirrors being used in large nighttime telescopes (Gemini, Subaru, VLT).

In addition to cooling the mirror from behind, there are other ways to control mirror seeing. Flushing the mirror with an ambient air flow is commonly used. Often this is done naturally by exposing the mirror to the wind flow. I suspect that many reports about seeing improvements of solar telescopes (for example the DOT; see Rutten (1999; these proceedings)) in the presence of wind over wind-still conditions is due to the flushing of the mirror seeing layer this way. In addition to removing the mirror seeing layer, the wind flow will also decrease the temperature difference by convective cooling. In the CLEAR concept (Beckers et al. 1998) an artificial flushing device is introduced, thus making this way of primary mirror seeing control independent of the existence of natural adequate wind flow.

As part of the Gemini program, engineers at NOAO have come up with another form of mirror seeing control: resistive heating of the metallic coating deposited on an undercooled mirror blank (Perona et al. 1998). It has the advantage over thermal control of the mirror blank only in that the time constant of the thermal control is very short (seconds versus about 15 minutes for a 75 mm thick mirror) so that good control is maintained under rapidly varying conditions like e.g. intermittent cloud cover.

If everything else fails, adaptive optics will remove any residual primary mirror seeing together with the external atmospheric seeing. Caution should be taken, however, to keep this residual mirror seeing low, less than the atmospheric seeing assumed for the design of the adaptive optics.

2.2. Prime focus heat stop

Siegmund (1998) designed a prime focus heat trap for CLEAR which removes the almost 10 KWatts of energy while introducing a minimum amount of seeing. After cooling, the residual heated air is removed by an air suction device. His design should allow primary f-ratios as fast as f/2. In addition CLEAR uses an off-axis parabolic mirror as primary mirror so that the heat trap is located outside the light beam. Any residual thermal disturbances in the air around the primary focus can therefore be prevented from entering the incident sunlight where such disturbances would cause a maximum seeing effect. Disturbances near prime focus therefore have a minimum effect on the light beam. However, even when entering the primary light beam there is little effect on the telescope point-spread-function since the heat stop disturbances are in the form of a plume which affects only part of the aperture, leaving most of the telescope aperture unaffected (Beckers 1994b).

2.3. Other sources of telescope seeing

Care should be taking to avoid other sources of telescope seeing. These could result from other surfaces of the telescope irradiated by sunlight and from instrumental heat sources. Telescope secondary mirrors also will heat up because of solar irradiation. With a 5 arcmin prime focus heat stop and a Gregorian secondary 1/6 of the primary mirror size the irradiation equals that of the primary, resulting also in 5K heating. That mirror therefore requires cooling or air flushing as well, although, because of the 6 times smaller beam diameter, the cooling requirements are about 6 times less than that of the primary mirror.

3. Wavelength Coverage

3.1. Photon flux

As already mentioned the desired wavelength coverage (0.30 - 30 μm) of the AST eliminates the use of refractive optics since the only large optics with acceptable optical quality (e.g. fused quartz, UBK7) absorb wavelength above 2.5 μm (except for the unlikely event that exchangeable/removable optics of the required sizes become available). The wavelength coverage requirement also reduces options for mirror coatings. Gold and silver are unacceptable because of their short wavelength efficiency. Silver coatings enhanced by UV enhancing coatings (Jacobson et al. 1998) might be an option if a lower cut-off wavelength of 0.35 μm would be acceptable, but straightforward aluminum coatings appear to be the only coating to cover the entire wavelength range specified. Al coatings are, however, not as efficient as other coatings, reflectivities being near 90% at visible wavelengths (and even less near 0.85 μm). The 10% absorption value for total solar energy for the primary mirror mentioned in section 2 is, however, the same for aluminum and silver coatings, the higher reflectivity at longer wavelength for silver being offset by the lower UV reflectivity.

3.2. Thermal emissivity

For nighttime telescopes IR thermal emissivity is a major issue. It makes the use of aluminum coatings undesirable. For solar disk observations it is only a very minor issue. Even in the Rayleigh-Jeans wavelength range a 10% telescope emissivity contributes only 0.5% to the radiation (assuming the same f-ratios for the detected solar and thermal telescope radiations). At 4 and 10 μm wavelengths it amounts to less: 0.4×10^{-6} and 230×10^{-6} of the solar disk brightness respectively. For off-limb observations at 10 μm this starts becoming of importance, however.

4. Polarization Accuracy

4.1. On-axis telescopes

The expression "polarization free telescopes" has been loosely used to mean on-axis telescopes like LEST and THEMIS. Without windows those telescopes would indeed be polarization free or neutral in the sense that the Müller matrix is purely diagonal at the 10^{-4} level (ignoring a small amount of de-polarization occurring because of curvature of the reflections on the telescope primary and other optics). That is the case for the DOT (Rutten 1999; these proceedings). At the 10^{-4} accuracy level the entrance window is not polarization neutral, however. Moreover its polarization effects are time-variable so that calibration is problematic. Nonetheless with care the required 10^{-4} accuracies appear to be achievable.

4.2. Off-axis telescopes

In contrast off-axis telescopes are not polarization neutral, even without a window. In addition to introducing linear polarization at the 10^{-2} level, they introduce a small amount of retardation. In the absence of windows the associated Müller matrix is, however, well-defined, time invariable, allowing calibration and correction for telescope polarization at the 10^{-4} level.

4.3. Other considerations for using off-axis telescopes

Were it not for the aperture and wavelength range requirements, which eliminate vacuum or helium filled telescopes, one would probably not want to seriously consider off-axis telescope optics because of their higher cost and the need for rather precise calibration of the instrumental Müller matrix. In addition, the decrease in field-of-view becomes a potential problem for fast primary mirrors but can be controlled for modest f-ratios (see section 8). However there are additional advantages for off-axis optical configurations in addition to better seeing control (in open telescopes) which may swing the balance in their favor.

They include: (i) the need for only two mirrors to the Gregorian focus rather than the larger numbers of mirrors needed for on-axis telescopes resulting from the requirement to transfer the light away from the main entrance beam (for example: for LEST there are ≥ 6 mirrors before the first instrumentation focus). More mirrors mean more light loss especially in the visible and near IR (850 μm Al low reflection band), a severe disadvantage when the "photon starvation" for diffraction limited solar spectroscopic imaging becomes a factor. (ii) an unobstructed telescope pupil resulting in less light losses and a better diffraction limited point- spread-function, (iii) a lower thermal emissivity where that may become important at far IR wavelength, (iv) less scattered light (see section 5), and (v) greater ease of mounting auxiliary telescope and instrumentation optics like atmospheric dispersion compensators, image derotators and polarization analyzers which otherwise may have to be placed in the incident light beam.

5. Scattered Light

5.1. The importance of scattered light control

Scattered light control should be part of any solar telescope design. The large extent of the solar disk easily introduces a few percent of scattered light which causes major complications especially in observations of sunspot structure and in observations of the chromosphere, prominences and possibly coronal structure off the solar limb. The precautions to minimize scattered light included in the CLEAR design (Beckers et al. 1998) address this issue head-on in contrast in the design of other solar telescopes where the window(s) and mirrors might be expected to introduce a significant amount of scattered light (for LEST I estimate $\sim 5\%$ scattered light in a 20 arcsec diameter sunspot umbra as compared with $< 0.5\%$ for CLEAR).

5.2. Why coronagraphic quality?

There is much scepticism in the solar community about the wisdom of the CLEAR design to include coronagraphic performance. Concerns range from beliefs that trying to do too many things with a single telescope would compromise its performance for the highest priority science (high angular resolution) to worries about the additional cost associated with super-polished optics and dust control, both features inherent in reflecting coronagraphs. The $\leq 10\%$ cost impact requirement given in section 1 is directly associated with the latter concern. The former concern should be addressed in a rational manner: no one has so far demonstrated such compromise in the CLEAR design.

Scientifically one often finds the view in the solar community that ground-based observations of the solar corona are outmoded, being overtaken by XUV and optical coronagraphic observations from space. That is correct where coronal structure is concerned, but wrong when magnetic field and line-of-sight velocity observations are needed. Zeeman observations of the magnetic field are not going to occur in the XUV and space-based optical coronagraphs with the apertures/sensitivities and angular resolution needed are not likely to occur soon.

The discovery by the HAO/MPI(Lindau)/NSO/Rhodes College team at the February 1998 solar eclipse (which occurred after the requirements in section 1 were defined) of the bright Si IX 3.932 μm coronal line predicted by P. Judge promises to strengthen the case for adding a coronagraphic capability to the AST substantially. Not only does it increase the sensitivity of coronal magnetic field observations, because of the increase in Zeeman splitting in the IR, but it also lessens the technical requirements for the AST since at the longer wavelengths things become less critical. I will comment on that next.

5.3. Comments on coronagraphy at IR wavelengths

At IR wavelengths scattered light control becomes easier because:

- the brightness of the dust aureole (dust either in the atmosphere, including aerosols, or on the optics) varies proportional with λ^{-2} resulting in a gain of 55× for the 3.932 μm line over the coronal green line (.53 μm).
- the brightness of the scattering of the optics/mirrors due to micro-roughness varies proportional to the square of the micro-roughness expressed in wavelengths, or like λ^{-2}. Alternately the micro-roughness requirements can be relaxed by a factor of to give the same aureole brightness, or 7.4× for the IR line over the green line. A micro-roughness of 2 nm RMS will therefore suffice for the IR line (rather than 0.3 nm RMS for the green line as assumed in the original CLEAR design). That corresponds to a standard polish quality astronomical mirror rather than to a super-polished mirror.
- the atmospheric Rayleigh Scattering decreases, of course, as λ^{-4}

On the other hand, the part of the aureole due to diffraction at the telescope periphery increases as $\lambda^3 D^{-3}$ (D = telescope diameter). This is the part which is mostly removed by the Lyot stop, but which still contributes significantly in the inner corona.. However, the larger telescope diameter of AST compensates the wavelength increase. This part of the aureole for a D = 300 cm telescope used at 3.932 μm equals that for a D=40 cm coronagraph operating at the green line. That is comparable to the aperture of the largest currently available visible light coronagraphs like the one at Sac Peak.

Finally one should consider the expected signal-to-noise ratios (SNR) for coronal observations. In case the aureole brightness IA dominates the coronal emission line signal IC one has:

$$SNR(:)ICD^2/(I_A D^2)^{0.5} = I_C D/\sqrt{I_S}, \quad (1)$$

so that a 7.5× larger aperture D (300 cm vs 40 cm) would offset a 56× increase in scattered light. In case IC dominates the noise one has:

$$SNR(:)(I_C D)^{0.5}, \quad (2)$$

and the aureole brightness does not matter.

6. Siting

Excellent seeing and large amount of sunshine are the highest priority requirements for siting the AST. A site with X% of excellent seeing should preferably have that seeing in a few long uninterrupted periods rather than in many short intervals so that long, uninterrupted observing sequences are possible. Moreover, for a unique national or international facility like the AST which will be used for many scientific observing programs, a site with a high probability of excellent seeing and sunshine are essential. Telescope operations which involve putting one or two programs on the telescope per year and then waiting for the occasional excellent seeing days are an option which should be avoided for the AST.

Secondary, but not unimportant, site considerations are: (i) Low sky brightness including the absence of airplane contrails. This is especially important if off-limb observations are made. (ii) Low water vapor. It is not clear yet how important this consideration is for solar observations since most or all of the presently known interesting solar IR lines are not contaminated by water vapor lines. (iii) Economy. Remote sites are more costly to develop and operate. (iv) Environmental considerations. The recent increasing awareness that our planet and its biosphere needs taking care of should be considered in site selection. (v) Staffing considerations. The recruitment of expert resident observing, maintenance and other support staff should be taken into account.

A recent NSO site surveys using the Seykora solar scintillation proxy (Beckers 1998a, Beckers & Rutten 1998) confirmed the already long known superiority of lake sites for solar high resolution observations (Evershed 1914; Zirin & Mosher 1988). Sceptics rightfully point at the exceedingly high quality solar observations obtained in mountain locations in the Canary Islands on sites, far away from lakes, as counter examples. So far limited scintillometer observations at those sites indeed suggest the occasional occurrence of excellent long duration seeing probably under favorable wind conditions, but with median conditions being worse than lake site conditions. More systematic site evaluations are needed, combining the scintillation proxy with direct optical image quality observations, emphasizing seeing statistics, (median, distribution and duration) and including other important seeing parameters like isoplanatic patch size (θ_0) and coherence time (τ_0), in addition to r_0, as well (Beckers 1998a,b).

7. Adaptive Optics

As is by now customary, I assume here that a functioning adaptive optics system will be based on wavefront sensing and control of subaperture sizes equal in diameter to the Fried parameter, r_0. This finite pupil resolution by itself results in a Strehl ratio of $\sim 70\%$. Limited temporal resolution and wavefront sensing errors could reduce the Strehl ratio further (Beckers 1993). Under these circumstances the number N of subapertures used equals;

$$N = (D/r_0)^2(\lambda/\lambda_0)^{2.4} \tag{3}$$

where λ is the wavelength for the adaptive optics system and λ_0 is the wavelength used in the r_0 definition (commonly 500 nm). N is also approximately equal to the number of lower Zernike polynomials which can be compensated. The cost and complexities of adaptive optics system is highly dependent on the size of N.

Brandt, Mauter & Smartt (1987) give for the Sacramento Peak Dunn Telescope a median r_0 of 8.6 cm and a 5 percentile value of $\sim$ 25 cm. The following table gives N for these conditions, for D = 300 cm, and for the wavelengths given in section 1 as well as for lake sites for which the median r_0 is suspected to be substantially larger (I assume 20 cm and the same distribution of occurrence at the 5 percentile level).

Table 1. Number N of Telescope Subapertures Needed for Adaptive Optics

	median r_0 = 8.6 cm	median r_0 = 20 cm
1.6 μm, 50%-tile	N = 75	N = 14
0.6 μm, 50%-tile	N = 93	N = 11

The adaptive optics system currently being developed for the Sac Peak Dunn Telescope (Rimmele 1999; these proceedings) has N = 20 and anticipates as its next step an expansion to N = 80. A successful completion of the N = 20 system will satisfy the technical requirements given in section 1 for the lake site assumption. Completion of the N = 80 system will satisfy the requirements for mountain sites similar to Sac Peak. It also approaches the adaptive optics need for diffraction limited imaging at lake sites at 0.6μm for median seeing. It requires N = 145.

8. Field-of-View

The 0.1 arcsec FWHM over 3 arcmin field diameter is quite easily satisfied even for off-axis, Gregorian telescopes. With this criterion the Gregorian FOV for CLEAR (Beckers et al. 1998) was larger than 5 arcminutes, as was the FOV for many of the other foci.

9. Conclusion

The Advanced Solar Telescope makes use of many of the major developments in telescope technologies which have occurred since the construction of the Richard B. Dunn Solar Telescope at Sac Peak in the late 1970's. Nighttime telescope technology has advanced in major ways in the last two decades, reaching apertures up to 10 meters and including many other sophisticated capabilities. The Dunn Telescope has been successfully copied a number of times with comparable or smaller apertures at a number of locations in the world, but except for the unsuccessful attempt to create LEST, solar telescope technology per se has progressed little. That made sense because it was the atmospheric seeing rather than the telescope aperture which limited the scientific objectives in observational solar physics.

With the implementation of solar adaptive optics technology foreseen in the near future, the telescope aperture becomes again the limiting factor. It is to be hoped that the solar community and its sponsors will have the courage to follow that advance with the creation of an Advanced Solar Telescope with the significant increase in capability over existing telescopes listed in section 1. That would be technical step comparable to the courageous effort by R.B. Dunn in building the telescope named after him during this workshop. Scientifically it would address many of the most important scientific questions described by Keller (1999; these proceedings).

References

Beckers, J.M. 1993, Ann. Rev. of Astron. Astrophy. 31, 13.

Beckers, J.M. 1994, in Infrared Tools for Solar Astrophysics: What's Next?, eds. J.R. Kuhn and M.J. Penn, 145.

Beckers, J.M. 1994, Proceedings SPIE 2199, 478.

Beckers, J.M. NOAO/ETS and NOAO/NSO Staff, 1998, SPIE Proceedings 3352, 588.

Beckers, J.M. 1998a, SPIE Proceedings 3352, 858.

Beckers, J.M. 1998b, in Proceedings of the Third Advances in Solar Physics Euro-Conference (Potsdam), in preparation.

Beckers, J.M., and Rutten, R.J. 1998, in New Astronomy Reviews, in print. 6.

Brandt, P.N., Mauter, H.A., and Smartt, R. 1987, A &A 188, 163.

Evershed, J. 1914, Kodaikanal Observatory Bulletin XLII.

Jacobson, M.R., Kneale, R.C., Gillett, F.C., Raybould, K., Filhaber, J.M., Camigilla, C., Laird, R., Kitchens, D., Shimshock, R., and Booth, D.C. 1998, Proceedings SPIE 3352, 477.

Keller, C.U. 1999, "The Advanced Solar Telescope, I. Scientific Rationale", this volume.

Perona, B., Hanson, E., Hagelbarger, D., and Rudeen, A. 1998, SPIE Proceedings 3352, 868.

Rimmele, T., Radick, R.R., Dunn, R.B., Richards, K., and Hull, B. 1999, "Solar Adaptive Optics at NSO", this volume.

Rutten, R.J. 1999, "The Dutch Open Telescope: Status and Prospects", this volume.

Siegmund, W. 1998, "Coronagraphic and Low Emissivity Astronomical Reflector (CLEAR): Heat Trap Design", SPIE Proceedings 3352, 572.

Zirin, H., and Mosher, J. 1988, "The Effects of Atmospheric Turbulence in Optical Astronomy", Solar Physics 115, 183

High Resolution Solar Physics: Theory, Observations, and Techniques
ASP Conference Series, Vol. 183, 1999
T.R. Rimmele, K. S. Balasubramaniam, R.R. Radick, eds.

High-Resolution Thermal Infrared Imaging of the Sun: A Pipe Dream?

Thomas R. Ayres

Center for Astrophysics and Space Astronomy,
University of Colorado, CB-389 (CASA), Boulder, CO 80309 USA

Abstract. I discuss the desired characteristics of an instrument to study the Sun in the 3–20 μm thermal infrared. I show that even the largest apertures under consideration for a next generation "Advanced Solar Telescope" fall short of the capabilities needed at the long wavelengths. I conclude that the design of AST should not be driven by access to the thermal infrared. Instead the IR would be better served by a special-purpose instrument; even a simple—perhaps temporary—modification of an existing or planned nighttime telescope.

1. Introduction

One of the central mysteries of solar physics—and indeed of *stellar* astronomy—is the process by which kinetic energy of the turbulent convection zone is transferred across dozens of scale heights to the outer atmosphere, deposited there as heat, boosting temperatures to 10^4–10^6 K. Decades of detailed reconnaissance of the solar surface and upper atmosphere have pointed to the central role of the spatially complex magnetic field, although there are thought to be heating processes—particularly in the largely field-free cell interiors—that are purely dynamical in nature. In parallel, solar modellers have moved away from thinking in terms of a "layered" atmosphere, to the more natural view of "structures." Growing awareness of the dynamic complexity of the coronal regions, motivated by the stunning images from *SOHO* EIT and now *TRACE,* has spawned new descriptive metaphors: for example, "the coronal junkyard" (attributed to Helen Mason); and my contribution, "battle in a rubberband factory."

1.1. The magnetic transition zone

Fig. 1 is a cartoon of the modern view. Within the new paradigm, one can identify a crucial region: the "magnetic transition zone" (MTZ) separating the high-β plasma of the deep photosphere, where the embedded magnetic field is carried about by the gas flows; and the low-β regions of the outer atmosphere, where it is the field that guides and controls the gas dynamics. The MTZ is of great interest because—as is the nature of "boundaries"—profound changes occur in the character of the atmosphere: thin, filamented, sparsely-distributed flux tubes below; broad, pervasive canopy, and tangled mess of coronal loops, above.

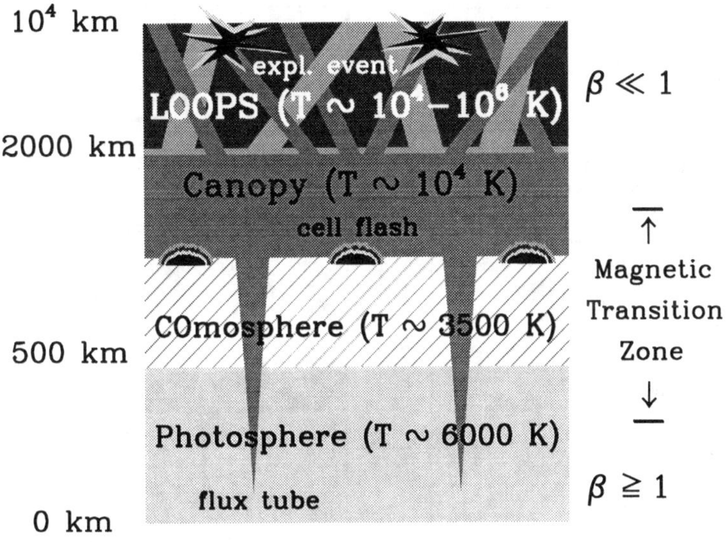

Figure 1. Cartoon of "magnetic transition zone." Height "0" is at $\tau_{5000\text{Å}} = 1$.

Perhaps coincidentally, the MTZ also corresponds to the pressure range in the outer atmosphere where acoustic waves radiating from the chaotic convective cells in the deep photosphere steepen and form shocks (as in the Carlsson-Stein [1997] 1-D radiation-hydro simulations): the MTZ not only is a critical interface between plasma-domination and magnetic-control, but it also defines a zone where the gas dynamics (away from strong fields) evolves from small-amplitude nearly-adiabatic disturbances, to strongly dissipative shock waves.

Remote sensing of physical conditions in the MTZ is a key observational goal for those who wish to understand the propagation of mechanical energy through the solar outer atmosphere, and the complexification of the magnetic "flora" as one proceeds from the deep photosphere out to the corona.

1.2. Far-UV space/time diagrams

One way to view the MTZ is through ultraviolet spectral imaging, in moderate excitation species (the 1300–1700 Å continuum, or first ions such as Si II and Fe II). This can be done at $\sim 2''$ resolution with *SOHO* SUMER. Fig. 2 illustrates an example, from an observation in May 1997. SUMER slit No. 2 ($300''$ in length, $1''$ in width), oriented N/S, was placed $+700''$ west of the central meridian ($\mu \sim 0.7$). Solar rotation carried the surface westward across it at a rate of 1 slit-width per 10 minutes, building up a drift-scan of transient and persistent surface features over time.

The intensity map (left panel) is based on the $\lambda 1321$ continuum; the Doppler velocities (right panel) are from a sharp Si II chromospheric emission at $\lambda 1309$. Each exposure was 60 s: the 350 consecutive frames spanned nearly six hours.

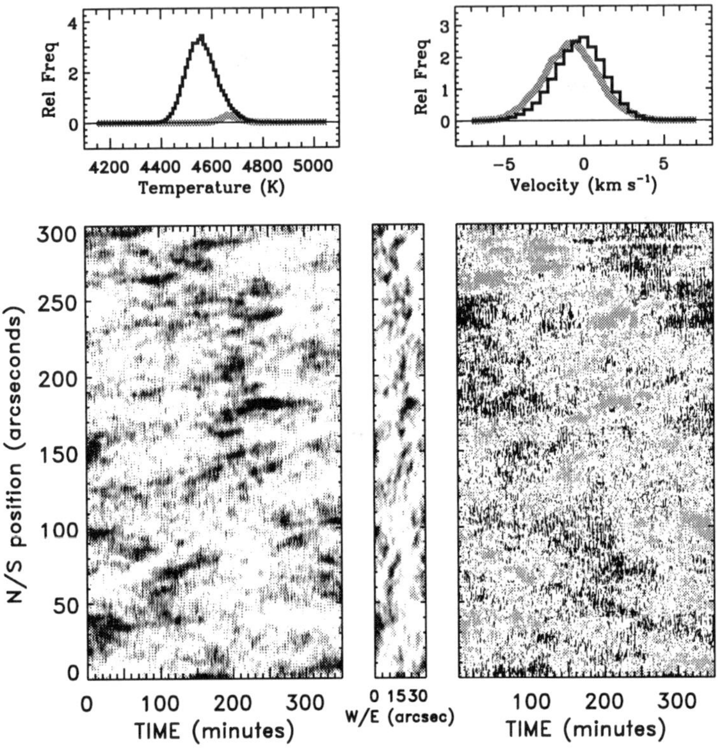

Figure 2. Far-UV space/time diagrams. Left: λ1321 continuum brightness (dark= high intensities; white= low intensities). Right: Si II λ1309 velocities (dark= redshifts; light-shaded= blueshifts).

According to classical layered models (Vernazza, Avrett & Loeser 1981), the λ1321 continuum forms on the high side of the initial temperature rise out of the $T_{\rm min}$ region. The Si II emission would form somewhat above that level.

The intensity image is peppered with short vertical streaks representing transient brightenings, and horizontally-elongated features produced by long-lived bright structures. The middle frame is a spatial map of persistent features obtained by filtering the space/time diagram to suppress the short-lived intensity fluctuations. The depiction is reversed with respect to normal convention, with W to the left and E to the right (to maintain the same orientation as the space/time diagram).

The velocity map is coded such that dark areas indicate strong line-of-sight redshifts ($v_{\rm los} > +2$ km s^{-1}), the lighter shaded areas are strong blueshifts ($v_{\rm los} < -2$ km s^{-1}), while the white background encompasses the smaller positive and negative velocities in the distribution. The velocity zero is based on a global average over the space-time diagram, after a systematic trend in the spatial dimension was removed.

The upper panels illustrate histograms of brightness temperatures derived from the calibrated intensities (left panel) and line-of-sight velocities (right panel). While the intensities in the image span more than a decade, the temperature histogram is nearly symmetric, with only a weak tail on the hot side; and the rms is only about 80 K. (Note: owing to NLTE effects, the λ1321 temperatures very likely are *lower* than the true kinetic temperatures in the continuum-forming gas.) In the ultraviolet, in the Wien limit, a great deal of intensity contrast can be produced by even a rather bland temperature field. The temperature histogram undoubtedly is dominated by oscillations, which at this level should be nearly adiabatic: the huge intensity swings are an insignificant energy drain, and don't affect the thermodynamics of the waves.

I constructed a mask to isolate the bright regions, above a selected threshold, which are persistent in time (as in the middle panel). The thick light-shaded curve in the temperature histogram depicts the population of the mostly network elements. These features are systematically 100–200 K warmer than the dominant temperature of the map, and occupy less than 10% of the scene.

I then applied the network bright-point mask to the velocity map. The light-shaded curve in the velocity panel shows the resulting histogram. The network points display a broad velocity distribution like the rest of the scene, but systematically skewed blueward (by about 1 km s^{-1}). The physical significance of the shift is uncertain.

What is clear, is that we can determine properties of the flow fields near the base of the MTZ through ultraviolet spectral imaging. However, the intensities in the ultraviolet are comparatively low (about five orders of magnitude less than visible intensities, when measured in instrumentally meaningful units of photons per Doppler width), and the contemporary small-aperture solar instruments in space are capable of only modest spatial ($1''$–$2''$) and temporal ($\sim$1 min) resolution. An instrumental remedy—large-aperture orbiting UV solar telescope—likely would be a billion dollar undertaking. Furthermore, the thermal emissivity in the far-UV is strongly biased towards hotter temperatures, and NLTE scattering effects are endemic. Thus, it is advantageous to look elsewhere in the spectrum to find more suitable diagnostics of plasma conditions and dynamics at the base of the MTZ.

1.3. Infrared thermal maps

There have been a number of previous discussions of the importance of the 4.7 μm fundamental vibration-rotation bands of carbon monoxide for inferring physical properties of the gas high in the solar photosphere (e.g., Ayres & Rabin 1996, and references therein). The lines are numerous and cover a wide range of excitation; they are strong—mostly because CO is abundant where conditions favor its formation; and collisional quenching of the excited states is rapid, so NLTE effects are unimportant. Altogether, CO is the diagnostic of choice for cool material high in the solar photosphere; occupying perhaps a unique position in that regard.

Fig.3 illustrates a space/time sequence of CO brightness temperatures and velocities, based on three strong absorptions: 2–1 R6, 3–2 R14, and 4–2 R23. The observations were obtained with the McMath-Pierce telescope and IR spectrometer, using a fixed slit at disk center, in 1996 May 10 under conditions of good seeing. The diffraction limit of the 1.5 m telescope is about $0.''8$. The N/S length of the slit is only about $\frac{1}{8}$-th that of the SUMER observations, and the

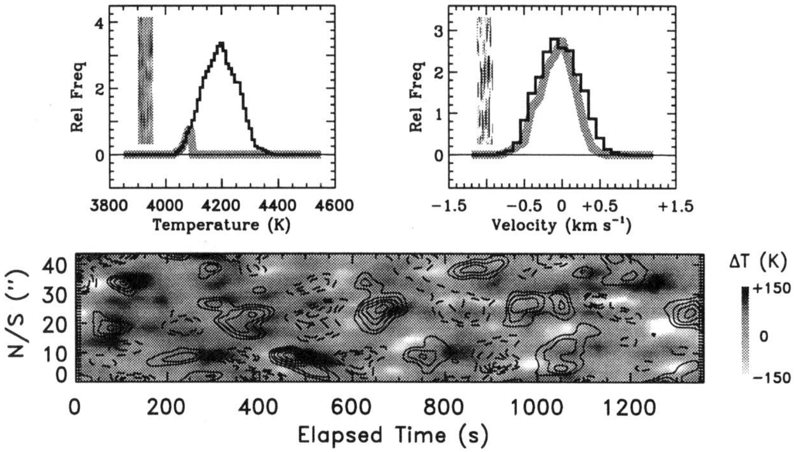

Figure 3. Space/time diagram for 4.7 μm CO lines.

frame cadence was about 20 times higher, yielding a (disk-storage-limited) sequence of only 25 min (still satisfactory for establishing temperature and velocity distributions for the typical 3–5 min waves at those altitudes).

Owing to the denser sampling of the IR sequence, I have depicted the velocities as an overlaid contour map: solid curves indicate redshifts; dashed curves, blueshifts. One clearly sees the alternating patches of Doppler shifts that signal the presence of 5-min p-mode "wavepackets." The temperature pattern loosely follows the compressions and rarefactions that accompany the velocity disturbances; namely, the compressional phase of the wave (velocity turning point following the maximum redshift) often shows an enhanced brightness temperature, while the other turning point (trailing the maximum blueshift) displays cooler temperatures expected for the near-adiabatic expansion phase.

The upper panels of Fig. 3 depict temperature (left) and velocity (right) histograms derived from the IR space/time diagram. The CO temperature centroid is a few hundred degrees cooler than its far-UV continuum counterpart, but very likely is a better gauge of the true kinetic temperatures where the strong CO lines become optically thick. Curiously enough, the rms width of the temperature distribution is similar to that of the far-UV continuum (although, again, the role of NLTE effects in the latter have not been evaluated). The CO velocity histogram, however, is much narrower than that of the far-UV Si II emission; by almost a factor of ten. This suggests that the Si II feature forms much higher up than the CO bands, at lower pressures where the wave amplitudes have grown considerably (at least in velocity).

As with the far-UV maps, I devised a mask for the CO space/time diagram; although now to isolate the *cooler* areas of the pattern. The selected population ($T < 4100$ K) is illustrated by the light-shaded curve in the temperature histogram; the corresponding velocity distribution is depicted in the right panel. Unlike the situation in Fig. 2, here we find that the "cool-patch" velocity distribution is somewhat narrower than the global histogram, and shows no evidence for any large systematic shifts. If the cooler areas are associated with the p-modes, one would expect a narrower line-of-sight velocity distribution, centered at $v = 0$, since the maximum rarefaction phase of the oscillation occurs at the velocity turning point.

On the other hand, if the cooler CO areas are produced by overshooting convective granules (Han Uitenbroek's 1999, this volume), one would expect to find systematic *blueshifts* associated with the cool regions at this level (disk center view)—with lifetimes comparable to those of granules ($\sim$ tens of minutes)—in order to produce the even cooler gas seen several scale heights above (in extreme limb views, and off-limb). Instead, the dominant velocity (and temperature) patterns are short-lived and oscillatory in nature.

The rectangular insets in the temperature and velocity panels illustrate a resampling of the rapid-cadence time sequence to the $\sim$ 1 min resolution of the far-UV maps. Although clearly the IR side is deficient in both spatial and temporal coverage, the temperature and velocity maps display the same "streaky" appearance as their far-UV counterparts; showing the importance of these types of measurements for evaluating the dynamical state at the lower boundary of the MTZ. Of course, the major limitation at present is the diffraction limit of even the largest available IR-capable solar telescope, the McMath-Pierce. We can't know, for example, whether there are finer-scale cold points below the current $\sim 1''$ horizon, whose properties might lead to a ready explanation for the extreme-limb darkening and off-limb emissions of the strong CO lines, without having to invoke a large volume of cool material in a pervasive "COmosphere" (see Ayres & Rabin 1996).

1.4. Translimb emissions of CO

I mentioned the off-limb emissions of CO in the previous discussion; they play a key role in understanding the thermal organization of the "belly" of the MTZ. Any cold material ($T < 4000$ K) at high altitudes—interlaced with the magnetically complex canopy fields—could be of fundamental importance owing to its large resistivity compared with the highly conducting gas inside the ionized flux tubes (J. Chae 1999, this volume). Fig.4 depicts an example of the best measurements of the CO translimb region obtained to date. The bottom panel depicts the coaddition of five frames selected from a sequence of 1000 by a limb-sharpness criterion; taken under calm conditions with good seeing at the North limb. Each of the "bumps" represents the frequency-resolved profile of the translimb emission of an individual CO line. (The image was divided by the average spectrum at $\mu \sim 0.2$ in order to normalize the intensities and suppress telluric absorptions.) In the upper panel, the dots depict frequency-resolved measurements of the off-limb extensions relative to the continuum; the thin solid curve at the top of the panel shows a high-resolution (FTS) *relative intensity* tracing of the disk-center absorption spectrum. At each frequency, the

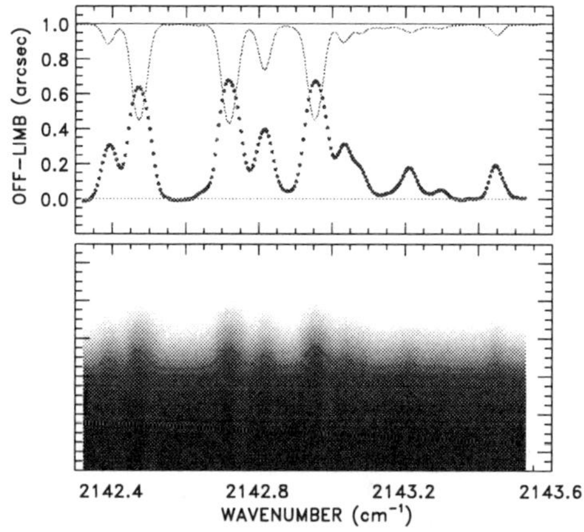

Figure 4. Translimb profiles of CO lines at 4.667 μm.

translimb intensities roll-off smoothly on a scale of $1''$–$2''$ representing the blurring effects of diffraction and scattering. The "subresolution" off-limb extensions are measured as *differential* displacements. On the other hand, higher angular resolving power would suppress any influence of the instrumental profile on the apparent off-limb extensions; this is particularly important for the weaker lines that provide key information concerning the "loading" of cool material along the long tangential sightline above the continuum limb (see Ayres & Rabin 1996).

2. Discussion

The foregoing examples should make it clear that with the existing McMath-Pierce infrared facility, we can obtain comparable or better spatial resolution at the base of the crucial MTZ than contemporary far-UV instruments, with much higher sensitivity. We could do even better from the ground, by going to apertures larger than the present 1.5 m maximum, at a tiny fraction of the cost (and risk) of a major space mission. How much aperture is needed? To answer that question, we must consider in more detail the two key issues: spatial resolution and sensitivity.

2.1. Infrared detectors and ambient thermal background

Before addressing these issues, I'll digress briefly to mention detectors and thermal background. In the ultraviolet and visible, camera technology is relatively mature, and high-quantum efficiency devices are readily available in increasingly large formats. In the thermal infrared, on the other hand, detector technology is much less advanced. Furthermore, even with the most sensitive devices, one must contend with the high external thermal background (sky plus telescope plus spectrometer, if the latter is not cooled) which contributes photon noise but no signal. While it is possible to approach the photon counting limit in the ultraviolet and visible, the accessible S/N in the mid-infrared—for the same photon flux—always will be much lower, if only because of the ubiquitous thermal background.

2.2. Spatial resolution

The angular resolving power of a telescope is

$$\delta\theta \sim 0.''2 \, (\frac{\lambda_{\mu m}}{D_m}) \ , \tag{1}$$

where $\lambda_{\mu m}$ is the observing wavelength in micrometers, and D_m is the primary mirror diameter in meters. For example, the planned *Solar-B* space-borne telescope is designed to achieve $0.''2$ resolution in the visible with a 50 cm mirror. Right away, you can see that to obtain comparable resolution at even the short end of the thermal IR (say, at the key 4.7 μm CO bands) requires a 5 m telescope. To image down to the horizontal photon mean-free path ($\sim H_p \sim 0.''1$ for structures that are $\tau \sim 1$ thick in their vertical dimension) adds another factor of two in aperture; i.e., 10 m for the CO bands.

2.3. Sensitivity

The specific intensity (photons per Doppler width) in the thermal infrared is down by about an order of magnitude compared with the peak of the radiant energy distribution near 4000 Å. That is not as bad as the far-UV case mentioned previously, but qualitatively points to the need for a larger collecting area in the IR to duplicate the sensitivity of an optical solar telescope of given diameter. The question is: how much?

Putting aside detector performance issues for the moment, let's assume that we have perfect photon-counting cameras for both the visible and 5 μm regions. We then can estimate the ratio of collecting areas which will yield the same sensitivity, given a specific remote sensing application. For the sake of argument, consider the most basic physical measurement one might wish to obtain, namely the detection of a given temperature difference ΔT on the solar surface with a specific S/N and angular resolution in an integration time t.

Consider a power-law approximation to the thermal emissivity (Planck function)

$$B(T) \sim b_\lambda \, (\frac{T}{5000 \text{ K}})^{a_\lambda} \ . \tag{2}$$

If we take a small temperature range near 5000 K (a typical photospheric value), the coefficients are as follows

λ	a_λ	b_λ
0.5 μm	6	5×10^{15}
5 μm	1.3	2×10^{15}

The b_λ values, above, are in units of photons cm^{-2} s^{-1} sr^{-1} per 1 km s^{-1} Doppler width. (Note: the visible and infrared photon emissivities at 5000 K are closer than the *observed* 0.5 μm and 5 μm intensities, because $\tau_C \sim 1$ occurs at a higher temperature in the visible than in the mid-IR.) The table emphasizes the fact that at the shorter wavelengths the thermal emissivity becomes a progressively steeper function of temperature, whereas in the mid-infrared the Planck function is essentially linear in T.

Now, for a specified temperature contrast $\Delta T/T$, we obtain the associated relative intensity contrast

$$\frac{\Delta B}{B} \sim a\,(\frac{\Delta T}{T})\;. \tag{3}$$

The corresponding signal is

$$\Delta\text{counts} \sim a\;B\;A_{\text{eff}}\;\delta\omega\;\Delta v_{\text{D}}\;t\;(\frac{\Delta T}{T})\;, \tag{4}$$

where A_{eff} is the effective area of the instrument, $(\pi/4)\,D^2\,\epsilon_\lambda$, the geometrical collecting area times the downstream efficiency (coatings plus instruments plus detector); $\delta\omega$ is the solid angle subtended by the camera pixels; and Δv_{D} is the equivalent Doppler bandpass of the spectral measurement. The noise (counting statistics, ignoring any backgrounds) is

$$\sigma_B \sim (B\;A_{\text{eff}}\;\delta\omega\;\Delta v_{\text{D}}\;t)^{1/2}\;, \tag{5}$$

i.e., the square root of the total counts. The S/N is the ratio of the former to the latter

$$\text{S/N} \sim a\;(B\;A_{\text{eff}}\;\delta\omega\;\Delta v_{\text{D}}\;t)^{1/2}\;(\frac{\Delta T}{T})\;. \tag{6}$$

To measure the same ΔT with the same S/N in the same t with the same angular resolution $\delta\omega$ (for spectral diagnostics of similar Doppler width) requires

$$(\frac{A_5}{A_{0.5}}) \sim (\frac{a_{0.5}}{a_5})^2\;(\frac{b_{0.5}}{b_5}) \sim 5\times10^1\;! \tag{7}$$

So, even if the system efficiency ϵ_λ is the same at 0.5 μm and 5 μm (and it is likely to be much *less* at the longer wavelength if only because of the thermal background), one still requires a $\sim 7\times$ larger diameter telescope in the IR to measure temperature contrasts with the same sensitivity as in the visible. One is forced to conclude that both resolution and sensitivity call for a $\sim 10\times$ increase in telescope aperture for thermal IR work. Another way of looking at it: the current McMath-Pierce telescope performs as well at 4.7 μm as a 15 cm instrument would in the visible!

The astute reader might be wondering at this point: why bother with the thermal infrared at all; can't we just work in the visible? The answer is that the only places in the visible where one can "see" as high an altitude as probed by the CO bands are the cores and inner wings of strong resonance lines, where NLTE effects and partial coherent scattering obscure the true sizes and temperatures of the radiating structures. The LTE CO lines, on the other hand, are much cleaner diagnostics of the physical conditions, but the price you pay is having to build a big telescope.

2.4. Is AST a possible solution?

Could the proposed Advanced Solar Telescope do the job? The facility presently under consideration is of 3 m class. With adaptive optics, such an instrument could provide—during periods of good seeing—resolution in the visible down to the $0.''1$ considered essential for breaking new ground in understanding small-scale solar phenomena. That performance could be achieved out to $\sim$1.5 μm, where important magnetically-sensitive lines fall (such as used by Doug Rabin's Near Infrared Magnetograph). With the large aperture, there also would be plenty of light to record fast-changing phenomena via narrow-band spectral imaging (Keller 1999,these proceedings). Out at 4.7 μm, however, a 3 m solar telescope would represent a much more modest improvement over the existing McMath-Pierce, achieving a diffraction-limited spatial resolution of only $0.''4$; still far away from the small scales where a great deal of crucial action might be occurring. While an "open" AST would yield better performance in the thermal infrared than the best existing solar IR telescope, it will not provide the order of magnitude improvement that likely is needed to boost our understanding to the next level.

2.5. Design considerations

The previous arguments point to a solar infrared telescope of aperture 5–10 m. Considering how difficult (and expensive!) it will be to design and build a properly-functioning 3 m AST, the concept of a much larger telescope—dedicated to the niche application of IR thermal imaging—sounds like a pipe dream. Perhaps. But, there are several considerations that I have not mentioned yet, which might make such a concept practical.

First, there already are (or soon will be) lots of telescopes around the world of 8–10 m aperture, albeit in the hands of the darksiders. Might it be possible to retrofit one of those nighttime facilities to permit it to be used profitably during the day? After all, the JCMT has been used to observe the Sun in the sub-mm band, with special precautions to protect the radio dish. I have not looked into the technical aspects (politics aside), but I believe that heating of the primary mirror (and potential damage to the coating) would be minimal (most of the $\sim$1 kW m^{-2} would be reflected, not absorbed), but clearly a specially designed secondary and heat dump would have to be provided to protect against the intense fluxes at the (typically fast) focus. Retrofiting a current nighttime telescope would be challenging (perhaps impossible politically), but the option is worth considering for a future facility for which solar access could be built in from the beginning. I have in mind a hypothetical nighttime IR-optimized telescope (not yet proposed) with an open, off-axis design that would lend itself naturally to the solar option (as in Jacques Beckers' original proposal for the CLEAR telescope).

A second important point is that while a sensitive high-resolution IR telescope (solar or otherwise) must be large, it need not be as expensive to build as a conventional optical facility. In particular, the tolerances for mirror curvature and surface roughness are a factor of 10 looser at 5 μm than in the visible, and that applies to the whole optical train as well. One can get away with a much lighter mirror, perhaps metal rather than glass, which in turn permits a lighter (therefore cheaper) support structure. Perhaps one could do away with the convential telescope tube by employing an articulated beam feed that would follow the gross motion of the solar image (from, say, a 10 m lightweight segmented mirror supported in a simple polar mount, or more flexible alt-az configuration), carrying a heat shield to dump most of the solar beam, with an agile secondary to correct for short-term image motion and more slowly changing mirror abberations. At the conclusion of observing, the arm could be retracted and stowed, allowing a low-profile enclosure (perhaps a clamishell like that used on the Dutch Open Telescope described by Rob Rutten 1999,these proceedings). The driving motivation for building such a telescope would have to be nighttime IR astronomy, but solar could piggyback with relatively modest additional cost.

A final possibility, that has been discussed for almost a decade now, is the "Big Mac" proposed by Bill Livingston and Bob Noyes. Their concept was to replace the 1.5 m Main mirror in the McMath-Pierce with a much larger aperture: up to 4 m diameter would fit in the existing tunnel. However, that design would cause enormous heating in the tunnel, and surely would lead to much worse internal seeing than already exists. I offer a modification; which is, I warn, not yet as well thought out as the original Big Mac design.

I would replace the heliostat by an off-axis main focussing mirror, with a fast enough f-ratio to form an image near the entrance to the tunnel, or somewhat inside. I believe there is room at the top of the McMath-Pierce for up to a 6 m heliostat, so a primary of that class probably could be accommodated. I would put a heat dump at the (short) focus to divert most of the light harmlessly away from the tunnel, passing only whatever few arcminutes of the beam the instruments downstream could profitably use. Suitable optics in the tunnel would magnify the image and transfer it to the Main spectrograph (or FTS) station at an appropriate plate scale.

One innovation is that the surface roughness of the main mirror could be controlled to be fully reflective at 5 μm, but mostly "dull" in the visible so that much of the incident optical light would be diffusely scattered, while the desired IR would be specularly reflected (I have seen such a design for a spacecraft instrument to measure $\sim$ 100 μm solar radiation). Heat loads at the beam dump would be reduced, but possible deleterious influences on the scattering wings of the instrumental spatial profile (e.g., for translimb work) would have to be evaluated carefully.

A final point to consider is that initially one might experiment with only a partially-filled aperture, in particular a *linear* mirror, say five or six 1×1 m segments carried perpendicular to the tunnel axis. Such a configuration would yield high spatial resolution in one dimension (East/West, in this case), but only modest resolution in the other. Such a design is well-suited for long-slit imaging of spatial structure on the disk, and for exploring the translimb zone at high resolution. Heat loads would be reduced by a factor of nearly five, and the supporting structure could be much simpler than for a monolithic 6 m. Costs certainly would be reduced, as well, perhaps to the level where such a project might be accomplished for only a modest investment. The main advantage of utilizing the McMath-Pierce is, of course, the good (i.e., dry) IR site, existing infrastructure, and full complement of high-performance infrared instruments.

3. Conclusions

I have three main conclusions: (1) it is astrophysically worthwhile to observe, at the very least, the 5 μm region of the solar spectrum—longer wavelengths still are pretty much unexplored, and indeed are awaiting development of solar IR telescopes of higher angular resolution and sensitivity; (2) the apertures required to achieve high spatial resolution and good sensitivity in the thermal infrared are so much larger than presently under consideration for the AST, that it does not make sense to shackle the design of the latter with the requirement to pass radiation longward of $\sim 3\ \mu$m; and (3) the mid-infrared is much less constrained—from a telescope design point of view—than the visible, so it is worthwhile to consider piggybacking solar IR capability on existing or planned nighttime facilities, or retrofitting the McMath-Pierce with specially-designed IR optics to take advantage of its existing complement of instruments. Failure to seize such opportunities, I fear, will cause infrared solar physics to languish; and a unique tool will be lost to explore physical conditions at the base of the vital—but enigmatic—magnetic transition zone.

Acknowledgments. I thank P. Lemaire, U. Schühle, and K. Wilhelm for their help with the SUMER side of the project, and to J. Chae for assistance with the calibrations; I thank D. Rabin for his help on the CO side. This work was supported by a grant from the National Science Foundation (AST–9618505) and *SOHO* Guest Investigator grant NAG5–6124.

References

Ayres, T. R., & Rabin, D. 1996, ApJ, 460, 1042 (AR)

Carlsson, M., & Stein, R. F. 1997, ApJ, 481, 500

Vernazza, J. E., Avrett, E., H.& Loeser, R. 1981, ApJS, 45, 635

Group Discussion

Chae : What line did you use for the production of the first Doppler shift map? Have you examined the possible effect of line blending?

Ayres : I used the SiII 1309 line, but did not examine the line blending.

High Resolution Solar Physics: Theory, Observations, and Techniques
ASP Conference Series, Vol. 183, 1999
T.R. Rimmele, K. S. Balasubramaniam, R.R. Radick, eds.

Scientific Objectives and Design Concepts of the Optical Telescope and its Observing Instruments aboard Solar-B

Y. Suematsu and the Solar-B working group

National Astronomical Observatory,
Mitaka, Tokyo 181-8588, Japan

Abstract. Solar-B is a Japanese solar observing satellite with a planned launch in 2004-2005. The satellite will carry an optical telescope with a magnetograph, a soft X-ray imaging telescope, and an EUV imaging spectrometer. It will provide a unique opportunity to study the physical connection between the photosphere and the corona. We describe some major scientific objectives and baseline designs of the observing instruments aboard Solar-B.

1. Introduction

Solar-B is the Institute of Space and Astronautical Science (ISAS) solar space mission to follow YOHKOH, which was called Solar-A by an ISAS prelaunch code name. YOHKOH, launched in August 1991, has provided exciting new results on the dynamical behavior of the corona and flares. A natural question then arose about the connection between these phenomena and changes in the photospheric magnetic fields. In order to answer this question, it will be necessary to observe the behavior of photospheric magnetic fields with a spatial resolution greater than $0.2''$, as suggested by high-resolution solar images, and coronal phenomena at the same time. This became a core principle in preparation of the Solar-B mission.

The Japanese Solar-B working group officially started in November 1992 at ISAS. Since then, many scientific as well as instrumentation studies have been done in collaboration with US and UK colleagues, and others. The Solar-B mission was formally approved by ISAS in 1997, and it is now scheduled to be launched by the ISAS M-V rocket into a circular, 600 km, sun-synchronous polar orbit during the Japanese fiscal year 2004 (starting in April 2004). The mission will be prepared and executed with multilateral international collaborations, including the US and the UK. We here describe some major scientific objectives for Solar-B, and its instrumentation to achieve them (see also Shimizu 1998).

2. Scientific Objectives: from YOHKOH to Solar-B

YOHKOH, SoHO and, more recently, TRACE (e.g. Title et al. 1998) have revealed the dynamical nature of the solar corona. For example, YOHKOH has shown that the corona is extremely dynamic, with restructuring (Tsuneta et al. 1992), rapid heating such as active region transient brightenings (Shimizu et al. 1992), nanoflares (Shimizu and Tsuneta 1997), and coronal flashes in polar coronal holes (Koutchmy et al. 1997), as well as mass accelerations such as X-ray jets from bright points (Shibata et al. 1992; Shimojo et al. 1996) being common phenomena. In addition, it turned out that the cores of many coronal bright points including small flares are too small in size to study their energy release mechanisms.

It is clear that magnetic fields play the most important role in coronal phenomena, and they are deeply rooted in the photosphere and controlled by plasma motions there and beneath. We think, therefore, that the next mission should be aimed at clarifying the physical connection between photospheric/subphotospheric dynamics (the driving engine) and the coronal active phenomena (the outputs), and at revealing the magnetic origins of the solar phenomena.

It is well known, however, that it is very difficult to observe the precise and detailed behavior of photospheric magnetic fields with ground-based instruments, because the fields are concentrated in sub-arcsecond structures that are much smaller than the atmospheric seeing-limited resolution. In particular, magnetic components transverse to the line of sight, which measure excess magnetic energy, are difficult to observe and cannot be measured with any degree of accuracy if the field is not spatially resolved.

Therefore, the next mission, Solar-B, should be capable of both better coronal imaging than YOHKOH/SXT and photospheric vector magnetic field measurements with sub-arcsecond resolution. Since unsolved problems in present-day solar physics are closely related to problems stemming from lack of spatial resolution and accuracy in vector magnetic field measurements, Solar-B will give us an opportunity to solve these for the first time. Its major objectives are to elucidate the nature of:

- coronal heating (including transient brightenings, X-ray bright points and jets, loop-loop interactions, dynamics of active regions, etc.),
- chromospheric heating (including the origin of spicules and their role in the transport of mass and energy between the chromsphere and corona, and the physics of the field free regions),
- eruptive events (including physics of flares and CMEs), and
- magnetic field generation and transport (including subsurface dynamo mechanisms, field emergence, formation of flux tubes, and evolution of active regions).

Among various mechanisms, magnetic reconnection seems to play the central role in the energy release of most solar dynamical phenomena. YOHKOH has provided strong evidence for magnetic reconnection over a wide range of size scales (e.g. Shibata 1997). As new magnetic flux erupts from the subphotosphere, previously emerged flux must be dissipated or ejected into the solar wind to maintain a steady state. Reconnection is likely to be a process by which the free energy contained in non-potential magnetic fields may be transformed rapidly into heating, mass motions, and/or structural changes of plasmas, and it likely plays the dominant role in the continual readjustment of the global solar magnetic field.

The corona may be heated by the dissipation of electric currents. These currents are generated by the slow, shuffling motions of magnetic flux bundles in the photosphere, and the subsequent twisting and braiding of the coronal magnetic field. In the corona, these currents are resistively dissipated in thin current sheets or produce localized impulsive heating by magnetic reconnection (e.g., microflares - Parker 1988).

Microflaring might also be responsible for heating the magnetized chromosphere. This idea requires a mixture of opposite polarities on sub-arcsecond scales in the heated areas (e.g. Rabin & Moore 1984). This might be the origin of spicules (for a brief review, see Suematsu 1998) and shocks which contribute to the heating of the chromosphere and the corona. Only a high-resolution vector magnetograph in space will reveal whether or not the requisite mixture of polarities is present, and whether the mixed-polarity fields contain enough free energy.

Therefore, understanding the detailed physical process of magnetic reconnection is one of the most important objectives of the Solar-B mission.

3. Baseline Solar-B Instrumentation

The major goal of Solar-B is a comprehensive understanding of the solar photosphere and corona as a system. To do this, we have to determine the evolution of the photospheric vector magnetic field, which is an indicator of free energy build-up. We also need to determine the topology, location, and timing of energy release. The determination of these boundary conditions will require high spatial resolution and high-cadence imaging.

Our approach is as follows. The photospheric and chromospheric measurements will be made with a 50cm-aperture optical telescope and its focal plane observation package, which contains both imaging and spectroscopic instruments. The coronal and transition region measurements will be made with an X-ray imaging telescope and an EUV imaging spectrometer.

The solar optical telescope (SOT) is designed to give diffraction limited images and to be capable of high precision polarimetry for vector magnetic measurements; we set the target spatial resolution at $\sim 0.2''$ and the S/N ratio at ~ 1000 to resolve fine-scale magnetic elements (e.g. Keller 1992). The strawman design of the optical telescope is an aplanatic Gregorian (1.2 m in length) with a heat dump oblique mirror at the primary focus (this mirror also works as a field stop of $400'' \times 400''$). A collimator comes after the secondary focus to yield a pupil image around a polarization modulator and a tip-tilt mirror just behind the primary mirror, and also to relax a positional tolerance of the following focal plane package. Table 1 summarizes the necessary components and their specifications to achieve the goal.

The focal plane package will have a monochromatic imager tunable to several spectral lines and continua useful for atmospheric diagnostics, and a spectrometer for magnetic sensitive lines (Table 2). In designing this package, we are fully utilizing heritage from solar high resolution instrumentation and precision spectropolarimeters.

The X-ray imaging telescope (XRT) will provide detailed information of high temperature material to permit study of both the dynamics of fine scale coronal phenomena, such as magnetic reconnection, and coronal heating mechanisms. Fabrication of the optics, whether grazing or normal-incidence, to meet the required angular resolution ($\sim 1''$) is well within the state of the art. Table 3 gives the specification for normal-incidence optics, which would permit imaging Doppler velocity measurements (Sakao et al. 1996).

The EUV imaging spectrometer (EIS) will provide detailed physical conditions of high temperature plasmas such as density, velocity and temperature. Note that the observing wavelength region was carefully selected to make both normal coronal and flare plasma diagnostics possible. Strawman specifications are given in Table 4 together with those for CDS/SoHO.

The optical telescope occupies a central part of the spacecraft structure. An optical bench enclosing the optical telescope structure is the base of accurate pointing (targeted $\sim 1''$) of the three independent telescopes (SOT, XRT and EIS) (Figure 2).

Table 1. Baseline components of the optical telescope and its focal plane instruments

Telescope	Aplanatic Gregorian with Heat Dump Mirror and Collimator
Polarization Modulator	Rotating Waveplate frequency=0.75 Hz, retardance=127°
Image Stabilizer	Tip-Tilt Mirror with Correlation Tracker
Two-Channel Filtergraph:	
narrow-band channel	Tunable Lyot (passband=100mÅ) or Fabry-Perot (passband=70mÅ)
wide-band channel	interference (passband= 3–10Å)
detector	2k×2k CCD, pixel=0.1″
Spectrograph	Littrow-type, Image Scanning Mirror
detector	frame transfer CCD

4. Summary

With the capabilities described above, Solar-B will allow us for the first time to observe the dynamics of the elemental, isolated magnetic flux tubes that are thought to be responsible for the activity observed in the upper solar atmosphere, and hence to study the detailed physical connection between the photosphere and corona. It should be noted that to gain the maximum scientific benefit from the mission, we will coordinate with theoretical studies based on numerical simulations and analytic methods to quantify the complex interactions between the magnetic fields, the radiation field and the ionized atmosphere. At the moment, Solar-B is scheduled to be launched in 2004-2005 into a circular, sun-synchronous polar orbit for continuous solar observation (Table 5).

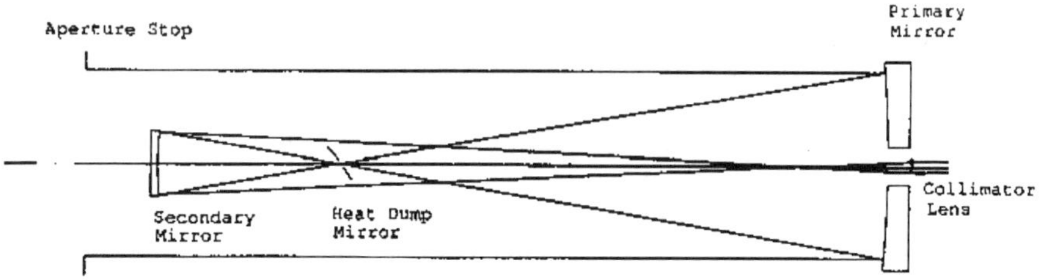

Figure 1. Layout of baseline Solar-B optical telescope: Aplanatic Gregorian with heat dump mirror and collimator.

Table 2. Focal plane instruments of the optical telescope and their primary observing spectral lines (g is the effective Landé g-factor) and continua. The spatial resolutions in the table are diffraction limited (λ/D) for 50 cm diameter. Abbreviations are as follows. ph: photosphere, ch: chromosphere, T-min: temperature minimum layer, v: velocity field measurements, h_v: horizontal velocity field measurements, and B: magnetic field measurements

wavelength (Å)	Filtergraph Broad-Band (3 – 10 Å)	Narrow-Band (100 mÅ) @ 6302Å	Spectrograph (25 mÅ) @ 6302Å	Spatial Resolution (arcsec)
3933 Ca II	ch heating		(ch, wave, heating)	0.16
4305 CH band	faculae, h_v			0.18
4508 clean cont.	deep ph h_v			0.19
4861 H I Hβ		ch, v		0.20
5173 Mg I b_2 (g=1.75)		T-min B		0.21
5250 Fe I (g=3)		ph B		0.22
5576 Fe I (g=0)		ph v		0.23
6302 Fe I (g=2.5)		ph B	ph, B with 6301Å	0.26
6563 H I Hα		ch, v		0.27

Table 3. Baseline design of X-ray telescope (normal incidence case).

		Remarks
Telescope	Ritchy-Chretian (20 cm diameter) with 5 sector coatings	
Wavelengths	Fe IX 171Å(0.9 MK)	for normal corona
(sensitive	Fe XIV 211 − 2 Å(1.8 MK)	Doppler velocity
temperature)	Fe XIV 211 + 2 Å(1.8 MK)	measurements
	Fe XII 193 Å(1.5 MK) & Fe XXIV 192 Å (16 MK)	for both corona & flare
	visible wavelength	for co-alignment
Detector	Back-illuminated CCD (2048×2048)	
Pixel size	1.24″	

Table 4. Specification of strawman EUV imaging spectrometer (EIS) for the Solar-B, with comparison to CDS aboard SoHO.

	EIS (Solar-B)	CDS (SoHO)
Optics	Normal incidence	Grazing Incidence
collective area	225 cm^2	34.3 cm^2
spatial res.	2″	3″
Wavelength Band	250–290 Å (He II – Fe XXIV)	308–381Å, 513–633Å
Detector	Back-Illuminated CCD	MCP+ Front-Illuminated CCD
Pixel size	2″, 20mÅ (@270Å)	1.68″, 70mÅ (@360Å)
FOV (slit)	~300″	240″
Method of Data Analysis	Line-ratio Line-profile	Line-ratio

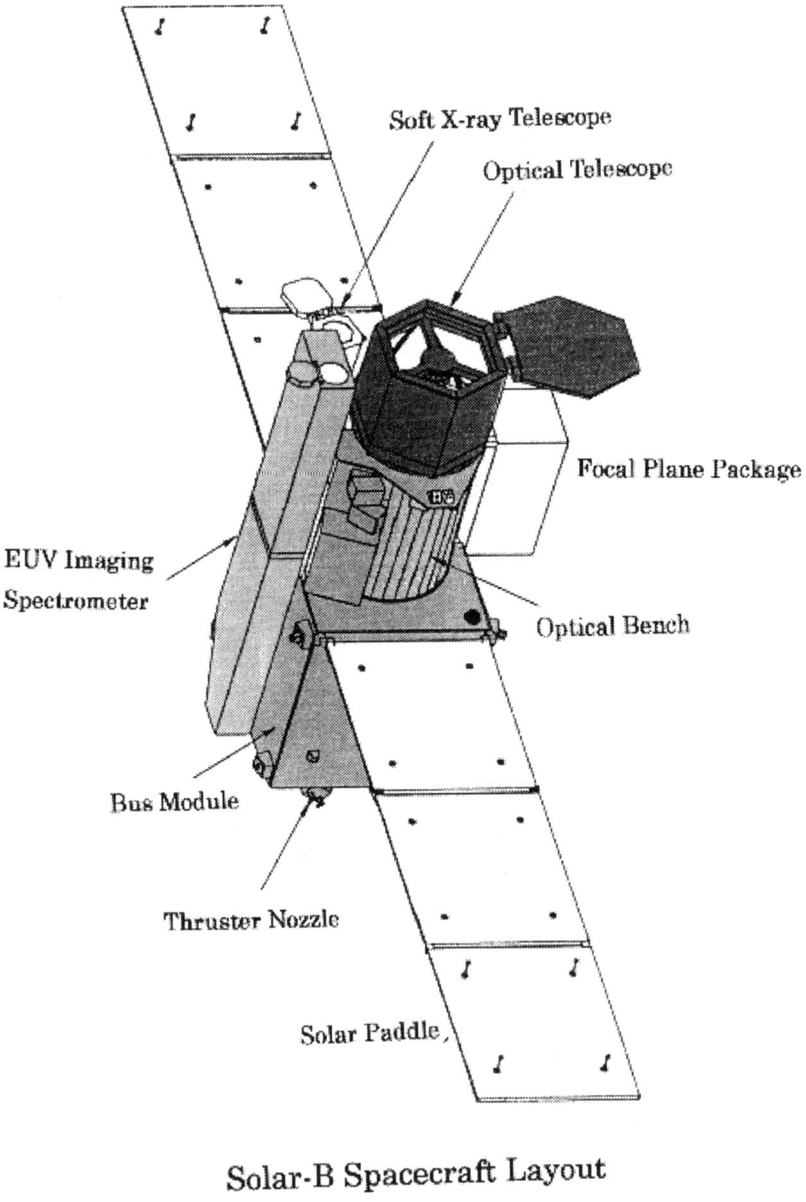

Figure 2. The Solar-B spacecraft layout.

Table 5. Overview of Solar-B satellite.

Orbit	600 km solar-synchronous polar orbit
Weight	$\sim$700 kg (dry) + $\sim$170 kg (thruster gas)
Launch	2004/8 or 2005/2
Lifetime	$\geq$ 3 years
Attitude Control	3-axis stabilized body control
Data Rate	$\sim$500 kbps
Amount of Data	$\sim$3 Gbits per orbit
Telemetry Rate	$\sim$5 Mbps

References

Keller C. U. 1992, Nature, 359, 307

Koutchmy S. Hara H., Suematsu Y. & Reardon K. 1997, A&A, 320, L33

Parker E. N. 1988, ApJ, 330, 474

Rabin D. & Moore R. 1984, ApJ, 285, 359

Sakao T. et al. 1996, SPIE, 2804, 153

Shibata K. et al. 1992, PASJ, 44, L173

Shibata K. 1997, in The Corona and Solar Wind near Minimum Activity, the Fifth SOHO Workshop, ESA SP-404, pp.103-112

Shimizu T. et al. 1992, PASJ, 44, L141

Shimizu T. & Tsuneta S. 1997, ApJ, 486, 1045

Shimizu T. 1998, in A Crossroads for European Solar and Heliospheric Physics: Recent Achievements and Future Possibilities, (eds. R. A. Harris et al.), ESA SP-417, pp.169-178

Shimojo M. et al. 1996, PASJ, 48, 123

Suematsu Y. 1998, in Solar Jets and Coronal Plumes, (ed. T. -D. Guyenne), ESA SP-421, pp.19-28 ESA

Title A. et al. 1999, in these proceedings

Tsuneta S. et al. 1992, PASJ, 44, L211

High Resolution Solar Physics: Theory, Observations, and Techniques
ASP Conference Series, Vol. 183, 1999
T.R. Rimmele, K. S. Balasubramaniam, R.R. Radick, eds.

The Large Vacuum Solar Telescope on the Way to High Resolution

A. V. Boulatov, V. M. Grigoryev, V. I. Skomorovsky

Institute of Solar-Terrestrial Physics, ISZF, P.O.Box 4026, Irkutsk, 664033, Russia

Abstract. The Large Vacuum Solar Telescope (LVST) is distinguished by its high theoretical optical parameters. We have carried out spectropolarimetric observations in various spectral bands, particularly in the HeI 10830Å line region (especially since 1997, when a new CCD system was obtained). The results cause us to conclude that only by means of adaptive optics will we be able to approach the theoretical resolution of the LVST. The possibility of using some kind of adaptive optics system is considered.

1. Introduction

One of the most important goals of solar astronomy is to understand the interaction of solar magnetic fields with the solar atmosphere. Supposing that such interactions occur on scales down to sizes near 50 km, resolution to $0.''05$ is required for studying such processes. Good resolution is required also for spectroscopic observations with narrow-band filters, since they require exposures longer than the temporal scale of the image motion that powerfully distorts these observations. Velocity fluctuation measurements in the upper atmosphere require repetitive scanning of one area for minutes to hours, and spatial registration must be maintained with high accuracy.

Modern ground-based telescopes are not capable of reaching such performance. In consequence of atmospheric turbulence, the resolution of ground-based solar telescopes is limited to an average value of about $1''$, with short periods of higher resolution in sites with the best seeing. In actual instruments it is important to suppress disturbances to the wave front (both inside the telescope and outside of it) that reduce resolution and consequently the information value of observations.

In the last decade progress in adaptive optics (AO) has raised the possibility of using such systems to improve the resolution of ground-based telescopes enough to study processes on the Sun at the angular scale required for their understanding.

The main difficulty in using AO, for solar observations particularly, is the small area over which it is possible to correct an image (isoplanatism), which in turn depends on the height distribution of turbulence in the Earth's atmosphere. It is very important for us to have the isoplanatic angle as large as possible in order to study extended scenes such as active regions and supergranular cells. For

Figure 1. View of the telescope.

imaging such extended objects, CCD arrays are used everywhere. For successful operation of AO systems with CCDs, where real-time correction of fast disturbances in the wave front arriving from the object is required, it is necessary to know the current and expected atmospheric parameters at each instant in order to optimize a disturbance compensation algorithm. This requires a careful study of seeing parameters. Under poor seeing an AO system simply will not be able to work.

In our article we will present some results of our efforts to achieve high resolution with the Large Vacuum Solar Telescope at Baikal Astrophysical Observatory.

2. Observations

2.1. Telescope

The Large Vacuum Solar Telescope (LVST) of the Baikal Astrophysical Observatory is situated near the shore of Lake Baikal in Listvianka village about 70km from Irkutsk. Its top is about 700 meters above sea level. The solar chromosphere and its fine structure are a main object of scientific studies at this telescope. In Fig 1 a view of the telescope is presented.

The LVST uses a single-mirror polar siderostat with a diameter of about 1 meter. The height of the siderostat tower is 25m. The diameter of the objective lens in the vacuum pipe is 76cm, the focal length is 40m, and the diameter of the solar image is 38cm. The theoretical spatial resolution is 0.18 arcseconds. The optical performance of the telescope is described by Skomorovsky & Firstova (1996). The telescope is equipped with a spectrograph. The good features of

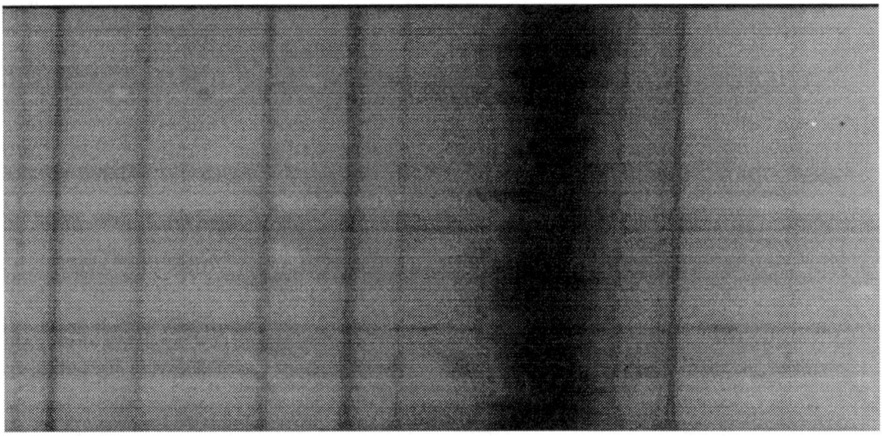

Figure 2. Raw CCD image of solar spectral lines in the H_α line region. The vertical size corresponds to 40 arc seconds

this spectrograph and particularly its diffraction grating allow the LVST to be used to study the fine structure of spectral lines.

2.2. Imaging system

Before 1995 imaging was performed only by photographic camera. In that year two CCD cameras were obtained: an ST-6 from Santa Barbara Instrument Group, and a special camera from a Russian manufacturer (St-Petersburg) with changeable camera heads (800×400 and 370×290). Toward the end of 1997 the first and still only commercial CCD system was installed in our observatory, a TEK 512×512 CCD from Princeton Instruments Company with a ST-130 controller. A new system with 2K×2K CCD camera will be installed in 1999.

During our observations the main objects of interest usually are flares and mustaches. As seen in Fig 2), we still do not reach optimal image quality.

3. Attempts to Improve Image Quality

What can we do to improve the solar image quality and its spatial resolution? Below we present short descriptions of our steps.

3.1. Image selection

Real-time selection of best frames with a seeing monitor has been used. The seeing monitor works on the basis of the contrast of granulation images (Karpinsky & Malyn 1983). Measurements have been performed at the Kislovodsk Solar Station, capturing exposures during the best quality moments with a photographic camera (Karpinsky & Bulatov 1983).

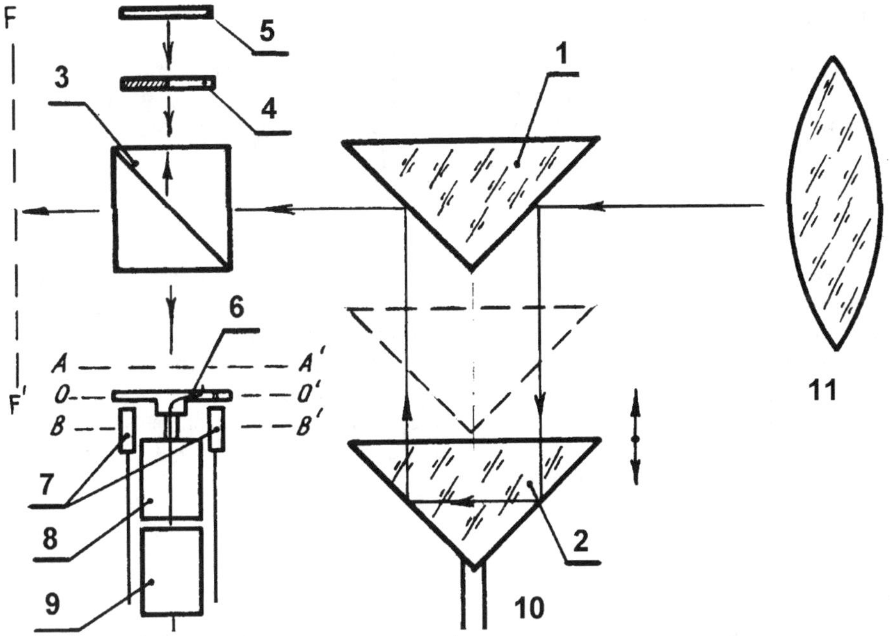

Figure 3. Optical scheme - focus stabilization system, indicating: 1 – fixed prism; 2 – movable prism; 3 – prism beamsplitter; 4 – shutter (mirror semi-circle); 5 – mirror; 6 – fiber; 7 – synchronizing elements; 8 – engine; 9 – photo-multiplier; 10 – rod to potentiometer; 11 – objective lens.

3.2. Focus stabilization

Combining the granulation contrast seeing monitor with the focus guiding system of the LVST has led to the development of an auto-focusing device. Fig 3 (Bulatov & Palachev 1988) is a sketch of the optical scheme. This defocusing compensation tool was designed to process both slow and comparatively quick variations of the focus position. The system worked only under very good seeing conditions. With poor quality images the correct position of focus was not defined. In summary, the system worked well in good seeing conditions, compensating both slow thermal and fast turbulence-induced variations in focus position.

3.3. Sunspot guiding

We have used a sunspot guider since 1987 for our observations (Druzhinin et al. 1988; Bulatov et al. 1991). A quadrant photodiode is used as a sensor. Regrettably, the inertia of the mirrors limited the image motion compensation to a frequency of 10 Hertz. We also met with performance problems for sunspots with asymmetrical form. A similar system at NSO/SP is the well-known spot tracker (von der Lühe 1988).

3.4. Site testing for large telescope project

At the same time we led the work to choose a place for the new large solar telescope. A suitable place was found in Central Asia (Bulatov et al. 1992) after investigation of seeing for several years (1987-1990) in different regions of the USSR.

3.5. Corrections inside the telescope

As a prototype for such a new generation telescope we considered the Large Vacuum Solar Telescope. In the LVST a thermal control system compensates the distortion of objective lens caused by non-uniform heating from the Sun. It is essential to control image motion, image rotation, and scattered light. The best method for correcting the image distortion is adaptive optics. As an initial stage we plan to install a correlation tracker, similar to schemes successfully working on several other solar telescopes (Von der Lühe et al. 1989, Rimmele et al. 1993, Ballesteros et al. 1996)

4. Current Status

In August 1998 a series of atmospheric turbulence measurements were conducted to evaluate the Fried parameter and isoplanatic angle on the top of the LVST. As is well known, these parameters must be taken into account in the design of an adaptive optics system or a tip-tilt mirror system like the correlation tracker. For more detailed information about seeing variations we usually monitor the image motion of the solar limb. An example of such a diagram is shown in Fig 4. This information also helps us define more optimal characteristics for future adaptive optics systems.

The following stage, planned for 1999, will be to test on the LVST a flat 3-cm active mirror driven by piezo ceramics. It will be part of a future adaptive optics system, which will have bimorph mirrors mounted on a base of piezo ceramics. The upper limit of the frequency response will be increased from 200 Hz to 2 kHz. A model of the AO system will be built with the servo supply system, allowing varying actuator configurations and contact points to the mirror. The computer model will calculate the accuracy of reproducing the Zernike polynomials, representing atmospheric aberrations. Work with such mirrors, mostly for night-time objects, has been led by the Tomsk Institute of Atmospheric Optics (Lukin 1994). We also hope for collaboration with Solar Adaptive Optics Teams from SVST, NSO/SP and IAC.

We are also going to improve observational data at the Large Vacuum Solar Telescope using filters for the BaII 4554$\AA$ (FWHM=0.08$\AA$) and H_β 4861$\AA$ (0.09$\AA$) lines, and a new two-bandpass birefringent filter for the HeI 10830$\AA$ (0.48$\AA$) and H_α 6563 (0.3$\AA$) lines (Kushtal & Skomorovsky 1998). Example of filtergrams are presented in Fig 5

Acknowledgments. I would like to thank the organizers and the European Office of Aerospace Research and Development (AFOSR) for supporting my participation in the Conference.

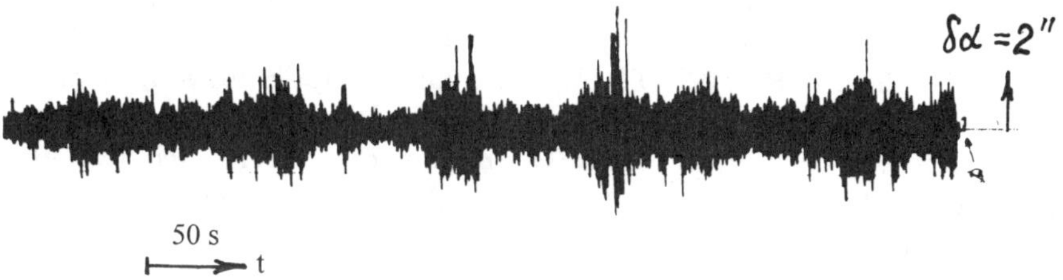

Figure 4. Solar limb image motion - temporal diagram.

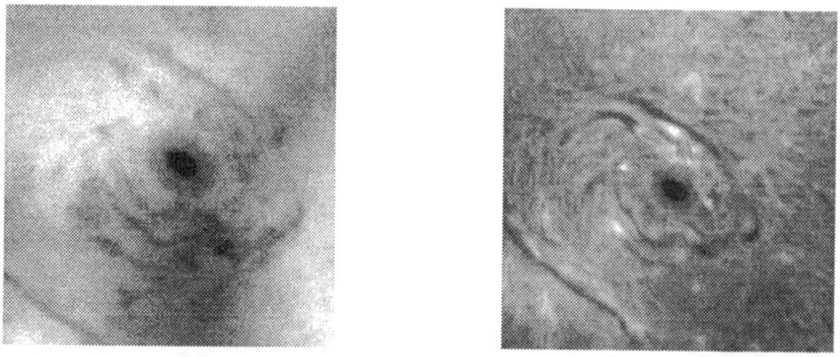

Figure 5. Filtergrams: left) obtained 8 July 1998 at 12:11 LT, centered on wavelength HeI 10830Å - 0.2Å, size = 205″; right) obtained 8 July 1998 at 12:29 LT, centered on wavelength H_α + 0.2Å, size = 266″

References

Ballesteros, E., Collados, M., Bonet, J.A., Lorenzo, F., Viera, T., Reyes, M., & Rodrigues Hidalgo 1996, A&AS, 115, 353

Bulatov, A.V., Palachev, & Yu.M. 1988, in "Atmospheric Instability and Adaptive Telescope", 125

Bulatov, A.V., Druzhinin, S.A., Pevtsov, A.A., Levkovsky, V.I., & Sattarov, I.S. 1991, IGAFS, 95, 15

Bulatov, A.V., Kovadlo, P.G., Sidorenkov, A.A., Prokopiev, A.A., Umarov, V.F., Khan, Yu.V., Skazhenyuk, V.I., Goshchdzhanov, M.M., & Raubaev, S.A. 1992, IGAFS, 99, 206

Druzhinin, S.A., Maslov, I.L., & Pevtsov, A.A. 1988, IGAFS, 83, 149

Karpinsky & Bulatov 1983, KGU Internal Report

Karpinsky, V.N., & Malyn, S.M. 1983, IGAFS, 64, 9

Kushtal, G.I., & Skomorovsky, V.I. 1998, in Solar Jets and Coronal Plumes, Proc. Internat. Meeting, Guadelupe, (ESA-SP 421, May 1998), 329

Lukin, V.P., 1994, in Proc. NATO ASI on Adaptive Optics for Astronomy, 59

Rimmele, T., Kentisher, T., & Wiborg P.H. 1993, in Real-Time and Post Facto Image Corection, Proc. 13th NSO/SP Summer Workshop, (ed. R. Radick)

Skomorovsky, V.I., & Firstova, N.M. 1996, Solar Phys., 163, 209

Von der Lühe, O. 1988, A&A, 205, 354

Von der Lühe, O., Widener, A.L., Rimmele, T., Spence, G., & Dunn, R.B. 1989, A&A, 224, 351

High Resolution Solar Physics: Theory, Observations, and Techniques
ASP Conference Series, Vol. 183, 1999
T.R. Rimmele, K. S. Balasubramaniam, R.R. Radick, eds.

Proposed 60-cm Solar Vacuum Telescope with Multi-Channel Facility of UPSO, Nainital, India

V. K. Verma

Uttar Pradesh State Observatory, Manora Peak, Nainital-263 129, India

Abstract. The proposed Solar Vacuum Telescope (SVT) at UPSO will have an equatorial mount and will be equipped with (a) an evacuated 60 cm (f/15) telescope, (b) an evacuated 35-cm refractor telescope and (c) two 15-cm refractor telescopes . The 60-cm (f/15) telescope will have a Gregorian mount, with four light beam channels to install four filters (6563 Å, 3933 Å, continuum at 5500 Å, and tunable filter in range of 4200 – 6000 Å, respectively). The 35-cm (focal length 900 cm) refractor vacuum tube telescope system will have two Lyot filters at wavelengths 5324 (passband $\sim$ 0.15 A) and 4861 A (passband $\sim$ 0.12 A), and KD*P crystals for simultaneous estimation of solar magnetic field. One of the 15-cm refractor telescopes will be equipped with full disk imaging in four differnt modes: 6563 Å, white light, 10830 imaging, and for a solar guider. The other 15-cm (f/15) refractor will be equipped with H-alpha (3 A) prominence monitor. The SVT of UPSO, NainiTal, India has sought funding from Government of Uttar Pradesh, India and about Rupees 10 crores (about US $ 2.5 million) has been allocated during coming 3 years.

At the image plane of each system, a CCD camera and processing computers will record the solar images. This telescope is mainly planned for solar activity observations.

1. Introduction

The Sun is the only star that can be observed with sufficient spatial, temporal and spectral resolutions, to investigate and understand physical structures and processes occurring in various regions of its atmosphere. Consequently, there are a number of important astrophysical problems which can be studied best or exclusively on the Sun.

Solar magnetic fields whose details are not understood require a major advance in the spatial resolution of magnetograms. It is now generally recognized that most or all aspects of solar activity - plasma heating, flares, acceleration of non thermal particles, high speed wind stream and other mass ejection, radio, hard X-ray and gamma ray emission are intimately related to magnetic fields.

The most violent phenomena are the solar flares which are still poorly understood. Radio and X-ray emissions from solar flares point to streams of relativistic electrons which call for an impulsive acceleration mechanism. The Hα emission line (6563 Å) provides spatial information very accurately to study time of electron flight information which is crucial for determining the energy release process in solar flares. Kampher and Magun(1983), and others have shown that there is a component of the Hα emission that shows impulsive variation and is coincident with microwave spikes. Keeping this problem in mind we have developed and tested a CCD Camera system for observing Hα images 2×2 arcminute^2 of the solar flares with a time resolution about 25 msec(Verma et al, 1996).

The motion of the filaments as eruptive prominences can be seen most clearly near the limb or in the lower corona. The study of prominence evolution has a great importance in understanding the energy storage and triggering mechanism.

To summarize, many of the processes occur in magnetic structures of a very small horizontal extent. Most recently, sunspot flux tubes of sizes 0.3 arcsecond or 250 km on the Sun have been observed by Swedish Telescope from Canary Island (Spain). Further, the observations obtained by solar optical/vacuum telescopes with the help of spectrographs, polarimeters and birefringent filters are important to understand the solar activity phenomena. The requirements for studying the above problems are:
(a) high spatial resolution
(b)sufficient resolution in height, i.e. observations in several spectral regions simultaneously
(c) adequate time coverage for following the evolution of a particular feature
(d) accurate polarization observations for magnetic field measurements.

To study above problems, one should have a solar optical telescope of apertures 50-cm to 100-cm and apporpriate backend instruments to understand the above mentioned problems. There are number of ground based solar telescopes having apertures ranging between 30 cm to 150 cm. A good account of the existing ground based solar telescopes is given by Stix (1989).

2. Scientific Objectives of Solar Vacuum Telescope

The principal scientific objective of SVT is to study the active Sun. Investigations of sunspots, faculae, plages, flares and prominences using photographic, photoelectric, CCD and magnetographic techniques with high temporal and spatial resolutions is planned with this telescope. These studies help us in identifying such changes which lead to flare eruptions. Also, observations revealing stretching, shearing and twisting of the magnetic field lines, due to increasing or decreasing magnetic energy content of the region can be studied. The scientific objectives of the programme are therefore as follows:

1. Registration of preflare, during-flare and post-flare changes during the course of evolution of flare.

2. Registration of the magnetic changes in spots with emphasis on such changes which lead to flare eruption. This will help us in understanding the physical mechanisms responsible for magnetic field changes in active regions and their effect on flare production.

3. To study the nature of the magnetic field structures in which the acceleration and transport of particles occur.

4. To study in detail the characteristics of microflares and their contribution to coronal heating.

5. To study in detail the eruptive prominences and their mechanisms.

3. Technical Specifications of the Proposed SVT

As mentioned in the preceding sections, our basic aim is to carry out detailed study of solar active phenomena namely, flares, prominences and structure of the sunspots etc. To achieve diffaraction limited 0.2 arcsec spatial resolution, the aperture of the proposed SVT should be at least 60 cm. In Fig. 1 we have shown the diagrams of the 60 cm SVT with multi-channel facility.

The proposed SVT has a fork type equatorial mounting, on which four optical tubes are set up; a 60-cm Gregorian tube, a 35-cm refractive tube for local region observations and two 15-cm refractive tubes for whole disk observations. The few mechanical details of 60-cm SVT with multi-channel facility are as follows:

- The 60-cm tube is mounted in the middle of the fork mounting.
- The 35-cm tube is firmly fixed with the 60-cm tube outside one fork.
- Two 15-cm tubes are mounted on a small fork firmly fixed with declination shaft outside another fork (with driving mechanism), and can be moved anti-synchronously against the main tube.
- Both 60-cm Gregorian and 35-cm refractive tubes can be vacuumed by a mechanical pump. The vacuum will be less than 10 mm of Hg for 10 days after evacuating.
- The main drive of the 60-cm SVT should in step of 0.1 arcsec so that the pointing accuracy og the telescope should accurate upto 0.1 arcsec.

The outline of optical specifications of 60-cm SVT which includes 60-cm vacuum Gregorian, 35-cm vacuum refractor tube and two 15-cm refractor tubes will be described ahead. The 60-cm (f/15) Gregorian vacuum tube with four channels to install four filters viz.6563 Å, 3933 Å, 5500 Å, and a tunable filter 4200 - 6000 Å. The optical ray diagram of the 60-cm vacuum Gregorian is shown in Fig. 2.

The main feature of the 60-cm vacuum Gregorian are as under:

- 60 cm f/2.5 free aperture primary mirror made of Zerodur.
- Two mirror Gregorian system with $0.2''$ angular resolution on $5'$ FOV.
- The effective focal length of 9 metres. The final images with a diameter of 8.5 mm corresponds to a FOV of $195''$.

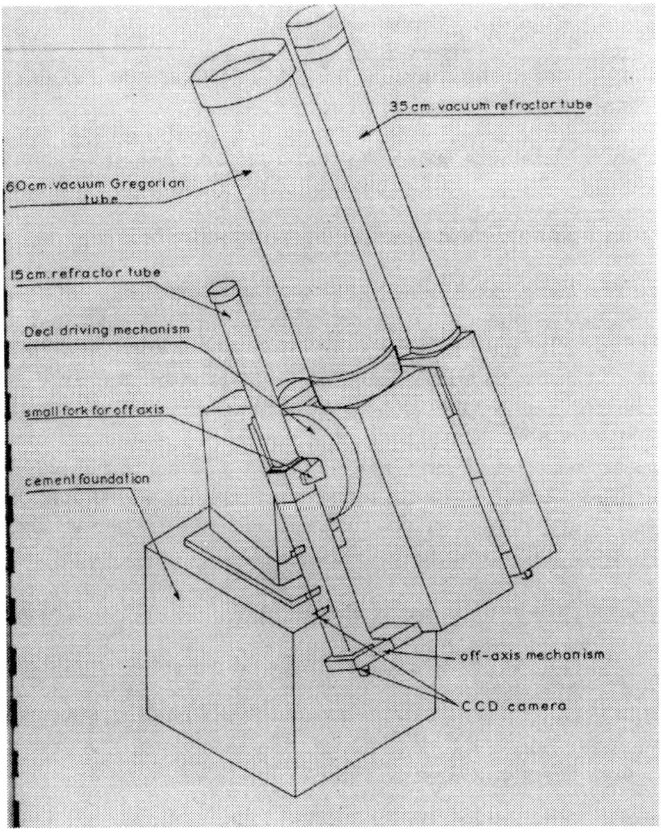

Figure 1. The schematic diagram of 60-cm SVT with multi-channel facility.

- Beam splitters would be dichroic or polarized prisms for access to the following wavelengths: 3933Å, 6563Å, 5500Åand a tunable filter of wavelength ranges 4200 -6000 Å.

The optical ray diagram of the 35-cm vacuum refractor telescope is shown in Fig 3.

The main feature of the 35-cm vacuum refractor are as under:

- A 38 cm entrance window and a F/8 doublet objective with a free aperture of 35 cm,
- An inclined metal diaphragm of $4' \times 6'$ at the focus for reflecting the unused sunlight out of the tube.
- A beam splitter two KD*P modulators and two Lyot filter will be used to estimate simultaneous vector magnetograms.

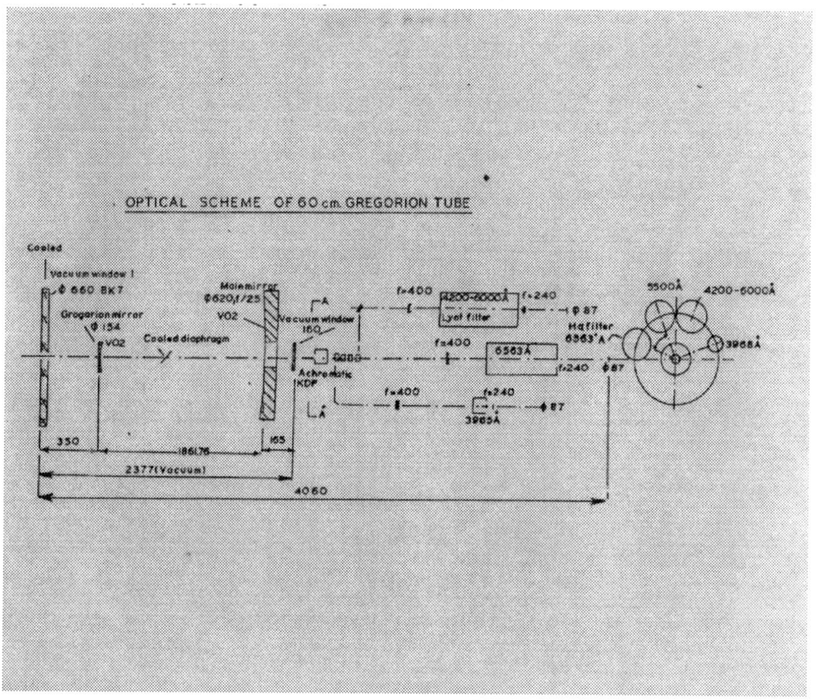

Figure 2. Optical ray diagram of 60-cm Gregory vacuum of SVT.

- The measurement of stokes parameter (I, Q, U, V) and radial velocities at photospheric and chromospheric height will be made simultaneously.

The optical ray diagram of two 15-cm refractor telescope are shown in Fig. 4.

The Ist 15-cm refractor telescope may have the four channels to installfour backend instruments and IInd will be equipped with prominence monitor. The specification of two 15-cm telescope are as under:

1. The 1st 15-cm refractor telescope may have a f/12 doublet objective with a clear aperture of 15 cm and with three beam splitters and relay optics providing to give four output to install four instruments.

2. Full-disk solar $H\alpha$ image of 10 mm dia. on a CCD using a $H\alpha$ Day Star filter.

3. Full-disk white light solar image of 10 mm dia. on a CCD.

4. Full-disk 10830 Å(Lyot filter) solar image of 28 mm dia. on a CCD.

5. 42 mm solar image monitored by two CCDs aimed on image limb at R.A. and Decl. directions.

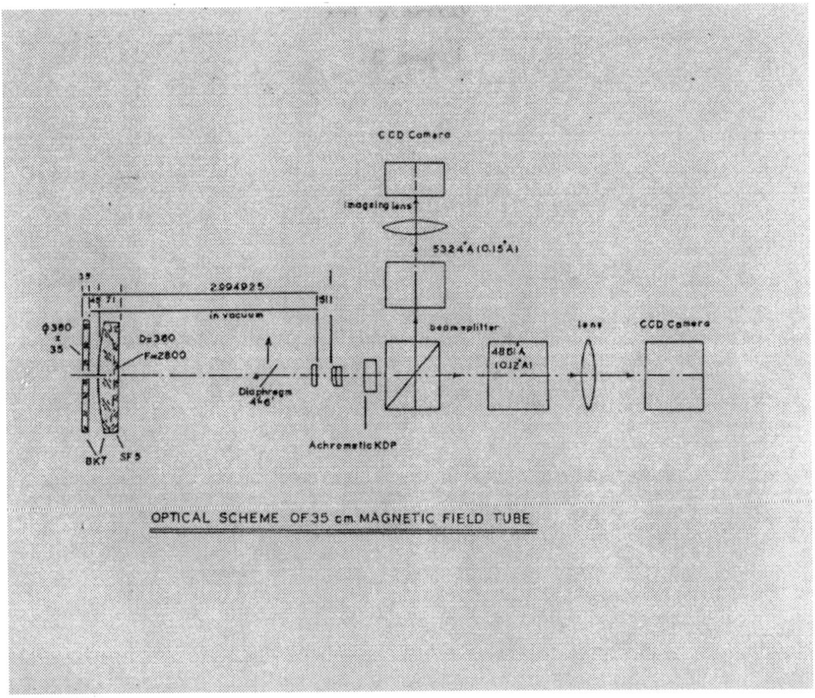

Figure 3. Shows the optical ray diagram of 35-cm vacuum refractor of SVT.

6. The IInd 15-cm f/15 ref ractive telescope will be used for prominence observations and backend instruments for the same will be installed.

The specifications of the 60-cm SVT with multi-channel facility are preliminary and may modified after extensive discussions, provided the suggestions are for improvement of the overall performence and output of science from the proposed telescope.

4. CCD Camera and Computer Systems

A total number of ten CCD systems and similar number of Sun Sparc computer will be required to be used at various focal planes of the SVT. A global position system (GPS) clock is essentially required to record time with atleast 1ms accuracy in header of each frame recorded through the CCD camera. The specifications of the CCD systems are given below:

1. *-CCD having 512 x 512 pixels, square pixel size of 15 μ 16-bit A/D convertor and a read out rate of 1-4 mega pixels/sec (6 Nos.)

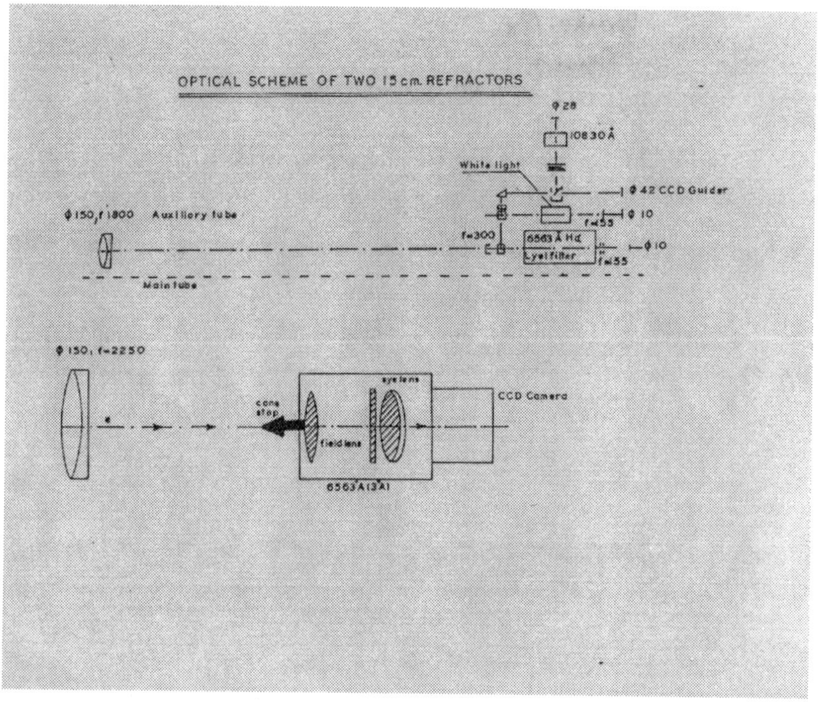

Figure 4. Shows the optical ray diagram of two 15-cm refractors of SVT.

2. *-CCD having 512 x 512 pixels, square pixel size of 24 μ, 16-bit A/D convertor and a read out rate of 1-4 mega pixels/sec (2 Nos.)

3. *-CCD having 2048 x 2048 pixels, square pixel size of 15 μ, 16-bit A/D convertor and a read out rate of 0.5-1.0 mega pixels/sec (1 No.)

4. *-CCD having 2048 x 2048 pixels, square pixel size of 24 μ, 16-bit A/D convertor and a read out rate of 1 mega pixels/sec (1 No.)

5. Building, Tower and Dome

The building for the SVT will be like a tower having a height of about 20 metres. The outline of building is shown in Fig. 5 and it should be RCC structure. At the top of the tower, a circular space having a diameter of 8 metre should be availble and at the centre of this, the SVT will be installed. The diameter of the motarized dome should be 8 metres and it should move freely on rails. The rail of the dome will be fixed on RCC wall of height about 2 metres above the top of the tower building. The dome must have a slit of 1.5 metre width through which telescope will point to the Sun. The temperature inside the dome will be

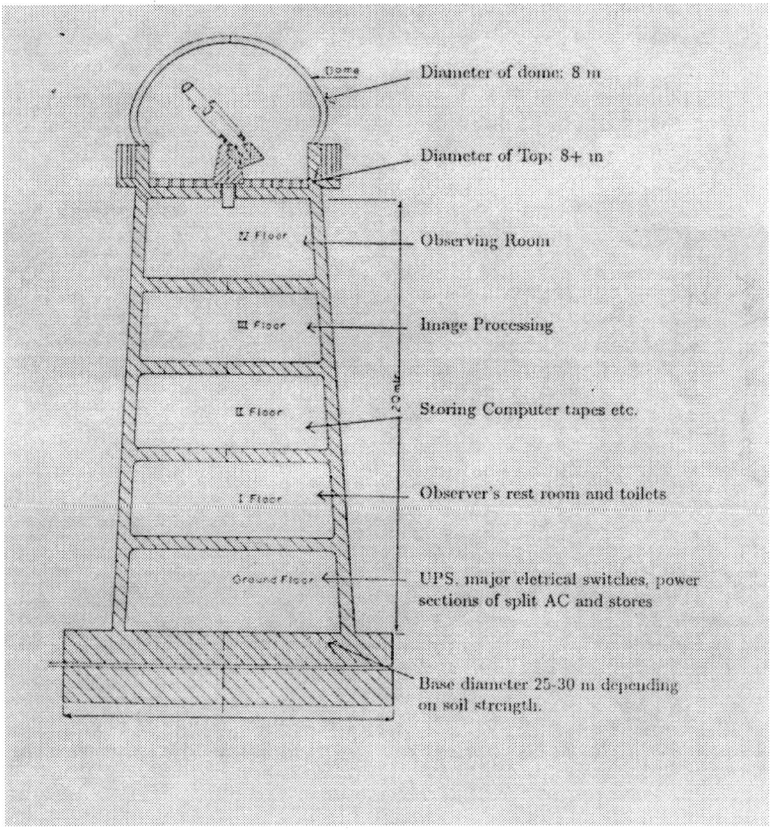

Figure 5. Shows the schematic diagram of tower and dome of 60-cm SVT.

maintained at -5 to -10 degree C less than the temperature outside the dome by a split AC system. This will help in stablizing temperature around the dome and the dome of the 60-cm SVT will not work as heat source. In addition to a staircase, tower building of the SVT must have a lift to carry a load upto 1000 Kg. The tower building of the SVT should have five stories and the base of tower building may have diameter between 25 to 30 metres depending on the soil strength. The ground floor of the tower may be used for uninterupted power supply, major electrical switches and power section of the split AC, and stores. Ist floor may be used for observer's rest room and toilets. IInd floor may be used for storing computer tapes etc. IIIrd floor may be used for image processing and IVth floor may be used for observing room.

References

Kampher, N., & Magun, A. 1983, ApJ, 274, 910.
Stix, M. 1989, The Sun; An Introduction, Springer-Verlag, Berlin, 56.
Verma, V. K., Uddin, W., & Gaur, V. P. 1996, J. Korean Astron. Soc., 29, S391.

Group Discussion

Yuanyong : Because of the low lande factor (g) of FEI 5324, it might be better (and would be possible to my knowledge) to choose some other magnetic-sensitive lines such as FEI 525O, etc.

Verma : I thank you for your suggestion and I will keep it in mind while finalizing the wavelengths selection for the magnetograph.

Keller : The 35-cm telescope for magnetic field measurements has a lot of glass in the aperture. A rough estimate gives a temperature induced retardation of about 40degrees for a 1k temperature gradient. I would advise you to minimize the thickness of the glass and choose glasses with small coefficients of thermal expansion and stress-optical coefficient.

Verma : Thank you Dr. Keller for your suggestion and we will keep it in mind when we go for final optical design selection of glass.

High Resolution Solar Physics: Theory, Observations, and Techniques
ASP Conference Series, Vol. 183, 1999
T.R. Rimmele, K. S. Balasubramaniam, R.R. Radick, eds.

Solar Adaptive Optics at the National Solar Observatory

T. R. Rimmele, R.B. Dunn, K. Richards

National Solar Observatory, Sunspot, NM-88349, USA

R. R. Radick

Air Force Research Laboratory, Sunspot, NM-88349, USA

Abstract. The National Solar Observatory at Sacramento Peak is developing adaptive optics (AO) for solar astronomy. The NSO AO program has recently achieved a major milestone in its effort to implement a low order adaptive optics system that corrects approximately 20 spatial modes. The control loop was closed for the first time on the bench, and the loop was again closed with solar granulation and small pores as the wavefront sensing targets during the system's first tests at the R.B. Dunn Solar Telescope (DST). We present the system design, give a status report of the project and show results from the first engineering run at the DST.

1. Introduction

This workshop has clearly demonstrated that high resolution observations of the Sun have become increasingly important for solving many of the outstanding problems in solar physics (see e.g. the keynote address by Thomas 1999, this volume) In order to study the physics of magnetic flux elements, or solar fine structure in general, spectroscopy and polarimetry of the fine structure is required. For exposure times of order 1 second, the resolution currently achieved in spectroscopic/polarimetric data on a more or less regular basis is typically 0.75 - 1.0 arcsec, which is insufficient to effectively study solar fine structure. The NSO is in the process of implementing adaptive optics at the DST, which will allow spectroscopy at or close to diffraction limit of 0."14 (500 nm), thereby increasing the scientific productivity of this high resolution facility.

Further motivation for the development of solar adaptive optics is provided by numerical MHD simulations, some of which were presented during this meeting. These simulations predict structure well below the resolution limit of current solar telescopes (Rast 1999, Sanchez-Almeida 1999, this volume). Observations of solar structures below the resolution limit of about 0."2 of current solar telescopes are needed to study the important physical processes that occur on such small scales.

A new generation solar telescope with a substantially increased aperture is needed. The LEST project is pursuing a 2.4-3m aperture telescope. NSO is planning to build an Advanced Solar Telescope with an aperture between 3

- 4m (Beckers 1999, Keller 1999, this volume) Plans to upgrade existing facilities such as the Swedish Vacuum Solar Telescope were also presented during this workshop (Scharmer 1999, this volume). Adaptive optics is essential to achieve consistent spatial resolution of $\leq 0.''1$ from ground based observatories and therefore is an essential component to all these new telescope projects. It is interesting to note that in nightime astronomy new exciting observational results are produced at large aperture telescopes equipped with adaptive optics. In many cases these ground based observations are either complementing observations from space (Hubble Space Telescope) or are crucial in interpreting the space based observations.

A first success with solar adaptive optics was achieved with the Lockheed segmented mirror system operated at the DST (Acton & Smithon 1992, Acton & Dunn 1993). The Lockheed system used a Shack-Hartmann wavefront sensor (WFS) based on an array of quad-cells. The quad-cell approach limits the system to use on solar pores and therefore severely restricts its scientific applications.

At NSO/SP we are currently developing solar adaptive optics for the DST. The system is based on a correlating Shack-Hartmann system, which uses arbitrary scenes, including solar granulation, as targets for wavefront sensing. As a first step the NSO is in the process of implementing a low order adaptive optics system at the DST. In the following we present the design of the low order AO system and describe some of the recent achievements of the NSO/SP adaptive optics program.

2. Low-order Adaptive Optics System Design

We have designed and are now in the process of implementing a low order AO system which will be capable of correcting of order 20 spatial modes. The design was driven by the following philosophy:

- parallel processing: The SH-WFS design is ideally suited for a parallel processing approach.
- one processor per channel: This results in a simple and clean architecture and allows the system to be scaled up to a higher order AO system.
- use of commercial components: The system consists mostly of commercially available off-the-shelf components. Expensive hardware development is avoided.

The aperture of the DST is 76 cm; the median seeing at the site is $r_0(500nm)$ = 8.7 cm (Brandt et al. 1987). The bulk of the scientific work in solar physics is done at visible wavelength, although observations in the near infrared (1-2 μ) are becoming increasingly important. This means that a AO system operating at the DST in median seeing conditions and in the visible requires of order 80-100 degrees of freedom. Since a point source is in general not available on the solar surface the wavefront sensor has to be able to sense the wavefront using solar granulation, a low-contrast extended source, as target. Several different approaches for solar wavefront sensors have been discussed, including curvature sensing (Kupke, Roddier & Mikey 1994, Fienup, Paxman & Thelen 1998) and

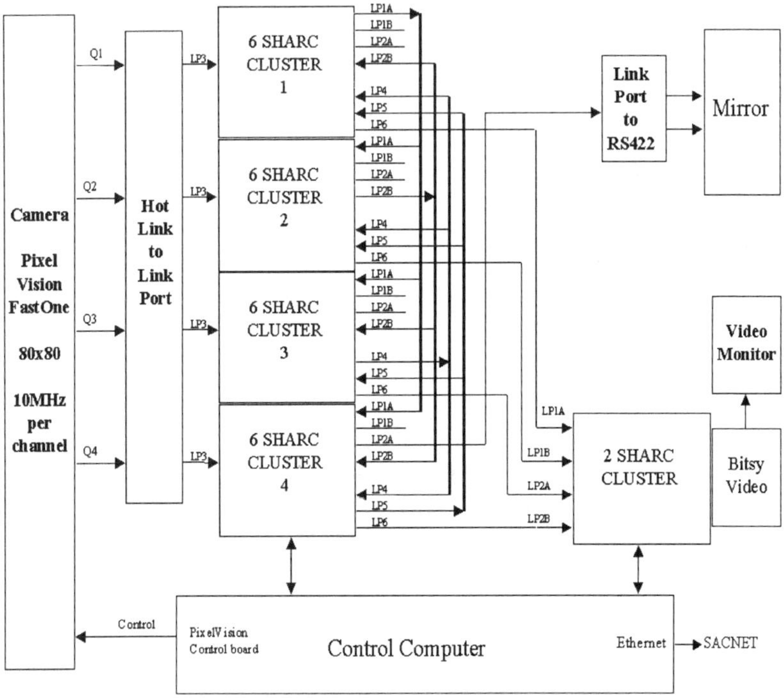

Figure 1. Hardware block diagram of the NSO low order solar adaptive optics system.

phase-diversity (Paxman, Thelen, & Seldin 1994, Paxman et al.1996, Loefdahl 1998, Scharmer & Blomberg 1999, this volume). The NSO previously pursued a wavefront sensing technique (v.d. Lühe 1988) that can be understood in terms of a generalized Foucault knife edge test, where a mask adapted to solar granulation is programmed on a LCD-display and placed in an image plane. However, our tests at the DST show that this approach suffers from severe S/N problems. Based on results of tests and simulations and our extensive practical experience with the correlating SH-WFS, and given the recent advances in digital processing that now makes this approach feasible for AO, we decided to use a correlating SH-WFS in our AO system.

Since the system will operate in day-time seeing conditions and at visible wavelength, a high system bandwidth is required. Temporal power spectra for the first few Zernike modes measured at the DST during typical seeing conditions using a correlating SH-WFS indicate that a system bandwidth of about 200 Hz or higher is desirable. To achieve this kind of bandwidth an update rate of $\geq 2kHz$ is required.

The NSO adaptive optics system consists of the following main components:

- A 97 actuator deformable mirror built by Xinetics Inc. (Ealey & Wellman 1994). The PMN actuators are operated in a stable temperature environment of about 20°C. Under these conditions the PMN actuators provide low ($\leq 1\%$) hysteresis and good linearity.

- A reconstructor based on commercial off-the-shelf DSPs

- A correlating Shack-Hartmann wavefront sensor, the principle of which has been described in detail by e.g. Rimmele & Radick 1998.

A hardware block diagram of the low order AO system is shown in Fig. 1. The detector is a 80x80 quad read-out CCD camera built by PixelVision. Inc. The camera is read out at a 10 Mhz rate/channel. At this rate the read noise is about 20-30e, which is much higher than the 3e read noise required for night time applications. However, since our solar WFS works with broad-band light, on the order of 10^5 photons can be easily collected within 0.5 ms and the photon noise dominates the read-noise. The basic processing unit is a SHARC DSP (AD21060) from Analog Devices. The processors are arranged in clusters of six DSPs which are mounted on a PCI bus board. The processor boards are off-the-shelf items. The data from the camera is transmitted to the processor units via the processor's LinkPort interface. The latency usually involved in transmitting data across the PCI bus is avoided. Some minor in-house custom design work is required in order to interface the camera outputs (HotLink) to the DSP processors (LinkPort). All processors are inter-connected via HotLink ports; data can be easily exchanged between processors. A separate board with two SHARC DSPs is used to perform support functions such as displaying Shack-Hartmann images, cross-correlations, and tilt-vector maps. These functions can be performed without interrupting or slowing the actual processing of Shack-Hartmann images.

The system is flexible enough so different wavefront sensor geometries can be implemented. We are currently working with a wavefront sensor with 24 subapertures. Our initial experiments, results of which are shown below, were carried out with a 800 Hz frame rate 128x128 CCD camera from DALSA Inc. The relatively slow read-out speed of this camera limits the system bandwidth to about 25-30 Hz. The update rate of the final system will be 2 kHz; we hope to achieve a closed loop bandwidth of 150-200 Hz The reconstruction algorithm currently used is based on a simple Gauss reconstructor. The size of the matrices involved in the reconstruction is relatively small and the reconstruction can be performed using the same DSPs used to compute the tilt vector maps.

Due to the parallel architecture expanding the system to a higher order AO system ($\approx$ 80 spatial modes) should be possible by adding more processors for processing the additional subapertures and a larger format CCD camera with more read-out channels. The system bandwidth is more or less limited by the update rate and servo delay achieved for a single channel.

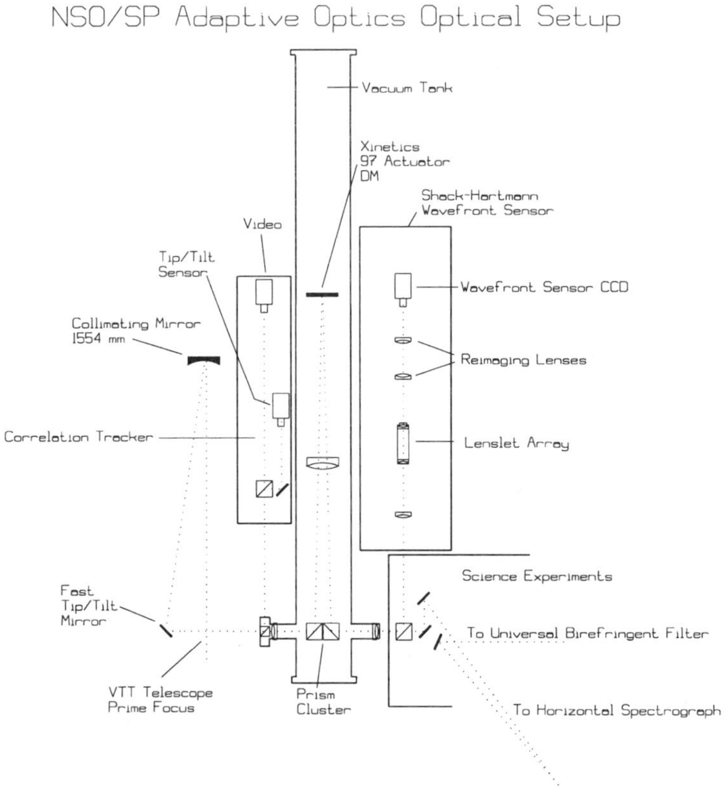

Figure 2. Optical layout of the NSO adaptive optics system at the DST.

3. Optical System

The optical layout used at the DST the adaptive optics experiments is shown in Fig. 2; it has been adapted from an earlier design (Dunn et al. 1992). A separate image motion compensation system provides tip/tilt corrected images to the adaptive optics system. Tip/tilt correction is achieved using a Mark II version of the correlation tracker described by v.d. Lühe et al. 1989, Rimmele et al. 1991). Both correlation tracker and the AO operate with broadband (10nm FWHM) visible light at about 500nm and only a few percent ($\leq 5\%$) of the light is directed to the correlation tracker and SH-WFS. A mirror directs the light to the DM where a 70 mm diameter image of the telescope's 760 mm entrance window is formed. A second mirror redirects the light onto an optical bench, where the science experiments are performed. Several post-focus instruments, such as a high resolution spectrograph, narrow-band Lyot- and/or Fabry-Perot filters, and CCD-detectors can be fed with the corrected beam. A Dyson interferometer is part of the optical setup and allows for real-time monitoring the DM's surface.

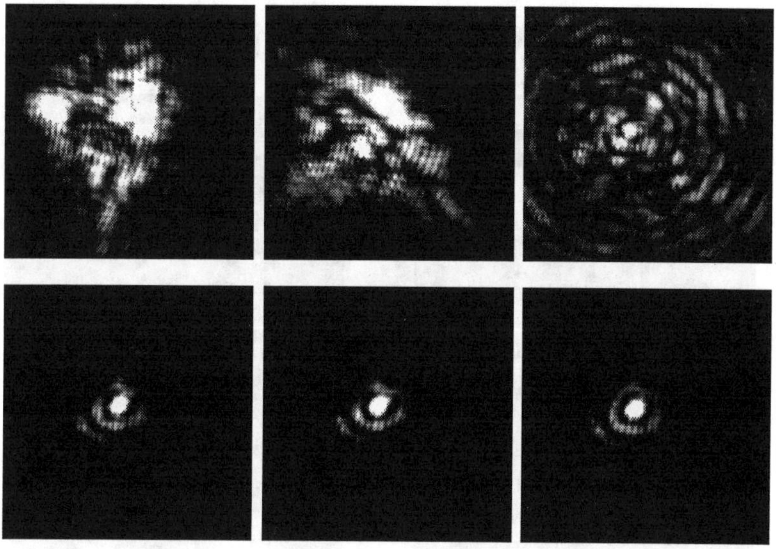

Figure 3. Examples of un-corrected and corrected images of a laser point source. The aberrations that were introduced in this example were Z5 (Astigmatism), Z5 (Ast45), and a fifth order spherical term (Z22).

4. First Results

At the time of the workshop we presented results from the first bench tests of the low-order AO system. The system was first tested using a laser point source as the wavefront sensing target, but the full cross-correlation algorithm was applied. Aberrations were introduced by either deforming the adaptive mirror before the servo loop was turned on or by using a rotating phase plate and heat sources to create time-varying aberrations. The AO system was able to correct aberrations up to about Zernike mode 20. Fig. 3 shows three examples of un-corrected and corrected images of the laser point source. The first diffraction ring of the Airy disk is clearly visible in the corrected images. Similar tests using extended targets imaged with a slide projector were also performed successfully.

A second major milestone was achieved shortly after. The loop was closed with solar granulation and small pores as the wavefront sensing targets during the system's first tests at the DST in early November 1998. For this initial observing run the servo loop was closed at an update rate of 800 Hz and a lag of more than 2.0 milliseconds, resulting in a system bandwidth that is inadequate to fully compensate the atmosphere except under the best seeing conditions. The loop bandwidth is currently limited by the frame rate and read-out speed of the wavefront sensor camera and to a lesser extent by the the processing hardware. These bottlenecks will be alleviated in the near future through the

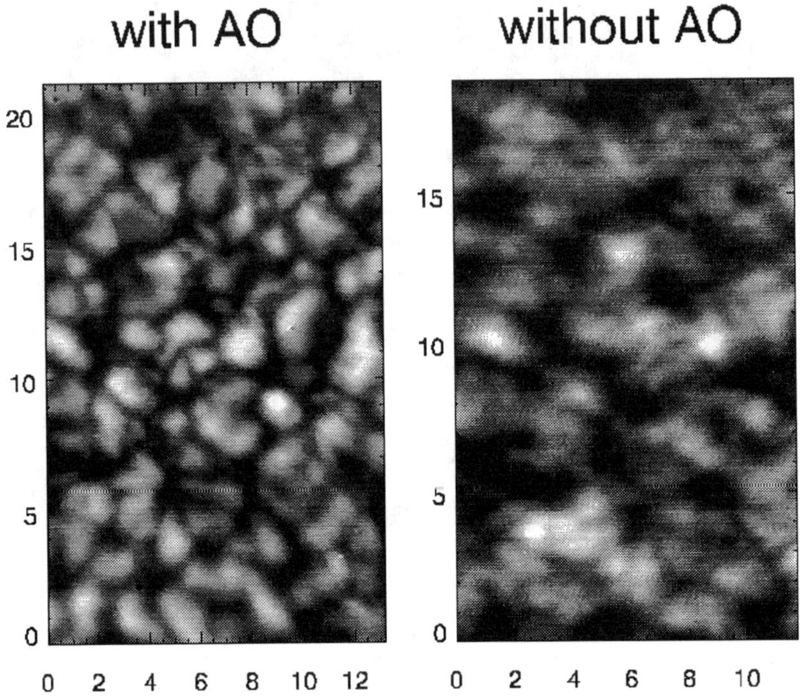

Figure 4. Example of corrected and uncorrected image of granulation at 500nm.

purchase of a faster camera and vendor improvements to the hardware. The small number of subapertures presents another performance limitation.

In the face of these difficulties, the partial atmospheric compensation that the system achieved under good, but not exceptional, seeing conditions at the DST was genuinely exciting: improved and more stable resolution and a reduction of residual jitter were both clearly evident in the corrected image. Fig. 4 shows an example of simultaneous corrected and uncorrected images of granulation at 500nm. These images were recorded using two synchronized video cameras and split-screen electronics. The image scale and the field of view for the two images are slightly different. The images were digitized from video tape using a frame grabber. The correction is slightly better in the center of the field-of-view, giving an indication of the size of the isoplanatic patch. However, a substantial improvement in image quality can be seen over the entire FOV, indicating the presence of a substantial amount of local turbulence close to the telescope aperture during these observations. More images and movies are available on the internet at www.sunspot.noao.edu/AOWEB.

5. Acknowledgments

We are grateful to Bill Hull for his technical support in developing the NSO adaptive optics system and to the observers at the DST for their excellent support during the engineering run.

References

Acton, D. S. & Dunn, R. B., 1993, Proc. SPIE , 1920, 348

Acton, D. S. & Smithson, R. C. , 1992, Applied Optics, 31, 3161

Beckers, J.M., 1999, "The Advanced Solar Telescope. II Technical Aspects", this volume

Brandt, P. N. , Mauter, H. A. , & Smartt, R.N., 1987, A&A, 188, 163

Dunn, R. B. ,Streander, G. W. , Hull, W., & Wilkins, L., 1992, Proc. SPIE, 1543, 88

Ealey, M. A. & Wellman, J. A. , 1994, Proc. SPIE , 2201, 680

Fienup, J. R. , Paxman, R. G., & Thelen, B. J. , 1998, Proc. SPIE, 3353, 930

Keller, C.U. , 1999, "The Advanced Solar Telescope.I Scientific Rational", this volume

Kupke, R., Roddier, F. J., & Mickey, D., 1994, Proc. SPIE, 2201, 519

Loefdahl, M. G., 1998, Proc. SPIE, 3353, p. 952

von Der Luehe, O., 1988, Optical Engineering, 27, 1078

von der Luehe,O., Widener, A.L., Rimmele,T. , Spence, G., & Dunn, R. B. , 1989, A&A, 224, 351

Paxman, R. G. , Seldin, J. H. , Loefdahl, M. G. , Scharmer, G. B. , & Keller, C. U. , 1996, ApJ, 466, 1087

Paxman, R. G. , Thelen, B. J. , & Seldin, J. H. , 1994 , Proc. SPIE, 2201, 1066

Rast, M. P., 1999, "Thermal Starting Plumes: Dynamics, Stability, and the Excitation of Solar Acoustic Oscillatons", this volume

Rimmele, T. & Radick, R.R., 1998, Proc. SPIE, 3353, 72

Rimmele, T., von der Luhe, O., Wiborg, P. H. , Widener, A. L. , Dunn, R. B. , & Spence, G., 1991, Proc. SPIE , 1542, 186

Sanchez-Almeida, J., 1999, "The Spectrum of Fluctuations Across Penumbral Filaments", this volume

Scharmer, G. B., & Blomberg, H.L., 1999, "Optimized Shack-Hartmann Wavefront Sensing for Adaptive Optics Systems", this volume

Scharmer, G.B., Ower-Petersen, M., & Korhonen, T., 1999, "The New Swedish Solar Telescope", this volume

Thomas, J.H., 1999, "The Case for High Resolution in Solar Physics", this volume

Group Discussion

Mathew : What is the time delay between the correction and sampling of the wavefront, is that a limitation?

Rimmele : The servo delay and the sampling frequency determine the system bandwidth. Our current AO system has a sampling rate of 800Hz and and an overall servo delay of about 2 ms resulting in a system bandwidth of roughly 25-30Hz. The bandwidth is limited mostly by the slow read-out rate of the Dalsa camera currently used. A faster camera will be implemented at a later stage. Our target is a system bandwidth of 200HZ.

Seldin: Given that the AO system requires that photons are diverted from science instruments, what is the impact on (photon-limited) scientific observations?

Rimmele : The AO operates with broadband light and will only receive a few percent of the available photons. v.d.Luhe comments that the optical passband of the system can be tuned on a per-observation basis.

High Resolution Solar Physics: Theory, Observations, and Techniques
ASP Conference Series, Vol. 183, 1999
T.R. Rimmele, K. S. Balasubramaniam, R.R. Radick, eds.

Correlation Tracking and Adaptive Optics Control Using Off-The-Shelf Workstation Technology

M. Shand

Compaq Systems Research Center, 130 Lytton Av., Palo Alto, CA 94301, USA

G. B. Scharmer, W. Wei

Royal Swedish Academy of Sciences, Research Station for Astrophysics, Stockholm Observatory, S-133 36 Saltsjöbaden, Sweden

Abstract. The microprocessors used in off-the-shelf workstations double in performance every eighteen months. The Swedish Vacuum Solar Tower (SVST) uses off-the-shelf workstations for all aspects of its on-line telescope control and data acquisition. Since 1995 workstation performance has been adequate for a correlation tracker of solar granulation controlling a tip-tilt corrector. In 1998 workstation performance permits the construction of a 19 subimage Shack-Hartmann wavefront sensor for adaptive optics and phase diversity based wavefront sensing is becoming feasible. It is argued that workstations provide a cost-effective, upgradable, low-risk and flexible means of instrument construction for solar adaptive optics systems and solar instrumentation in general.

1. Introduction

The Swedish Vacuum Solar Telescope (SVST) in La Palma is widely recognized for its high spatial resolution and excellent image quality. Initially used by a small Swedish community, there has been an increasing demand of observing time from an international community over the past ten years. Associated with this potential for high-resolution imaging, there has been a demand also for a variety of instruments to support such observations. Initial efforts to develop these instruments were heterogeneous and based on a variety of more or less custom made systems. It soon became evident that the staff and budget of the SVST were unable to support and develop this instrumentation further and that a drastically different approach was needed if such efforts were to be successful. During this period, workstations were in increasing use by staff at the SVST. Such machines were becoming ever more powerful and it became natural to investigate whether such technology could be used also for real-time control of instrumentation and for recording data, since this would reduce the requirement for training and maintenance know-how to a minimum.

At the same time the research laboratories of Digital Equipment Corporation (now part of Compaq Computer) were investigating the use of reconfigurable interface and coprocessor boards based on field programmable gate arrays (FPGAs). These boards provide a means of interfacing workstations to high bandwidth external devices that is particularly cost effective in application that are low volume or whose requirements are ill-defined or evolving. Digital and SVST began a collaboration in 1993 that has led to the construction of a number of control and data acquisition devices, all built around the same basic components of off-the-shelf workstations and reconfigurable interface boards.

These instruments include:

- A telescope pointing control system.
- A large frame CCD image acquisition system with on-line frame selection and continuous image preview (Shand 1995).
- A solar polarimetry system based on ferroelectric liquid crystals, built by the Instituto de Astrofisica de Canarias (Martinez Pillet et al. 1999, this volume).
- A solar polarimetry system based on two electrically tunable retarders with a single CCD time-multiplexed to receive image data across multiple polarization states (Shand & Moll 1998).
- A low latency correlation tracker for active tip-tilt mirror control and image stabilization (Shand, Wei, & Scharmer 1995).

Of the devices constructed so far, the most computationally demanding has been the correlation tracker. This device is also the basis for a more ambitious device, a full Adaptive Optics (AO) system controlling a deformable mirror. The AO system is crucial if the SVST is to realize the potential gains in spatial resolution from its upgrade to a larger telescope with 95cm aperture. Initially the SVST AO system will be targeted to a 19 electrode mirror, correcting the first 15 Zernike aberrations, but it is built to scale to mirrors with larger numbers of electrodes as more powerful microprocessors become available at the time of installation of the new telescope.

This paper describes the computational aspects of the current correlation tracker and plans for the AO system. Section two provides motivation for the use of general purpose microprocessors for the correlation tracker computation and compares them to custom developed hardware and specialized Digital Signal Processors (DSP). Section three discusses the existing correlation tracker, the sum of absolute differences algorithm used in the inner loop of its computation and how, due to steady improvements in general purpose microprocessors, its performance has increased since it was first deployed. Section four looks at the 19 subimage Shack-Hartmann wavefront sensor we are currently constructing. Section five examines the performance of general purpose microprocessors on FFT computations and considers the prospects for real-time phase diversity wavefront sensing in a future AO system.

2. Comparison of General Purpose Microprocessors, DSPs, and Custom Digital Hardware Implementations

Moore's Law has accurately predicted the increase in silicon density of the last 25 years and looks likely to continue to do so for at least the next 10 years. Every 18 months the number of transistors that can be packed into a given area of silicon doubles. This relentless progress of the underlying manufacturing technology has led to ever more powerful semiconductor devices whose computational power is growing exponentially over time.

Moore's law has several consequences for any data acquisition or control system based on a fixed choice of semiconductor technology:

- The system soon becomes obsolete with respect to a system based on newer semiconductor devices. Where development times are longer than a few months, the underlying technology in a system may already be nearing obsolescence before initial deployment.

- Fixed technology systems tend to be expensive or uneconomic to upgrade.

- As the constituent parts become obsolete they may become difficult to obtain, thereby affecting the maintainability of the system.

To maximize the return on our engineering investment, we want to build data acquisition or control systems that can easily migrate to new semiconductor devices as they become available, both to avoid the pitfalls of fixed technology systems, and to realize the performance or cost advantages the newer devices bring. The limited scope of our engineering resources and the difficulty of building custom hardware that can be easily migrated to new semiconductor devices dictate the use microprocessor based solutions wherever possible. Beyond our compute performance needs, the properties we require in our hardware platform are portability, high performance I/O and scalability.

2.1. DSPs and portability

DSPs are a form of microprocessor that offer an internal organization and specialized instruction set primitives targeted toward the operations that arise in digital signal processing and real-time control. As such they may seem the natural choice for any *processor* based data acquisition or control system. However DSP solutions entail many of the disadvantages of custom hardware development that lead to early obsolescence with no upgrade path to new technology. To maximize performance, DSPs tend to expose the internal architecture of the processor to a much greater extent than does a general purpose microprocessor. As a consequence, it is difficult to write efficient compilers from languages like C to DSPs and so they are often programmed in assembly language. Furthermore, there is limited compatibility in program code of successive generations of DSPs due to the internal architecture changes that accompany new generations.

General purpose microprocessor on the other hand are careful to maintain compatibility between successive generations at the binary machine code level and between different architectural families through the provision of high-level language optimizing compilers which allow efficient applications to be written in a highly portable manner. Meanwhile advances in computer architecture such as caches and superscalar pipelined execution units have rendered general purpose microprocessors as efficient per clock tick as DSPs while the clock speed improvements of general purpose microprocessors have been outpacing DSPs (Stewart et. al. 1992).

2.2. High Performance I/O Mechanisms and Real-Time Scheduling

The key feature of DSPs which general purpose processors typically lack is flexible high-speed low-latency I/O facilities for direct interfacing with external instruments. For SVST these facilities are provided by Compaq's *Pamette* reconfigurable interface technology (Compaq 1998).

The Pamette technology collects several FPGAs (field programmable gate arrays) together on a I/O expansion card and provides a runtime environment for applications to configure and control these FPGAs. Using FPGAs for computation—reconfigurable computing—has recently become a popular idea. Certainly a number of image processing tasks could be done on the FPGAs, but we resist this temptation. By keeping the computation in the host microprocessor, our systems get faster with each new workstation release and they are simpler to develop and enhance. The FPGAs' role is restricted to marshaling data and meeting the most stringent of our real-time requirements.

For less demanding real-time scheduling, modern workstation operating systems provide the ability to assign real-time process priorities and agressive scheduling policies. On fast microprocessors these facilities are adequate for sub millisecond response.

2.3. Scalability

It is important to make the distinction between systems that must scale to deliver increased performance in the future versus systems that must scale to deliver a range of performances in a fixed technology. In the former case semiconductor progress will often provide the necessary performance increases. Moreover, much of the development of such systems can be carried out on hardware with less performance than is required in the operational system, in anticipation of newer, more powerful hardware that we can predict will become available before the scheduled deployment of our system. In the latter case systems must be constructed to allow a variable number of physical devices to be applied to the computation. Much care is required in the design of such systems if communication between cooperating devices is not to become a bottleneck. Fortunately, our scalability needs are almost always of *future* scalability.

3. Existing Correlation Tracker

The existing correlation tracker for active mirror control and image stabilization (Shand, Wei, & Scharmer 1995) became operational in August 1994 and entered routine use in the 1995 observing season. This system receives a continuous stream of 128 × 128 images from a high frame rate (800Hz) Dalsa CCD and computes the most probable shift between successive images. The shift is used to drive the servo controls of an active mirror that compensates for perceived image motion due to seeing. The comparison between successive images is done in the workstation Alpha microprocessor (Sites 1992). The system can operate in cross correlation mode for tracking on the solar granulation pattern, or in quad cell mode with large pores or sunspots. Sub-pixel resolution is achieved by quadratic interpolation of the discrete correlation function on a 3x3 grid centered on the point of maximum correlation.

```
// V is an 8 element vector held in a 64-bit word, v[i] is the ith element of V
// R = reference image = r[7], r[6], .. r[0]       0 <= r[i] < 128
// B = new image       = b[7], b[6], .. b[0]       0 <= b[i] < 128
// R is preprocessed to be biased by 128
// A = a[7], a[6], .. a[0]      a[i] = r[i]+128, 128 <= a[i] < 256
C = A - B                     // c[i] = 128+r[i]-b[i], 0 < c[i] < 256
H = C & 0x8080808080808080    // extract top bits of bytes
X = H ^ 0x8080808080808080    // x[i] = 128 IFF r[i]-b[i] < 0
L = X >> 7                    // l[i] = 1 IFF r[i]-b[i] < 0
Y = X - L                     // y[i] = 127 if r[i]-b[i] < 0, else 0
Z = H | Y                     // z[i] = 127 if r[i]-b[i] < 0, else 128
C = C ^ Z                     // if r[i]-b[i] < 0, one's compl. else sub 128
C += L                        // add one to one's complemented bytes to form
                              // two's complement.  Now c[i] = abs(r[i]-b[i])
C += C >> 8                   // add odd bytes to even bytes, can guarantee
                              // no overflow since abs(r[i]-b[i]) < 128
C &= 0x00ff00ff00ff00ff       // clear odd bytes, this accommodates
                              // overflow on subsequent adds
C += C >> 16                  // add c[0] += c[2], c[4] += c[6]
C += C >> 32                  // add c[0] += c[4]
                              // Bits 0..15 of C = sum(abs(a[i]-b[i])) : i=0..7
```

Figure 1. 8 byte parallel sum of absolute differences

The correlation tracker is particularly demanding from a computational viewpoint. When it first became operational the fastest workstation available was barely adequate for this application and the algorithm demanded careful implementation. Rather than treating byte-wide pixels individually, wherever possible we pack eight pixels into a 64-bit word and treat them in parallel in the 64-bit wide Alpha CPU integer datapath. Cross-correlation compares various translations of an incoming image against a reference image to determine the translation that yields the closest possible match. The comparison of incoming pixels against the corresponding reference pixels is done using a sum of absolute differences. Computing absolute values normally uses conditional operations on each data element however a branchless sequence of unconditional boolean and arithmetic instructions can compute the same function. Applying this unconditional sequence to several data elements in parallel more than offsets its extra cost in basic operations. For the SVST cross-correlation implementation the input pixels are scaled to seven bit quantities to accommodate a guard bit on each pixel. The resulting algorithm is illustrated in pseudocode in Fig. 1.

When the correlation tracker first became available it took 2 ms to compute all possible shifts up to 4 pixels of a 24×24 pixel window. With the passage of time, new more powerful workstations have become available which have allowed the system to be enhanced, with very little engineering effort, to treat all possible shifts up to 6 pixels of a 48×48 pixel window while reducing the total computation time to below 1 ms.

The latency of the tip-tilt system currently installed at SVST is totally dominated by the 20-25 ms response time of the DC motors which drive the tip-tilt mechanism and limit the overall servo bandwidth to low frequencies. The total latency from the CCD and computations is approxmately 2.7 ms (0.6 ms from exposure, 1.2 ms from read-out, 0.9 ms from computations).

4. Shack-Hartmann Wavefront Sensor

The Shack-Hartmann wavefront sensor consists of a field-lens to re-image the entrance pupil of the telescope onto an array of micro-lenses. Each micro-lens gives a sub-image of the target seen through that micro-lens. The displacement of each such sub-images gives a measure of the average gradient of the wavefront over the corresponding micro-lens, which allows the wavefront to be reconstructed. When used with extended targets, such as solar fine structure, the Shack-Hartmann wavefront sensor requires cross-correlations to estimate the image displacements. Present performance of the cross-correlation computation is such that cross-correlations applied to subimages from a Shack-Hartmann wavefront sensor can process all possible shifts up to 3 pixels of a 16×16 pixel window of 19 simultaneous subimages on a 600 MHz Alpha 21164 based workstation in 0.9 ms. This time includes gain and dark correction of the live images and the matrix multiplications required to convert the computed x, y positions to aberrations and the aberrations to voltages. Using techniques described in Shand et al. (1995) we can overlap most of these computations with the receipt of the incoming image so that the computations finish 0.25 ms after the CCD image is fully delivered to workstation memory. This is adequate to enable adaptive optics control with a deformable mirror using solar granulation as a target for wavefront sensing.

Shack-Hartmann wavefront sensors for solar telescopes larger than the SVST requires a larger format CCD than our existing correlation tracker. In collaboration with the Kiepenheuer-Institut für Sonnenphysik (KIS), we have interfaced a Reticon 256×256 by 1 kHz CCD to our Pamette interfaces and have been able to sustain continuous transfer of 64 MBytes/second of image data directly to microprocessor memory. In this collaboration, one of the KIS contributions has been to develop the hardware needed to output the voltages that control the high-voltage amplifiers of the adaptive mirrors. The SVST uses a bimorph 19-electrode mirror from Laplacian Optics whereas the KIS system currently uses the 19-segment adaptive mirror developed at Lockheed-Martin. The software is developed in order to allow arbitrary types of adaptive mirrors and it is expected that at least one of the adaptive mirrors will be tested at more than one solar telescope.

4.1. Multimedia Instruction Set Extensions

In response to the growing importance of image and video processing in general purpose microprocessors, most major microprocessor architectures have added *multimedia* extensions. These extensions are dedicated instructions to perform parallel manipulation of multiple small integer data items (typically pixels), packed into single machine words. Of particular interest in our context is the new Alpha pixel error instruction (PERR). This instruction computes the sum of absolute differences of eight pixels presented in two 64-bit input operands—exactly like our 15 instruction sequence in Fig. 1, except that PERR operates on a full eight bits per pixel and takes just one machine cycle for the full sum of absolute differences calculation over 8 pixels. Overall, the use of PERR offers a one-off performance boost of about 5 times for the Shack-Hartmann computation—equivalent to a little more than 3 years worth of the regular improvements we can expect from Moore's law. Current compilers do not in general generate code that uses multimedia extensions since they are incapable of automatically recognizing situations where these primitives can be used. So, today, the use of primitives such as PERR does reduce the portability of our code, but these primitives can be embedded seamlessly in C code to resemble function calls and we can easily revert to fully portable code (with some loss in performance) if the need arises subsequently.

For the basic 19 subimage Shack-Hartmann sensor described above we do not rely on PERR. Its existence gives us the latitude to increase the bandwidth of the system or the number of subimages such as is required by adaptive optics systems on the next larger Swedish solar telescope. The total latency of the present system is approximately 2.7 ms when not using PERR instructions or overlap of computations and read-out. Even without these two mechanisms, the total latency is dominated by the CCD for the present system.

5. Prospects for Real-Time Phase Diversity

The growth we have seen in correlation tracker performance is due to increases in microprocessor integer performance, but microprocessor floating point performance has been growing at a similar rate. On a 600 MHz Alpha 21164 we can perform a 64×64 FFT in 0.31 ms. If further refinements can be achieved in the phase diversity method proposed by Scharmer (1999, this volume), current floating point performance may be adequate to build a prototype phase diversity based wavefront sensor. We are actively pursuing this possibility.

Meanwhile the next generation Alpha microprocessor (Alpha 21264) offers at least 60% more floating point performance than the current Alpha 21164 at comparable clock speed, and it has been announced that Alpha 21264 derivatives running at 1 GHz will be available by the year 2000. So, even without further algorithmic advances, real-time phase diversity based wavefront sensing will soon be in reach of general purpose microprocessors. At present, the major question is not computational speed but how to establish initial lock with the system when aberrations are large.

6. Discussion

Since 1993, all new instruments at SVST have been built around a common platform of off-the-shelf workstation and reconfigurable interface board. In accordance with Moore's law, successive workstation models have exhibited steady performance improvements such that today our correlation tracker code running on the latest model workstation can perform Shack-Hartmann wavefront sensing on 19 subimages where the same code, when first operational on the fastest workstation of that time, was only fast enough to control a simple tip-tilt corrector. We have been able to exploit these performance improvements because newer workstations are compatible with preceding models in terms of operating system and programming language support. Although we must update workstations to increase performance the application code we have engineered remains largely unchanged.

To date, all our work has been based on the Alpha microprocessor primarily because Alpha has been the consistently highest performing microprocessor since its introduction in 1992. We have found Alpha's 64-bit integer datapath to be valuable in treating several pixels in parallel (see for example Fig. 1), however most major microprocessor will complete their transition to 64-bits in the coming years. Although we expect to continue to use Alpha, nothing in our approach is particularly tied to this processor. Our code, based on the C language and POSIX standards, could be ported without undue effort to a slower microprocessor with the only cost being some loss in performance.

At the present time, the best choice of wavefront sensor for solar telescopes is not obvious. We are proceeding with a Shack-Hartmann sensor, but are at the same time investigating phase diversity based approaches. By basing our wavefront sensor on a general purpose microprocessor with good overall performance, the same platform will be able to implement either technique. We may find that phase diversity, while working in principle, requires processing of 128 $\times$ 128 pixel images and is therefore not currently capable of adequate real-time operation. In such a case, we will be able to use the existing system to investigate performance limitations and, using Moore's law, predict with a fair degree of certainty when a microprocessor will be available that will allow real-time operation. When that microprocessor does become available, it will be a simple matter to upgrade to it with our already tested phase diversity code.

References

Compaq 1998, http://www.research.digital.com/SRC/pamette

Martinez Pillet, V., Collados, M., Sanchez Almeida, J., Gonzalez, V., Cruz-Lopez, A., Manesau, A., Joven, E., Paez, E., Diaz, J. J., Feeney, O., Sanchez, V., Scharmer, G. B., & Soltau, D. 1999, this volume

Scharmer, G. B. 1999, this volume

Shand, M. 1995, IEEE Workshop on FPGAs for Custom Computing Machines, (eds P. Athanas and K. Pocek), IEEE Computer Society Press, Los Alamitos, CA, ISBN 0-8186-7086-X

Shand, M., Wei, W., & G. B. Scharmer 1995, SPIE Proceedings Volume 2706-16, (ed. John Schewel)

Shand, M. & Moll, L. 1998 IEEE Workshop on FPGAs for Custom Computing Machines, (eds K. Pocek and J. M. Arnold), IEEE Computer Society Press, Los Alamitos, CA

Sites, R. L. 1992, Digital Technical Journal, Vol. 4, No. 4.

Stewart, L. C., Payne, A. C., & Levergood, T. 1992, CRL Technical Report 92-10, http://www.crl.research.digital.com/

High Resolution Solar Physics: Theory, Observations, and Techniques
ASP Conference Series, Vol. 183, 1999
T.R. Rimmele, K. S. Balasubramaniam, R.R. Radick, eds.

Optimized Shack–Hartmann Wavefront Sensing for Adaptive Optics and Post Processing

Göran B. Scharmer, Henrik L. A. Blomberg

Royal Swedish Academy of Sciences, Stockholm Observatory, SE–133 36 Saltsjöbaden, Sweden

Abstract. Shack–Hartmann wavefront sensors are frequently used for control of adaptive optics to compensate atmospheric seeing and in post-processing to restore aberrated images. In this work, it is pointed out that equations can be derived which allow errors from cross-talk with arbitrarily many Zernike aberrations and from random noise in the measured positions of the sub-images to be minimized in a statistical (average) sense, assuming that the average variances of the atmospheric aberrations are known. We give the equation which determines the "optimum" matrix needed to estimate Zernike coefficients from measured displacements of micro-lens images. This matrix minimizes the total wavefront error by properly balancing errors from cross-talk, noise and amplitudes for the estimated Zernikes.

We have calculated such matrices for various configurations with 7–44 micro-lenses and optimized for cross-talk with more than 300 Zernike aberrations. Monte Carlo simulations are used to verify the superior performance of this approach compared to methods which rely on singular-value decomposition methods.

1. Introduction

The Shack–Hartmann wavefront sensor is probably the most common type of wavefront sensor for night-time and solar telescopes. Its main drawback is that it measures average gradients of the wavefront over sub-areas of the pupil, corresponding to the location of the micro-lenses, rather than the wavefront more directly. This makes the Shack–Hartmann wavefront sensor sensitive to cross-talk with higher-order aberrations. Algorithms to process Shack–Hartmann data frequently employ least-squares fit methods to determine wavefronts (e.g. Takato, Iye, & Yamagachi 1994, Dai 1994), usually expressed in Zernike coefficients. Due to the strong cross-talk with higher-order Zernikes, such methods tend to give large residual wavefront errors and allow only a relatively small number of Zernike coefficients to be reliably determined.

Methods based on least-squares fit of measured x, y positions to a set of aberration coefficients ignore that known statistics of atmospheric Zernike coefficients can be used to optimize the wavefront sensor algorithm, including the effects of cross-talk with aberrations not corrected by the AO system or post-processing algorithm. The first work to consider optimal wavefront estimation and correction in a unified manner, taking into account the effects of mirror geometry, noise and wavefront statistics, appears to be that of Wallner (1983). In this paper, we present a simplified approach and demonstrate that the residual wavefront errors in a Shack–Hartmann wavefront sensor can be strongly reduced by optimizing with respect to seeing and with respect to random errors in the measured relative positions of the sub-images. In Section 2, we give the optimum matrix for estimating aberration coefficients and in Sections 3 and 4 we describe simulations and results of simulations which have been made in order to verify the performance of the method.

2. Optimization of Wavefront Estimates

Let us write:

$$x_m = \sum_{k=4}^{\infty} A_{mk} \cdot c_k, \qquad m = 1, \ldots, 2M, \tag{1}$$

where x_m is the vector of x, y positions of all M micro-lens images, and c_k are the true (exact) aberration coefficients. Ideally, c_k should be Karhunen–Loeve coefficients, but Zernikes should work nearly as well. The matrix A_{mk} contains the average wavefront gradients per unit Zernike for each micro-lens m and all aberrations k and is assumed known.

We wish to find a pseudo-inverse matrix B_{nm} such that we can minimize the RMS error in the measured wavefront, i.e.

$$\tilde{c}_n = \sum_{m=1}^{2M} B_{nm} \cdot (x_n + N_m), \qquad n = 4, \ldots, Z_{\max}. \tag{2}$$

Here $\tilde{c}_n$ is the estimated Zernike coefficient and B_{nm} is the matrix we wish to determine. N_m is the random error in x, y position at micro-lens m.

We now determine B_{nm} such that the average variance of the error in the measured wavefront is a minimum. This corresponds to minimizing L, where

$$L = \sum_{n=4}^{Z_{\max}} \left\langle (c_n - \tilde{c}_n)^2 \right\rangle. \tag{3}$$

Combining equations (1)–(3) gives

$$L = \sum_{n=4}^{Z_{\max}} \left\langle \left(c_n - \sum_{m=1}^{2M} B_{nm} \cdot \left(\sum_{k=4}^{\infty} A_{mk} \cdot c_k + N_m \right) \right)^2 \right\rangle. \tag{4}$$

This can be rewritten as

$$L = \sum_{n=4}^{Z_{\max}} \left\langle \left(c_n - \sum_{k=4}^{\infty} \beta_{nk} \cdot c_k + \sum_{m=1}^{2M} N_m \cdot B_{nm} \right)^2 \right\rangle,$$

where $\beta_{nk} = \sum_{m=1}^{2M} B_{nm} \cdot A_{mk}$ represents the contribution of c_k to $\tilde{c}_n$.

In evaluating this, we assume that:

1. All N_m are statistically independent and with the same variance $\langle N^2\rangle \equiv \langle N_m^2\rangle$.

2. The covariance between *different* aberrations is zero, $\langle c_n c_k\rangle = 0$ for $n \neq k$. This is true for Karhunen–Loeve polynomials and nearly true for Zernikes.

We obtain

$$L = \sum_{n=4}^{Z_{\max}} \left\{ \langle c_n^2\rangle + \sum_{k=4}^{\infty} \beta_{nk}^2 \cdot \langle c_k^2\rangle - 2\beta_{nn} \cdot \langle c_n^2\rangle + \langle N^2\rangle \cdot \sum_{m=1}^{2M} B_{nm}^2 \right\} \tag{5}$$

By separating out the diagonal terms β_{nn}, we can interpret L:

$$L = \sum_{n=4}^{Z_{\max}} \left\{ (1-\beta_{nn})^2 \langle c_n^2\rangle + \sum_{k\neq n}^{\infty} \beta_{nk}^2 \cdot \langle c_k^2\rangle + \langle N^2\rangle \sum_{m=1}^{2M} B_{nm}^2 \right\} \tag{6}$$

The first term contains contributions from errors in the estimated amplitudes of the $Z_{\max}$ aberrations, the second represents errors due to cross-talk from other aberrations c_k to c_n, the third are errors from random errors in the measured x, y positions. Minimizing the total wavefront error, therefore requires trading these three errors by choosing appropriately the matrix elements B_{nm}.

Inserting the definition of β_{nk} into equation (5), we obtain

$$\begin{aligned} L \;=\; & \sum_{n=4}^{Z_{\max}} \left\{ \langle c_n^2\rangle + \sum_{k=4}^{\infty} \Big(\sum_{m=1}^{2M} B_{nm} \cdot A_{mk} \Big)^2 \cdot \langle c_k^2\rangle + \right. \\ & \left. - 2 \sum_{m=1}^{2M} B_{nm} \cdot A_{mn} \cdot \langle c_n^2\rangle + \langle N^2\rangle \cdot \sum_{m=1}^{2M} B_{nm}^2 \right\} \end{aligned} \tag{7}$$

We now find the matrix elements $B_{n'm'}$ such that L has a minimum, i.e. $\partial L/\partial B_{n'm'} = 0$ for all values of n' and m'. This gives

$$\sum_{k=4}^{\infty} \sum_{m=1}^{2M} B_{n'm} \cdot A_{mk} \cdot A_{m'k} \cdot \langle c_k^2\rangle - A_{m'n'} \cdot \langle c_{n'}^2\rangle + \langle N^2\rangle \cdot B_{n'm'} = 0 \tag{8}$$

Minimizing L is therefore equivalent to minimizing the wavefront variance for each individual Zernike aberration. We define

$$\alpha_{m'm} = \sum_{k=4}^{\infty} A_{m'k} \cdot A_{mk} \cdot \langle c_k^2\rangle, \tag{9}$$

which is the correlation function between the different micro-lenses of atmospheric sub-image movements. This matrix contains only known quantities. We can then rewrite equation (8) as

$$\sum_{m=1}^{2M} B_{n'm} \alpha_{m'm} + \langle N^2\rangle \cdot B_{n'm'} = A_{m'n'} \cdot \langle c_{n'}^2\rangle, \tag{10}$$

which is a linear set of equations determining B_{nm}. This approach minimizes the expected RMS error in the *estimated wavefront*, which is preferable to making least-square fits to the *data* (minimizing the residual errors in the measurements). Note that the matrix determining B_{nm} is the *same* for all n, only the right hand side is different. This matrix is not singular and can be inverted or solved for with standard matrix methods.

The solution to equation (10) depends only on the ratio of $\langle c_n^2 \rangle$ to $\langle N^2 \rangle$. Since the former is proportional to $(D/r_0)^{5/3}$ we can always be near optimum in varying seeing conditions by pre-determining matrices for a small number of values of γ,

$$\gamma = \frac{(D/r_o)^{5/3}}{\langle N^2 \rangle} \tag{11}$$

and using the matrix which best corresponds to the current seeing conditions.

3. Implementation and Simulations

To compare the performance of the proposed method with that of least-squares fits using the singular value decomposition (SVD) method, we have made simulations relevant to the diameter of the Swedish Vacuum Solar Telescope (47.5 cm). To mimic the use of cross-correlations to measure the relative displacement of the sub-images, we have measured average wavefront gradients over each microlens and subtracted the average gradient of a "reference" micro-lens, located at the center of the pupil when possible. The assumption that cross-correlations measure average wavefront gradients has not been verified but obviously needs to be done.

We used an SVD routine by Press, Vetterling, Teukolsky, & Flannery (1992). To use that SVD routine requires specification of the ratio between the smallest and largest non-rejected singular value. We refer to this ratio as the SVD parameter. This parameter was adjusted by trial and error for each micro-lens configuration so that the obtained RMS wavefront error was as small as possible.

To mimic seeing, we simply generated random Zernike coefficients and renormalized their amplitudes according to the variances of Noll. This ignores the (small) co-variance between Zernike of the same azimuthal degree but different radial degree, but is consistent with the derivation made above. For more accurate optimization of performance, Karhunen–Loeve polynomials should have been used, but this is expected to be of small importance. In the simulations, we generated random wavefronts including the first 325 Zernikes. We calculated the gradient of the wavefront numerically at each pixel within the pupil and averaged this gradient over the micro-lenses. The average gradients were then scaled to give the relative image motion in arc-sec for each sub-image and random errors of 0.00, 0.02 or 0.05 arc-sec RMS were added. From these sub-image motions, the low-order Zernike coefficients were then estimated using the method presented here and the SVD method described above.

We have used a number of different micro-lens geometries and configurations. Here, we report on results with three different configurations using 7, 19 and 44 micro-lenses, respectively, as shown in Figure 1.

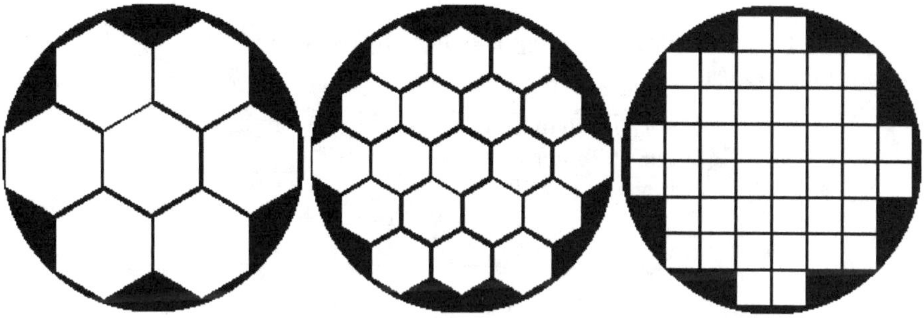

Figure 1. Three micro-lens configurations with 7, 19 & 44 micro-lenses.

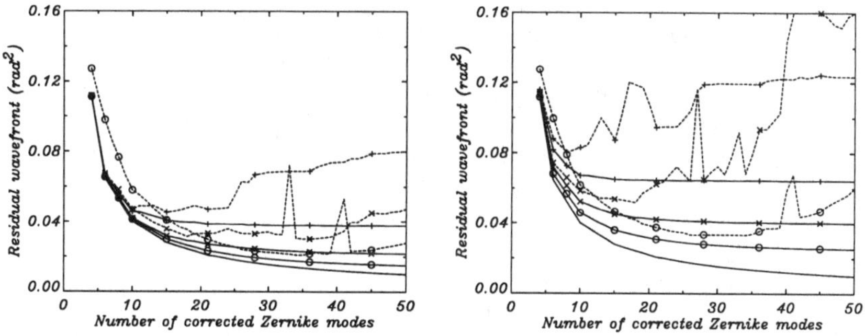

Figure 2. The residual wavefront error for three different micro-lens configurations and assuming $D/r_0 = 1$. Left: residual wavefront variance for zero random error in the measured x, y positions. Right: corresponding curves for 0.″05 RMS error in the measured positions. Full-drawn curves: configurations with 7 (+), 19 (×) and 44 (⊙) micro-lenses and perfect wavefront correction (no symbols) with the optimized method described in the paper. Dashed curves: results with the SVD method using the same configurations.

4. Results and Conclusions

The results are given in Figs. 2–4 and summarized in Table 1. Fig. 2 shows the residual wavefront variance when trying to measure Zernike coefficients up to some order. It is clear from these plots that the optimized matrix gives a smooth behavior and that the wavefront residual always decreases when increasing the number of estimated Zernike coefficients. It is also clear that the optimized matrix consistently gives smaller residuals than with the SVD method which gives a more erratic behavior, even after tuning the SVD parameter. We also conclude that 0.05 arc-sec RMS error in the measured x, y positions gives a significant increase of the wave-front rms in excellent ($D/r_0 = 1$) seeing conditions.

Table 1. Comparing the equivalent number of perfectly corrected Zernike modes.

RMS x, y error	Number of micro-lenses	SVD	Optimized
0	7	9	11
	19	14	20
	44	21	31
0."05	7	5	6
	19	8	10
	44	13	17

Table 1 summarizes the results by giving the equivalent number of perfectly corrected Zernikes. The table shows that we can in principle measure nearly one Zernike coefficient per micro-lens with the optimized method when the random x, y errors are negligible.

In Figure 3, we show the total (top) wavefront variance for the configuration with 19 micro-lenses as dashed curves and the curves corresponding to perfect correction, obtained from Noll (1976), as full-drawn curves. The results with the optimized matrices are shown on the left and those obtained from the SVD method are shown on the right. In the lower two plots, we show the individual contributions to the wavefront variance for the two methods. The contributions are calculated from Eq. (6) and are from the un-corrected Zernikes (long dashes), amplitude errors of the corrected Zernikes (full drawn), cross-talk (short dashes) and noise (dotted). The main difference between the curves of the optimized and SVD matrices are that with the SVD method, the wavefront errors are totally dominated by noise and cross-talk whereas the amplitude errors are zero when less than 25 Zernikes are corrected. With the optimized matrix, the contributions from noise and cross-talk are strongly reduced at the expense of small amplitude errors. When attempting to correct many Zernikes, noise dominates completely with the SVD method whereas with the optimized method, the higher Zernikes are effectively switched off giving (small) amplitude errors as the dominating source of wavefront errors. This is demonstrated more clearly in Fig. 4 which shows the relative amplitudes for each Zernike when trying to determine the first 50 Zernikes with the optimized (left) and SVD (right) methods. For $D/r_0 = 3$, the noise is negligible and the optimized matrix gives maximum gains, limited only by cross-talk. For $D/r_0 = 1$, noise deteriorates the estimates of Zernikes higher than approximately 18 and the optimized matrix reduces these amplitudes accordingly. With the SVD method, there is no information in the inverted matrix on cross-talk with un-corrected Zernikes or the effects of noise. The SVD method underestimates the amplitudes of the lower aberrations and amplifies noise by not reducing the amplitudes of the higher Zernikes to the point of giving useless wavefront estimates.

In excellent seeing conditions, noise dominates the estimates of the higher Zernikes, which are effectively switched off by the optimized matrix. This pre-

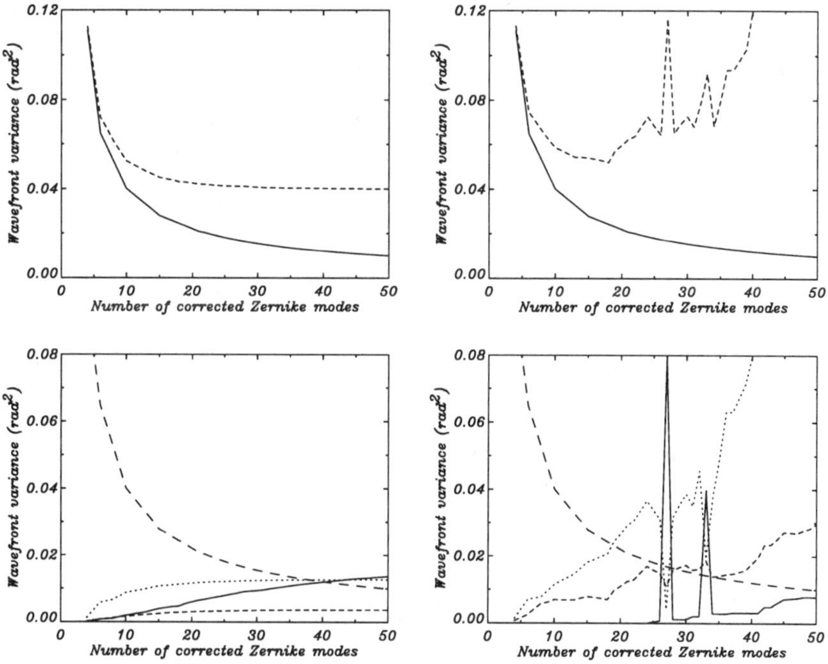

Figure 3. Top plots: residual wavefront variance as function of the number corrected Zernike modes (dashed curves) for the optimized (left) and SVD (right) methods. Full drawn curves: perfect correction of the same number of Zernikes. Bottom plots: individual contributions to the wavefront variance. These are from uncorrected Zernikes (long dashes), amplitude errors of the corrected Zernikes (full drawn), cross-talk from other Zernikes (short dashes) and errors from noise in the measured x, y positions of the micro-lens sub-images (dotted). Calculations were made with 19 micro-lenses, for $D/r_0 = 1$ and assuming $0''\!.05$ random x, y errors.

vents noisy estimates of Zernike coefficients from degrading performance of the AO mirror or the post-processing algorithm in excellent seeing, which is clearly a desirable property of the method.

In conclusion, it appears that the method for estimating Zernike coefficients from Shack–Hartmann wavefront sensors described here is superior to matrix inversion using the SVD method. By taking advantage of known statistics of atmospheric aberrations and by estimating the random errors in the measured positions of the sub-images, it is possible to make optimum wavefront estimates which minimize the errors from cross-talk and noise. The use of an optimized matrix for estimating wavefronts from the Shack–Hartmann wavefront sensor is similar to the use of optimum filters in image processing. The sources of "noise" with the Shack–Hartmann wavefront sensor are both cross-talk from many higher-order aberrations and random errors in the measured positions of

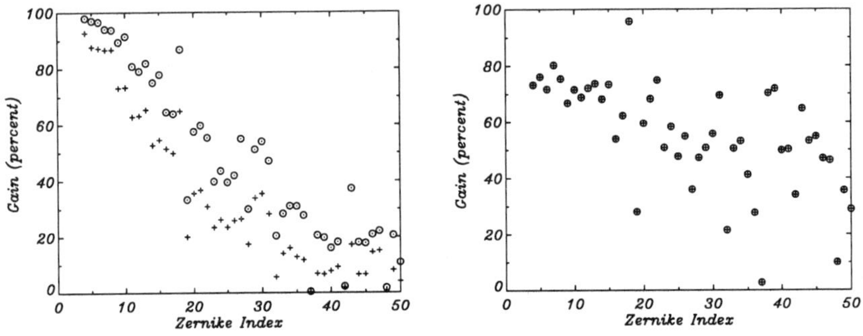

Figure 4. Relative amplitudes for each Zernike mode when trying to estimate 50 Zernikes with 19 micro-lenses. Left: optimized method. Right: SVD method. Calculations were made assuming $0.''05$ RMS errors in the measured x, y positions for $D/r_0 = 1$ ($\odot$) and $D/r_0 = 3$ ($+$).

the sub-images. The calculations presented here, show that these sources of errors can be reduced to reasonable levels.

For real-time applications, there is no difference in the computational effort of using the optimized matrix or that obtained with the SVD method. In both cases the matrices are obtained from calculations which are made off-line and in advance.

In this work, we have assumed that the Shack–Hartmann wavefront sensor measures the average gradient of the wavefront over each micro–lens. For use with extended targets, such as solar fine structure, cross-correlations are used to provide such measurements. We have not investigated various cross-correlation algorithms to verify or modify this assumption, but this clearly needs to be done. We suspect that different cross-correlation algorithms will show very small differences for low-order Zernikes but that important differences will exist for high-order Zernikes which, as we have demonstrated, contribute severe cross-talk. Optimizing the wavefront estimates requires adequate understanding also of cross-talk and errors from the cross-correlation algorithm.

Acknowledgments. Mats Löfdahl is thanked for adapting the SVD routine and giving us access to his subroutine library. Dan Kiselman and Mats Löfdahl are thanked for comments on the manuscript.

References

Dai, G., 1994, in Adaptive Optics in Astronomy (eds. Mark A. Ealey & Fritz Merkle), Proc. SPIE 2201, pp. 562

Noll, R. J., 1976, JOSA, 66, 207

Press, W. H., Vetterling, W. T., Teukolsky, S. A., & Flannery, B. P. 1992, Numerical Recipes in C, (2nd Edition)

Takato, N., Iye, M., & Yamaguchi, I. 1994, PASP, 106, 182

Wallner, E. P., 1983, JOSA, 73, 1771

High Resolution Solar Physics: Theory, Observations, and Techniques
ASP Conference Series, Vol. 183, 1999
T.R. Rimmele, K. S. Balasubramaniam, R.R. Radick, eds.

An Introduction on Chinese Efforts for High Resolution Observations of the Sun

Binxun Ye

Beijing Astronomical Observatory, Chinese Academy of Sciences, Science Park, Beijing 100012, China

Abstract. The efforts for making high resolution solar observations recently in China, such as Space Solar Telescope (SST) project and its test-bed experiments in a Balloon Solar telescope (BST), Large Infrared Solar telescope and its site-testing campaign, Active compensation optics and other endeavors are introduced.

1. History

Back to early 70's, at Kunming where now Yunnan Observatory locates, Prof. Ding Youji got an idea of hunting high resolution solar images under the disturbing atmospheric seeing: he wrapped himself and a Leica camera body with a large black cloth around the telescope focal plane, as a photographer did in old times. By his eye (certainly his brain) to judge the best moments to push down the shutter button, which make 1/2000 sec exposures. He obtained many of very fine resolution images of the sunspots(Ding,1973), a really surprise to the people at that time. One of the best is still hanging on the wall of McMath-Pierce Telescope today.

Based on his idea and the yields later in solar physics, in early 80's people in Yunnan Observatory designed and built a pair of so-called Solar Fine Structure Telescopes(Wu and Li,1983), each of which contains two 26 cm refractors for filtered Hα imaging and wide-band images. Those two Solar Fine Structure Telescopes were stationed in wide longitude separations: one in west China and another in the east coast of the country. They played had very important roles for more than a decade, especially as main working horses in the solar maximums. A significant mount of papers generated from the observing results of them.

Then, there came the fine structure magnetic fields measurements. For almost a decade, Ai and his colleagues had succeed in building a vedio magnetograph of 35cm aperture(Ai,1995). Further developments had been carried on to yield an invention of multi-channel solar magnetograph which can produce nine magnetograms of different wavelengths, hence of the different heights on the Sun.

Those have become the history. But efforts are still continuing for high-resolution observations of the Sun both in the space and the ground.

2. Space Solar Telescope (SST)

SST is the largest space science project ever made in China. The scientific targets it aims : " Through coordinated, high spatial and temporal resolution observations of transient and steady state solar hydrodynamic and magnetohydrodynamic process over 2-D real time polarizing spectrum, UV, X-ray and Hα image, and continuous time evolution to achieve a break through advance in solar physics." (Ai, 1996). Actually, Its 0.1 arcseconds high angular resolution will allow people to reach those critical scientific goal the modern solar physics is longing for, such as :

- Exploring 3-Dimensional structure of vector magnetic fields and velocity fields at flux-tube scale;
- Exploring the fine structures of solar atmospheres, especially the heating mechanism of the chromosphere and corona;
- Studying the build up, storage, triggering and release of energy in solar flare and radio spike;
- Studying of fine evolution of solar active region, especially that of sunspots and prominence;
- Studying of various solar transient phenomena associated with solar terrestrial space environment;
- Providing various parameters serving the purpose of forecasts of solar activity and associated calamities.

The major payload of SST is large optical telescope, the largest solar telescope ever sent to space. The diameter of its primary mirror is one-meter with F-ratio of 3.5:1. Via attached 2-Dimensional Real Time Polarized Spectrograph, eight magnetograms can be obtained simultaneously from the data generated at eight 0.005nm narrow band channels with eight 1024×2048 frame-transfer CCD cameras and operational circuits. The pointing and tracking accuracy of the telescope will be maintained to ±0.02 arcseconds by various measures (including correlation tracker). The telescope can reach 0.1 arcseconds resolution over about 2 arcminutes field of view. The photons from the line center arrive the focal plane range from 2.46×10^4 to 2.41×10^5 per pixel (0.075 arcseconds square) depends the wavelengths. The optical configuration of this main optical telescope and 2-D Real-time Polarized Spectrograph is shown on Fig. 1.

Other payloads on the SST includes:

- Four 15cm EUV telescopes working at 129,180,284 and 304 Åwith 0.5 arcseconds resolution over 9 arcminutes FOV;
- A Coronagraph;
- A Group of spectrographs ranging from soft X-ray, hard X-ray to γ-ray;
- A Full disk Hα & white light telescope;
- A Guiding telescope.

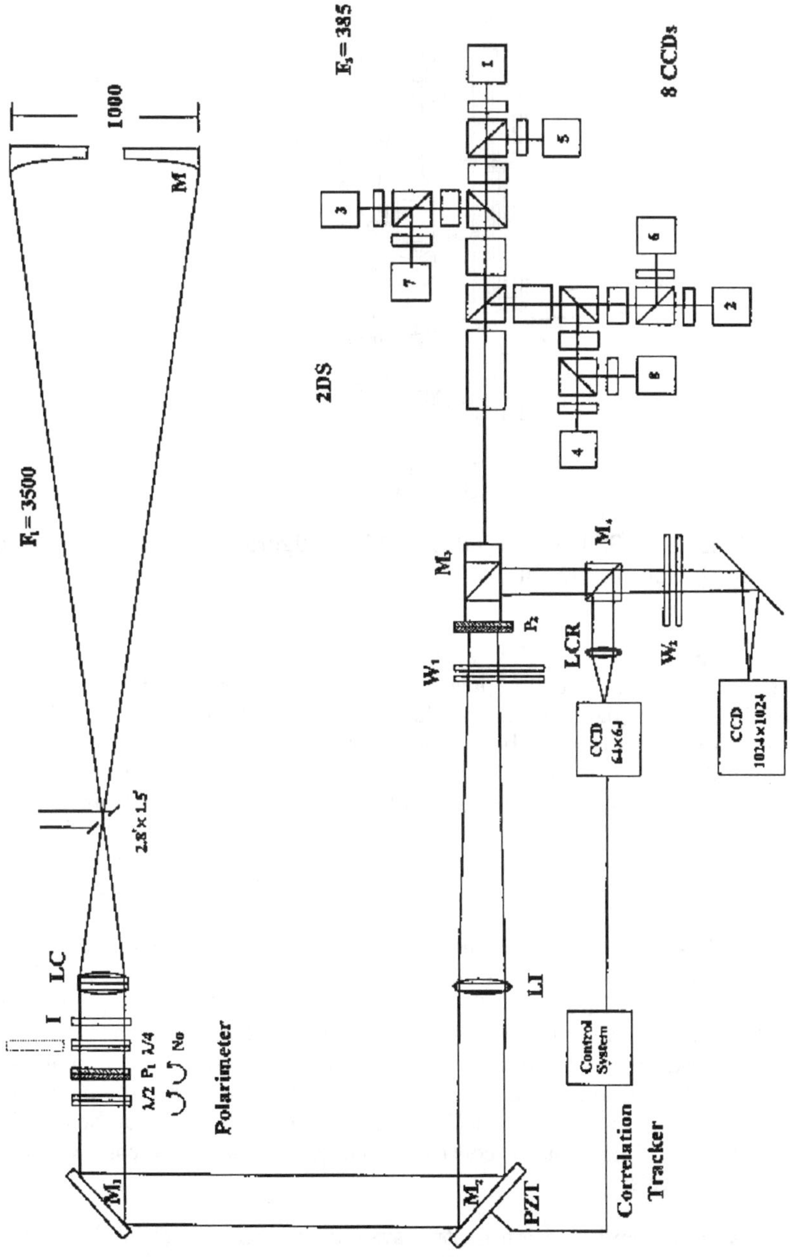

Figure 1. The optical configuration of this main optical telescope (courtesy G. Ai)

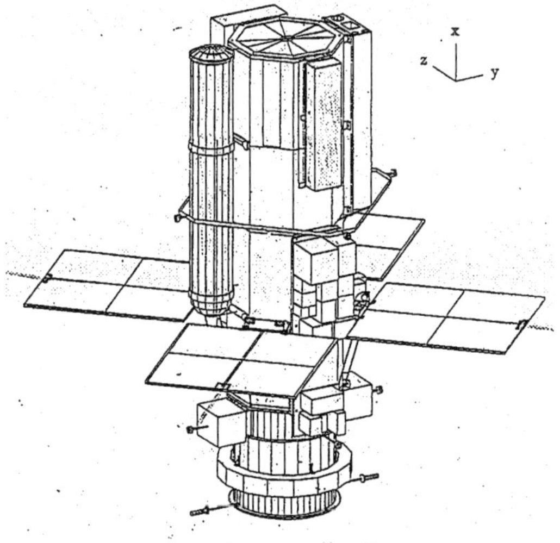

Figure 2. SST Spacecraft on orbit configuration (three dimensional view), Courtesy Q.Hu

The SST satellite (Fig. 2) weighs about 2.5 tons with the size of 5×2×2 meters and consumes 600 w power (800W peak). The satellite will be on a 730 km high round (0.003e) sun synchronous orbit. It is 3- axis stabilized to 12 arcseconds. Data rate is about 3 GB/day. Telemetry will be made via L-band for uplinks and S-band for down links as there will be large mount of image data to be transmitted to the ground station daily. The satellite is scheduled to be launched in 2004 by CZ-4B rocket from the Taiyuan Space flight center , northern China for a 3 years mission life. SST project was conceived in early 90's and began phase A study in 1994 with German colleague's participation in 1995. later in 1997, it entered Phase B schedule. Recently, a balloon solar telescope(Fig. 3) has been built as a test-bed for experimenting SST optics , attitude control, and some other electronics. The Balloon Solar Telescope has almost the same optical configuration as SST except the primary is 80 cm in diameter. The gondola weighs about 1.4 tons with the size of 5×3×2.5 meters. The telescope structure was made by titanium alloy and put in equatorial mounting. The gondola will be lifted by a 300,000 cubic meter balloon to 32,000 meter elevation for a 8 hours flight based on 600W power consuming from the on board Lithium-ion batteries. The attitude control of the gondola and telescope will be made over three levels:

1. Azimuth and rotation control system by controlling CMG (Control Momentum Giro) with 50 N.M.S and decoupling mechanism;

2. Equatorial tracker with linear CCD guider;

3. Correlation tracker by piezoelectric and tip-tilt mirror.

Figure 3. The Balloon Solar Telescope (Courtesy Xiangmin Ye)

Table 1 shows the characteristics of those levels:

Table 1. Characteristics in Three Controlling Levels

LEVELS	SENSITIVITY CONSTANT	TIME ACCURACY	TRACKING RANGE	CONTROL
1	±0.5 arcmin	300 ms	±6 arcmin	±180°
2	± 1 arcsec	100 ms	±12 arcsec	±18 arcmin
3	±0.01 arcsec	1 ms	±0.025 arcsec	±15 arcsec

3. Large Infrared Solar Telescope and Its Site-testing Campaign

Except McMath-Pierce Solar Facility, This Large Infrared Solar Telescope will have the world largest aperture primary mirror for solar observations. Besides the IR capability, this telescope also aims high-resolution observations of the Sun. The optical configuration and building layout (Fig. 4) reminds people the LEST design, a modified Gregorian system with active compensation optics. The primary mirror is an F/2.4 parabolic. The secondary is F/8.8 elliptic. The third is also elliptic with f number 45. Polarize Analyzer is put in the symmetric optics. A multi-channel spectrograph was considered as a first instrument for obtaining Stocks parameters from the spectra. Fig. 6 shows the details of the spectrograph. As a ground-based support facility for the SST, this telescope will work at 0.36 2.50μm for the magnetic and velocity fields with about 0.3 0.5 arcseconds spatial resolution. For this grade of resolution, to find a compatible site for the telescope is a real challenge. A site-testing campaign has been carrying on for several years. Now, Huxian Lake (Fig. 5) is selected for final concentration. The instruments used for evaluation the site characteristics called Daytime-DIMM (Differential Image Motion Monitor), designed and built by Liu Zhong of Yunnan Observatory. The concept is quite new for daytime r_0 measurements but has been shown as a powerful tool in site testing. Fig. 6 gives

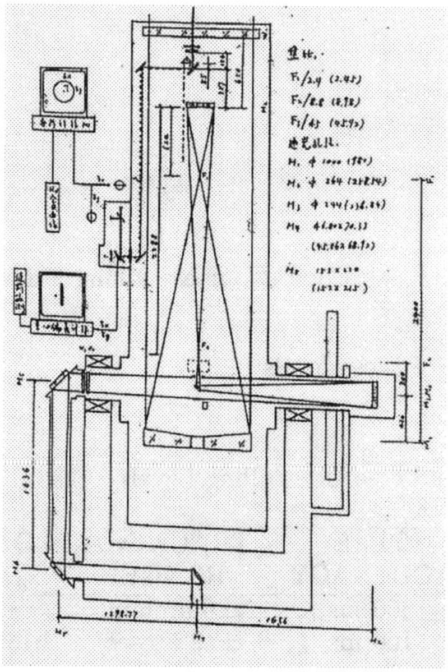

Figure 4. Optical configuration of Large IR Solar Telescope (Courtesy L. Sheng)

Figure 5. Huxian Lake Site (Courtesy Z.Liu)

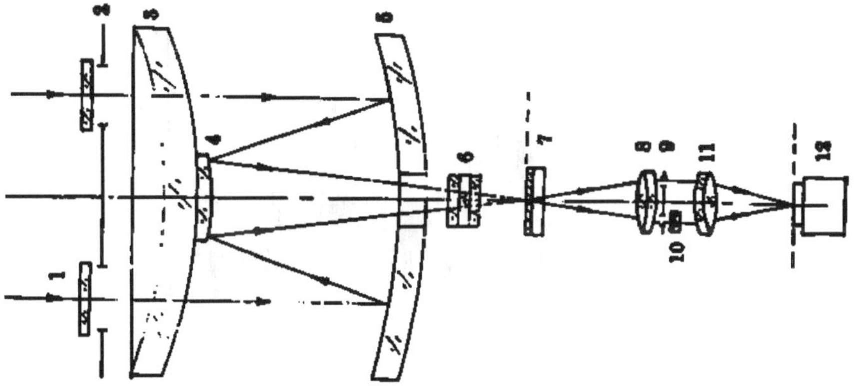

Figure 6. DIMM system configuration

the drawing of the system, where 1 is filters for entrance pupils, 3 5 consists a telescope, 6 is a filter group, 7 is slit, 8 is collimator, 9 is the differential masking, 10 is mask, 11 is imaging lens and 12 is a CCD camera. r_0 is calculated from the equations(Liu,1996):

$$\sigma_l^2 = 2\lambda^2 r_0^{-\frac{5}{3}}\left[0.179D^{-\frac{1}{3}} - 0.097d^{-\frac{1}{3}}\right]$$
$$\sigma_t^2 = 2\lambda^2 r_0^{-\frac{5}{3}}\left[0.179D^{-\frac{1}{3}} - 0.145d^{-\frac{1}{3}}\right]$$

where D and d are the diameter of the sub-apertures and their separation, σ_l^2 and σ_t^2 are the differential image motions measured in longitudinal and transversal directions.

Fig. 7 is a sample of the result from a measurement at the site.

4. Active Compensation Optics and other Endeavors

Besides going to space and better sites on the earth, third effort has been making is doing active compensations in the telescope optics to reduce the effect of earth's atmospheric seeing. Though these kinds of the effort are still in the experimental stage, significant progresses, especially in Correlation Tracking techniques has been made (Deng,1997). Other efforts for high resolution observations should be mentioned are Real Time Shift & Add Magnetic Field Measurements (by Deng, Yuanyong), A plan for using Orthogonal Transfer CCDs for real time correction of inclination of the wavefronts.

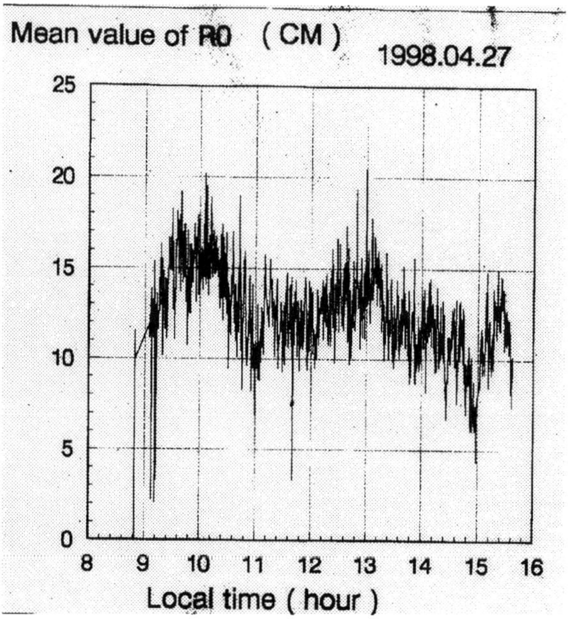

Figure 7. A sample of the result from the measurement taken by Daytime DIMM

References

Ding, Y.J.,1973: Acta Astron. Sinica ,10, 127

Wu,M.C,Li,Z.K,1983: Publication Of Yunnan Obs. No.4

Ai,G.X.,1995: Optical Technology, 1995/9, p1–19

Ai,G.X.,et al,1992: Proposal of Space Solar Telescope

Brungs,W.,Schmidt,W.K.H,Zhang, R.W.,1996: Space Solar Telescope Assessment Study Report

Sheng,L.X.1998: 1-Meter IR Solar Telescope (preprint)

Liu, Z.,1996: "Image Quality Degradation in the Turbulent Atmosphere, Parameters and their Measurement", Thesis for Master Degree, Yunnan Observatory

Deng,J.,1997: "Development of Correlation Trackers for Solar Telescope", Proceedings of CAS 97' Symposium on Telescope and Instrumentation, ed. X.Cui P.128

Deng,Y.,1998: poster paper of this workshop

Group Discussion

Denker : What would you use to measure r_0 and to what wavelength does this value refer to?

Ye : Idealogically, the method for obtaining r_0 by using a Daytime-DIMM has no principle difference from the DIMM concept, though Mr. Zhong Liu made quite a lot of developments for it in his Master's degree work (cf. Zhong Liu, 1996: "Image Degradation in the Turbulence Atmosphere-Parameters and Measurements", thesis for Master, Yunan Observatory CAS). for the wavelength, I'm pretty sure Liu made the measurements at 5500A.

Jain : My question is related to your proposed IR telescope. You mentioned that it will be a vacuum telescop. I understand a similar vacuum telescope, the SUST, already exists at the Kunming Observatory. Will the proposed IR telescope be similar to the SUST?

Ye : No, it will be totally different. The SUST you mentioned was a 50cm aperture facility which was installed in the Phoenix Hill near Kunming City.

Jain : When is your Solar Space Telescope expected to be launched? Further, you have not shown the results of optical tests of the system carried out through balloon experiment.

Ye : The SST(Space Solar Telescope) is scheduled to be launched in 2004 from Taiyuan Space Flight Center, Northern China by a CZ-4B rocket. We have made the balloon telescope optics test in lab at Prounhased Observations but unfortunately we're suffering a problem with the air controller near our balloon flight site. We've decided to move our flight site away from Beijing and make the flights early next year.

High Resolution Solar Physics: Theory, Observations, and Techniques
ASP Conference Series, Vol. 183, 1999
T.R. Rimmele, K. S. Balasubramaniam, R.R. Radick, eds.

High Resolution Longitudinal Magnetic Field Measurements Using a Fabry-Perot Lithium Niobate Filter Based Video Magnetograph

S. K. Mathew, A. Bhatnagar, C. Debi Prasad, & A. Ambastha

Udaipur Solar Observatory, P. O. Box: 198, Udaipur, INDIA

Abstract. A tunable Lithium Niobate ($LiNbO_3$) Fabry-Perot filter (FP) (passband 165 mÅ at 6122 Å) based Video magnetograph has been designed and fabricated. This instrument is capable of providing near simultaneous observations of photospheric longitudinal magnetic field, chromospheric Hα, and photospheric CaI observations using the same telescope and back-end set-up. The magnetic field measurements are made by using the polarization properties of the Zeeman components of the photospheric CaI line at 6122 Å (Landé g factor of 1.75). The CaI line has been chosen due to its low temperature sensitivity and no blend with other solar or atmospheric lines. A variable electro-optic quarter wave retarder, KD*P (Potassium di-Deuterium Phosphate) along with a linear polarizer is used for analyzing the circular polarization of the Zeeman components. The filter tuned at 140 mÅ away from the line center in the blue wing is found to give the best linear response for the field strength up to 1500 Gauss. A field of view (FOV) of $\sim$ 4 x 3 arc-min on the solar disk is imaged using a 699 x 288 pixel Cohu CCD camera in synchronous with the KD*P modulation. The $\lambda/4$ modulation is achieved by applying $\pm$2100 volts to the KD*P to obtain alternate frames of oppositely circular polarized images. These images are stored in separate frame buffers of an image acquisition system. To achieve high signal to noise ratio, a large number of images (maximum 256) are added in the respective frame buffers and then the difference between the left and the right circularly polarized images is obtained. This difference is related to the magnetic field strength. On comparing the video magnetograms (VMG) obtained at Udaipur Solar Observatory (USO) on 09 April 1997 at 09:32 UT with those taken by SOHO/MDI at 09:41UT, it was found that all the magnetic features matched very well in both the magnetograms. In this paper we present the details of the instrument and examples of observations.

1. Introduction

A filter magnetograph is a more suitable instrument for observing rapid changes in the magnetic fields associated with the solar dynamic phenomena compared to the spectrograph based magnetographs (Zirin 1995). The filter magnetographs operating at Marshall Space Flight center, Big Bear Solar Observatory (USA) and Huairou solar station (China) have shown the importance of these instruments in flare study. However the birefringent filters used in these magnetographs can not be tuned rapidly to different wavelengths. Keeping in mind the above factors, we have designed and fabricated a voltage tunable $LiNbO_3$ Fabry-Perot filter based video magnetograph, which can obtain near simultaneous photospheric magnetograms, Hα filtergrams and continuum images of an active region. Since these images are made by using the same imaging set-up, the problem of registration and inter calibration are minimized.

The Fabry-Perot (FP) etalons are extensively used in astronomy (Meaburn 1970; Vaughan 1989). The availability of high finesse FP etalons with large free spectral range made it possible to use them in solar observations (Rust, Burton, & Leistner 1987; Bonaccini & Stauffer 1990; Debi et al. 1998). Compared to the birefringent Lyot filters, which are extensively used in solar observations, the $LiNbO_3$ FP etalons have the added advantage of fast voltage tunability which is very convenient for multi-wavelength observations.

The USO video magnetograph (VMG) is the first and only such instrument operational at the Indian longitude and fills an important gap for continuous coverage of solar magnetic field observations. The solar video magnetograph operating at USO could provide valuable data for synoptic studies of evolution of active regions. In this paper we present the details of the instrument and examples of observations.

2. Basic Principle of Solar Magnetograph

The principle of Zeeman splitting of a spectral line, is employed to measure the solar magnetic field. For the line-of-sight field the σ_1 and σ_2 components of the Zeeman triplet are left and right circularly polarized. For the transverse magnetic field the σ_1 and σ_2 components are linearly polarized parallel to the field and the unshifted π component is linearly polarized perpendicular to the field. The σ_1 and σ_2 components are separated in wavelength from the unshifted line,

$$\Delta\lambda_H = 4.67 \times 10^{-13}\, g\, \lambda^2\, H \qquad (1)$$

where H is the field strength in Gauss and g is the Landé factor of the spectral line, λ and $\Delta\lambda$ are in Å units. For weak magnetic fields, the direct measurement of Zeeman splitting is not possible even with very high resolution spectrographs. Therefore the polarization properties of the Zeeman components is used to measure the magnetic field. Since the σ_1 and σ_2 components are shifted in wavelengths, choosing the left or right polarized light i.e., σ_1 or σ_2, entering the system corresponds to shifting of the lines. This principle is used in our video magnetograph. We tune the narrow band filter on the blue wing of the magnetically sensitive CaI 6122 Å line. It is found that tuning the filter at 140 mÅ away from the line center gives acceptable linearity for magnetic field measurement up to ± 1500 Gauss. By switching an electro-optic variable $\lambda/4$ retarder, the FOV of solar photosphere are imaged alternately in σ_1 and σ_2 components. A number of images, up to 256, are added for each σ_1 and σ_2 components separately in order to achieve the acceptable signal to noise ratio. The difference between the averaged σ_1 and σ_2 images is related to the longitudinal magnetic field.

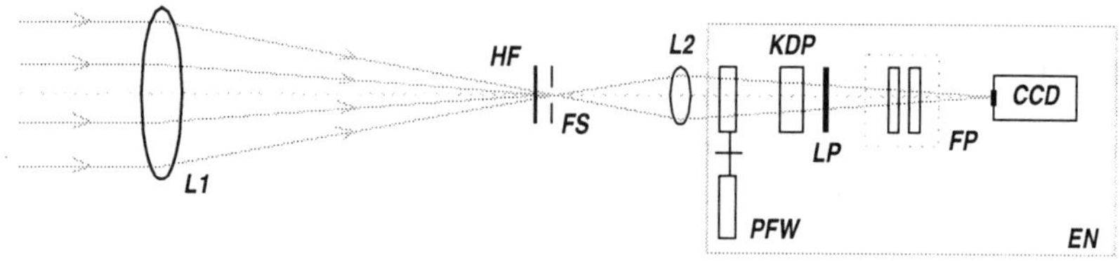

Figure 1. Optical layout of USO video magnetograph. L1 - 15 cm objective lens, HF - heat filter, FS - field stop, L2 - relay lens, PFW - pre-filter wheel, KDP - KD*P electro-optic modulator, LP - linear polariser, FP - Fabry-Perot etalon filter, CCD - CCD video camera, EN - Wooden enclosure.

3. Instrumentation

3.1. Optical layout

The optical layout of the video magnetograph along with light feed is given in Figure 1. The objective lens **L1** ($\phi = 15\,cm$, $f/15$) makes a solar image of 22 mm diameter at the focal plane where a field stop **FS** is placed. A heat filter **HF** is used to reduce the heat load on the optics and CCD camera saturation by blocking the IR radiation. A portion of the image is enlarged by a factor of 2.7 by lens **L2** to yield a 60 mm image (*f/40*). The pre-filter wheel **PFW**, **KD*P** modulator, linear polarizer **LP** and **FP** etalon filter are placed in the telecentric beam following the lens **L2**. The fast axis of the **KD*P** crystal makes an angle of 45° with the linear polarizer **LP**. Finally the image is recorded by a **CCD** camera. To avoid scattered light and ambient temperature variations, all the components are mounted on the optical bench, which is enclosed in a wooden box **EN**.

3.2. Fabry-Perot Narrow Band Filter

A 60 mm aperture, high finesse voltage tunable $LiNbO_3$ FP etalon acquired from CSIRO Australia, is used as a narrow band filter to isolate a portion of the wing of the selected absorption line. The FP etalon is made by using a $LiNbO_3$ substrate of 0.175 mm thickness, both sides of which are polished to $\lambda/200$ and coated with high reflective SiO_2 and Ta_2O_5 films. The resultant etalon has a reflectivity of 93% over a wavelength range of 5000 to 6700 Å. The tunability of the etalon is achieved by applying high voltage across the $LiNbO_3$ wafer which varies its refractive index. A conductive *ITO* (Indium Tin Oxide) coating is deposited for the application of electric field across the crystal. The high voltage terminals made of gold wires are bonded to the *ITO* coating with the help of silver epoxy. The details of the FP parameters are given in Table 1. In order to avoid the drift of the FP band pass due to a change in the ambient temperature, the etalon is enclosed in a constant temperature oven. The oven temperature is maintained at 43° C with a stability of ± 0.05° C which provides a wavelength stability 5 mÅ. To isolate 6122 Å (CaI) and 6563 Å (Hα) lines two narrow band interference filters of passband 2.6 Å and 3.5 Å respectively are placed before the FP, and the voltage on the FP is accordingly changed to obtain VMG in 6122 Å line and Hα filtergrams. These two pre-filters are enclosed in separate temperature controlled ovens and mounted on a computer controlled filter wheel, in order to make magnetic field and chromospheric observations.

Table 1. *Narrow band filter parameters*

Etalon Properties			
Substrate material	:	Z-cut $LiNbO_3$	
Substrate diameter	:	73.6 mm	
Usable aperture	:	60mm	
Etalon thickness	:	0.175 mm	
Reflectivity	:	0.95 at $\lambda = 6120$ Å	
Voltage sensitivity	:	0.45 Å per 1000 Volts	
FP filter parameters		CaI 6122 Å	Hα 6563 Å
Free Spectral Range(Å)	:	4.22	5.05
Passband (Å)	:	0.16	0.22
Finesse (FSR/FWHM)	:	26	23
Voltage applied (Volts)	:	−1200	+2500
Pre-filter parameters			
Central wavelength (Å)	:	6122.23	6563.98
Bandwidth (Å)	:	2.68	3.58
Peak transmission (%)	:	33.50	23.03
Ambient temperature (^{0}C)	:	43.0	43.0

3.3. Circular Polarization analyzer

A KD*P electro-optic quarter wave plate and a linear polarizer constitute the circular polarization analyzer for measuring the longitudinal magnetic field. The modulator uses a thin (< 3 mm) Meadowlark KD*P crystal which makes these cells suitable for using them in convergent light beam slower than $f/20$ (West 1989). The fast axis of the KD*P crystal is aligned at 45° to the transmission axis of the linear polarizer and housed in an insulated enclosure. The KD*P crystal is converted to $\pm\lambda/4$ retarders (at $\lambda = 6122$Å) by applying ±2100 Volts. In order to operate the KD*P modulator a fast switchable high voltage power supply was made, which can provide ±2100Volts on the application of low voltage TTL pulses at the input. The TTL pulses are obtained from the centronics port of the image acquisition system and synchronized with the image frame acquisition, such that alternately captured video frames contain left or right circularly polarized images. The left and right circular Zeeman components are converted in to two mutually perpendicular linear polarizations

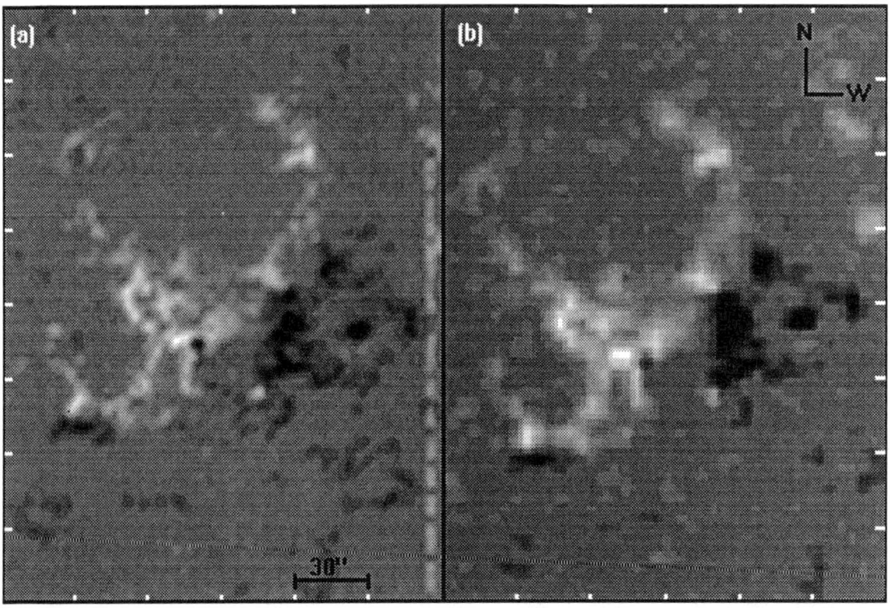

Figure 2. The magnetograms obtained for comparison on 09 April 1997 by: **a)** USO video magnetograph at 09:32 UT, **b)** SOHO/MDI at 09:41 UT.

depending on the sign of the applied voltage. One of the components is blocked by the linear polarizer allowing the selection of the image corresponding to left or right circular polarization in the emerging beam.

3.4. Detector - Cohu 4710 series CCD camera

The detector used in the video magnetograph is a Cohu CCD monochrome camera with an image sensor chip TC277 from Texas Instruments. This camera provides high resolution images with sensitivity as low as 0.25 lux, zero geometric distortion and no lag or retention of images. The CCD used is a frame transfer device with $9.2 \times 16.8\,\mu m$ pixels arranged in 699×576 array, with half masked for image storage and the other half is exposed to light. The resulting image area is 6.4×4.8 mm on the CCD chip. CCIR scanning system is employed for the image read out. The vertical blanking pulses after each video field (two vertical blanking pulses for each video frame) in the CCIR video output are detected through software and used for the synchronization of the KD*P switching and the entire data acquisition and reduction process. In the present optical set-up an area of $\sim 4 \times 3$ arc-minute of the solar disk is imaged by the camera, which gives a resolution of 0.3×0.6 arc-seconds/pixel.

3.5. Image acquisition and control system

An Innovision Inc. workstation, based on a Motorola MC68030 single board VMEbus computer, in conjunction with Imaging Technology Series 150 image

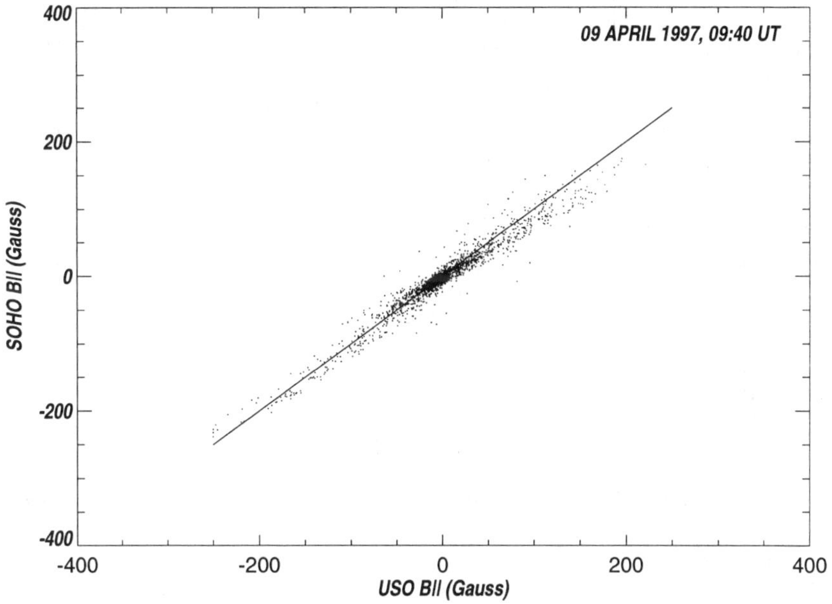

Figure 3. Scatter plot made between calibrated USO and SOHO magnetograms.

processing modules makes the complete data acquisition and control system. The componnents of the image processing modules are: one analogue-to-digital interface (ADI) unit, two frame buffers (FB) and one arithmetic and logic unit (ALU); all connected to the VMEbus of the host computer. The frame buffers FB0 and FB1 contains the image storage required for the real time processing of the data. Each FB consists of a single 512 by 512 by 16-bit frame store (FRAME A) and two 512 by 512 by 8-bit frame stores (B1 and B2). ALU is a pipelined image processor which provides real time image processing capabilities when used with ADI and FB. The images from FRAME A and FRAME B are processed by the ALU and the result is stored in FRAME A, which is finally transferred to the host computer.

The software was developed using C-language and the ITEX 150/151 image processing library (Imaging Tech, Inc) for real time processing and control of the video magnetograph (Mathew 1998). In a single acquisition cycle the operations to obtain a selected number of magnetograms, CaI 6122 Å and Hα images are performed. In order to produce magnetograms the Series150 modules are set up for continuous acquisition and accumulation of the images. The ALU is programmed for adding the incoming video data on to the pervious image already present in the FRAME A. The alternate incoming video images are routed to FRAME A of FB0 and FB1 respectively. This is synchronized with the KD*P high voltage switching such that the alternate frames correspond to the left and right circular polarizations. This process can be repeated for a preselected number of frames (maximum of 256 frames) for an increased signal to noise (S/N) ratio. The operations involved in a single cycle of image acquisition, which consists of a magnetogram by averaging 256 frames (12 seconds), CaI and Hα images takes less than one minute for completion. The delay between the CaI and Hα observation is mainly attributed to the change over speed of the pre-filters, which takes $\sim$ 15 seconds, while the frequency response of the FP etalon is much higher (Rust 1987).

3.6. Calibration and comparison

The video magnetograms were calibrated by using the profile line slope method (Varsik 1995), which is based on the weak field approximation (Jefferies & Mickey 1991; Cauzzi et al. 1993). The Stokes parameter V can be expressed as,

$$B_{||} = k_{cal} V \tag{2}$$

$$k_{cal} = (-\frac{dI}{d\lambda} \mu \lambda^2 g)^{-1} \tag{3}$$

where $\mu = e / 4\pi mc^2$; e and m are the charge and mass of the electron, λ_0 the central wavelength of observation, g the Landé factor of transition for absorption line, $B_{||}$ is the longitudinal magnetic field and k_{cal} is the calibration factor. I the Stokes intensity obtained in the absence of magnetic field and V is the intensity obtained by subtracting a single pair of video frames taken in opposite circular polarization. $dI/d\lambda$ is the measured line slope by tuning the filter from -40 mÅ to -170 mÅ from the line center in the steps of 25 mÅ. The measured slope $dI/d\lambda$ which is found to be 0.047, gives the value of $k_{cal} = 2059$ using Eqn. 3. This constant is used in Eqn. 2 for the conversion of measured Stokes V intensity into the magnetic field values in Gauss. The measured longitudinal magnetic field is the observed average field over the spatial resolution element and limited by the solar seeing. Further as the spectral resolution of the FP etalon is small, the accuracy of the calibration by this method is rather low.

Fig.2a and b show USO and SOHO/MDI magnetograms obtained around the same time on 09 April 1997. The SOHO/MDI magnetogram is a part of the calibrated full-disk data (Scherrer et al. 1995). It may be noticed that all the features observed in both the video magnetograms match well. After accurate registration of the two VMGs and degrading the spatial resolution of USO VMG to match the SOHO/MDI, a scatter diagram has been plotted in Fig.3. This plot demonstrates the validity of the calibration procedure, except for a small deviation of the slope of the scatter line from unity. This deviation may be attributed to the use of different spectral lines and the difference in spatial resolution in the two instruments.

4. Example of Observations

The USO video magnetograph is being used for near-simultaneous observations of longitudinal magnetic field, photosphere, and Hα chromospheric activity. Fig.4 shows an example of a typical magnetogram and the corresponding image in the blue wing of CaI 6122 Å line. During the best seeing conditions, magnetic

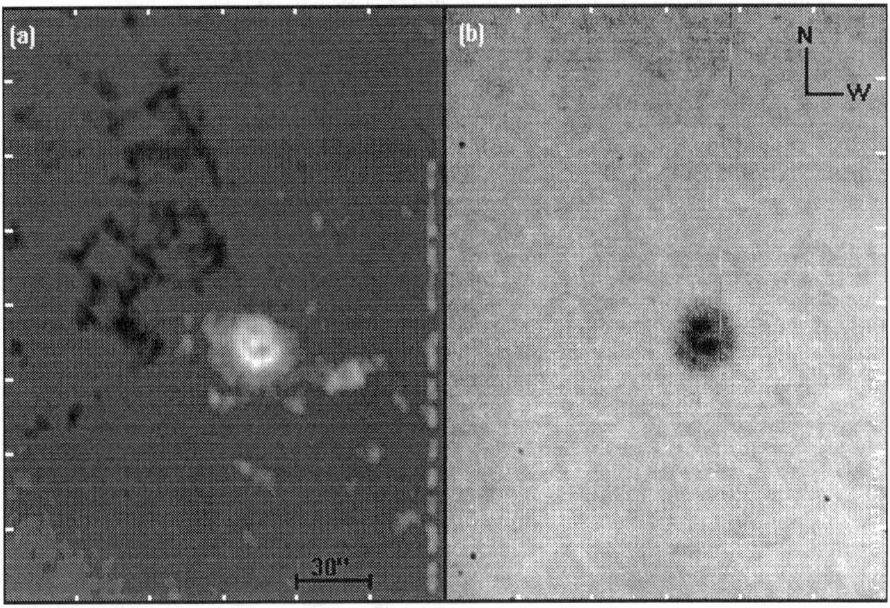

Figure 4. A typical: **a)** video magnetogram and **b)** CaI 6122Å image.

features of about 2 - 4 arc-seconds could be observed. The minimum detectable field was found to be 15 Gauss as obtained from the background noise level.

References

Bonaccini, D., & Stauffer,F. 1990, A&A. 229, 817

Cauzzi, G., Smaldone, L.A., Balasubramaniam, K.S., Keil, S.L. 1993, Sol. Phys. 146, 207

Debi Prasad, C., Mathew, S.K., Bhatnagar, A., Ambastha, A. 1998, Exp. Astro. 8, 2

Jefferies, J.J., Mickey, D.L. 1991, ApJ. 372, 694

Mathew, S.K. 1998, *A study of Solar Magnetic and Velocity Fields*, Ph.D Thesis

Meaburn, J. 1970, Ap&SS. 9, 206

Rust, D.M., Burton, C.H., Leistner, A.L. 1987, Soc. Photo-Opt. Inst. Eng. Conf. Proc. 627

Scherrer, P.H., Bogart, R.S., Bush, R.I., et al. 1995, Sol. Phys. 162, 129

Varsik, R.C. 1995, Sol. Phys. 161, 207

Vaughan, J.M. 1989, The Fabry-Perot interferometer, Adam Hilger

West, E.A. 1989, Soc. Photo-Opt. Inst. Eng. Conf. Proc. 1166, 434

Zirin, H. 1995, Sol. Phys. 159, 203

High Resolution Solar Physics: Theory, Observations, and Techniques
ASP Conference Series, Vol. 183, 1999
T.R. Rimmele, K. S. Balasubramaniam, R.R. Radick, eds.

LPSP & TIP: Full Stokes Polarimeters for the Canary Islands Observatories

V. Martínez Pillet, M. Collados, J. Sánchez Almeida,
V. González, A. Cruz-Lopez, A. Manescau, E. Joven,
E. Paez, J.J. Diaz, O. Feeney, V. Sánchez

Instituto de Astrofísica de Canarias, 38200-La Laguna, Tenerife, Spain

G. Scharmer

Royal Swedish Academy of Sciences, SE-133 36 Sweden

D. Soltau

Kiepenheuer-Institut für Sonnenphysik, Freiburg, D-79104 Germany

Abstract. Two Stokes polarimeters have seen first-light at the solar telescopes on the Canary Islands (Spain), one for visible wavelengths and one for the infrared region. Both polarimeters measure all four Stokes parameters simultaneously using a modulation scheme based on ferro-electric liquid crystals. The polarimeters use a polarizing beamsplitter (Savart plate) that produces a double image useful to reduce seeing induced crosstalk and to double the number of detected photons. The visible polarimeter (La Palma Stokes Polarimeter, LPSP) uses a CHIL-CCD camera with a readout rate of 60 Hz. For the Tenerife Infrared Polarimeter (TIP) a NICMOS-3 array is used at a rate of 8 Hz. The LPSP is designed to work at the Swedish Vacuum Solar Telescope (La Palma) and it aims to obtain high spatial resolution (0.5 ") maps of the polarization signals. TIP will be used at the German telescopes on Tenerife (Vacuum Tower Telescope and Gregory Coude Telescope) and it will be one of the few full Stokes polarimeters available working in the infrared. Both polarimeters obtain 2D maps by scanning the spectrograph slit through the region of interest. First-light results are presented.

1. Introduction

In recent years full Stokes spectral profiles obtained with the Advanced Stokes Polarimeter (ASP, Elmore et al., 1992) have led to an increase in our understanding of photospheric magnetic fields (see, e.g., Lites et al., 1996; Sánchez Almeida et al., 1996; Westendorp et al., 1997). Only through the analysis of spectrally resolved profiles of all four Stokes parameters over a photospheric map were these discoveries possible.

At the Instituto de Astrofísica de Canarias (in collaboration with the Kiepenheuer-Institut für Sonnenphysik and the Royal Swedish Academy of Sciences), we have been involved in the development of two Stokes polarimeters for the solar telescopes at the Observatories on Tenerife and La Palma (Canary Islands, Spain). Progress reports were given elsewhere (Collados 1994; Sánchez Almeida et al., 1997). Both polarimeters are similar to the ASP in the sense that they make use of the telescope' spectrograph and two dimensional information is obtained by scanning the slit of the spectrograph. However, the final solutions adopted for the polarimeters differ in many ways from those adopted in the ASP. The most important is the use of two ferroelectric liquid crystals (FLCs) to modulate the polarization information instead of a rotating retarder. This contribution describes in some details the properties of the two instruments. The Tenerife Infrared Polarimeter (TIP) has been designed for use at the German Vacuum Tower Telescope (VTT) on Teide Observatory (Tenerife). TIP can be used in the spectral range of 1 μm up to 2.3 μm. It can also operate at the Gregory Coude Telescope (GCT) on Tenerife but with no slit scanning option. The La Palma Stokes Polarimeter (LPSP) is designed for the Swedish Vacuum Solar Telescope (SVST) and it will work at visible wavelengths. Both, the SVST and the GCT have plans to increase their aperture up to 1 meter to increase the light gathering power and/or the spatial resolution of our polarimeters.

2. A Description of LPSP & TIP

It is well known that polarimeters that only use spatial modulation of the polarization information through the use of a beamsplitter are limited by the accuracy with which the flat-field is known. Pixel-to-pixel errors in the flat-field are translated into different (pixel-to-pixel) crosstalk with Stokes I. Because always the polarization signals are small compared with the intensity levels, this flat-field induced crosstalk fixes the final signal-to-noise (S/N) of the polarization profiles (Q, U and V). Typically the noise in the polarization profiles (compared to the continuum intensity) achieved by this method is hardly any better than 10^{-2}. It is for this reason that modern polarimeters modulate the polarization information in time, measuring the different intensities involved in the polarization analysis on the same resolution element. In this way, it is relatively simple to achieve $(S/N)^{-1} \approx \sigma_{stokes}/I_c \approx 10^{-3}$ while keeping good spatial resolution (<1 arcsec). Additionally, with time modulation polarimeters, better S/N can be achieved by simply using larger exposure times and/or integrating over extended regions of the solar surface.

LPSP & TIP modulate the polarization information in time through the use of FLCs. Two temperature controlled FLCs are used for the modulation of the four Stokes parameters. We explain how this is actually made in Sec.3. Here, we just mention that all four Stokes parameters are measured almost simultaneously not sequentially. After the two FLCs, both polarimeters have a calcite beamsplitter (a Savart plate) that makes the linear polarization analysis in two orthogonal directions. This beamsplitter is used here to reduce the seeing induced crosstalk (in a similar way as the ASP; see Lites, 1987).

In Fig. 1, we show the optical scheme for the LPSP integrated in the SVST. With some modifications it is also a valid scheme for TIP. A linear polarizer (TCsS in Fig. 1) can be put at the entrance of the telescope to calibrate the instrumental polarization produced by the entrance and exit windows and

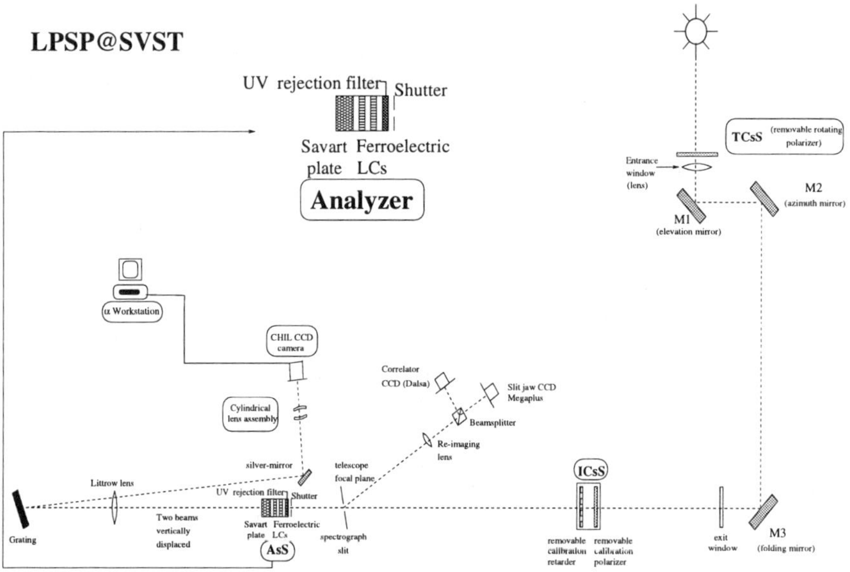

Figure 1. Optical layout of the LPSP at the SVST on La Palma. TCsS: Telescope Calibration sub-System, ICsS: Instrumental Calibration sub-System, AsS: Analyzer sub-System.

the three flat mirrors M1, M2 and M3. The rotation of the linear polarizer is controlled from the LPSP workstation. The instrumental calibration unit (ICsS in Fig. 1) is located in between the exit window and the entrance slit. For TIP it is inside a vacuum chamber before all the correlation tracker optical components. The task of the ICsS is twofold. First, by generating known states of polarization, it calibrates the performance of the analyzer (in particular it calibrates the FLCs). The result of this calibration is the polarimeter modulation matrix (see section 3). Second, it takes care of all the instrumental polarization effects induced by the optical elements between the ICsS and the analyzer. The ICsS is composed of a linear polarizer and a quarter wave retarder. During an instrumental calibration, the linear polarizer is fixed at some orientation and the retarder is rotated from 0° (with respect to the linear polarizer orientation) to 355° every 5°. This gives a total of 72 data points useful to characterize the analyzer. The analyzer itself (AsS) is located behind the slit to reduce the amount of light that goes through the FLCs. The AsS is composed of a shutter (for dark current measurements), a UV rejection filter to protect the FLCs, a half wave FLC followed by a quarter wave FLC and, finally, a Savart plate acting as a beamsplitter. Light, then, enters the spectrograph and after dispersion it goes through re-imaging optics to finally reach the CCD. Both TIP & LPSP use re-imaging optics that adjusts the pixel size to the right sampling size and, in both cases, to increase the amount photons detected by each pixel. TIP uses two spherical lenses that reduce the focal length by a factor of ≈2. LPSP uses two cylindrical lenses that reduce the focal length in the spatial direction by a factor of 2, keeping the same focal length in the spectral direction.

A 12 bit CCD camera with a readout rate of 60 Hz is used for LPSP (the same camera used by the ASP). The CCD camera is described in Elmore et al. (1992). For TIP a 16 bit NICMOS3 infrared camera is used. This camera is run at 8 Hz in double readout mode (Collados et al., 1997). Data is transfered from the detectors to a DEC-alpha workstations where a PCI Pamette FPGA-based board writes the images in the computer memory. This is made using a DMA interface that does not require time from the workstation CPU (Shand et al., 1998). The Pamette board is also in charge of the synchronization between the FLCs driving voltages and the exposing of the detectors.

All the components of the polarimeter described above are controlled by a DEC-alpha workstation (400 MHz) through a user's interface developed at the IAC (called `polar`). The interface allows the user to set all the parameters needed for the slit scanning mode. This mode is performed by controlling the correlation tracker CCD camera. At each slit position, the data is written into the hard disk in FITS format.

Table 1. Numbers for LPSP & TIP

Property	LPSP	TIP
1.- Nominal λ_o	6300 Å	1.56 μm
2.- Usable Range	4500-7500 Å	1.-2.3 μm
3.- Spectral Resolution	40 mÅ	60 mÅ
4.- Field-of-View (FoV)	60"×60"	40"×40"
5.- Spectral Range at λ_o	4.3 Å	7.4 Å
6.- Time to scan FoV [a]	30 minutes	10 minutes
7.- Accumulations for $S/N \approx 10^3$	67	10
8.- Exposure time	16.6 ms	50 ms
9.- Exposure time for $S/N \approx 10^3$	4.45 s	5.00 s
10.- Number of pixels	755×242	256×256
11.- Pixel size [b]	8.5 μm×19.75 μm	40 μm×40 μm
12.- Pixel size[c]	0.157"×17.95 mÅ	0.37"×29.0 mÅ
13.- Modulation frequency[d]	15 Hz	2 Hz

[a]Computed for 0.16 "/step for LPSP and 0.4 "/step for TIP.

[b]Spatial direction×spectral direction.

[c]Including the magnification factors of the re-imaging optics.

[d]Number of independent determinations of the 4 Stokes parameters per second.

Table 1 summarizes the properties of both polarimeters. To measure the Stokes parameters our modulation scheme (see section 3) uses 4 images corresponding to 4 different states of the FLCs. With these 4 images, a determination of the four Stokes parameters can be made but the S/N will be far from the 10^3

goal (due to the short exposure time). To increase the S/N of the final Stokes parameters a number of images, for each of the 4 modulation states, is accumulated (i.e., added). This number (N_A) is given in entry 7 of Table 1. It fixes the total time the polarimeter spends at each slit position. For example, LPSP uses a CCD camera at 60 Hz and thus, the exposure time of each individual image is 16.6 ms (entry 8). Because 4 different modulation states are used and then N_A images are accumulated for each state, the "effective" exposure time per slit position is 0.0166×4×67=4.45 seconds. This number is given in entry 9 of Table 1.

3. Ferroelectric Liquid Crystals as Polarization Modulators

FLCs are electro-optical devices with a fixed retardance and an optical axis that can be tuned (in a few microseconds) between two orientations as a function of the input voltage. The driving signal for tuning the FLCs is a simple DC compensated square wave working between ±10 volts. The FLCs change the orientation of the optical axis in response to the positive-to-negative or negative-to-positive voltage ramps. If one state has an orientation of the optical axis of, say, θ, the other state will have an orientation of $\theta+\Delta\theta$. $\Delta\theta$ is a smooth function of temperature. For our FLCs it equals 45° when working at temperatures in the range 25-30°. The larger the temperature the smaller the value of $\Delta\theta$. At a fixed temperature, FLCs have, then, only one free parameter, the orientation of the optical axis in one of the states, θ.

In both polarimeters, light goes first through an FLC with a retardance close to half wave and, then, through a second FLC with a retardance close to quarter wave. Suppose that the free orientations we have set up for each one of the FLCs are $\theta_{\lambda/2}$ and $\theta_{\lambda/4}$, respectively. Then, as a function of the input voltage, the Mueller matrix of the analyzer is:

$$\mathbf{M}^i(\theta_{\lambda/2}, \theta_{\lambda/4}) = \mathbf{M}_{\pm}\mathbf{M}^i(\theta_{\lambda/4})\mathbf{M}^i(\theta_{\lambda/2}), \qquad i = 1, 2, 3, 4 \tag{1}$$

where $i = 1, 2, 3, 4$ correspond to different combinations of the angles of the optical axis of the two FLCs (i=1: $\theta_{\lambda/2}$, $\theta_{\lambda/4}$; i=2: $\theta_{\lambda/2} + \Delta\theta$, $\theta_{\lambda/4}$; i=3: $\theta_{\lambda/2}$, $\theta_{\lambda/4} + \Delta\theta$; i=4: $\theta_{\lambda/2} + \Delta\theta$, $\theta_{\lambda/4} + \Delta\theta$). $\mathbf{M}_{\pm}$ represents the Mueller matrix of the beamsplitter. Each of the $i = 1, 2, 3, 4$ modulation states has a different Mueller matrix and the first rows of these matrices determines the output intensity as a function of the incoming Stokes vector:

$$I^i = M^i_{1,1}I + M^i_{1,2}Q + M^i_{1,3}U + M^i_{1,4}V, \qquad i = 1, 2, 3, 4 \tag{2}$$

Each one of the I^i is a different linear combination of the input polarization where all Stokes parameters play a role, but with different weighting factors (the $M^i_{1,j}$ $i, j = 1, 2, 3, 4$). It is because of this property that the Stokes parameters are measured almost simultaneously in our polarimeters. Using these Mueller matrix elements one can define the modulation matrix, $\mathbf{A}$, as:

$$I^i = \sum_{j=1}^{4} A_{ij}S_j, \qquad i = 1, 2, 3, 4 \tag{3}$$

where we use the notation $(I, Q, U, V) = (S_1, S_2, S_3, S_4)$. Equations (2) and (3) imply that $A_{ij} = M^i_{1,j}$ for all values of i and j. To increase the S/N, the detectors measure the four intensities I^1, I^2, I^3 and I^4 a number N_A of times and the polarimeter computer adds them in each of the modulation states giving the accumulated intensities $\bar{I}^i$, $i = 1, 2, 3, 4$. The Stokes parameters are, then, obtained from the demodulation matrix $\mathbf{B}$:

$$S_i = \sum_{j=1}^{4} B_{ij}\bar{I}^j, \qquad i = 1, 2, 3, 4 \tag{4}$$

where $\mathbf{B} = \mathbf{A}^{-1}$. From equation (4), one sees that the uncertainties in the $\bar{I}^i$ (photon noise, readout noise) are translated into uncertainties of the final Stokes parameters. Indeed, the S/N of the Stokes parameters (normalized to the continuum intensities) can be shown to be given by (Sánchez Almeida et al., 1994; Martínez Pillet, 1997):

$$(S/N)_i = (s/n)\epsilon_i\sqrt{4N_A}, \qquad i = 1, 2, 3, 4 \tag{5}$$

(s/n) represents the signal to noise of each of the individual intensities I^i, which is limited by the number of photons available and the detector (for LPSP $(s/n) \approx 120$). The other three factors are related to how polarimetry is made. The modulation efficiencies, ϵ_i, are given by:

$$\epsilon_i = (4\sum_{j=1}^{4} B_{ij}^2)^{-1/2} \quad i = 1, 2, 3, 4 \tag{6}$$

The efficiencies are a function of the demodulation matrix coefficients and they are directly related to the properties of the FLCs (retardances, orientations and temperature). At a temperature of 25°, we have measured for LPSP: $\epsilon_2 = 0.45$, $\epsilon_3 = 0.45$, $\epsilon_4 = 0.67$. For TIP, the same values (at a temperature of 19°) were: $\epsilon_2 = 0.65$, $\epsilon_3 = 0.5$, $\epsilon_4 = 0.47$. As it can be seen, LPSP is less efficient for measuring Q and U than V. This is due to the use of an almost perfect half wave and quarter wave retardances for the visible FLCs. (If one wants to match all three efficiencies, the retardance of the first FLC should have been in the range [130-160] degrees.) For comparison, the efficiencies, defined in a similar way as equation (6), for the ASP at 6300 Å are $\epsilon_2 = 0.41$, $\epsilon_3 = 0.41$, $\epsilon_4 = 0.61$.

The modulation efficiencies are very sensitive to the retardances of the FLCs which are not achromatic. To use both polarimeters at wavelengths other than the nominal (see Table 1) a different couple of FLCs should be installed and aligned in the analyzer. For LPSP two pairs of FLCs are available, one for the range 4500-5800 Å and another for 5800-7500 Å. TIP needs three pairs of FLCs for each of the spectral bands: J ([1,1.3] μm), H ([1.5,1.8] μm), K ([2,2.3] μm).

4. First Light Results

During the summer of 1998, both TIP & LPSP have seen first light. All critical parts of the polarimeter were tested and the performance was quite satisfactory. Communication with the telescopes and 2D scans were performed under the

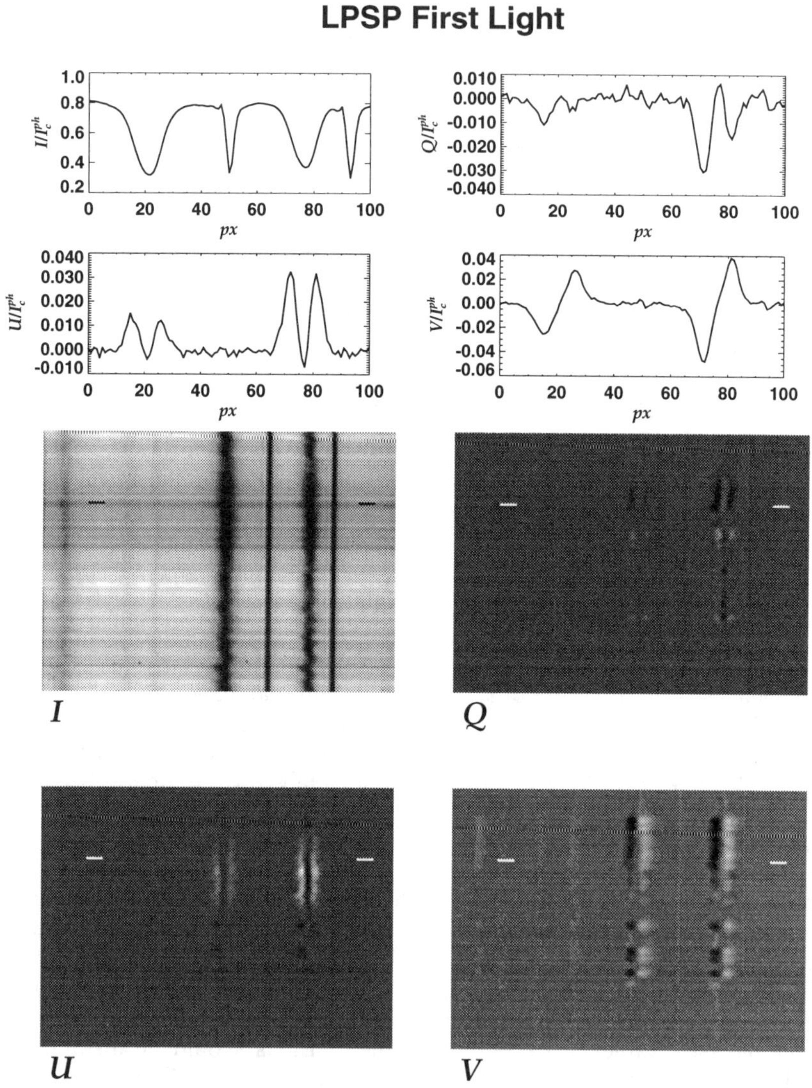

Figure 2. LPSP first light profiles with the slit crossing the penumbra of a sunspot. The 4 plots on the top show the profiles of the point marked with horizontal bars on the images. The FoV is about 60 arcsec.

parameters specified on Table 1. Polarization modulation was achieved with efficiencies similar to those measured at the laboratory. The temperature control of the FLCs was the only important part that was not ready for first light tests and they were run at room temperatures (around 18°). This made the efficiencies for LPSP smaller than the values quoted above. But even at these low temperatures, LPSP reached $(S/N)_Q \approx (S/N)_U \approx 500$ and $(S/N)_V \approx 1000$.

Fig. 2 shows an example of the Stokes parameters from the LPSP with the slit crossing the penumbra of a sunspot. The telescope crosstalk was corrected using a preliminary model based on measurements of the instrumental polarization of the SVST. Work is being made to improve this model with better data (from the TCsS, see Figure 1), but already the results are quite satisfactory.

The performance of TIP & LPSP is quite promising. Both polarimeters are open for use at the telescopes on the Canary Islands by contacting us or the Institutions running the telescopes (to contact us visit http://www.iac.es/cat/diurno/IMPRESO.html).

Acknowledgments. This work has been funded by the Spanish DGES (project PB 95-0028) and CICYT (project ESP95-1023-E).

References

Collados, M., Joven, E., Fuentes, F.J., et al., 1997, in 1st Advances in Solar Physics Euroconference, Advances in the Physics of Sunspots, ASP Conference Series Vol. 118 (eds. B. Schmieder, J.C. del Toro Iniesta, M. Vazquez), 361

Collados, M., 1994, JOSO Annual Report, 96

Elmore, D. F., Lites, B.W., Tomczyk, S., et al., 1992, Proc. SPIE, 1746, 22

Lites, B.W., 1987, Appl.Optics, 26, 3838

Lites, B.W., Leka, K.D., Skumanich, A., Martínez Pillet, & V., Shimizu, T., 1996, ApJ, 460, 1019

Martínez Pillet, V., 1997, Optical Design of a Visible Polarimeter, IAC internal report

Sánchez Almeida, J., Collados, M., & Martínez Pillet, V., 1994, Modulation Schemes for the Polarimeter of the SVST, IAC internal report

Sánchez Almeida, & J., Landi Degl'Innocenti, E., Martínez Pillet, & V., Lites, B.W., 1996, ApJ, 466, 537

Sánchez Almeida, J., Collados, M., Martínez Pillet, V., et al., 1997, in 1st Advances in Solar Physics Euroconference, Advances in the Physics of Sunspots, ASP Conference Series Vol. 118 (eds. B. Schmieder, J.C. del Toro Iniesta, M. Vazquez), 366

Shand, M., Scharmer, & G., Wei, W. 1998, these proceedings

Westendorp, C., del Toro Iniesta, J.C., Ruiz Cobo, B., Martínez Pillet, V., Lites, B.W., & Skumanich, A. 1997, Nature, 389, 47

Group Discussion

Bala: Are you satisfied with the surface qualities of the LC filter?

Martinez-Pillet: At the optical laboratories of the IAC we have measured our homogenity better than 1 degree for both the retardance and the orientation of the fast axis. Besides any in homogenity will be taken into account by doing a pixel dependent calibration along the slit.

High Resolution Solar Physics: Theory, Observations, and Techniques
ASP Conference Series, Vol. 183, 1999
T.R. Rimmele, K. S. Balasubramaniam, R.R. Radick, eds.

Deep Integration Magnetograms with Image-Selection and Cross-Correlation

Yuanyong Deng, Bin Zhang, Yihua Yan

Beijing astronomical Observatory, Chinese Academy of Sciences, Beijing 100012, P. R. China

Abstract. Using a PC computer with 512MB memory, we can record, in real-time, large amount of images. Applying image-selection and cross-correlation method to these data, very "deep" integration vector magnetogram can be obtained. As a test, we compare the extrapolation of this kind of vector magnetogram with the observed chromospheric field. We hope these "deep" integration magnetograms can be used to calculate vector current.

1. Introduction

The electric current in solar atmosphere is an important parameter to describe active phenomena of the Sun. Unfortunately, it is very difficult to directly measure the current field. A potential way to deduce this parameter comes from the Maxwell equation in case of force-free assumption:

$$\vec{j} = \nabla \times \vec{B} \tag{1}$$

or

$$(j_x, j_y, j_z) = (\frac{\partial B_z}{\partial y} - \frac{\partial B_y}{\partial z}, \frac{\partial B_x}{\partial z} - \frac{\partial B_z}{\partial x}, \frac{\partial B_y}{\partial x} - \frac{\partial B_x}{\partial y}) \tag{2}$$

Because the magnetic field components, (B_x, B_y, B_z), are observable, the current field can be determined. However, there still exist some difficulties. The most difficult is that the sensitivity of transverse magnetogram is much lower than that of longitudinal magnetogram.

By a simple estimation with spectral line FeI 5250.2, J. Evans (1966) formulated the relation between the observed magnetic signals (S) and the field strength (B) as:

$$S_L = 7.1 \cdot 10^{-4} B_L \tag{3}$$

$$S_T = 1.5 \cdot 10^{-7} B_T \tag{4}$$

Where, L and T denote longitudinal and transverse component, respectively. From these two equations one can see that the transverse component, B_x and B_y, must exceed the longitudinal one, B_z, by a factor of about $70\sqrt{B_z}$ in order to produce equal signal in the magnetograph. That means, if a magnetograph has 10G sensitivity for longitudinal field measurement, its transverse sensitivity will be about 221G. Thus, when one inserts the vector field measured with this instrument into Equation (2), the calculated current field will have poor quality and reliability because of the much lower sensitivity of the transverse field.

In order to get high sensitivity, one can use the deep integration method to observe the transverse field, because the sensitivity is proportional to the square-root of the integration time. In the above example, the integration time should be 400 times longer than the time taken in the observation of longitudinal field! Obviously, this is impractical. However, it is a basic requirement to improve the transverse sensitivity by several times in order to calculate the current field more reliably. One of the disadvantages of long time integration observation is the reduced spatial resolution due to the turbulence of earth atmosphere. This may cause the disappearance of small-scale mixed-polarity field. Thus, it is necessary to apply high resolution observational methods in the "deep" integration observation.

In Huairou Solar Observing Station (HSOS), Beijing Astronomical Observatory (BAO), we are developing a high capacity, real-time data collection system based on a PC computer that will allow very deep integration magnetograms to be obtained within several minutes. Because each frame is recorded, post-facto image-selection and image registration can be applied in order to reduce the effect of the earth's atmosphere.

2. Data collection system

Fig.1 shows the sketch of the data collection system. In the post processing module, each image will be written to computer disk for further analysis. This part has been finished. The real-time image-selection and correlation module is being developed.

We use a PULNIX-6701AN CCD camera. Some of its specifications are listed in table 1.

Table 1. Some Specifications of CCD Camera

Camera	TM-6701AN
Imager	Progressive scanning interline transfer
Pixel	648 × 484
Cell size	$9.0\mu \times 9.0\mu$
Sync.	Internal/external
Electronic shutter	1/32000 to 100ms

Following the A/D conversion, the digital signals are stored in computer memory. Some specifications of the computer and signal processing are listed in Table 2.

Then, all frames are written from memory to computer disk or other media. It takes about two minutes to write the data to disk. In this mode all frames

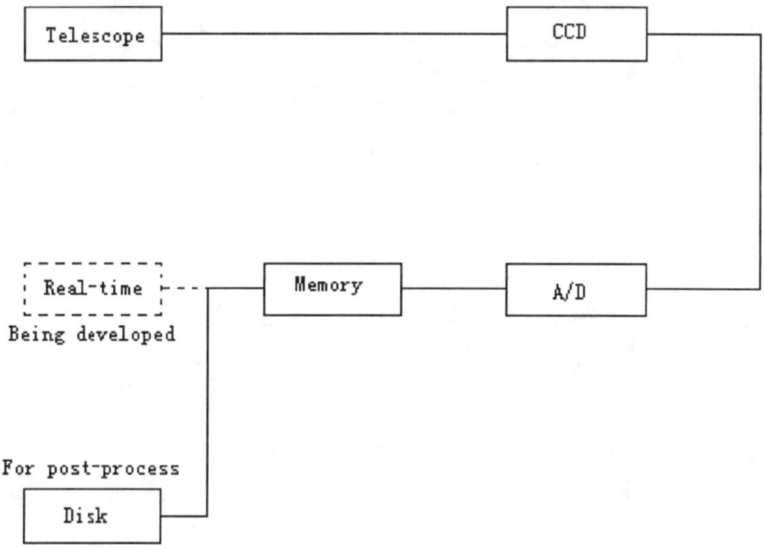

Figure 1. Sketch of Real-time Data Collection System

Table 2. Some Specifications of CCD Camera

Computer	Pentium-300
Extended memory	512MB
A/D accuracy	8bits
Image size	Changeable
	640 × 480 pixels (Max.)
Pixel resolution	$0.''35 \times 0.''35$
Frame number	Changeable
	4096 frames for 256 × 256 pixels
Total time	Changeable
	100s for 4096 frames, 20ms exposure

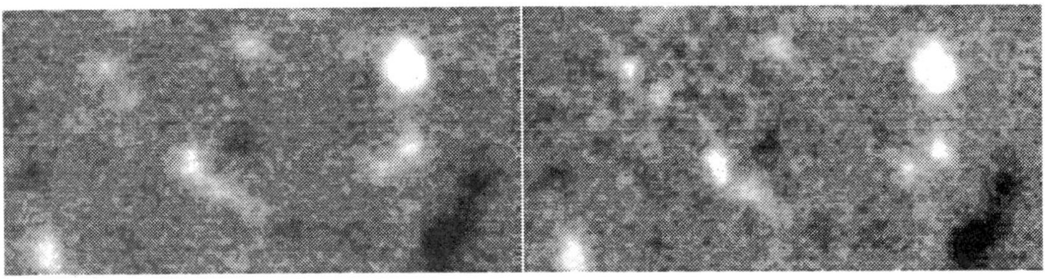

Figure 2. Comparison between direct and post-facto processed magnetogram

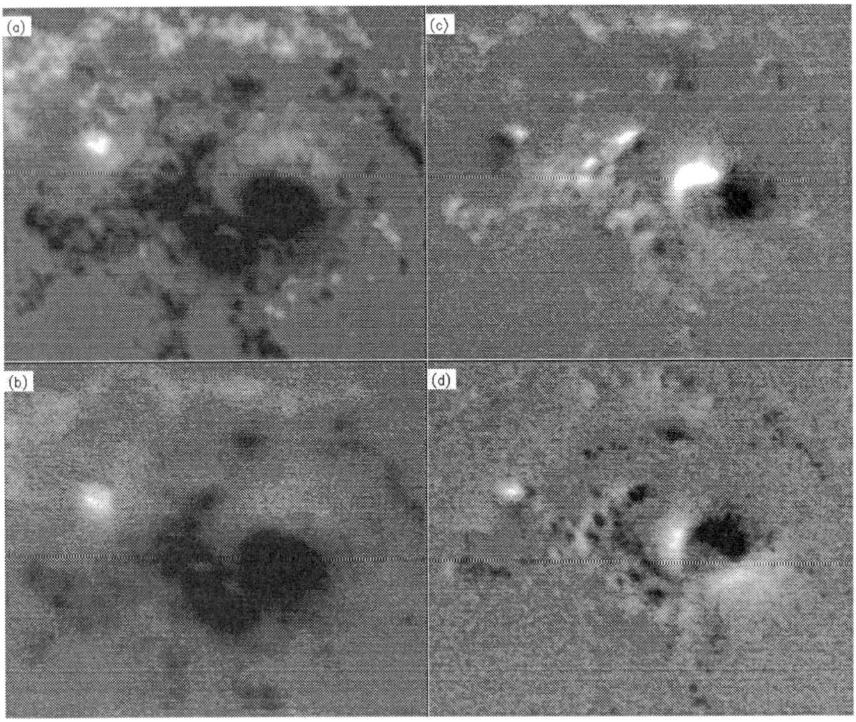

Figure 3. "Deep" integration vector magnetogram

are stored, and thus, post-facto cross-correlation and image-selection can be used. Sometimes one may just try to get a magnetogram in real-time. For this purpose, our system requires an online accumulation capability. This function is under development.

Fig.2 shows an example of a magnetogram in a quiet region. Fig.2a is the magnetogram obtained by direct accumulation. Fig.2b is the result after image-registration and image-selection have been applied. The improvement of spatial resolution and signal-to-noise ratio is obvious.

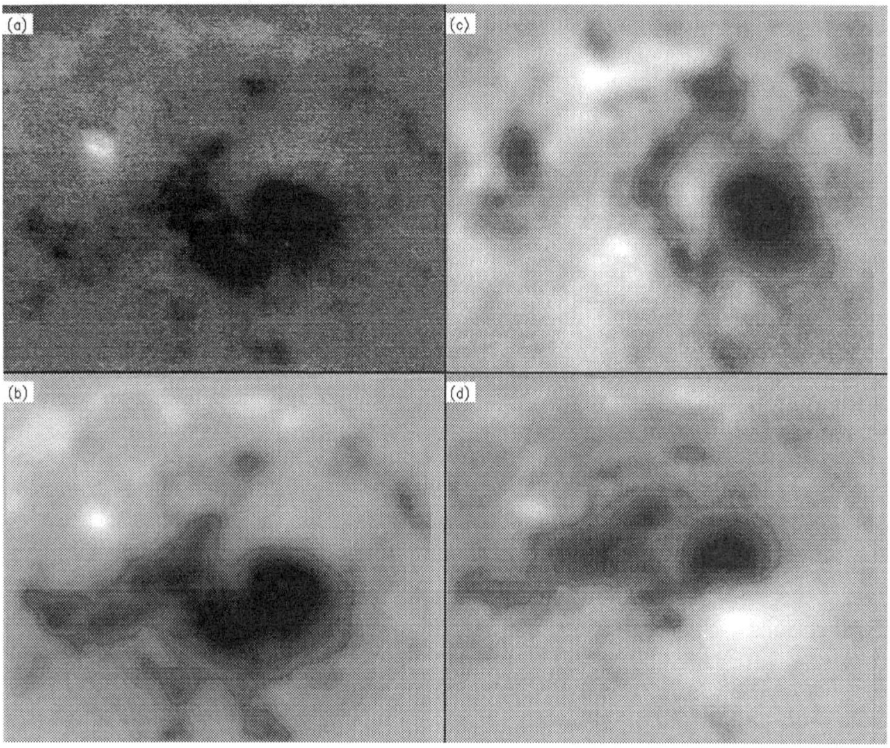

Figure 4. Comparison between observed and extrapolated chromospheric field

3. Observations

On Aug.28, 1998, we carried out a test observation. The target was a sunspot at north-west of solar disc. The exposure time was 20ms, and the field of view was about 2.′24 × 1.′68 (384 × 288 pixels). The number of frames summed was 1024 for the photospheric longitudinal field (FeI 5324.185), and about 6000 for photospheric transverse and chromospheric longitudinal (H_β) field. In this case, the sensitivity of one transverse field is about 60G. These magnetograms are shown in Fig.3, where, a, c and d are V, Q and U of photospheric magnetic field, while b is the chromospheric longitudinal field measured at H_β.

4. Discussions

Fig.4 is a comparison between the observed chromospheric longitudinal field (Fig. 4(a)) and the extrapolated field at height of H_β formation from the observed photospheric vector field. Fig. 4b, c and d represent the extrapolated components, B_z, B_x, and B_y, respectively. The extrapolated field is calculated from the photospheric vector magnetogram using the boundary element method based on a force-free magnetic field. The boundary element code was developed by Yan and Sakurai (1997). From Fig. 4 we can see that the extrapolated and observed result are consistent, both for the sunspot or for the network field (upper part in each panel).

Vector magnetograms at two different altitudes are needed to calculate the vector current from Eq. 2. Unfortunately, our magnetograph just works at one chromospheric line, H_β, which is not the best candidate for measurements of the chromospheric vector magnetic field. In addition, the present experiment measures just the longitudinal chromospheric magnetogram. It is possible to measure the vector field at H_β by careful selection of spectral position. We will continue this experiment. On the other hand, a rough estimation of the vector current can be deduced from the extrapolated field.

Acknowledgement. This work is supported by the NSF of China and the Chinese Academy of Sciences. We thank Profs. Ai G. X. And Wang J. X. for their suggestions and comments. One of the authors, Deng, would like to thank the Asian Office of Aerospace Research and Development, AFOSR for financial support; and the SOC and LOC of this workshop for their kind assistance.

References

Evans,J., 1966, Proc. Rome Meeting on Solar Magnetic Fields, Vol. II, M. Cimino (eds.), P223

Yan,Y., & Sakurai ,T., 1997, Solar Phys., 174, 65

High Resolution Solar Physics: Theory, Observations, and Techniques
ASP Conference Series, Vol. 183, 1999
T.R. Rimmele, K. S. Balasubramaniam, R.R. Radick, eds.

High Resolution Polarimetry With a Balloon-Borne Telescope: The Flare Genesis Experiment

P. N. Bernasconi, D. M. Rust, G. A. Murphy, H. A. C. Eaton

JHU/Applied Physics Laboratory, 11100 Johns Hopkins Road, Laurel, MD 20723-6099, USA

Abstract. The Flare Genesis Experiment consists of an 80 cm balloon-borne Ritchey-Chrétien telescope which will provide 0.″2 spatial resolution vector magnetograms. Wavelength and polarization state selection are accomplished respectively with a lithium niobate Fabry-Perot etalon, and a liquid crystal modulator. This paper describes recent modifications to the experiment, mainly to the optics and to the pointing system. The aim is to reach, in the next flight, the diffraction limit of the telescope and to assure 0.″1 pointing accuracy during the time needed to record a full set of vector polarimetric measurements (typically 10 seconds). The telescope optics have been extensively tuned and tested at the McMath-Pierce Solar Telescope at Kitt Peak during winter and spring 1998, while the pointing system has been tested in Palestine (Texas) in August/September 1998. Some test results are presented.

1. Introduction

There is a consensus in the solar physics community that to better understand the origin and evolution of solar activity, greatly improved magnetic field observations are needed. Higher resolution, greater stability during observations and long runs without interruptions are expected to reveal key features of the magnetic energy buildup and release needed to explain solar activity.

The major limiting factor in ground-based observations is the thermal turbulence of the Earth's atmosphere (seeing) that distorts the wavefront of the light coming from the Sun. In recent years various techniques have been developed to overcome this problem. One method is to calculate the wavefront distortion that occurred during the data recording and to subsequently reconstruct the undistorted images (speckle interferometry and phase diversity). Another way is to compensate the wavefront distortion in real-time by means of adaptive or active optical systems. Finally, the best way to eliminate the seeing degradation effects is to place a telescope above the Earth's atmosphere. This can be done either by installing the telescope in a spacecraft or by using a stratospheric balloon.

The Applied Physics Laboratory (APL) of the Johns Hopkins University has launched an effort to fly a balloon-borne 80 cm aperture telescope in the Antarctic stratosphere. It will be able to image and measure the vector magnetic field in the solar photosphere and chromosphere over long periods of time with an unprecedented spatial resolution of $0.''2$.

The primary scientific objective of the FGE is to understand the mechanisms which lead the fibrous magnetic fields at the solar surface to erupt in solar flares. In addition the FGE is well suited to study the evolution of the magnetic structure and helicity (twist of field lines) of active regions and filaments, to investigate the properties of small-scale magnetic elements and their contribution to solar luminosity variations and coronal heating, and to study solar oscillations at the smallest scales.

2. The Instrument

The FGE characteristics and capabilities have already been extensively described and discussed in Rust et al. (1996, 1993), and Murphy et al. (1996). Here we will just summarize its main features.

The telescope is an 80 cm Ritchey-Chrétien with an F/1.5 primary mirror made of ULE and honeycombed to a weight of just 50 kg. Its support and spider arms are made of graphite-epoxy and provide high thermal stability over a wide temperature range. The secondary mirror is made of single-crystal silicon which provides optimum heat transfer with minimal thermal distortion. The heat is removed from the secondary by means of a set of copper heat straps that connect the back of the secondary to the spider arm heat shields that also operate as radiators. In order to reduce the heat load on the primary mirror, a "heat dump" mirror was installed in front of it (see M3 in Fig. 1), which passes the light only from a small region on the Sun while the rest is reflected out the front of the telescope. The FGE telescope at floating altitude (~ 32 km) can achieve its diffraction-limited spatial resolution of $0.''2$.

Attached to the back of the mirror cell is a 1.2 m diameter pressure vessel (OPV, see Fig. 1), where most of the post focus optics is housed, as well as some of the electronics. The main post focus instrument is a vector magnetograph, whose major components are a polarization analyzer composed of two liquid crystal modulators and a linear polarizer, a tunable 75 mm diameter lithium niobate Fabry-Perot etalon (Rust 1994), and a 10 bit 1532×1024 pixels CCD camera. The temperature controlled polarization analyzer is located directly behind the prime focus to minimize instrumental polarization (see Fig. 1). The Fabry-Perot filter has a passband of about 160 mÅ and, in conjunction with a series of narrow-band blocking filters (~ 1.5 Å passband), it can be tuned to any of the four spectral lines: 6563 Å (Hα, $g = 1.045$), 6122 Å (Ca I, $g = 1.75$), 6302 Å (Fe I, $g = 2.5$), or 6249 Å (continuum). The image scale at the CCD camera is $0.''1$ per pixel which gives a field of view of about $150'' \times 100''$. The vector magnetograph will provide time series of vector magnetograms at various wavelengths, and also of vector velocity and intensity in the photosphere and chromosphere. Depending upon on the tradeoffs in data processing, the magnetic sensitivity will be $1 - 10$ G for the longitudinal component and $50 - 100$ G for the transversal component. The on-board tape recorder is provided with 9 tape cassettes, each capable of storing 6000 images, for a resulting data storage capability of 54000 images.

To maintain an image at the FGE focal plane stabilized down to the level of 0.″1 during the exposures, despite the oscillation of the gondola, we adopted an Image Motion Compensator (IMC). Its components are a Lateral Effect Diode (LED), installed at the focus of the pointing telescope which gives the pointing errors in elevation and azimuth with a sensitivity of 0.″05, and an agile tip-tilt mirror mounted right after the polarization analyzer where an image of the pupil is formed (see Fig. 1). The mirror is moved to compensate for the pointing errors as read by the LED.

The telescope is mounted to the gondola frame on its elevation axis which goes through the center of gravity of the telescope. It is moved on this axis by a motor whose stator is connected to the gondola and its rotor to the telescope. The pointing in azimuth is achieved by turning the entire gondola using a Reaction Wheel (RW) mounted on the top of the structure. The gondola is rotated by spinning up or down the RW while its motor is connected to the gondola (see Sec. 4.2).

The electronics is housed in four pressure vessels mounted on a mezzanine above the telescope. The vessels which are maintained near 20° C, allow commercial-grade components to be used, thereby considerably lowering the costs.

3. First Flight

From January 7 to 26, 1996, FGE had its first Antarctic flight, one of the longest balloon flights yet in Antarctica. The pointing system remained locked on the Sun for the entire flight. With the exception of the first 18 hours of flight, when the telescope was in the line-of-sight, and during an 11 hours flight beneath the payload with an LC-130 aircraft on January 10, the telescope operated autonomously. For the entire flight a 30″ solar target tracking was maintained.

On January 26, the FGE payload came down in an area inaccessible to large cargo aircraft. However, a team did reach the site in a small airplane and did recover the data tapes. The rest of the payload had to remain on the Antarctic ice until December 1996, when, with the aid of the U. S. and French Antarctic programs, we recovered the gondola and the telescope. The overall condition of the payload was good. The solar panels were destroyed during the hard landing, and the gondola aluminum frame was bent, but the telescope itself was in good condition. At APL the telescope tube and mirror cell were carefully measured and no damages or warping were detected. We started then to rebuild the payload for a second flight.

The analysis of the flight data revealed that the image resolution in the first flight was limited to only about 1″. However, the major FGE systems performed as expected and we feel that the basic engineering approach, in particular the mechanical, thermal and electrical system design, is sound.

4. Design Modifications and Test Results

Even though the first flight of FGE proved its feasibility, the scientific return was relatively small for mainly two reasons. First, because of a problem in focusing and alignment the image resolution was considerably lower than the expected

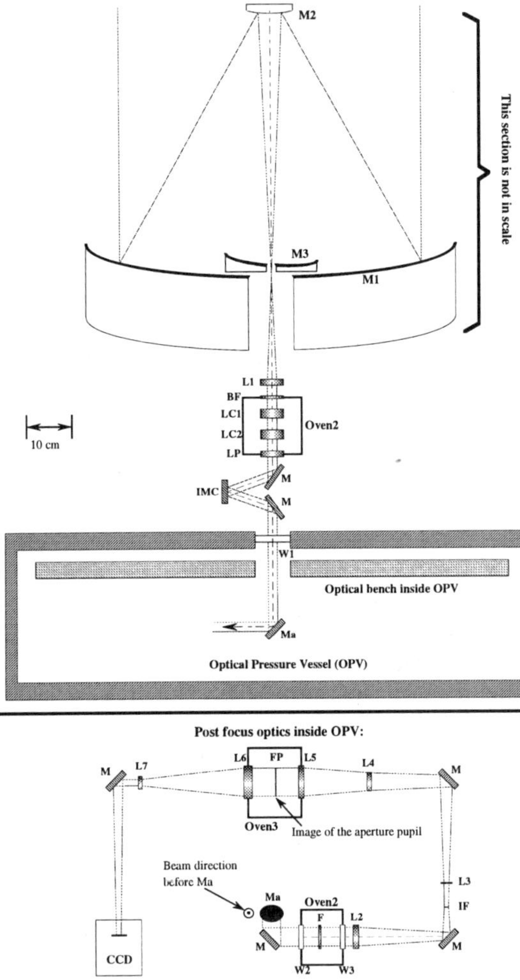

Figure 1. Scheme of the FGE optics: M = beamfolding mirror, M1 = primary mirror (ULE, ϕ = 80 cm, F/1.5), M2 = secondary mirror (silicon), M3 = heat dump mirror, L1-L7 = objective lenses, BF = broadband filter (500 Å passband centered at λ = 6328 Å), LC1-2 = liquid crystal modulators, LP = linear polarizer, IMC = image motion compensator mirror, W1 = OPV window, Ma = first beamfolding mirror inside the OPV, W2-3 = flat windows of Oven2, F = narrow-band filter (0.16 Å passband, a filter wheel is used to switch between different filters), IF = intermediate image focus, FP = Fabry-Perot etalon.

0.″2. Second, the pointing was not accurate enough to enable the Image Motion Compensator to stabilize the image jitter down to 0.″1 during the exposures. In the next two sections we will describe the modification we made to improve the FGE performance.

4.1. Optics

The original optical design was very complex (see Murphy et al. 1996), and it was very difficult to exactly align all the optics. Therefore we decided to significantly simplify the optical path. In Fig. 1 the new optical path is shown. In particular, we removed the Target Selector Telescope (TST), which was mounted parallel to the main telescope and formed a full disk image also on the CCD camera inside the OPV. The images from the TST were analyzed by the computer on-board to determine regions of interest on the Sun to direct the observation program while out of contact with the ground. Both, the main telescope beam and the TST beam were feeding the same detector, and it turned out to be very hard to correctly coalign these two beams.So we decided to eliminate the TST. During the next flight the coordinates of the regions of interest will be transmitted to the on-board computer via a low speed radio link making use of the Inmarsat or NOAA/Argos data packets (see Murphy et al. 1996).

In order to test the new FGE optics with sunlight, the telescope was shipped to Kitt Peak (Arizona) and installed at the NOAO/NSO McMath-Pierce Solar Telescope. The latter was used to feed the FGE telescope with a stable, 1.5 m diameter parallel solar beam. In this way we could separate the problems of pointing from those of achieving optimum optical performance. On June 21, 1998, we had good seeing conditions and we were able to record the filtergram shown in Fig. 2, and the longitudinal magnetogram of the same active region displayed in Fig. 3. We believe that the filtergram in Fig. 2 is seeing-limited. Other interferometric tests confirm that the telescope and the focal plane optics are operating near the diffraction limit of 0.″2.

The longitudinal magnetogram was obtained by combining the two polarization images $I + V$ and $I - V$ recorded consecutively with the Fabry-Perot passband sitting in the blue wing of the Ca I 6122 Å line. The magnetogram is not very accurate since the two images have been taken with a time interval of a few seconds between each other, and due to the seeing, they are not exactly the same (no destretching procedure was applied to the filtergrams). In flight there will be no seeing and this problem will disappear. In addition many consecutive polarization measurements will be added together in the post flight data processing, significantly increasing the magnetic sensitivity.

4.2. Pointing system

Even though the rms azimuthal pointing error during the first flight was $\sim 30''$ (in elevation the rms error was about a factor 2 smaller) we encountered periodic pointing disturbances with error peaks up to $100''$ (see the top diagram in Fig. 5). These spikes were too large to be compensated by the IMC mirror. The origin of the disturbances was found in the Momentum Transfer Unit (MTU), which is used to transfer excess angular momentum to the balloon.

The left image of Fig. 4 shows a scheme of the MTU employed during the first flight. The MTU was originally designed by the Harvard/Smithsonian

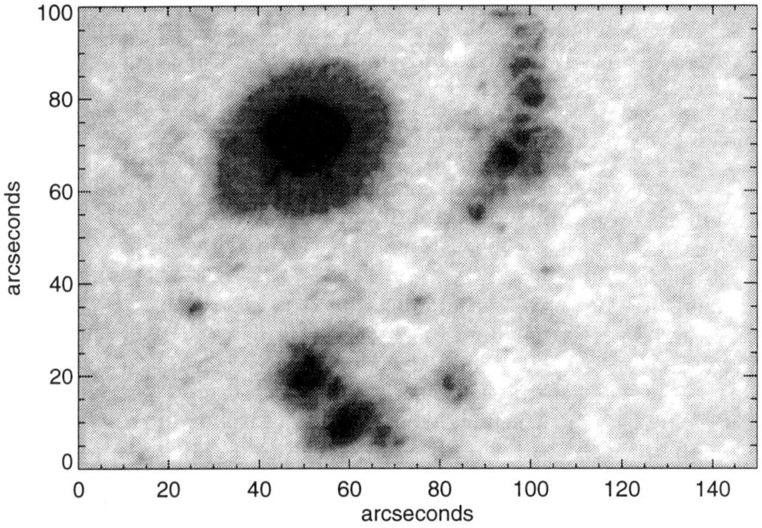

Figure 2. Filtergram of a sunspot group located at $\theta = 37.5°$ in the NW solar quadrant, recorded on June 21, 1998 with the FGE CCD camera. The Fabry-Perot passband was positioned in the core of the Ca I 6122 Å line.

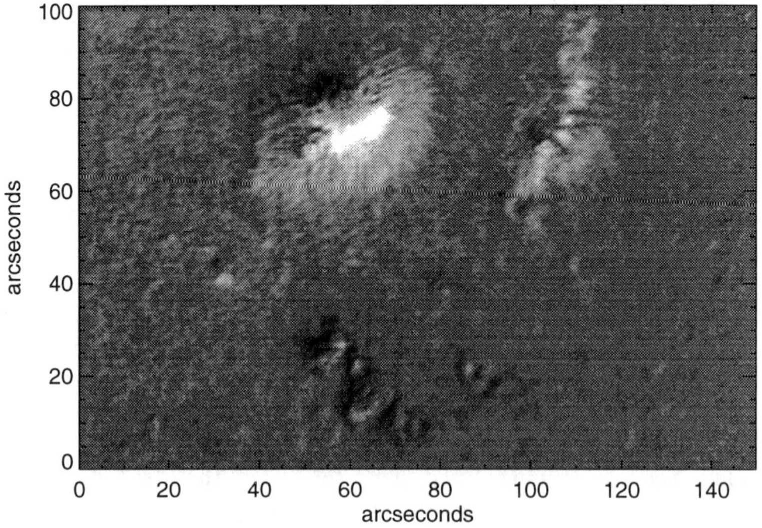

Figure 3. Longitudinal magnetogram of the same active region shown in Fig. 2. It represents the degree of circular polarization as recorded in the blue wing of the Ca I 6122 Å line. White (black) means field pointing towards (away from) the observer.

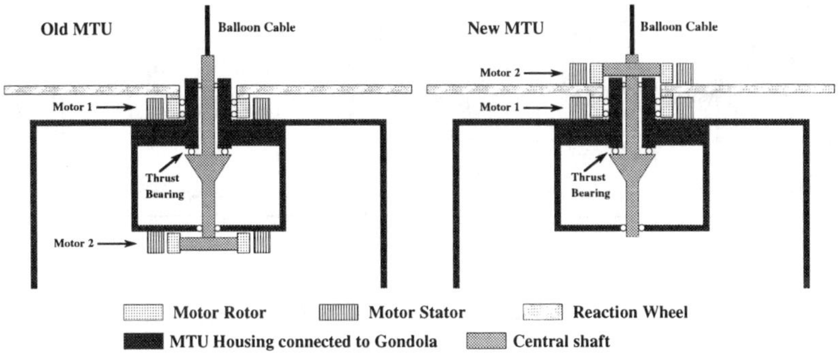

Figure 4. Simplified diagrams of the Momentum Transfer Unit used during the first flight (left) and the improved present one (right).

Center for Astrophysics. All the weight of the gondola is resting on the thrust bearing, transferring the weight to the central shaft which is connected to the balloon. Motor 1 has its rotor connected to the RW and its stator to the gondola. When a current is applied to Motor 1 a torque is produced and gondola and RW are accelerated, spinning in opposite directions. In this way the telescope can be steered in azimuth. If the RW starts to rotate at high speed (above 1 revolution per second) it becomes difficult to apply large torques that would speed it up further, and the the servo system will not be able to provide the necessary steering torques. To slow it down, an additional torque opposite to the spin must be applied. This additional torque cannot be transferred to the gondola since this would compromise the pointing. It must be transferred to the balloon. This is achieved with the aid of another motor (Motor 2) whose stator is connected to the gondola and whose rotor is connected to the central shaft. Motor 2 generates a torque exactly opposite to the one needed to slow the wheel down. This is transferred to the balloon cable. However, despite all efforts, it was no possible to match exactly the torques produced by the two motors. Consequently, an error torque is applied to the gondola during momentum dump. The top diagram in Fig. 5 shows that the high disturbances occur when momentum is transferred from the RW to the balloon cable.

To reduce this effect as much as possible we designed a new MTU, which is shown in the right image of Fig. 4. Motor 2 has its rotor still connected to the central shaft, but its stator is now connected directly to the RW instead of the MTU housing. In this configuration only Motor 1 can apply torques to the gondola. Motor 2 will operate as before whenever the RW needs to be slowed down.

We tested the new MTU at the NASA Scientific Balloon Facility (Texas) by hanging the gondola from a crane and by letting the telescope point at the Sun. The azimuthal pointing errors are shown in the bottom diagram in Fig. 5. The improvement in pointing performance is evident. Even though we must stress that the test conditions on the ground are much less difficult than the ones during

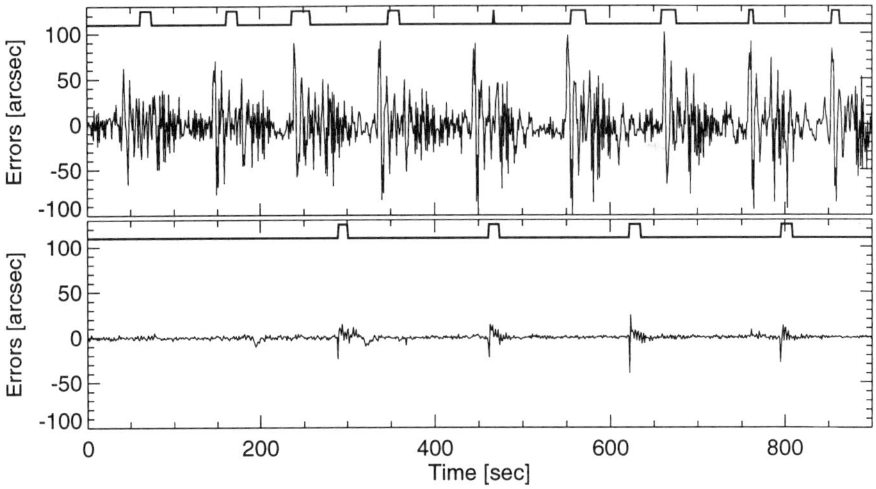

Figure 5. Azimuthal pointing errors in a typical time frame during the first flight (top), and during the ground tests with the new MTU in nearly ideal environmental conditions (bottom). The blips in the thick lines at the top of each diagram show the times when momentum was transferred from the reaction wheel to the balloon cable.

the flight. Now the rms error in azimuth is below 2″, and the disturbances during momentum dumps are considerably smaller and briefer. These kind of pointing errors can be removed by our IMC.

5. Conclusions

The FGE is a system for obtaining uninterrupted high resolution solar observations. The experience gained from the first flight led us to simplify the optical system and modify the pointing system. The next flight is scheduled for December 1999 during the peak of the solar activity in the present cycle. More information and the status of the project can be obtained by visiting our Web site: http://sd-www.jhuapl.edu/FlareGenesis.

Acknowledgments. We thank NOAO/National Solar Observatory for their help and for giving us the chance to test our telescope at their facilities at Kitt Peak. This work was supported by NSF under grant OPP-9615073 and NASA under grant NAG5-4955.

References

Murphy, G. A., Rust, D. M., Strohbehn, K., Eaton, H. A. C., Keil, S. L., Keller, C. U., and Wiborg, P., 1996, Proc. SPIE, 2804, 141

Rust, D. M., 1994, Optical Eng., 33, 3342

Rust, D. M., Hayes, J. R., Lohr, D. A., Murphy, G. A., and Strohbehn, K., 1993, Johns Hopkins APL Tech. Digest, 14, 358

Rust, D. M., Murphy, G. A., Strohbehn, K., and Keller, C. U., 1996, Sol. Phys., 164, 403

High Resolution Solar Physics: Theory, Observations, and Techniques
ASP Conference Series, Vol. 183, 1999
T.R. Rimmele, K. S. Balasubramaniam, R.R. Radick, eds.

Development of an Instrument for the Observations of Solar Hα Flares With Time Resolution of 25 ms

V. K. Verma

Uttar Pradesh State Observatory, Manora Peak, Nainital-263 129, India.

Abstract. We have developed a tested an instrument for solar flare observations based on CCD camera for observing Hα images of the flares with time resolution 25 ms. In the 512 $\times$ 512 pixel mode, images of CCD camera at 2 Mpixels/sec can be recorded at the rate of more than 5 frames/sec while 100 $\times$ 100 pixels mode, images can be obtained 40 frames/sec. The 100 $\times$ 100 pixels image of CCD camera scales to 65 $\times$ 65 $arcsec^2$ of the solar disc.

1. Introduction

The major effort of solar physics during the last two solar cycles has been directed towards observations of flares with high time resolution. The study of high time resolution data will lead us in understanding the rapid fluctuations of the emission of radiation from solar flares which in turn will provide a promising approach for probing the plasma processes that are responsible for solar flares. According to Kiplinger(1983) the high-time resolution hard X-ray bursts spectrometer (HXRBS) instrument aboard Solar Maximum Mission (SMM) recorded hundreds of X-ray flare spikes of less than 1 s duration with rise and decay times of some tens of milliseconds. The time resolution of the HXRBS instrument on SMM satellite was 1 ms while the minimum time resolution of Hα observation so far obtained 100 ms (Kiplinger et al.,1988). The Hα emission line (λ6563 Å) can provide spatial information more accurately as compared to HXR. Moreover, there is a good temporal correlation between the Hα and hard X-ray time profiles. Correlations of subsecond X−ray variations with UV and MW variations have placed important constraints on models of solar flares. MW images obtained with the Very Large Array (VLA) radio telescope, USA has a good spatial resolution $<$ 1 arcsec but relatively poor time resolutions (about 3s). UV images obtained with the SMM had a good time resolution of 32 ms , but at the expense of observing at only one arcsec or a few pixels image. Clearly, there is a great need for new correlative studies of solar flares to achieve high time resolution $<$ 10 ms and high spatial resolution $<$ 1 arcsec simultaneously in order to resolve effects such as those resulting from electron time−of−flight. These effects are crucial for determining causal relationship among emitting components and in identifying the flare energy release process or processes. In CCD observations study of Hα fluctuations, Kampher and Magun (1983) have shown that there is a component of the Hα emission which shows impulsive variations $<$ 2s and is coincident with microwave (MW) spikes. Flares observed at MW wavelengths also show spikes in their light curves similar to HXR spikes. However, observations with a time resolution of only 1.4 s, the coincidences could only be established to within 2s. Wusler and Kampher(1987) found that time evolution of total Hα intensity during impulsive phase shows fast fluctuations and spikes which are correlated with MW and HXR spikes. These spikes indicate chromospheric heating by fast electrons and peak of spikes originate from different kernels. By studying the chromosphere, with sufficiently high temporal resolution and comparing the time profile of Hα emissions with (MW) and hard X−ray (HXR) data one can study about the flare triggering mechanism in the solar atmosphere. The most important electron time of flight effect that are crucial for understanding flare triggering phenomenon can be determined by combined study of high temporal resolution Hα , MW and HXR observations.

Even on the global scene, there is a definite dearth of high time resolution Hα data of solar flares. There are few instruments which can record flare data with time resolutions less than 1s. The Institute of Applied Physics, Univ. of Bern, Switzerland has a digital CCD camera which can record flare Hα data with a temporal resolution of 0.3s (Greater,1990). Earlier, Kiplinger et al (1988) have reported that they have developed a high speed Hα camera system which can record Hα flare data with a time resolution 0.1s. Keeping above facts in mind we have tested and installed a solar flare observing system which can record a flare on 100x100 pixels area at an interval of 25 ms.

In the present paper we are reporting about an instrument " Flare Observing System " (FOS) which has been developed and installed at UPSO, Naini Tal , India , can record solar flare images at a interval of 25 ms.

2. Scientific Objective of Flare Observing System

The high speed CCD camera developed for observations of solar flares may record flare images at a time interval of 25 ms. The main aims of high speed CCD camera as under:

(i) To observe Hα flare data with time resolution of 25 ms. The recorded digital CCD images will be analysed for obtaining FTP and flare contours of the flare as a whole and of each flare kernel separately.

(ii) The FTP of Hα emissions will be compared with the MW and HXR emission data to understand the flare triggering mechanism and also the initial energy release in the chromosphere and the corona.

(III) From the FTP of Hα flare data we will also study the periodicity of rapid changes in Hα flares which in turn help us in understanding the energy release processes and other mechanism during flares.

3. Details of Flare Observing System

In Fig. 1 we have shown the schematic diagram of a 15-cm Coude' refractor telescope and back-end CCD camera system for solar flare observations.

The instrument developed for the solar flare observations employs a CCD camera includes camera controller, Sun Sparc station-20 computer a a GPS

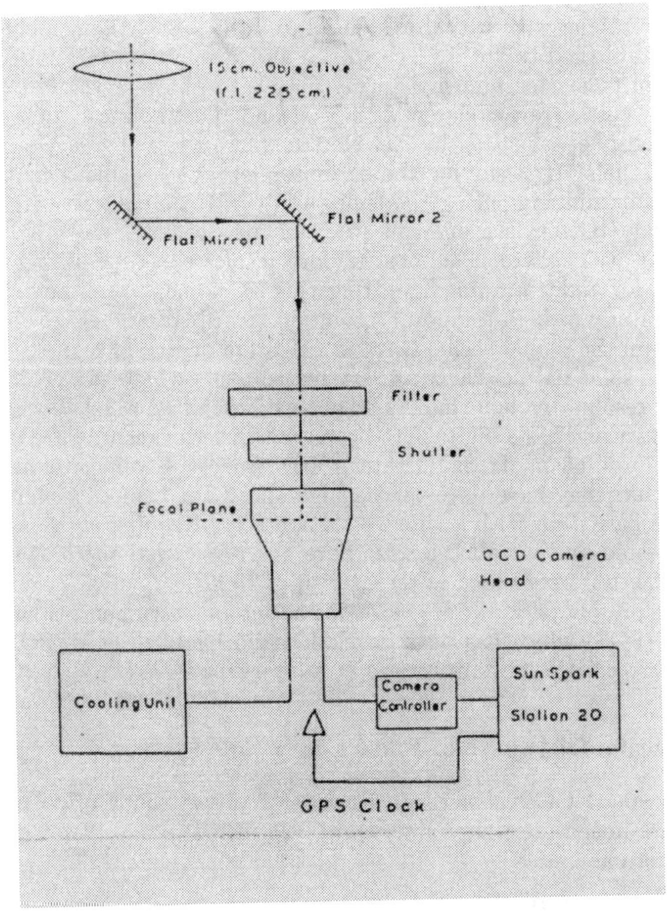

Figure 1. Shows the schematic diagram of a 15-cm Coude' refractor telescope with back-end instruments for solar flare observations.

(global position system) clock for accurate timings. At the back of Halle Hα filter with 0.5 Åpassband a the shutter the CCD camera system is installed at focus of the refractor telescope. This CCD camera system employs a CCD chip EEV37 (512 × 512 pixels, pixel sizes 15μ^2). The camera controller of the system has a variable read out rate from 0.5 to 4.0 Mega pixels/sec with 12 bit (4096 grey levels) resolution. At present, the CCD system has maximum read out rate of 2M pixels/sec. The software, which controls the working of camera controller is installed on Sun S c Station 20, 50 MHz computer having 7.3GB disk capacity a 14 GB Cartridge drive. A GPS clock is connected to Sun S c 20 computer to record time with an accuracy of 1 ms in header of the each flare image. The CCD chip of CCD camera is cooled to -25 ° C by a liquid circulator unit connected to head of the camera system. The high speed CCD camera system has been

Figure 2. Shows the photograph of 15-cm refractor telescope with backend instruments for flare observations

tested for observations of solar Hα flares. During high speed read out mode CCD camera works in frame transfer mode a it also works as shutterless CCD camera. To record flare images in high frame rate mode of the CCD camera, a program "Flare" is written. This program can record any number of frames in form of a single large file (say 1000 frames) on the external hard disk of the computer provided, there is sufficient space in the disk. From this big file we can extract similar number of frames in FITS format using "roi2ic" program. Few steps of Flare a roi2ic program are given in next section.

4. Computer Program for High Speed Flare Observations

As described in earlier section that high speed CCD camera can record up to 40 frames/sec of 100x 100 pixels of image of CCD camera. These program are written based on various routines of Isee software supplied by Inovision Corp., USA. This high speed CCD recording is done with help of computer program and few steps of these program are given below:"

flare program

Fast acquisition program for Solar imaging using tccd routine

"This program creates one big raw datafile which can be subsequently"
"converted into series of images in Inovision's roi format and then"
"converted to FITS format. Refer to icimport for creating individual"
"roi format and roi2fits for FITS conversion"

Hit any key to continue .....

"Enter image X size (default 512) = "

"Enter image Y size (default 512) = "

"Enter image X origin (default 0) = "
"Enter image Y origin (default 0) = "

"Enter expose time in msecs (default 15)= "

"Enter filename = "

"Below is space available on home0, home1 and home2 disks. Please check how many images you can save on any one disk"

"Enter home0 or home1 or home2 = "

"Enter number of images to be acquired (default 1000) = "

starting acquisition.....

done

"This window vanishes in 30 secs...."

roi2ic program

" Conversion program to be used after using flare program using icimport routine "

"This program creates images from one big raw datafile created from flare"
"convert these images to FITS format using roi2fits routine. "

"Hit any key to continue ....."

"Enter big source file from /export/home0, or /export/home1 or /export/home2 (with full path) = "

"Below is the space on file system "

"Required space should be greater than space on the destination dir "

"Enter destination filename to /export/home0, or /export/home1 or /export/home2 "

"Enter path with dirs other than one the source dir = "

"Enter image X size (default 512) = "
"Enter image Y size (default 512) = "
"Enter the time interval in secs to be set in the header = "
"Enter number of images acquired = "

"This script will create frames images from flare file" with created file name as extension 1, 2, 3, ..
"creating images ...."

done
" This window vanishes in 30 secs......."

The FOS instrument generates a stream of flare data in FITS format at rate about 4MB/sec. The flare data recorded on disk of computer is later on transferred to 14 GB Exabyte tape for storing solar flare data. In Fig. 2 we have shown the photograph of 15-cm refractor telescope with back-end instruments, which includes Halle Hα filter, CCD camera head, camera controller and liquid cooling unit.
The FOS instrument of UPSO is equipped with Halle Hα Filter which is giving good Hα images. To accommodate Halle Hα filter at the back-end of telescope we have enlarged a solar image to the twice of the initial size i.e. from 21 mm to 42 mm. Earlier the FOS instrument was equipped with DayStar Hα filter (Verma et al., 1996) Earlier with the DayStar filter the 512 X 512 pixels image of CCD camera could record upto 5 frames/sec at read out rate 2 Mpixels/sec and 100x100 area of image could upto 40 frames/sec. The 100x 100 pixels of CCD camera with the DayStar filter corresponds to 130 x 130 arcsec square of s olar disk while after accommodating Halle H-alpha filter 100x 100 pixels of image of CCD camera corresponds to 65x 65 arcsec square of the solar disk. The quality of Hα image with DayStar was not up to the mark therefore we have replaced the same with Halle Hα filter. The Hα image obtained through the FOS instrument equipped with Halle Hα filter is shown in Fig. 3.

The FOS instrument was used for the flare observations for about 50 days (observing duration between 3-5 hrs) but unfortunately we could not record a single solar Hα flare during above said period therefore we are unable to present flare time profile (FTP) with time resolution of 25 ms. At moment the camera controller is undergoing repair work and flare observations will restart after camera controller fault is corrected. In Table 1 we have shown the performance of the solar Hα CCD camera with various pixel areas. The spatial resolution of the system is 0.65 arc-sec per pixel or 1.3 arcsec which can be considered good for the solar flare observations.

The FOS instrument developed by us, based on 12 bit CCD camera is the fastest Hα flare monitoring system and is about four times faster than the system of Kiplinger et al (1988) which may be based on 8 bit (256 grey level) CID camera. The observing cost of this system is about \$ 1 per Gigabyte flare data which appears to be quite low for such high quality Hα flare data.

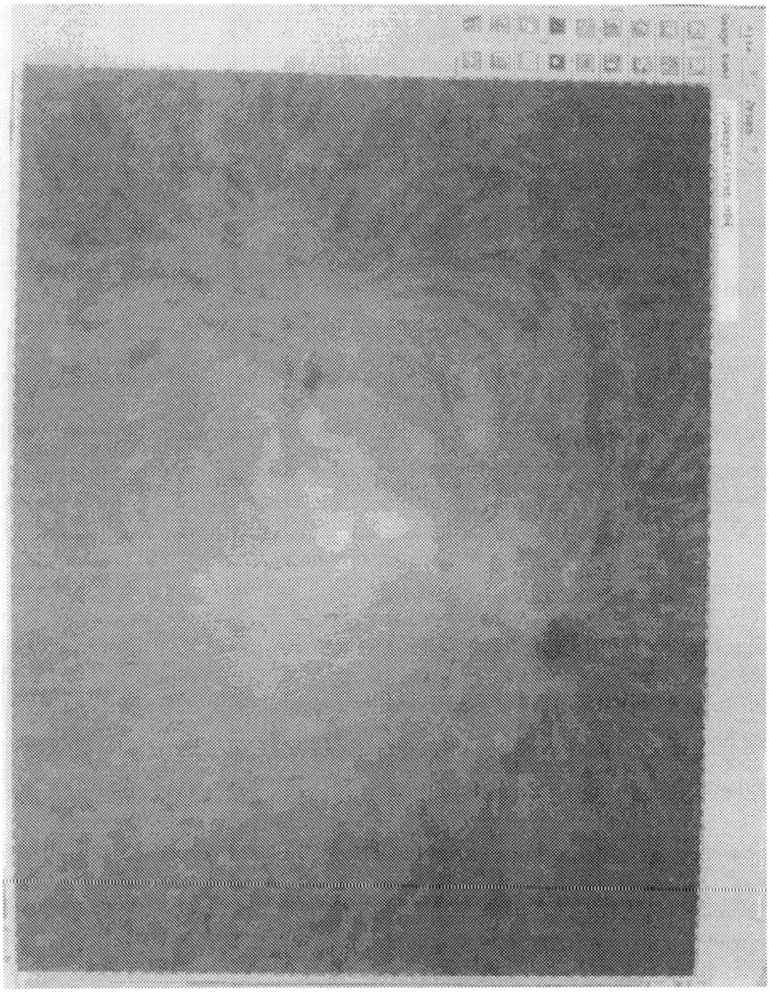

Figure 3. Shows the Hα image obtained through the FOS instrument,

References

Greater, M. 1990, Solar Phys., 133, 337.

Kampher N., & Magun, A. 1983, ApJ,274, 910.

Kiplinger A. L., Dennis B. R., Emslie A. G., Frost K. J.,& Orwig L. E. 1983, ApJ,265, L99.

Table 1. The performance of the High Speed Solar flare CCD camera with various pixels.

Image Area (pixels)	frames per sec.
100 × 100	40
200 × 200	20
512 × 512	5

Kiplinger A. L., Dennis B. R., Orwig L. E., & Chen P. C. 1988 , Flare Research at the Next Solar Maximum (Eds. R. C. Canfield & B. R. Dennis) p.214.

Verma, V. K., Uddin, W., & Gaur, V. P. 1996, J. Korean Astron. Soc., 29, S391.

Wuls, & J. P., Kampher, N 1987, Rapid Fluctuations in Solar Flares, NASA CP 2449, 301.

High Resolution Solar Physics: Theory, Observations, and Techniques
ASP Conference Series, Vol. 183, 1999
T.R. Rimmele, K. S. Balasubramaniam, R.R. Radick, eds.

Two-Dimensional Phase-Diverse Stokes Spectrometer

Robert J. Rutten

Sterrekundig Instituut, Utrecht, The Netherlands

Abstract. Optical fibers may be used to rearrange a two-dimensional field of view into a slit-like configuration feeding a spectrometer. In theory, such input reconfiguring permits two-dimensional Stokes spectrometry with phase-diverse/speckle restoration to the diffraction limit and represents an alternative to adaptive optics for small telescopes. In practice, the question is whether current fiber technology achieves such reconfiguration without loss of angular resolution at acceptable spectrograph speed and light loss. A feasibility study is undertaken for application at the Dutch Open Telescope.

1. Introduction

The simple idea behind 2D fiber spectrometry is to use optical fibers to reformat a two-dimensional input field into a linear spectrometer slit arrangement (Fig.1). For example, 2K×2K CCD's in the spectrum plane might be fed with 45×45 fibers sampling a 5×5 arcsec^2 field of view at 0.1 arcsec pixel resolution. Adding liquid-crystal polarization encoding before the fiber pickup would permit multi-line full-profile Stokes polarimetry. Adding an extra camera out of focus would permit phase-diverse/speckle reconstruction per spectral element.

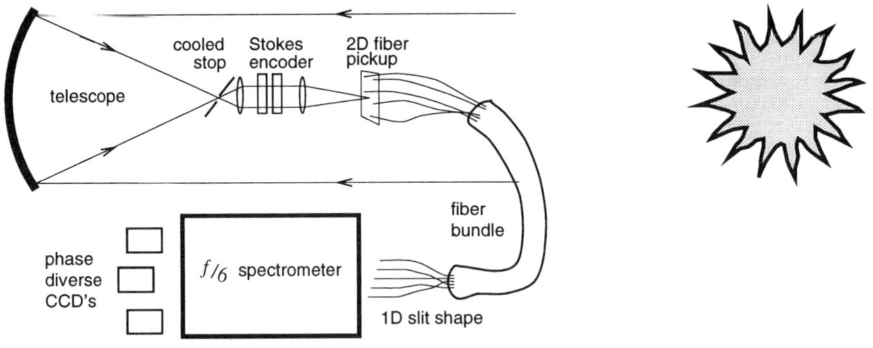

Figure 1. Scheme for a solar fiber spectrometer at the DOT. The fibers reformat a square input field into a linear slit-like arrangement. Stokes encoding is done before the fiber pickup, for example with liquid crystal modulation. Phase diverse registration is realized by having one CCD register the continuum out of focus.

With phase-diverse/speckle registration, the scheme represents a poor-man's alternative to adaptive optics using post-detection computer processing rather then real-time wavefront monitoring and correction. In both cases detailed spectral information including Stokes vector analysis may be obtained with angular resolution down to the diffraction limit. In both cases the price paid is that the angular field is very small — although with off-line processing it may exceed the isoplanatic patch through tessellation into independent subfields and separate (parallel) restoration.

The scheme differs intrinsically from the now ubiquitous application of fibers in nighttime astronomy (see the conference proceedings edited by Bardeen 1988 and by Gray 1993). There, fibers are usually used to reduce confusion between close stars or galaxies, each fiber accepting a field patch larger than the seeing disk. In the solar case, the fibers should map the focal plane with angular resolution at the diffraction limit, an order of magnitude smaller than the seeing disk. The resolution elements are also geometrically small because solar telescope diameters are an order of magnitude smaller than nighttime telescope diameters.

The problem is in the fibers. The scheme requires very thin fibers to permit fast input beams that do not suffer too much from beam spreading (focal ratio degradation) within the fibers. Otherwise, the spectrometer size grows out of bounds or the light loss becomes unacceptable.

2. Solar 2D spectrometry

The intrinsically intricate structuring of the solar atmosphere and its rapid changes require fast 2D field registration. In addition, the even faster wavefront perturbations added by our own atmosphere, even at good sites such as La Palma, are intrinsically 2D in character (or even 3D) and so require 2D (or 3D = phase diverse) registration at sampling times no longer than 10 ms in order to be hardware corrected in real time or software restored post detection.

On the other hand, the intricacy of solar phenomena and processes requires all the diagnostic constraints provided by precise multi-line Doppler mapping and Stokes magnetometry. These require spectral resolution of narrow bands within selected spectral lines. Height discrimination also requires narrow bands within selected spectral lines. The more spectral elements, the larger the number of lines, the wider the line formation variety, the preciser the Stokes polarimetry, the better the diagnosis and the astrophysical interpretation. High spectral resolution and high spectral variety require a spectrometer, intrinsically a 1D instrument, and long exposure times, exceeding the 10 ms seeing freezing limit by one or two orders of magnitude.

Let me review some examples. The extreme of spectral resolution and spectral variety are of course reached by the NSO/Kitt Peak FTS (Brault 1978) in broad-band (solar atlas) mode, at nearly complete sacrifice of both temporal and spatial resolution.

The best example of precise Stokes spectrometry is the HAO ASP here at the DST (eg. Elmore et al. 1992, Lites et al. 1994, Lites 1996), which is currently the most accurate and productive magnetograph. It uses detailed Stokes profile information from 2 to 5 spectral lines with different Landé factors simultaneously to derive the vector field with an elaborate calibration procedure (Skumanich et al. 1997, cf. Westendorp Plaza et al. 1998). The French-Italian THEMIS telescope on Tenerife aims to perfect this technique using yet more lines, of order 10 simultaneously (Mein & Rayrole 1991, Semel 1994).

The best example of height tomography obtained by combining spectral detail from diverse lines in the optical part of the spectrum is Beckers' classic HIRKHAD setup of the Echelle Spectrograph at the DST, registering Ca II H & K, Na I D_1, Hα and the Ca II infrared lines next to each other on a single detector (70 mm film at the time) in order to span most of the optically-accessible solar atmosphere (Beckers et al. 1972).

The filter instruments are exemplified by the versatile Lockheed SOUP tunable filter (e.g. Title et al. 1989) and the superb G-band imaging at the Swedish Vacuum Solar Telescope (e.g. Loefdahl et al. 1998).

The advantages and disadvantages of these various techniques are obvious. It takes a spectrometer to obtain varied line profile information simultaneously, whereas it takes a filtergraph to measure many spatial elements simultaneously. As a result, solar observing strategies have concentrated on filter instruments emphasizing surface patterning at the cost of measurement precision, and on spectrometers emphasizing quantitative measurement at the cost of 2D spatial coverage and resolution. Each strategy has its pros and cons; neither suffices by itself. How to combine the advantages of 2D registration with spectrometer-type spectral resolution and spectral variety?

Classical spectrometer slit stepping as in spectroheliography builds up the solar scene too slowly, or samples only a narrow strip if a higher cadence is required (for example, only 11 arcsec with the ASP in the dynamics study of Lites et al. 1999). In addition, there is no easy method for resolution restoration (cf. Keller & Johannesson 1995).

Rapid tuning of a Fabry-Perot, with a co-tuned birefringent or Fabry-Perot prefilter, delivers a large field with good spectral resolution but is usually limited to spectral elements in just one spectral line. Examples are the Fabry-Perot at the Sac Peak DST (e.g. Bonaccini et al. 1990, Rimmele et al. 1995), the MISC at the German VTT (Bendlin et al. 1992, Kneer 1997, Al et al. 1998, Stolpe & Kneer 1998), and the IPM on THEMIS (Bonaccini et al. 1989). Taking simultaneous wide-band images enables off-line wavefront restoration with the multi-channel deconvolution method of Keller & von der Lühe (1992; cf. Tritschler et al. 1997).

The MSDP instruments (see Mein 1991, Roudier et al. 1991, Mein 1992), of which one will be the major post-focus instrument of THEMIS, represent a hybrid compromise between spectral and spatial resolution. They trade most of one spatial dimension against multiple (well-defined but fairly wide) spectral channels. This is done by projecting a relatively wide entrance slot onto the CCD's, with intermediate slits selecting spectral passbands within sufficiently wide spectral lines. They are subsequently de-dispersed by a reversed ("subtractive") grating pass in order to regain full spatial resolution (Evans 1956). The end result consists of narrow more-or-less monochromatic strip maps per bandpass, typically measuring 10–30 arcsec by 100-300 arcsec.

A fiber-fed spectrometer similarly sacrifices field size for spectral information and spatial simultaneity while maintaining 2D registration. Compared with the MSDP, it obtains full spectral resolution and spectral variety by sacrificing most of the second spatial dimension as well, down to about 5×5 $arcsec^2$ for diffraction-limit pixels. Its strength is that it permits full spectral detail including Doppler imaging and Stokes vector magnetometry in multi-line tomography for its small 2D field. Addition of phase-diverse/speckle restoration of atmospheric wavefront degradation makes it do so at the diffraction limit in reasonable seeing, which implies gathering high-resolution sequences of desirable duration (hours) reasonably frequently rather than just on very rare fluke super-seeing days. There are obvious science applications for such capability — essentially the same as for adaptively corrected spectrometry.

3. Solar fiber spectrometry

The basic scheme of a solar fiber spectrometer is shown in Fig.1. In principle, it resembles the multi-fiber spectrometer inputs now regularly used in nighttime applications. Most of these achieve low focal-plane filling because they are intended to pick up discrete objects avoiding confusion, but some integral-field fiber pickups are used as well (Vanderriest 1993).

In practice, the solar implementation is quite different. In the solar case, the fibers should sample image elements at the diffraction limit of a small-diameter telescope instead of just picking up stars of galaxies in the image plane of a large telescope. Even in integral-field spectrometry with the VLT at its expected best-seeing resolution, the required 0.2 arcsec pixels correspond to as much as 100 μm in the Nasmyth foci. Solar diffraction-limited resolution elements are far smaller geometrically than the seeing disks at large nighttime telescopes. Direct prime-focus pickup therefore requires very thin fibers, with diameters of order 1–10 micron only, at the edge of fiber technology if not beyond.

Thicker fibers may be used after magnification, but that brings the problem of beam spreading known as focal ratio degradation. Fibers do not conserve étendu (the product of beam cross-section and solid angle) or intensity (measured per unit cross-section and unit solid angle), but produce fast output beams with speed of order $f/6$ for beams with slower input speed, whatever it is, due to microbending within the fibers (e.g. Ramsey 1988, Carrasco & Parry 1993). The Dutch Open Telescope, for example, has prime focus at $f/4.4$ with 1 μm pixel size per 0.1 arcsec diffraction-limit resolution element. Magnification to 10 μm or 100 μm pixels to match fibers that are possibly or readily available produces input at $f/44$ or $f/444$, respectively, but the output yet requires an $f/6$ spectrometer which grows with the fiber diameter — the effective slit length growing to well over 20 cm for 2K fibers of 100 μm— to impossible size. Thin fibers are therefore required, but their cladding produces increasing fill fraction loss and necessitates an array of square microlenses feeding the fibers individually.

If these problems are solved it seems relatively straightforward to add nematic liquid-crystal Stokes encoding before the fiber pickup (cf. Sanchez Almeida et al. 1997, Del Toro Iniesta et al. 1997) and to add phase-diverse/speckle restoration as demonstrated by Paxman et al. (1996) by registering a (sufficiently wide) continuum segment out of focus. The individual narrow-band spectral channels may then be restored through the multi-channel deconvolution technique developed by Keller & von der Lühe (1992).

Phase-diverse/speckle restoration makes a fiber-fed spectrometer directly comparable to an adaptive-optics fed spectrometer. The advantages are the cheapness of off-line processing over real-time wavefront measurement and control, and the potentially larger field size which at 0.2 arcsec resolution grows beyond the isoplanatic patch with fiber arrays and CCD cameras beyond 2K spatial pixels. The disadvantages, apart from the fiber problems discussed above, are the need to acquire sufficient speckle samplings without swamping by read-out noise and the diminishing effectiveness of speckle restoration at larger aperture due to decrease of speckle S/N (A. Title, private communication). Adaptive optics is clearly the way to go for future large-aperture solar telescopes — short of putting them in space — but a highly attractive niche for multi-line Stokes spectrometry at 0.2 arcsec resolution over fairly long duration at reasonable frequency seems available for computer-restored fiber spectrometry with small telescopes at good sites.

4. Solar fiber spectrometers

A fiber feed has recently been put into use at the spectrograph of the Tatranska Lomnica Observatory in Slovakia (Kucera et al. 1990) and is described by Kucera *et al.* (1997, 1999). It is not an integral-field feed but is designed to sample a fairly wide solar scene containing a sunspot at 23 discrete locations with 200 μm fibers for photometric spectrometry. Four additional fibers add iodine absorption lines for wavelength calibration.

During the present NSO workshop, Dr. Haosheng Lin of NSO/Sacramento Peak informed me that he has submitted a proposal for a fiber-fed near-infrared spectrometer along the lines discussed here, and was so kind to show me its details. It concerns a 16$\times$16 integral-field prototype of which the bundle size is set by the presently available cameras at NSO/Sacramento Peak, in particular the 256$\times$256 NICMOS 3 array. The tentative design uses microlenses, 200 μm fibers with 10 μm cladding, and an accompanying high-speed $f/6$ spectrometer with a 20$\times$40 cm^2 grating. The latter addition makes clear that fiber feeds require custom spectrometers that are much faster than solar spectrometers usually are. An obviously worthwhile goal is to achieve high-resolution 2D Stokes magnetometry at 1.6 μm.

Presently, the Netherlands Foundation for Research in Astronomy (NFRA) is embarking on a feasibility study for a fiber-fed spectrometer at the Dutch Open Telescope (DOT, see my description elsewhere in these proceedings). The requirements are harder than for Dr. Lin's proposal since it is intended for application in the visual with much smaller resolution elements (0.1 arcsec instead of 0.5 arcsec) and much larger detector arrays. Hence the need for a feasibility study. The DOT is an on-axis parabolic reflector with nearly axial symmetry and should enable precise Stokes magnetometry with relatively simple time-independent calibration. Also, it has no diurnal image rotation thanks to its parallactic mount. These properties make a DOT fiber spectrometer particularly attractive.

An obvious advantage of a fiber feed that is essential for application at the DOT is that the spectrometer may be detached from the telescope. The DOT is an open telescope with prime focus at its top (Fig.1) where there is neither sufficient space nor a sufficiently non-harsh environment to place any spectrometer, let be a large fast one. Using a fiber bundle as feed, the spectrometer may be placed in an enclosure just below the telescope, or perhaps on the ground below the telescope, or perhaps even in the nearby building of the Swedish Vacuum Solar Telescope from which the DOT is operated.

References

Al N., Bendlin C., Kneer F., 1998, A&A 336, 743

Bardeen S. C. (ed.), 1988, Fiber Optics in Astronomy, Astron. Soc. Pac. Conf. Series 3

Beckers J. M., Mauter H. A., Mann G. R., Brown D. R., 1972, Solar Phys. 25, 81

Bendlin C., Volkmer R., Kneer F., 1992, A&A 257, 817

Bonaccini D., Falchi A., Cauzzi G., Smaldone L. A., Falciani R., 1990, Astrophys. Space Sci. 170, 117

Bonaccini D., Righini A., Cavallini F., Ceppatelli G., 1989, A&A 217, 368

Brault J. W., 1978, in G. Godoli, G. Noci, A. Righini (eds.), Future solar optical observations: needs and constraints, Procs. JOSO Workshop, Osservazioni e Memorie Oss. Astrof. Arcetri, Florence, p. 33

Carrasco E., Parry I. R., 1993, in P. M. Gray (ed.), Fiber Optics in Astronomy II, Astron. Soc. Pacific Conf. Series 37, p. 392

Del Toro Iniesta J. C., Martinez Pillet V., Gonzalez Escalera V., 1997, in ASP Conf. Ser. 118: 1st Advances in Solar Physics Euroconference. Advances in Physics of Sunspots, p. 356

Elmore D. F., Lites B. W., Tomczyk S., Skumanich A. P., Dunn R. B., Schuenke J. A., Streander K. V., Leach T. W., Clambellan C. W., Hull H. K., Lacey L. B., 1992, Procs. SPIE 1746, 22

Evans J. W., 1956, Sky & Tel. 15, 436

Gray P. M. (ed.), 1993, Fiber Optics in Astronomy II, Astron. Soc. Pac. Conf. Series 37

Keller C. U., Johannesson A., 1995, Astronomy and Astrophysics Supplement Series 110, 565

Keller C. U., von der Lühe O., 1992, A&A 261, 321

Kneer F., 1997, in B. Schmieder, J. C. del Toro Iniesta, M. Vázquez (eds.), Advances in the Physics of Sunspots, Procs. First Adv. in Solar Physics Euroconf., Astron. Soc. Pac. Conf. Series 118, p. 329

Kucera A., Rybak J., Minarovjech M., Novocky D., Saniga M., 1990, Astrophys. Space Sci. 171, 279

Kucera A., Rybak J., Wöhl H., 1997, in M. Saniga (ed.), JOSO Annual Report 1996, Astron. Inst. Slovak Ac. Sciences, Tatranska Lomnica, p. 68

Kucera A., Rybak J., Wöhl H., Hanslmeier A., 1999, in B. Schmieder, A. Hofmann, J. Staude (eds.), Solar Magnetic Fields and Oscillations, Procs. Third Adv. Solar Physics Euroconf., Astron. Soc. Pac. Conf. Series, in press

Lites B. W., 1996, Solar Phys. 163, 223

Lites B. W., Martínez Pillet V., Skumanich A., 1994, Solar Phys. 155, 1

Lites B. W., Rutten R. J., Berger T. E., 1999, ApJ submitted

Lofdahl M. G., Berger T. E., Shine R. S., Title A. M., 1998, ApJ 495, 965

Mein P., 1991, A&A 248, 669

Mein P., 1992, in ESA, Solar Physics and Astrophysics at Interferometric Resolution p 153-155 (SEE N93-10901 01-90), p. 153

Mein P., Rayrole J., 1991, Advances in Space Research 11, 151

Paxman R. G., Seldin J. H., Loefdahl M. G., Scharmer G. B., Keller C. U., 1996, ApJ 466, 1087

Ramsey L. W., 1988, in S. C. Barden (ed.), Fiber Optics in Astronomy, Astron. Soc. Pacific Conf. Series 3, p. 26

Rimmele T. R., Goode P. R., Harold E., Stebbins R. T., 1995, ApJ 444, L119

Roudier T., Vigneau J., Mein P., Malherbe J. M., Coutard C., Hellier R., Muller R., 1991, A&A 248, 237

Sanchez Almeida J., Collados M., Martinez Pillet V., Gonzalez Escalera V., Scharmer G. B., Shand M., Moll L., Joven E., Cruz A., Diaz J. J., Rodriguez L. F., Fuentes J., Jochum L., Paez E., Ronquillo B., Carranza J. M., Escudero-Sanz I., 1997, in ASP Conf. Ser. 118: 1st Advances in Solar Physics Euroconference. Advances in Physics of Sunspots, p. 366

Semel M., 1994, in R. J. Rutten, C. J. Schrijver (eds.), Solar Surface Magnetism, NATO ASI Series C 433, Kluwer, Dordrecht, p. 509

Skumanich A., Lites B. W., Martinez Pillet V., Seagraves P., 1997, ApJS 110, 357

Stolpe F., Kneer F., 1998, A&AS 131, 181

Title A. M., Tarbell T. D., Topka K. P., Ferguson S. H., Shine R. A., the SOUP Team, 1989, ApJ 336, 475

Tritschler A., Schmidt W., Knolker M., 1997, in ASP Conf. Ser. 118: 1st Advances in Solar Physics Euroconference. Advances in Physics of Sunspots, p. 170

Vanderriest C., 1993, in P. M. Gray (ed.), Fiber Optics in Astronomy II, Astron. Soc. Pacific Conf. Series 37, p. 338

Westendorp Plaza C., Del Toro Iniesta J. C., Ruiz Cobo B., Martinez Pillet V., Lites B. W., Skumanich A., 1998, ApJ 494, 453

High Resolution Solar Physics: Theory, Observations, and Techniques
ASP Conference Series, Vol. 183, 1999
T.R. Rimmele, K. S. Balasubramaniam, R.R. Radick, eds.

Tridimensional Spectroscopic Observations of the Sun with a Microlens-Array Spectrograph

Y. Suematsu, H. Imai

National Astronomical Observatory,
Mitaka, Tokyo 181-8588, Japan

K. Yoshimura, T. Ishigaki, S. Ueno, T. Hayashi, and H. Ohtani

Department of Astronomy, Kyoto University,
Kitashirakawa, Sakyo-ku, Kyoto 606-8502, Japan

Abstract.
We describe a design concept and the observational limitations of the microlens-array (MLA) spectrograph with which simultaneous spectroscopy of two-dimensional fields (i.e., tridimensional spectroscopy) can be realized. As an example, we describe the Mitaka MLA spectrograph, which forms simultaneous Hα spectra covering 10 Å sampled at 0.13 Å at 1.4″× 1.4″ resolution over a 57″× 57″ field of view.

1. Introduction

Spectroscopy of two-dimensional fields is necessary for observations of most solar phenomena. This is usually done with spectroheliographs, or with scanning filter systems incorporating tunable Lyot filters or Fabry-Perot filters. However, these observational methods are not truly two-dimensional spectroscopy: they require some time to complete the necessary data set, and hence cannot give us accurate information, especially for transient phenomena.

The technique for producing a spectrum at each spatial element over an extended, two-dimensional field is called tridimensional (or integral field) spectroscopy, since there are two dimensions in space and one in wavelength. The scientific advantages of this technique for studies of localized and transient solar surface phenomena such as flares are obvious. Because light from the full extent of the structures is sampled simultaneously, one can make both monochromatic images and line profile analyses across two-dimensional fields.

One of the techniques for realizing tridimensional spectroscopy is the microlens-array spectrograph (Bacon 1995; Courtès et al. 1988; Ohtani et al. 1994; Ohtani 1995). An advantage of this technique is that we can adapt existing telescope-spectrograph systems simply by replacing the slit with a microlens-array and inserting a few optical components. Basically, the tiny pupil images made by the microlenses work as a multi-slit system (cf. the image slicer described by Weitzel et al. 1996). It should be pointed out that this technique may be incorporated easily and advantageously into high resolution imaging

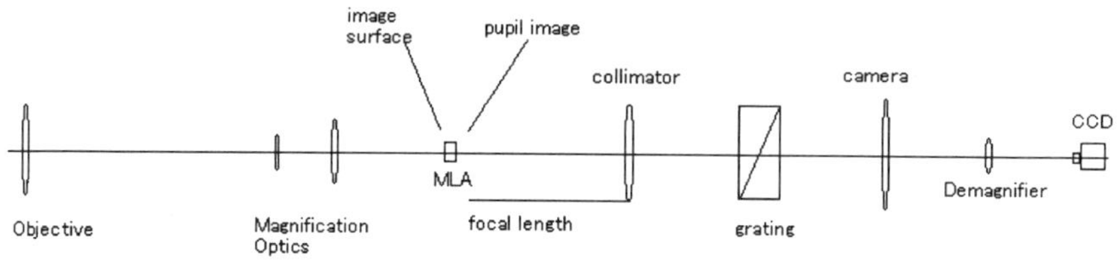

Figure 1. Layout of microlens-array (MLA) spectrograph. The MLA is substituted for the spectrograph slit.

procedures such as adaptive optics or post-facto image processing, and into polarimetric observations.

We will next present the design concepts and the observational limitations of a microlens-array (MLA) spectrograph implemented by converting an existing solar telescope-spectrograph system, and introduce the example that has been developed at Mitaka, NAOJ.

2. MLA Spectrograph Design

We assume utilization of an existing telescope-spectrograph system, because most solar observatories are equipped with them. We replace the spectrograph slit with an MLA (see Fig. 1).

The tiny pupil images formed by the microlenses work as a multi-slit. These pupil images should be equal to or larger than the optimum slit width of the spectrograph. Because the pupil image array is projected onto a detector through the spectrograph, the amount of information (space × wavelength) available is limited by the aperture area of one microlens. The maximum spectral length (Å) available is given by

$$\alpha \frac{\Delta\lambda\, D_{MLA}^2}{D_{pupil}}, \tag{1}$$

where D_{MLA} is the size of a square microlens, D_{pupil} is the diameter of the pupil image formed by a microlens, $\Delta\lambda$ is the dispersion (Å/mm) on the detector, and α is a filling factor specifying the separation of neighboring spectra in the direction perpendicular to the dispersion.

In order ensure an appropriate pupil size, we need to insert magnifying optics between the telescope objective and the MLA. To maximize the spectral length, we rotate the MLA around the optical axis by a fixed angle, as shown below. We also need to use a blocking filter of appropriate bandwidth to prevent spectral overlap from adjacent pupil images.

It is desirable for spectral line analysis that the spectral length be at least 20 times larger than the line width of the target spectral line. Using equation (1), we can then determine the necessary aperture size or pitch of the microlenses.

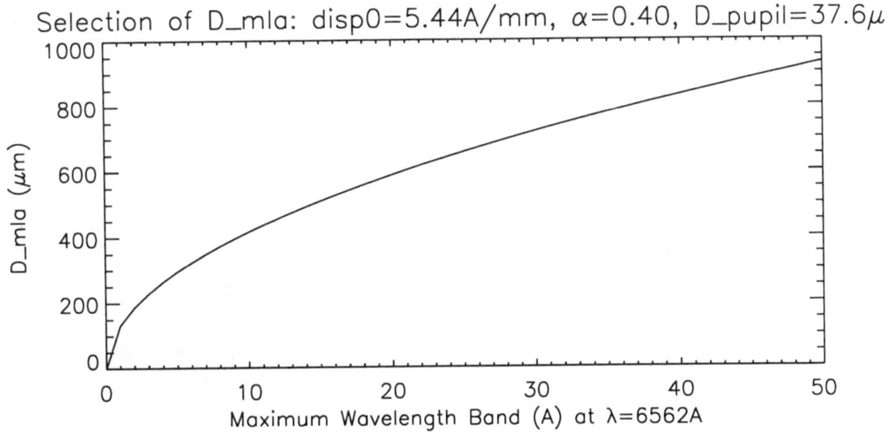

Figure 2. The relationship between the aperture size of one microlens and the achievable spectral length. The dispersion of the spectrograph is 5.44 Å/mm, the diameter of the pupil image produced by a microlens is 37.6 μm, and the filling factor for the spectrum height is 0.4.

Because the focal ratio of each microlens should be equal to or larger than that of the spectrograph, the focal length of the microlenses can then be decided.

Next, we consider the configuration of the spectra array on the detector. If we suppose that one spectrum has a height of j pixels, then the actual diameter h of the pupil on the detector is given by

$$h = \alpha \, j \, p, \tag{2}$$

where p is the pixel size, j is the given integer, and α is a filling factor less than unity. If the separation between neighboring pupil images is $n_x \, j \, p$ and $n_y \, j \, p$, along and perpendicular to the dispersion, respectively, the pitch L of the pupil array on the detector is given by

$$L = j \, p \, \sqrt{n_x^2 + n_y^2}. \tag{3}$$

Note that n_x and n_y should both be integers, ensuring that the spectra array fills the detector area as cleanly as possible. If n_x and n_y have no common measure, then the distance L_{sp} between two neighboring pupils aligned in the direction of dispersion is

$$L_{sp} = j \, p \, (n_x^2 + n_y^2). \tag{4}$$

Since the distance L_{sp} is the spectral length available, we select n_x and n_y to make L_{sp} as large as possible. Then the rotation angle θ of the MLA around the optical axis is given by

$$tan \, \theta = n_y / n_x. \tag{5}$$

Once α, j, n_x, n_y, and the specification of the MLA are selected, we can determine the magnification of the enlarger optics in front of the MLA and demagnifier optics in front of the detector.

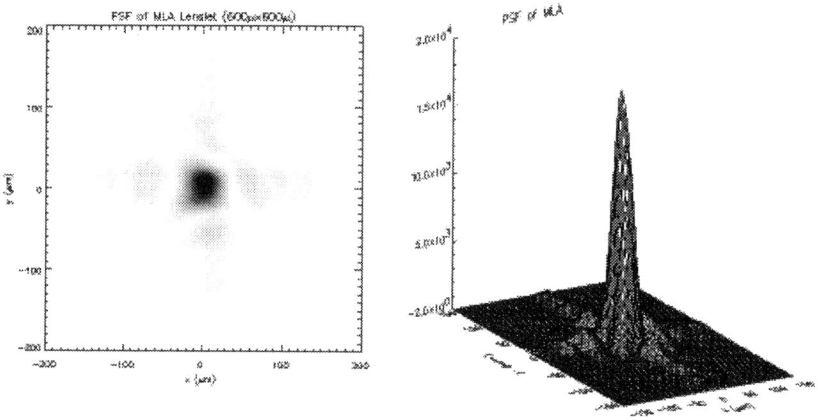

Figure 3. Example of the point spread function (PSF) of one microlens. The wave front error of this microlens is estimated to be about $1/20\lambda$ rms.

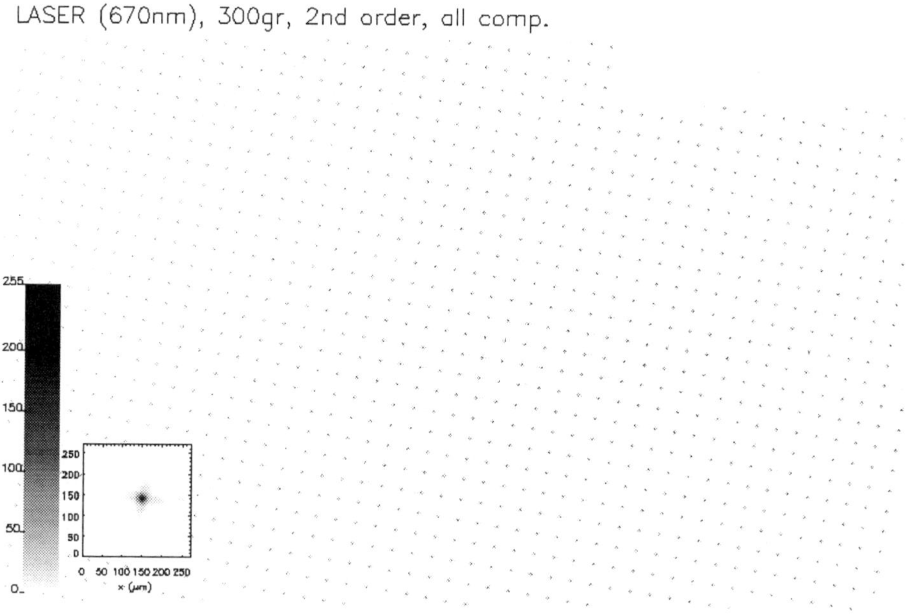

Figure 4. A (negative) image of a monochromatic point source (λ670nm laser) made with the MLA spectrograph. This illustrates the PSF of the entire system.

In this way, we can uniquely determine the specification of the necessary optical components.

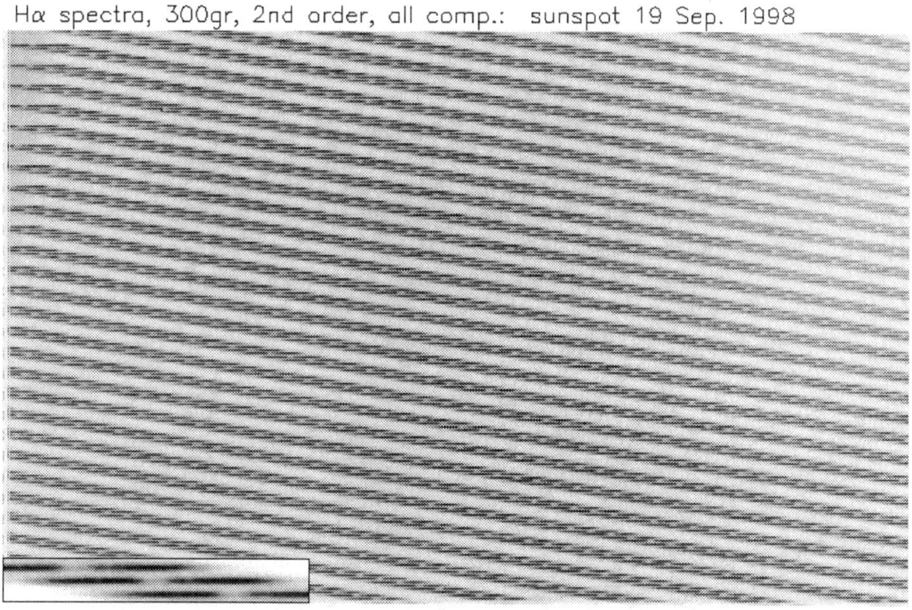

Figure 5. 2D Hα spectra (negative image) of the sun with the Mitaka MLA spectrograph. The area on the upper right corresponds to a sunspot. Enlarged spectra are shown in the lower-left corner; we can see Hα and a few atmospheric absorption lines.

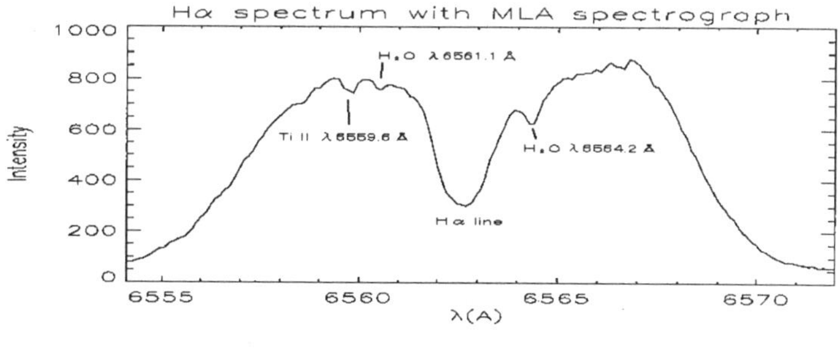

Figure 6. Example of Hα spectrum from the Mitaka MLA spectrograph.

The MLA spectrograph has a few drawbacks. When the focal ratio of telescope is equal to that of spectrograph ($F_{telescope} = F_{spectrograph}$, which is typical for most existing telescope-spectrograph systems), the spatial sampling by the MLA becomes under-sampled when we follow the procedure described above. We can derive the relation

$$\Delta x \geq 1.22 \frac{\lambda}{D}, \tag{6}$$

where Δx is the pitch of the MLA in radians, λ is the wavelength, and D is the diameter of the objective. This problem can be avoided if we have a spectrograph whose focal ratio is half as large as that of the telescope. Generally, the spectral sampling is also under-sampled.

Table 1. Telescope and Spectrograph at Mitaka

Sun Tracking	Coelostat	
Telescope	Refractor	ϕ12 cm, f460 cm (F/38.3)
Spectrograph	Evans-Fasty type	
	Grating	600 grooves/mm
		width 8.5cm, height 6.7cm
	Collimator mirror	ϕ15cm, f300cm
	Camera mirror	ϕ15cm, f300cm

3. The Mitaka MLA Spectrograph and a Sample of Hα Observation

We here describe the microlens-array spectrograph at Mitaka, NAOJ, which aims at improving flare observation in the Hα line. Although analysis of the Hα line profile in flare kernels would be useful for determining the energy transport mechanism from the corona to the chromosphere, the observations so far have been limited basically to one spatial dimension with conventional spectroheliographs (Ichimoto & Kurokawa 1984; Shoji & Kurokawa 1995; Wülser et al. 1994), or to one wavelength within the line with tunable filtergraphs (Kitahara & Kurokawa 1990).

Table 2. MLA Selection

Format	Square aperture, 50×50 array
Pitch	600μm × 600μm
Focal length	28.2 mm
Substrate	P.M.M.A.
Manufacturer	NALUX Co., LTD

Since the MLA spectrograph can produce simultaneous line profiles in a two-dimensional field, we can study more accurately the energy transfer mechanism in the two-dimensional field, and even in three dimensions when X-ray imaging observations from YOHKOH are available. When optimized, the Mitaka MLA spectrograph produces simultaneous Hα spectra, 10 Å wide, sampled at 0.13 Å for a $1.4'' \times 1.4''$ spatial element over a $57'' \times 57''$ field of view.

Table 1 gives the characteristics of the telescope and spectrograph. Using the relation given in equation (1), we chose the aperture size of a microlens to be 600μm (F/47) (see Fig. 2). The specifications of the MLA are summarized in Table 2, and a point spread function for one microlens of the purchased MLA is shown in Fig. 3. The optical quality of the microlenses and their uniformity were found to be very good. The necessary additional optical components are described in Table 3.

Table 3. Newly Inserted Optical Elements

Magnifier optics in front of MLA	Telecentric (convex + convex), Magnification of 18.6 $\times$
Demagnifier optics before CCD	Plano convex, Magnification of 0.365 $\times$
Blocking Filter	Center: Hα line, Bandpass: 10 Å
MLA rotation	9.46deg
CCD Camera	Format: 1536x1024, Pixel size: 9μm

To test imaging quality, we fed the monochromatic light of a laser (λ670 nm) through the objective and examined the aberration and scattering properties of the whole MLA spectrograph system (Fig. 4). We found that the main degradation originates from diffraction of the microlenses; the point-like image in Fig. 4 is very close to the PSF shown in Fig. 3.

An example of a solar observation with the MLA spectrograph is given in Fig. 5, which shows simultaneous Hα spectra of about 10 Å spectral length in a two dimensional field. In the figure, we can see a sunspot on the upper right side of the negative image. In the spectra, we can see a few atmospheric absorption lines around the Hα line. A representative profile of the spectrum is given in Fig. 6. To recover the spectral profile, we had to remove the transmission curve of the blocking filter. A full calibration scheme for the spectra is being developed at Mitaka.

Table 4. Currently-available Observation Scheme

Spatial sampling	$1.44''$
FOV	$57'' \times 57''$
Wavelength sampling	0.134 Å
Wavelength band	10 Å centered on the Hα line
Cadence	$\geq$2.5 sec (CCD readout limit)

References

Bacon R. 1995, in 3D Optical Spectroscopic Methods in Astronomy, (eds. G. Comte & M. Marcelin), ASP Conf. Ser. 71, 239

Courtès G. et al. 1988, in Instrumentation for Ground Based Optical Astronomy, (ed. L. B. Robinson), Springer-Verlag, New York, pg 267–274

Ichimoto K. & Kurokawa H. 1984, Solar Phys., 93, 105

Kitahara T. & Kurokawa H. 1990, Solar Phys., 125, 321

Ohtani H. 1995, in 3D Optical Spectroscopic Methods in Astronomy, (eds. G. Comte & M. Marcelin), ASP Conf. Ser. 71, 239

Ohtani H. et al. 1994, SPIE, 2198. 229

Shoji M. & Kurokawa H. 1995, PASJ, 47, 239

Weitzel L. et al. 1996, A&AS, 119, 531

Wülser J.-P. et al. 1994, ApJ, 424, 459

High Resolution Solar Physics: Theory, Observations, and Techniques
ASP Conference Series, Vol. 183, 1999
T.R. Rimmele, K. S. Balasubramaniam, R.R. Radick, eds.

Phase-Diversity Data Sets and Processing Strategies

R. G. Paxman, J. H. Seldin

ERIM International, P.O. Box 134008, Ann Arbor, MI 48113-4008

C. U. Keller

National Optical Astronomy Observatories/National Solar Observatory, P.O. Box 26732, Tucson, AZ 85726-6732

Abstract. Phase-Diverse Speckle (PDS) is a data-collection and processing technique that blends phase-diversity and speckle-imaging concepts. PDS has been successfully used to achieve near diffraction-limited resolution in ground-based imaging of solar granulation. Variants of PDS that involve narrow-band, spectroscopic, and polarimetric data provide more informative observations. We present preliminary results of a narrow-band time-series reconstruction. A short-lived continuum bright point with a peak intensity at a factor of 2.1 above the mean intensity in the continuum was observed. We also describe other candidate PDS data sets and their respective scientific goals.

1. Introduction

The method of phase-diverse speckle (PDS) provides fine spatial and temporal resolution images of evolving solar phenomena from ground-based observations. PDS has been demonstrated on broad-band solar-granulation data. However, more compelling science investigations require narrow-band observations. We report on a generalization of PDS to accommodate narrow-band data. In the next section, we briefly review the method of PDS. In section 3 we describe data collection and the attendant processing procedure leading to co-spatial and co-temporal continuum and narrow-band reconstructions. An unusual and short-lived continuum bright point with a peak intensity at a factor of 2.1 above the mean intensity in the continuum was observed. In section 4 we discuss advanced PDS imaging modalities for collecting and restoring polarimetric, Doppler, and spectroscopic images to resolutions approaching the diffraction limit.

2. Phase-Diverse Speckle

Phase-diverse speckle is a relatively new imaging modality that has been used to overcome image degradation arising from atmospheric turbulence. PDS blends the strengths of short-exposure speckle data and phase-diversity imaging. In phase-diversity imaging (Gonsalves & Chidlaw 1979, Paxman et al. 1992), two or more images are simultaneously collected. Typically, the first image is a

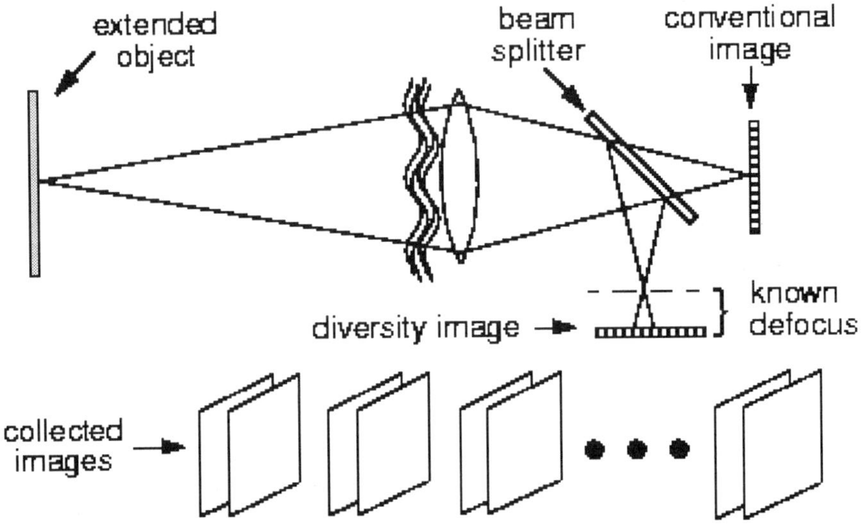

Figure 1. Phase-diverse speckle data-collection scheme.

conventional focal-plane image that is degraded by unknown aberrations. A second (diversity) image that is intentionally defocused by a known amount is also collected with the aid of a simple beam splitter. Note that the diversity image is influenced by both the unknown aberrations and the known amount of defocus. The goal in phase diversity is to identify an object and an aberration function that are consistent with both collected images, given the known phase diversity.

In many scenarios, PDS data are richer and more informative than simple phase-diversity data. PDS data are recorded when a pair of short-exposure images (conventional and diversity) are collected for each of multiple aberration realizations. A schematic of the PDS data collection is given in Fig. 1. We require that the object being imaged is essentially unchanged over the multiple realizations. The principle of maximum likelihood can be used to jointly estimate a single unaberrated object and an aberration function for each realization associated with the data set (Seldin & Paxman 1994, Paxman et al. 1996). Often the object estimate is of most interest but the aberration estimates can give diagnostic information about fixed aberrations and turbulence statistics.

PDS is attractive for ground-based solar astronomy for the following reasons: (1) the optical hardware needed is simple, (2) far fewer realizations (relative to speckle imaging) are needed to achieve near diffraction-limited results, thus providing enhanced temporal resolution in time-series studies, (3) the need for speckle calibration is obviated, and (4) PDS can be generalized to accommodate narrow-band, polarimetric, and spectroscopic observations. In addition, PDS has been shown to effectively correct residual aberrations in images of satellites *after* adaptive-optics (AO) compensation (Seldin et al. 1997). We view the complementary use of AO and PDS as a most compelling strategy for ground-based solar observations.

3. Narrow-Band Reconstructions

In this section, we show how PDS can be generalized to reconstruct faint, narrow-band images, report on scientifically intriguing time-series observations made with this generalized method, and discuss advanced processing strategies that we are currently investigating. On 1995 June 22, a PDS set of white-light images were collected simultaneously with narrow-band images using the 76-cm Richard B. Dunn Solar Telescope at the National Solar Observatory on Sacramento Peak in Sunspot, New Mexico. Results from processing the phase-diverse continuum images were reported previously (Seldin & Paxman 1996), where the focus was on the temporal evolution of fine-resolution continuum bright points and time-series processing strategies. The restored time series of a plage depicted the highly dynamic photosphere at scales below 0.3 arcsec, close to the diffraction limit of the telescope (0.18 arcsec). Others have successfully used PDS to study evolving solar phenomena in differing bands (Löfdahl et al. 1998). The method we discuss here is distinct in that the point-spread functions (PSF's) estimated from the phase-diverse images in the broad-band channels are used to restore the images obtained behind the narrow-band filter. As a consequence, spectral resolution (optical bandwidth) of filtergrams can be significantly improved. The methodology developed here demonstrates the general utility of using PDS to recover fine-resolution information from fairly noisy narrow-band data.

At each exposure interval, we collected a triplet of images consisting of a phase-diverse pair of images in a 5-*nm* passband centered at 656.3 *nm* and a third, narrow-band (0.05-*nm* optical bandwidth) image in the wing of the Hα line. For the data discussed here, the broad-band and narrow-band images were taken at a common center wavelength. The methodology outlined here can also be used to retrieve a near diffraction-limited image at an arbitrary narrow-band wavelength (Paxman & Seldin 1993). With the scientific goal of studying the relationship between filigree and continuum bright points, data were recorded with three cameras in parallel at a rate of seven frames per second per camera using the Zurich Imaging Stokes POLarimeter I (ZIMPOL I) system. Two high-speed CCD cameras were used for the phase-diversity data, and a third was placed behind the Universal Birefringent Filter (UBF) in the wing of Hα. An exposure time of 15 *msec* was used, and we assumed that the turbulence did not evolve significantly during an exposure. At most observing sites, this assumption is valid for exposure times between 10 and 20 *msec*, and we have obtained good results using 20 *msec* exposures (Paxman et al. 1996).

Images of a plage were collected for 13.5 minutes, using real-time frame selection to retain the sharpest 3 out of every 50 triplets acquired. A set of 300 image triplets was obtained during this period of observation, and this set was divided into disjoint groups of 10 triplets each. Isoplanatic image patches with a width of approximately 3 arsec were then extracted. Examples of these patches are shown along the left side of Fig. 2, with the phase-diverse (focus/defocus) image pairs shown at the top and the Hα patches below. The processing of a 10-

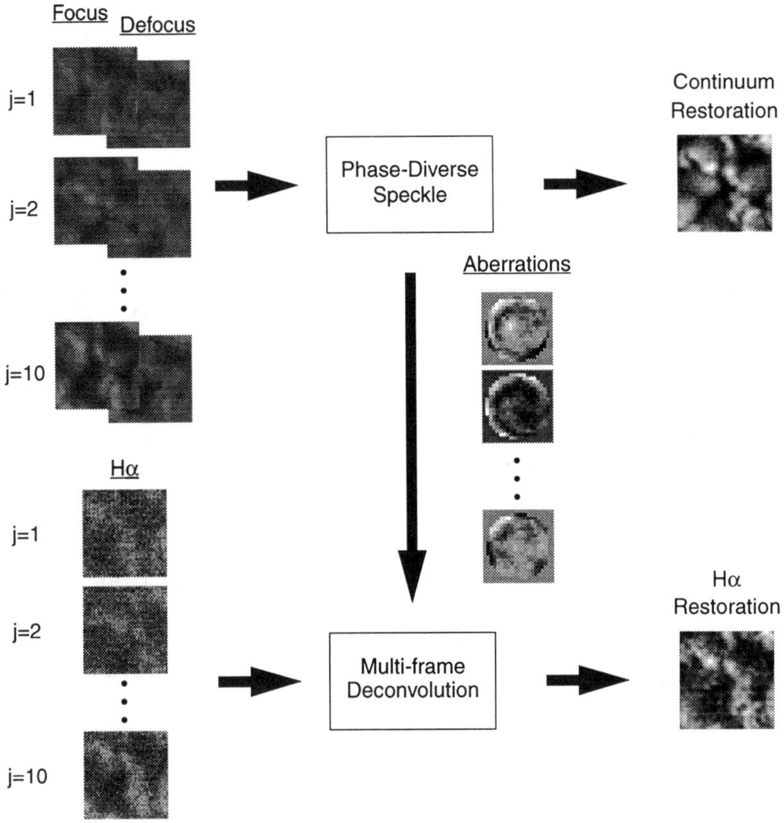

Figure 2. Three-channel, multi-realization image restoration is depicted. Ten phase-diverse image pairs (top left) are processed with the PDS algorithm to produce maximum-likelihood estimates of the continuum object (top right) and the 10 realizations of aberrations (center). The aberration estimates are input to a multi-frame deconvolution step that is used to form a reconstruction (bottom right) from the 10, co-temporal narrow-band images taken in the wing of Hα (bottom left).

realization set, which spans a period of about 27 seconds of observation time, is depicted in this figure. We consider the scene to be approximately constant over this 27-second window, and the phase-diverse pairs are processed to produce a set of 10 aberration estimates (middle) and a single reconstructed continuum image (upper-right). The aberration estimates are parameterized initially with the first 45 orthonormal Zernike polynomials on the circular aperture of the Dunn Solar Telescope. This is a fairly low-order parameterization that helps to establish the overall structure of the phase aberration, but which has been shown in simulations to lead to the under-estimation of the variance in the restored object (Seldin et al. 1995). In an attempt to avoid this problem, we use the low-order Zernike phase representation for the first portion of the PDS processing, then finish by representing the phase with a separate parameter for each of the 250 pixels used to model the pupil.

The 10 phase-diversity aberration estimates yield 10 PSF's that are used to deconvolve the narrow-band data. The multi-frame deconvolution is done in a single step to produce the restoration shown at the bottom right of Fig. 2 The narrow-band object estimate is given by

$$\hat{f}_n = \mathcal{F}\left\{\frac{\sum_j D_j \hat{S}_j^{\,*}(u)}{|\hat{S}_j|^2 + \epsilon}\right\}, \tag{1}$$

where D_j is the Fourier transform of the jth narrow-band image, $\hat{S}_j$ is the estimated optical transfer function (OTF) for the jth realization of turbulence, ϵ is a regularization parameter that suppresses high-frequency artifacts, and $\mathcal{F}\{\cdot\}$ is a Fourier transform. This estimator is known to give the regularized, maximum-likelihood object estimate under a Gaussian noise model and the assumption that the OTF's are known exactly. In practice, the jth OTF is estimated by using the jth PDS aberration estimate and is defined as the Fourier transform of the jth PSF estimate (Goodman 1968). The formation of the narrow-band estimate, $\hat{f}_n$, from Eq. (1) is direct; it does not involve an iterative optimization procedure and is, thus, very fast. The two-step process depicted in Fig. 2 is similar in concept to the staged procedure used in deconvolution with wavefront sensing (Gonglewski et al. 1990, Primot et al. 1990).

The narrow-band data shown in the figure are quite noisy and suffer from a time-varying, horizontal striping artifact. Thus, the restoration at the bottom right of the figure has a mottled appearance in the low-intensity areas. Fortunately, the scientific goals are related to bright points in the wing of Hα, and the high-intensity areas of interest are more faithfully restored. The restorations are tiled together to produce a larger image product, and a time series of full-field restorations in both pass bands are shown in Fig. 3. Fig. 3(a) contains a series of 12 consecutive continuum restorations in a 9-arcsec by 13-arcsec field of view. Each full-field restoration is formed by tiling 35 overlapping 3-arsec by 3-arcsec images. The time sequence rasters from the upper left to the lower right of Fig. 3(a). The restorations have significantly more than twice the contrast (defined as the ratio of the standard deviation of an image to its mean) as the original, aberrated data frames, and fine-resolution details approaching the diffraction limit of 0.18 arcsec can be observed. The corresponding narrow-band restorations are shown below in Fig. 3(b), where it is easy to associate fine-resolution bright points with corresponding regions in the white-light restorations. The high degree of correlation between narrow-band and continuum features in the areas bordering the pores enhances our trust in the features recovered from the noisy, narrow-band data. The ultimate goal of this analysis is to study the association between continuum and narrow-band bright points, and in this series we observe the evolution of an extremely bright region in the upper-middle portion of the continuum images. The bright area reaches its peak intensity at a factor of 2.1 above the mean intensity in the ninth restoration (3rd row, far left). Continuum bright points at this intensity have been observed in the past (Koutchmy

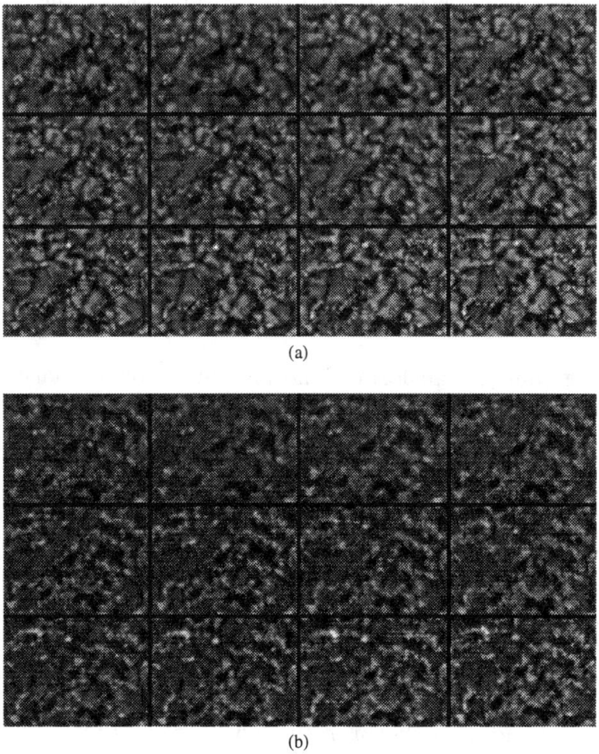

Figure 3. Time series of full-field continuum and narrow-band restorations. (a) A series of 12, 9 x 13-arcsec continuum restorations. The frames are spaced in time by approximately 27 sec and raster from the upper left to the lower right. A bright point evolves in the upper-middle portion of the field, reaching a maximum of 2.1 times the continuum mean in frame 9. (b) The corresponding restorations of data taken in the wing of the Hα line share many of the fine-resolution features along the borders of pores. The brightest features are actually found to the left of the region where the brightest point in the continuum restorations is observed.

1977), but only infrequently. It is interesting to note that this same area is not, in fact, the brightest region in the Hα data. A high-intensity region to the left of this area that forms during the same period of time is actually brighter.

Additional data collected during this observation run will be processed in the future, with the goal of enhancing our understanding of the spatial and temporal correlation of broad-band/narrow-band bright points. Refined approaches to accelerating the processing and enhancing the consistency of the restored resolution will also be considered. The time-series processing approach used to generate these restorations does not use the temporal history of object estimates to aid the formation of new estimates. The entire sequence of data was divided into disjoint groups, and restorations were formed independently for each group. As an alternative to this approach, we have implemented what we refer to as a "sliding window" estimation technique that takes advantage of previous object estimates to help form new estimates (Seldin & Paxman 1996). We believe that this will be an important method to use to overcome the loss of resolution in the reconstructions during intervals of poor seeing. In addition to time-series processing strategies, we will explore sub-fielding approaches to accelerate the restoration of large fields of view. One strategy is to estimate PSF's from small and sparsely located patches. A reduced patch size implies a computational savings in the iterative PSF estimation procedure. The estimated PSF's can then be used in a direct (non-iterative) step to deconvolve larger isoplanatic patches prior to mosaicing.

4. Advanced PDS Imaging Modalities

In addition to single-channel narrow-band observations, PDS can be generalized to make polarimetric, Doppler, and spectroscopic solar observations. It is our goal to demonstrate each of these advanced PDS imaging modalities by making scientifically useful observations, thereby validating the utility of each modality. We briefly touch on acquisition requirements and scientific goals for each of these imaging modalities.

PDS polarimetry can be accomplished by collecting 4 image channels with the aid of 3 ZIMPOL cameras. Two cameras are used to collect phase-diversity image pairs in the continuum. A third ZIMPOL camera can be used to modulate between left and right circular polarization at 42 KHz. The two polarization images (channels) will be spatially encoded in an interleaved fashion on this third camera. As with the single narrow-band processing, the broad-band PDS images will allow us to estimate a series of PSF's that will in turn allow us to perform multi-frame deconvolution on the polarimetric images. PDS polarimetric data will aid in studying the evolution of small-scale magnetic fields.

PDS Doppler observations can also be obtained with 3 ZIMPOL cameras that collect 4 image channels. For this modality, phase-diversity image pairs are collected in the continuum while a third camera is used to time-multiplex between 2 wavelengths using a Fabry-Perot spectrometer. Alternatively, the UBF could be used to measure both wavelengths. Under this scheme, the 2 wavelengths could be encoded in opposite polarization states so that the ZIMPOL system will automatically interleave the two images. As before, PSF's estimated from the diversity channels will be used for multi-frame deconvolution in the narrow-band channels. PDS Doppler reconstructions can be used to study the evolution of shock waves in granulation and active regions. In addition, we hope to study small-scale p-mode waves.

PDS spectroscopy resembles speckle spectroscopy (Keller & Johannesson 1995) except that PSF estimates derive from PDS rather than from speckle imaging. PDS spectroscopy requires the collection of PDS data in the continuum while simultaneously collecting a data cube (angle-angle-wavelength) using a slit spectrograph. The acquisition timing must insure that the slit traverses the scene on a time scale that is slow relative to the evolving turbulence but fast relative to the evolving solar phenomena. Keller has worked out the theory of deconvolving such data, given the PSF estimates (Keller & Johannesson 1995). We plan to use PDS spectroscopy to observe velocity, temperature, and magnetic-field variations at the diffraction limit. This modality will also be useful for studying waves in magnetic elements and the Evershed effect in sunspot penumbrae.

Finally, these advanced PDS modalities complement the use of adaptive optics (AO) in observations. From a signal-to-noise perspective, pre-detection correction afforded by AO is preferable to post-detection correction accomplished with PDS. However, owing to wavefront-sensing errors, deformable-mirror fitting errors, and servo-lag errors, AO systems never achieve perfect correction. Thus PDS retains an important role in correcting residual aberrations after compensation. Furthermore, AO can only correct over one isoplanatic patch. PDS can be used in conjunction with AO to extend the size of corrected images for observations that require larger fields of view.

5. Conclusions

PDS has been generalized to accommodate narrow-band observations. We have reported on data collection, the associated processing methodology, and the resulting time-series reconstruction. A short-lived continuum bright point with a peak intensity at a factor of 2.1 above the mean intensity in the continuum was observed. Advanced PDS imaging modalities for collecting polarimetric, Doppler, and spectroscopic images at resolutions approaching the diffraction limit were discussed. PDS modalities open a new window for the investigation of small, rapidly evolving solar phenomena.

Acknowledgments. We acknowledge the efforts of Cassandra Hoye and Kevin Auslander of ERIM International for developing time-series processing and mosaicing tools. Also, Dr. Guenther Elste of the University of Michigan assisted in the collection of the data. The bulk of this research was supported by the NSF Advanced Technologies and Instrumentation Program, NSF Grant No. AST-9817796. The data collection was supported by the NASA space Physics Program Office, NASA Grant No. NAGW-4069. The National Solar Observatory is one of the National Optical Astronomy Observatories, which are operated by the Association of Universities for Research in Astronomy, Inc. (AURA) under cooperative agreement with the National Science Foundation.

References

Gonglewski, J. D., Voelz, D. G., Fender, J. S., Dayton, D. C., Spielbusch, B. K., & Pierson, R. E. 1990, Appl. Opt., 29, 4527

Gonsalves, R. A. & Chidlaw, R. 1979, Proc. SPIE, 207, 32

Goodman, J. W. 1968, Introduction to Fourier Optics, McGraw-Hill

Keller, C. U. & Johannesson, A. 1995, A&AS, 110, 565

Koutchmy, S. 1977, A&A, 61, 397

Löfdahl, M. G., Berger, T. E., Shine, R. A.,& Title, A. M. 1998, ApJ, 495, 965

Paxman, R. G., Schulz, T J., & Fienup, J.R. 1992, J. Opt. Soc. Am. A, 9, 1072

Paxman, R. G. & Seldin, J. H. 1993, Proc. SPIE, 2029, 287

Paxman, R. G., Seldin, J. H., Löfdahl, M. G., Scharmer, G. B., & Keller, C. U. 1996, ApJ, 467, 1087

Primot, J., Rousset, G., & Fontanella, J. C. 1990, J. Opt. Soc. Am. A, 7, 1598

Seldin, J. H.& Paxman, R. G. 1994, Proc. SPIE, 2302, 268

Seldin, J. H., Paxman, R. G., & Elste, G. H. 1995, BAAS, 27, 950

Seldin, J. H. & Paxman, R. G. 1996, Proc. SPIE, 2804, 166

Seldin, J. H., Reiley, M. F., Paxman, R. G., Stribling, B. E., Ellerbroek, B. L., & Johnston, D. C. 1997, Proc. SPIE, 3170, 2

High Resolution Solar Physics: Theory, Observations, and Techniques
ASP Conference Series, Vol. 183, 1999
T.R. Rimmele, K. S. Balasubramaniam, R.R. Radick, eds.

Orthogonalization of Basis Functions for Diagonalized Wavefront Sensing

Mats G. Löfdahl

Royal Swedish Academy of Sciences
Stockholm Observatory, SE–133 36 Saltsjöbaden, Sweden

Abstract. Phase diversity problems are usually solved by gradient search techniques or by non-linear least-squares optimization methods. In the latter case, when the wavefront is expanded in Zernike polynomials, a system matrix with significant off-diagonal elements is obtained. We can approximately diagonalize the system matrix by orthogonalizing the basis functions used to expand the wavefront. Using only the diagonal of the $M \times M$ resulting matrix, reducing the need for time-consuming inner products from $\sim M^2$ to $\sim M$, closed loop simulations converge to the same error level as full-matrix solutions. We show results with real solar data as well as with simulations corresponding to a closed loop adaptive optics system.

1. Introduction

Wavefront sensing methods for real-time control of adaptive optics (AO) systems need to combine accuracy and sensitivity, large capture range, low cross-talk with unwanted aberrations and speed with ease of alignment and operation. For certain applications, such as with solar telescopes, wavefront sensors operating on extended objects are needed.

Phase diversity (PD) methods have been described in several publications (e.g. Löfdahl & Scharmer 1993, 1994;Kendrick, Acton & Duncan 1994; Paxman, Schulz, & Fienup 1992) since the concept was first presented by Gonsalves & Chidlaw (1979). PD wavefront sensors are very simple optically and require only a simultaneous focused and defocused image. PD sensors are also easy to calibrate and can be used with extended scenes as well as point sources. However, existing PD implementations are computationally expensive.

Our group is investigating different schemes for reducing the amount of computation for real-time applications of PD, while retaining the necessary accuracy of the estimated wavefronts (Löfdahl, Duncan, & Scharmer 1998b; Scharmer 1998). In this paper we show that the number of costly inner-products (IPs) can be reduced from $\sim M^2$, where M is the number of modes, to $\sim M$ by using only the diagonal of the $M \times M$ system matrix. In this context, Zernike polynomials (Noll 1976) are shown *not* to be an optimum choice of basis functions for expansion of the wavefront. Better results are obtained when the basis functions are orthogonalized with respect to the optical setup and PD algorithm used.

Section 2 describes how each iteration in our algorithm can be viewed as a least squares fit to a wavefront and how the system of normal equations can be approximately diagonalized by orthogonalizing the basis functions used in the fit. In Section 3 we test, with real data as well as simulations, a number of basis function sets orthogonalized with this procedure. We discuss the results in Section 4.

2. Phase Diversity and Diagonalization

2.1. Phase diversity as a least-squares fit

PD wavefront sensing is usually formulated as a non-linear optimization problem where an error metric is minimized so that wavefront and object estimates are optimized in a maximum likelihood (ML) sense (Paxman, Schulz, & Fienup 1992). A detailed description of our algorithm is given by Löfdahl & Scharmer (1994). When additive Gaussian noise is assumed in two phase-diverse images, D_1 and D_2 in the Fourier domain, the joint ML estimates of wavefront and object are given by the minimization of an error metric,

$$L = \|E\|_2^2, \tag{1}$$

that is the squared norm of an error function,

$$E = H \cdot \frac{D_1 \widehat{S}_2 - D_2 \widehat{S}_1}{\sqrt{|\widehat{S}_1|^2 + |\widehat{S}_2|^2}}, \tag{2}$$

where H is a low-pass noise filter and $\widehat{\cdot}$ denotes the current estimate. The estimated OTFs, $\widehat{S}_k$, are functions of the estimated wavefront, $\widehat{\phi}$, through the phase of the auto-correlated pupil transmission function. Note that L is independent of the estimated object, which greatly reduces the number of parameters. The wavefront is expanded in basis functions, usually Zernike polynomials $\{Z_m\}_{m=1}^{M}$,

$$\widehat{\phi} = \sum_{m=1}^{M} \alpha_m Z_m. \tag{3}$$

The norm in Eq. (1) is the Euclidean distance, defined through

$$\|A\|_2 = (A \mid A)^{1/2}, \tag{4}$$

where the IP can be written as

$$(A \mid B) = \sum_u A(u) B^*(u) \approx \sum_{x \in \mathcal{X}} \mathfrak{F}^{-1}\{A\}(x) \cdot \mathfrak{F}^{-1}\{B\}(x). \tag{5}$$

Here, complex conjugate is denoted with $*$ used as a superscript and the inverse Fourier transform operator as $\mathfrak{F}^{-1}$. $\mathcal{X}$ is a centered field of view, surrounded by a guard band that prevents large wrap-around errors from convolutions over finite subfields from influence the result (Löfdahl & Scharmer 1994). Parseval's identity and related theorems assure that with no guard band the two sums are strictly equal.

We linearize the problem and find the optimum wavefront by iteratively minimizing $\|E + \sum_m^M \alpha_m \cdot \partial E/\partial \alpha_m\|_2^2$. The normal equations for this minimization can be written as

$$\begin{pmatrix} \left(\frac{\partial E}{\partial \alpha_1} \mid \frac{\partial E}{\partial \alpha_1}\right) & \cdots & \left(\frac{\partial E}{\partial \alpha_1} \mid \frac{\partial E}{\partial \alpha_M}\right) \\ \vdots & \ddots & \\ \left(\frac{\partial E}{\partial \alpha_M} \mid \frac{\partial E}{\partial \alpha_1}\right) & & \left(\frac{\partial E}{\partial \alpha_M} \mid \frac{\partial E}{\partial \alpha_M}\right) \end{pmatrix} \begin{pmatrix} \delta\alpha_1 \\ \vdots \\ \delta\alpha_M \end{pmatrix} = \begin{pmatrix} \left(-E \mid \frac{\partial E}{\partial \alpha_1}\right) \\ \vdots \\ \left(-E \mid \frac{\partial E}{\partial \alpha_M}\right) \end{pmatrix}, \tag{6}$$

where we refer to the $M \times M$ matrix as the system matrix. The solution vector, $\delta\boldsymbol{\alpha}$, is corrections to the current estimate of the wavefront expansion coefficients, $\boldsymbol{\alpha} = (\alpha_1 \ldots \alpha_M)^{\mathrm{T}}$. Our experience with these matrices is that the off-diagonal elements are significant. Thus, Zernike polynomials are not orthogonal for this problem. Normally, we use singular value decomposition (SVD) methods in order to deal with near-singular matrices.

2.2. Phase diversity as a wavefront fit

Transforming Eq. (6) into equations involving wavefronts and basis functions allows us to use vector-space methods to improve the properties of the matrix. Because $\partial E/\partial \alpha_m$ is a function of Z_m, we can formally express the IPs of $\partial E/\partial \alpha_m$ as new IPs of Z_m directly, so that

$$\langle Z_{m_1} \mid Z_{m_2} \rangle = \left(\frac{\partial E}{\partial \alpha_{m_1}}(Z_{m_1}) \,\middle|\, \frac{\partial E}{\partial \alpha_{m_2}}(Z_{m_2}) \right) \tag{7}$$

When the aberrations are small, the error function derivatives can be written as

$$\frac{\partial E}{\partial \alpha_m}(Z_m) = HF \cdot \frac{S_1 \frac{\partial \widehat{S}_2}{\partial \alpha_m}(Z_m) - S_2 \frac{\partial \widehat{S}_1}{\partial \alpha_m}(Z_m)}{\sqrt{|\widehat{S}_1|^2 + |\widehat{S}_2|^2}}, \tag{8}$$

where we have used $D_k = FS_k$ and thus ignored noise effects in the signal-dominated regime where H is ~ 1. Using the first sum in Eq. (5), we obtain

$$\langle Z_{m_1} \mid Z_{m_2} \rangle = \sum_u H^2 \cdot |F|^2 \cdot \frac{\left(S_1 \frac{\partial \widehat{S}_2}{\partial \alpha_{m_1}} - S_2 \frac{\partial \widehat{S}_1}{\partial \alpha_{m_1}} \right)\left(S_1^* \frac{\partial \widehat{S}_2^*}{\partial \alpha_{m_2}} - S_2^* \frac{\partial \widehat{S}_1^*}{\partial \alpha_{m_2}} \right)}{|\widehat{S}_1|^2 + |\widehat{S}_2|^2}. \tag{9}$$

With this notation, Eq. (6) can be rewritten as

$$\begin{pmatrix} \langle Z_1 \mid Z_1 \rangle & \cdots & \langle Z_1 \mid Z_M \rangle \\ \vdots & \ddots & \\ \langle Z_M \mid Z_1 \rangle & & \langle Z_M \mid Z_M \rangle \end{pmatrix} \begin{pmatrix} \delta\alpha_1 \\ \vdots \\ \delta\alpha_M \end{pmatrix} = \begin{pmatrix} \langle \phi_E \mid Z_1 \rangle \\ \vdots \\ \langle \phi_E \mid Z_M \rangle \end{pmatrix}. \tag{10}$$

The Zernike polynomials together with the new IP and its corresponding norm, $\|\cdot\|_\phi = \langle \cdot \mid \cdot \rangle^{1/2}$, form a Hilbert space. Therefore, we can treat the problem as a least-squares fit of ϕ_E directly, minimizing $\|\phi_E - \widehat{\phi}\|_\phi^2$.

We note that ϕ_E is unknown but also that we never have to actually calculate the IPs at the right-hand side of Eq. (10) directly. Also, in general ϕ_E is not equal to the real wavefront, ϕ, because ϕ_E has to allow for errors due to the joint estimation of the object. This is a desired property for image restoration but not necessarily for wavefront sensing.

2.3. Orthogonalization

A least-squares fit is best performed if, instead of Zernike polynomials, we are using an orthonormal set of basis functions because the off-diagonal elements in the system matrix disappear and the diagonal elements are unity. If we denote the orthogonal basis by $\{\psi_m\}$, we can write down the solution directly as

$$\widehat{\phi} = \sum_{m=1}^{M} \beta_m \psi_m = \sum_{m=1}^{M} \frac{\langle \phi_E \mid \psi_m \rangle}{\langle \psi_m \mid \psi_m \rangle} \psi_m. \tag{11}$$

In reality, the diagonalization will not be perfect, particularly when the assumption of small wavefronts is violated. In spite of this we can expect advantages with the new basis functions. The equation system should at least become more well-conditioned. With a closed loop system, the diagonalization becomes better as the wavefront estimate is improved and as long as we obtain convergence, final accuracy is not compromised.

Zernike polynomials are orthonormal with respect to a third, distinct IP,

$$[Z_i \mid Z_j] = \frac{1}{\pi} \iint_{r \leq 1} Z_i Z_j \, r dr \, d\theta \tag{12}$$

but not to the new IP in Eq. (7) that is relevant to PD. It is, however, easy to orthogonalize a set of linearly independent functions using the modified Gram–Schmidt orthogonalization process[1].

We want the results to be comparable to those obtained with Zernike polynomials, in the sense that for any M, the solution space spanned by the M first Zernike polynomials should also be spanned by the same number of functions in the new basis set. A solution in terms of the new basis is then easy to express in Zernike polynomials. We use as the linearly independent input basis to the Gram–Schmidt process the highest order terms of the Zernike polynomials, in the same order.

Set 1 in Table 1 is orthogonalized with respect to a sampled version of Eq. (12), appropriate for 128×128-pixel subfields of the data discussed in Section 3.1. Note the difference between Zernikes and Set 1, which one might expect to be similar. The differences decrease when the pupil sampling is denser, which indicates that this orthogonalization process not only catches mathematical differences, but also numerical misrepresentations in the pupil geometry. Since the real IP used in the PD code is subject to the same sampling misrepresentation, this alone is a reason to consider changing basis functions.

---

[1]Mathematically equivalent to the original Gram–Schmidt process but better numerically (e.g. Dahlquist et al. 1974).

Table 1. Example basis functions and the IPs used for orthogonalization. The symbols in parentheses are used in Figs. 1 and 3.

Set	Description	Example polynomials
Z (−)	Zernike polynomials.	$Z_{11} = 13.4(r^4 + 1.00r^2 + 0.17)$ $Z_{14} = 3.16r^4 \cos 4\theta$ $Z_{15} = 3.16r^4 \sin 4\theta$
1 (+)	Sampled Euclidean IP.	$\psi_{11} = 13.8(r^4 + 0.99r^2 + 0.16)$ $\psi_{14} = 3.26r^4 \cos 4\theta$ $\psi_{15} = 3.27(r^4 \sin 4\theta + 0.19r^4 + 0.15r^2 + 0.02)$
2 (×)	Point source, $H(60\%)$, diffraction limited MTFs.	$\psi_{11} = 3.33(r^4 + 0.90r^2 + 0.12)$ $\psi_{14} = 0.89r^4 \cos 4\theta$ $\psi_{15} = 0.89(r^4 \sin 4\theta + 0.22r^4 + 0.18r^2 + 0.02)$
3 (□)	Granulation power, $H(100\%)$, diffraction limited MTFs.	$\psi_{11} = 306(r^4 + 0.92r^2 + 0.13)$ $\psi_{14} = 104r^4 \cos 4\theta$ $\psi_{15} = 104(r^4 \sin 4\theta + 0.12r^4 + 0.09r^2 + 0.01)$
4 (△)	Granulation power, $H(60\%)$, diffraction limited MTFs.	$\psi_{11} = 314(r^4 + 0.92r^2 + 0.13)$ $\psi_{14} = 105r^4 \cos 4\theta$ $\psi_{15} = 105(r^4 \sin 4\theta + 0.13r^4 + 0.09r^2 + 0.01)$
5 (◇)	Granulation power, $H(30\%)$, diffraction limited MTFs.	$\psi_{11} = 407(r^4 + 0.95r^2 + 0.14)$ $\psi_{14} = 135r^4 \cos 4\theta$ $\psi_{15} = 134(r^4 \sin 4\theta + 0.15r^4 + 0.12r^2 + 0.02)$
6 (▽)	Granulation power, $H(60\%)$, average MTFs from data.	$\psi_{11} = 461(r^4 + 0.94r^2 + 0.14)$ $\psi_{14} = 184r^4 \cos 4\theta$ $\psi_{15} = 185(r^4 \sin 4\theta + 0.09r^4 + 0.06r^2)$

2.4. Approximate orthogonalization

The new IP in Eq. (9) cannot be used directly for the orthogonalization procedure, since it involves the observed images. Those are unknown when the orthogonalization is performed, prior to running the real-time wavefront sensing system. However, it is possible to construct a similar, approximate IP, where expectation values are substituted for the unknown data. We will now attempt to customize a set of basis functions for a particular data set, see Section 3.1. Example polynomials are shown in Table 1. Note that PD is insensitive to modes 1, 2, and 3, so we use the Euclidean IP whenever any of those is involved. Therefore the first three functions are the same in all sets 1–6 and all normalizations are with respect to $\psi_1 \equiv 1.0$.

The influence of noise in the images is limited to the filter H^2. As an expectation value, this factor corresponds to a circular filter with a large radius for low-noise, small-aberration data and a smaller radius for higher noise and more severe aberrations. The average equivalent circular filter radius (Löfdahl et al. 1998a) of H for these data is $\langle U_H \rangle = 60\%$ of the diffraction limit with a standard deviation of 2.4%. We use a circular filter with a radius of 60% of the diffraction limit for Sets 2, 4, and 6. We use 30% for Set 3 and 100% for Set 5, respectively ,for comparison.

The power spectrum, $|F|^2$, is reasonably well known for solar granulation. For these particular data, Paxman et al. (1996) estimate the power spectrum out to about 70% of the diffraction limit of the SVST. We use an extrapolation of the flat part to the diffraction limit for Sets 3–6. Set 2 is made with a unit power spectrum corresponding to a point source for comparison.

The remaining factor involves OTFs and OTF derivatives. The estimated quantities, $\widehat{S}_k$ and $\partial\widehat{S}_k/\partial\alpha_m$, should be calculated using zero wavefronts, because that is the initial value we typically use. Estimated, averaged, circularly symmetrical MTFs are substituted for the data OTFs, S_k, for Set 6.

3. Testing

In this section we describe some experiments that were performed in order to compare the different basis sets in Table 1. They were tested with real solar data and with simulated data that were made to correspond in quality to the real data. The PD algorithm was employed using Eq. (11), i.e. only the diagonal elements of the system matrix. For comparison, the same experiments were repeated using the full matrix and SVD disconnected. A fixed, circular noise filter, H, with a radius of 60% of the diffraction limit was used. Because PD data is insensitive to global wavefront piston and tilt, basis functions starting from $m = 4$ (focus) were used. A 70×70-pixel area, $\mathcal{X}$, was used for summations.

3.1. Real data

Solar granulation data were collected using real-time frame selection with the 47.5 cm Swedish Vacuum Solar Telescope (SVST) in La Palma on April 27, 1993 (Löfdahl & Scharmer 1994). A 5.4 nm wide interference filter centered at 470 nm was used. The quadratic diversity corresponds to a phase shift of 1.0 wave peak-to-peak. The image scale is 0.″07/pixel. The data set used here was recorded with 20 ms exposure time during 6 s of good seeing at 14:19 UT. A 128×128-pixel subfield was extracted from the full field of view (FOV).

The PD algorithm was iterated as usual in post-processing.

When the full matrix is used, the convergence is very similar for all basis sets as shown in Fig. 1. The average of the RMS of the wavefront deviations are shown as functions of iteration. The convergence is essentially the same for all bases except for the Zernikes, that appear to go better. This is an artifact, however, because the deviations are calculated with respect to previously calculated wavefronts using Zernikes, optimum H, and SVD parameter $w_c = 10^{-4}$ (for real data there is no ground truth). This appears to favor Zernikes solutions but the average metrics show no corresponding effect. In fact, Zernikes yield marginally *worse* metrics. Essentially, these data confirm that the different bases span the same vector space of wavefronts.

When only the diagonal is used, it is evident in both wavefront deviation and metric that the new basis sets are better than Zernikes. Even Set 1 gives better results than Zernikes. A closer look reveals that Sets 2–6 converge to approximately the same level. Sets 3 and 4 give slightly faster convergence. Because the first iteration is most important for real-time applications, we show in Fig. 2 scatter plots of RMS wavefront residuals and error metrics obtained using the diagonal only, comparing Zernike polynomials and Set 4. It is clear that in most cases the new basis yields better results.

The experiments were repeated with $M = 21$. The full matrix results were slightly better but similar to the $M = 15$ cases. Using only the diagonal, Sets 2–4 gave the best results. Sets 1, 5, and Zernikes were intermediate. Set 6 gave the worst results.

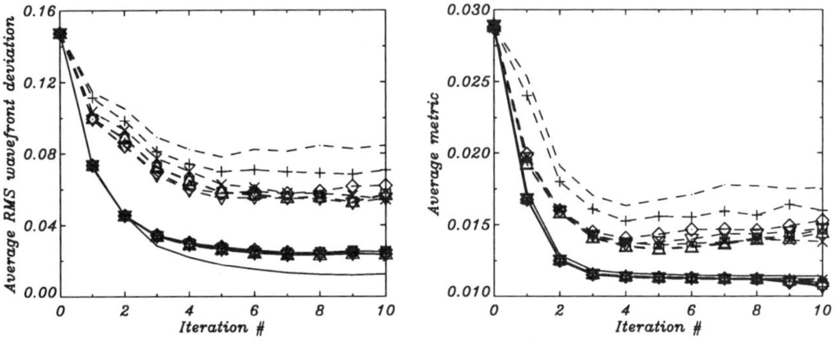

Figure 1. Real data results, 100 realizations, $M = 15$. The left tile shows the average RMS of wavefront deviations in waves. The right tile shows the average error metric, $\langle L \rangle$ (note that the vertical axis is offset from the origin). Symbols from Table 1. Full lines: full system matrix. Dashed lines: diagonal only.

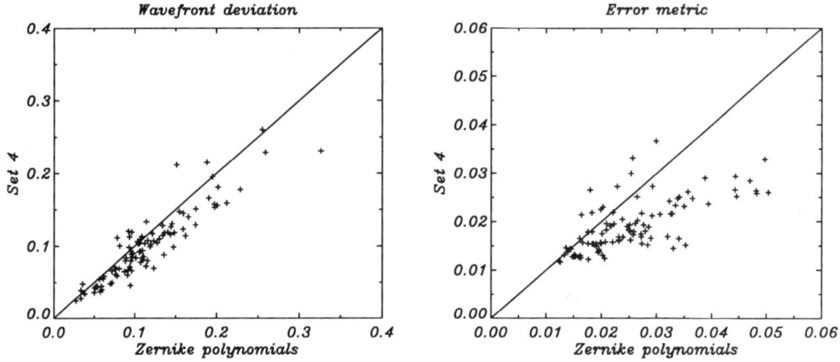

Figure 2. Real data results after a single iteration using only the diagonal of the system matrix. The left tile shows the RMS of the wavefront deviations in waves. The right tile shows the error metric, L. Horizontal axis: Using Zernike polynomials. Vertical axis: Using orthogonalized basis 4.

3.2. Closed loop simulation

Twenty random Kolmogoroff-type wavefronts were generated and scaled to approximately match the real data in the previous section. PD data were made by convolving a previously restored image with the corresponding PSFs using a 256×256-pixel FOV and then cropping to 128×128 pixels.

The estimates from a single iteration of the PD algorithm were used. 0.8 times the estimated wavefront was subtracted from the input wavefront, and new PD images were made from the residual wavefront. This process was repeated 15 times per original Kolmogoroff wavefront. For the purposes of this discussion, we define convergence as cases when the RMS of the residual wavefront after 15 corrections is less than RMS of the input Kolmogoroff wavefront. We calculate

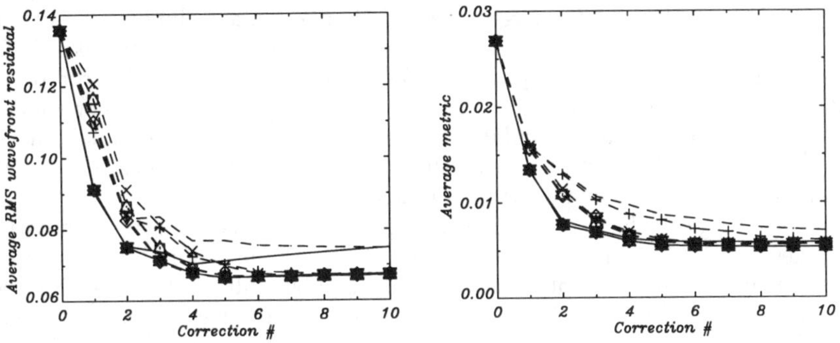

Figure 3. Closed loop results for 17 converging cases, $M = 15$. The left tile shows the average RMS wavefront residuals in waves (note that the vertical axis is offset from the origin). The right tile shows the average metric, $\langle L \rangle$.

RMS only of the part of the wavefront that *can* be expanded with polynomials $m \leq M$, thus not counting the high-order tail. With $M = 15$, all loops but 3 converge (2 with full matrix, set 1) and the average wavefront residuals and error metrics can thus be compared.

The results of these experiments are shown in Fig. 3. When the full matrix is used there is essentially no difference between the new Sets 1–6, while Zernikes converge more slowly in corrections 3 and 4 and then starts to diverge slightly, more than the diagonal-only Zernikes. This, we believe, is a sign of the ill-conditioning that we normally deal with using SVD.

When only the diagonal is used, convergence is more varied. Set 1 gives the best first correction, Set 2 the worst. Sets 5 and 6 converge slightly faster than Sets 3 and 4. Sets 1–6 all converge to the same residual wavefront RMS level as when the full matrix is used, while Zernikes converge to a noticeably worse level. The one or two extra corrections needed do not matter because diagonal-only solutions are much faster.

The same experiments were also done with $M = 21$. Too many cases diverged for diagrams like those in Fig. 3 to be meaningful. However, the number of diverged cases is in itself a measure of how good a basis set is. For the full matrix experiment, The Sets Z, 1–6 diverged 5, 6, 6, 6, 6, 6, 6 times, respectively. When only the diagonal was used, 14, 13, 12, 6, 6, 4, 8 diverged, respectively. It is hard to see any significant trend in the full-matrix data but it is clear that the diagonal-only experiments worked better using the new IPs, particularly using granulation power.

4. Conclusions

The new basis sets are made using a Gram–Schmidt orthonormalization algorithm based on IPs that incorporate various effects that are present in the real problem but not in the analytical IP corresponding to Zernike polynomials.

We note that Set 1 is more successful than Zernike polynomials, although the only difference is that the IP is a sampled version of the Zernike IP. The conclusion is that pupil sampling can be important sources of error or sub-optimal fits. We conjecture that sampling alone can be a significant source of error in transfer functions calculated using differently sampled Zernike functions (or any other basis functions) with the same expansion coefficients, e.g. when deconvolving images using different subfield sizes.

Also, we note that all the Sets 2–6 give as good as or better results than Zernikes and Set 1. This means that although taking pupil sampling into account is important, allowing for other effects when orthogonalizing the basis functions can help further. Image scale, object power, amount of diversity, noise statistics, as well as the exponentiation and correlation involved in making OTF derivatives, can be taken into account with the procedure described in this paper. Including granulation power, as in Sets 3–6, gives better results than assuming unit power, as in Set 2. It is harder to argue the merits of any of the varieties of H radius and average transfer functions in Sets 3–6 over the others. There may be other effects that should be included before a clear optimum can be found, like a noise term in Eq. (8).

We emphasize that the lesson learned is also applicable to other modes of wavefront sensing, e.g. with Shack–Hartmann data.

Acknowledgments. I have enjoyed many helpful and interesting discussions with Göran Scharmer. The SVST is operated by the Royal Swedish Academy of Sciences in the Spanish Observatorio del Roque de los Muchachos of the Instituto de Astrofísica de Canarias (IAC).

References

Dahlquist, G., Björck, Å., & Anderson, N. 1974, Numerical Methods, Prentice-Hall

Gonsalves, R. A. & Chidlaw, R. 1979, in Applications of Digital Image Processing III, Proc. SPIE, (ed. A. G. Tescher), pp. 32

Kendrick, R. L., Acton, D. S., & Duncan, A. L. 1994, in Image reconstruction and restoration, Proc. SPIE 2302, (eds. T. J. Schulz & D. L. Snyder), pp. 312

Löfdahl, M. G., Berger, T. E., Shine, R. A., & Title, A. M. 1998a, ApJ, 495, 965

Löfdahl, M. G., Duncan, A. L., & Scharmer, G. B. 1998b, in Adaptive Optical System Technologies, Proc. SPIE 3353, (eds. Bonnaccini & Tyson), pp. 952

Löfdahl, M. G. & Scharmer, G. B. 1993, in Real Time and Post Facto Solar Image Correction, Proc. 13th NSO/SP Summer Workshop, (ed. R. Radick), pp. 89

Löfdahl, M. G. & Scharmer, G. B. 1994, A&AS, 107, 243

Luenberger, D. G. 1969, Optimization by Vector Space Methods, John Wiley & sons

Noll, R. J. 1976, JOSA, 66(3), 207

Paxman, R. G., Schulz, T. J., & Fienup, J. R. 1992, JOSA A, 9(7), 1072

Paxman, R. G., Seldin, J. H., Löfdahl, M. G., Scharmer, G. B., & Keller, C. U. 1996, ApJ, 466, 1087

Scharmer, G. B. 1998, this volume.

High Resolution Solar Physics: Theory, Observations, and Techniques
ASP Conference Series, Vol. 183, 1999
T.R. Rimmele, K. S. Balasubramaniam, R.R. Radick, eds.

Object-Independent Fast Phase-Diversity

G. B. Scharmer

Royal Swedish Academy of Sciences
Stockholm Observatory, SE–133 36 Saltsjöbaden, Sweden

Abstract. We propose the use of an object-independent error metric to process phase-diversity data. This error metric has several advantages over that proposed earlier by Gonsalves and Chidlaw. In particular, it compresses information about aberrations to a small number of pixels in the Fourier and image planes, allowing efficient noise suppression and fast computation. In addition, it allows gradients of the error metric with respect to aberration parameters to be calculated without involving the observed data i.e. without contamination from noise and sub-fielding effects.

By taking advantage of known statistics of atmospheric aberrations, it is possible to formulate phase-diversity techniques to minimize expected RMS errors in the wavefront instead of RMS errors in fits to the data. This is illustrated by describing a technique for real-time phase-diversity, which is fast enough to drive an adaptive optics system for a meter-class or smaller telescope with a low-cost work station. Simulations are presented which illustrate that cross-talk with higher-order aberrations is reduced to very low levels. This technique is limited to small aberrations. The form of the proposed error metric suggests possibilities for rapidly estimating also larger aberrations to initialize the wavefronts in post–processing or to establish initial lock with an adaptive optics system.

1. Introduction

Phase diversity (PD) techniques are attractive for wavefront sensing partly because of the simplicity of the setup and the ease of calibration. With these techniques, wavefront sensing is made from images which can be used to directly verify the performance of an adaptive optics (AO) system and the focused image can be used in a straight-forward and optimum way for image stabilization. For post-processing, the object can be restored from the same images as are used for wavefront sensing. In contrast to Shack–Hartmann wavefront sensing, PD uses diffraction theory and therefore has the potential of more accurate wavefront sensing, in particular with extended objects. The disadvantage of PD techniques is that the measured intensities of the images used for wavefront sensing depend in a non-linear way on the aberrations.

Present PD techniques are largely influenced by the work of Gonsalves & Chidlaw (1979), who derived an error metric which is optimum for additive white noise. Later work by Paxman, Schulz, & Fienup (1992) formulated PD problems using maximum likelihood theory and extended the work of Gonsalves & Chidlaw (1979) to include e.g. Poisson noise and many diversity channels.

Recently implemented PD techniques, such as that of Löfdahl & Scharmer (1994) and Paxman, Schulz, & Fienup (1992) are computationally expensive for real-time applications and even post-processing applications require large computational resources. In our opinion, a more important drawback of these methods is that they provide the best estimate of the wavefront for a given noise model and effectively assumes that the residual errors are due to noise. In reality, residuals are present also because of cross-talk from high-order aberrations not included in the assumed expansion of the wavefront. These methods ignore that known statistics of atmospheric aberrations can be used to optimize the wavefront sensing algorithm, in particular including the effects of cross-talk with aberrations not corrected by the AO system or post-processing algorithm.

In this presentation we report on recent attempts to develop fast and more accurate PD techniques. We propose an error metric which is independent of the object, which allows regularisation of data in both the image and Fourier planes and retrieval of wavefront information with a minimum of computation. The proposed error metric allows its gradients with respect to aberration parameters to be calculated without using the observed data i.e. without corruption from noise and sub-fielding effects. The use of object-independent error metrics has been proposed earlier by Kendrick, Acton, & Duncan (1994) in combination with neural network algorithms. In the present work, it is shown that equations can be derived which allow errors from cross-talk with high-order Zernike aberrations and from noise to be minimized in a statistical (average) sense, consistent with the requirement of AO systems and post-processing techniques. Finally, a very fast method for real-time wavefront correction with PD is presented.

2. Error function and metric

Gonsalves & Chidlaw (1979) proposed that the wavefront be estimated by minimizing an error metric which can be written as

$$L = \sum_u |E|^2, \tag{1}$$

where E is an "error function" given by

$$E = \frac{D_k \widehat{S}_0 - D_0 \widehat{S}_k}{\sqrt{|\widehat{S}_0|^2 + |\widehat{S}_k|^2}}. \tag{2}$$

In these equations, u represents the two Fourier variables, D_0 and D_k are the Fourier transforms of the focused and the diversity images, S_0 and S_k are the transfer functions of the corresponding images and $\widehat{\ }$ represents an estimated quantity. Here, we propose to use instead the error function

$$\mathcal{E} = \left(\frac{D_k S_{d0} - D_0 S_{dk}}{D_0 S_{d0} + D_k S_{dk}} - \frac{\widehat{S}_k S_{d0} - \widehat{S}_0 S_{dk}}{\widehat{S}_0 S_{d0} + \widehat{S}_k S_{dk}} \right) \sqrt{{S_{d0}}^2 + {S_{dk}}^2}, \tag{3}$$

where S_{d0} and S_{dk} are the transfer functions which are obtained when there are no telescope or atmospheric aberrations present. These will be referred to as the "aberration-free" transfer functions. The proposed error function has two parts: one containing only the observed data and the aberration free transfer functions and one part containing the transfer functions corresponding to the estimated aberrations, but no data. The proposed error function can be rewritten as

$$\mathcal{E} = \frac{(D_k \widehat{S}_0 - D_0 \widehat{S}_k)\left(S_{d0}{}^2 + S_{dk}{}^2\right)^{3/2}}{(D_0 S_{d0} + D_k S_{dk})(\widehat{S}_0 S_{d0} + \widehat{S}_k S_{dk})}. \tag{4}$$

This is similar to the error function of Gonsalves & Chidlaw (1979) in being proportional to $(D_k \widehat{S}_0 - D_0 \widehat{S}_k)$, but has a different weighting factor. The proposed error function in Eq. (3) has several attractive properties:

- The error function is in principle independent of the object, which can be seen by replacing D_0 by FS_0 and D_k by FS_k where F is the Fourier transform of the object.
- The part containing the data is fixed and the same for all iterations.
- The part containing the estimated transfer functions does not contain observed data.

Because $\mathcal{E}$ is basically independent of the object, we can explore methods for extracting wavefront information from $\mathcal{E}$ which should be applicable to a wide range of objects. Of particular importance is that the part of $\mathcal{E}$ which contains the data is in the form of an error in a transfer function. When taking the inverse Fourier transform of this part of $\mathcal{E}$ we obtain the corresponding error in a point spread function. In Fig. 1, we show both the Fourier and image-plane representations of $\mathcal{E}$ as calculated from slightly oversampled 128×128 pixel focused and defocused granulation images aberrated by approximately 1/6 wave RMS. It is obvious from this figure, that all information about aberrations in the images is concentrated to a small area in the image plane, corresponding to the first diffraction rings. Any processing to extract wavefront information needs to involve only a small number of pixels in the image plane. This is equivalent to saying that the information about the transfer functions has much lower resolution in the Fourier plane than information about the images from which wavefront information is extracted. Therefore, regularisation (elimination of noise) is possible both in the image plane and in the Fourier plane. This allows noise to be eliminated from the data in a pre-processing stage, which involves compressing the wavefront information into an array of the smallest possible dimension. Fig. 1a shows the real (top) and imaginary (lower) part of $\mathcal{E}$ in the Fourier domain. The error function is a smooth function overlayed by noise spikes at isolated Fourier frequencies. This is not noise from the images but noise from artifacts related to the sub-fielding and apodization procedures. This noise disappears in the compressed error functions shown Fig. 1b. In Fig. 1c is shown the compressed Fourier plane representation of the same error function but calculated directly from the corresponding transfer functions, without using any data. This demonstrates the small influence of sub-fielding and that we

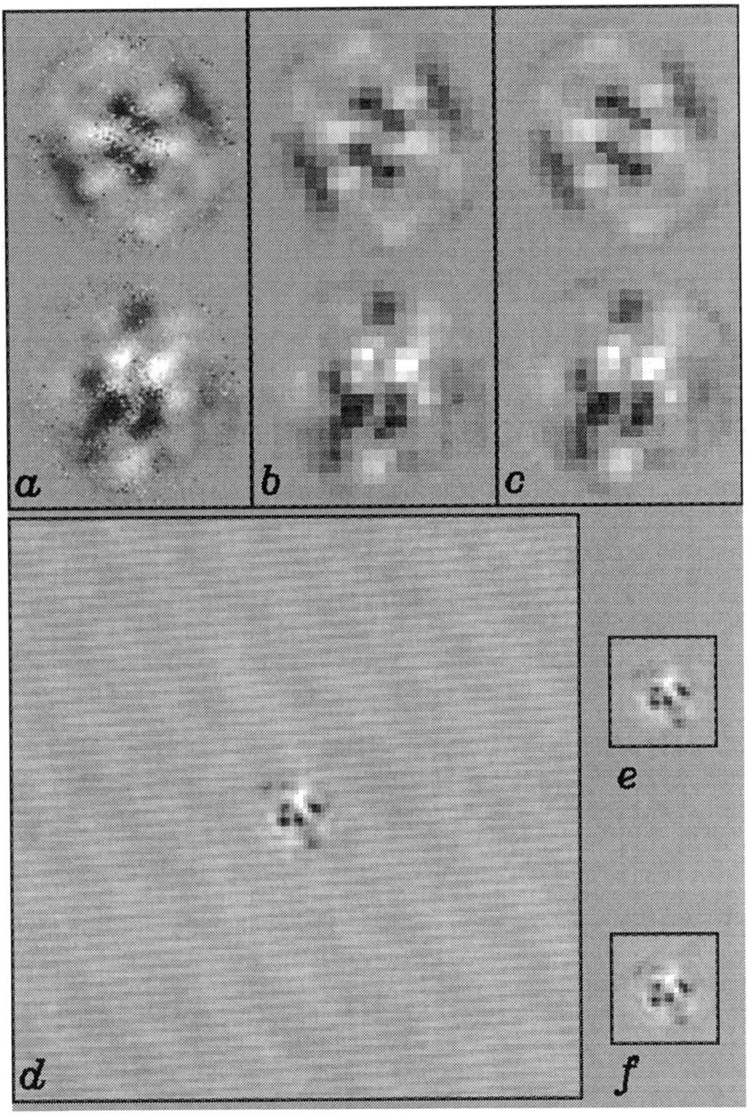

Figure 1. Representations of a sample error function calculated from focused and defocused granulation images aberrated by approximately 1/6 wave rms. The real (top) and imaginary (bottom) part of the Fourier plane error function, $\mathcal{E}$, is shown in **a**. The absence of information on aberrations over most of the corresponding function in the image plane **d**, implies that most of the information in $\mathcal{E}$ is redundant, requiring only low resolution. In **b** is shown $\mathcal{E}$ compressed by binning over 5×5 pixels and in **e** the corresponding image plane representation. **c** and **f** correspond to the same functions but calculated directly from the corresponding transfer functions, without involving image data.

can use the theoretically calculated transfer functions to calculate gradients of the error function with respect to aberration parameters, without involving the observed data which are normally contaminated by noise.

In the lower part of Fig. 1 are shown image plane representations of $\mathcal{E}$. The large panel shows the uncompressed error function which is contaminated by fringe-like patterns from the noise spikes at individual Fourier frequencies. Shown are also Fourier plane representations of the error function compressed by binning over 5×5 pixels, using the data (e) and theoretical transfer functions (f). Obviously, the compression does not destroy any information. On the contrary, it removes the Fourier fringes shown in the uncompressed error function. It is also evident that the compressed image plane representations contain no information about the aberrations in most of the pixels and that therefore further compression is desirable in order to make the Fourier plane representations non-redundant.

We also note that the part of the error function containing data in Eq. (3) only involves the aberration-free transfer functions and is, unlike for the error metric of Gonsalves & Chidlaw (1979), the same for all iterations. For determining wavefronts from small sub-fields of large images, such as is normally required when observing the Sun, this is advantageous. Earlier work by Löfdahl & Scharmer (1994) clearly demonstrated the severe effects of sub-fielding and apodization on the error function and the derived aberrations for small sub-fields. For extracting wavefront information with the here proposed error function, we can first perform the implied convolutions using FFT:s of the large, unsegmented images to get separately the numerator and denominator of $\mathcal{E}$. These two (large) functions are then transformed back to the image plane. Because the convolutions are made using the large images, there will be no errors from sub-fielding within the field of view of these large images. Finally, when extracting wavefront information over small fields, we extract subfields from these two image plane functions, apodize and calculate the forward Fourier transforms to extract the wavefront information. This approach minimizes the errors from sub-fielding.

Methods for solving PD problems normally require evaluation of the gradients of the error function with respect to the estimated aberration coefficients. For the error function proposed in Eq. (3), the gradients are entirely independent of the data. This means that noise from the data or errors from sub-fielding and apodization cannot corrupt such gradients. Our previous experience indicate that gradients contaminated by noise and sub-fielding is the reason for slow convergence and failure to converge when many aberrations are used to represent the wavefront. The proposed error metric should eliminate this problem. Finally, the proposed error function should allow very fast methods methods for real-time wavefront correction. This can be done for small aberrations by pre-calculation of the partial derivatives of $\mathcal{E}$ with respect to the aberration coefficients for the aberration-free case. For larger aberrations, this can probably be done by pre-calculation of tables of error functions and comparing with the observed data.

3. An optimum filter approach to fast real-time phase diversity

In the PD implementation of Löfdahl & Scharmer (1994) wavefronts are determined by solving iteratively non-linear matrix equations in a way similar to standard non-linear least squares techniques described in Press et al. (1992). The curvature matrix is calculated from inner products involving the partial derivatives of the error function, $\partial E/\partial c_n$. The RHS of the matrix equation is the inner product of $\partial E/\partial c_n$ and E.

We expect that for small aberrations, the partial derivatives $\partial \mathcal{E}/\partial c_n$, which are independent of the observed data, can be calculated in advance for the aberration-free case and that these derivatives can be used as fixed derivatives for controlling an adaptive mirror in good seeing. For such closed-loop operation, the estimated transfer functions $\widehat{S}_0$ and $\widehat{S}_k$ are set equal to their aberration-free values, S_{d0} and S_{dk}, causing the wavefront sensor to drive the mirror to eliminate the aberrations in the focused image.

We conclude that it should be possible to estimate small aberrations c_n from an inner product of $\mathcal{E}$ and pre-calculated functions H_{nu}, i.e. from

$$\widehat{c}_n = \sum_u H_{nu}\mathcal{E}_u, \tag{5}$$

where $\mathcal{E}_u$ is the error function and H_{nu} are the pre-calculated functions, different for each aberration n. Both of these functions are Hermitian, which ensures that the inner product produces a real quantity. This is in the form of a set of filters H_{nu} acting on the error function. We now determine H_{nu} to be optimum filters in the sense of producing an estimate of c_n which gives the smallest possible RMS error. This corresponds to choosing H_{nu} such that

$$L_n = \langle (c_n - \sum_u H_{nu}\mathcal{E}_u)^2 \rangle \tag{6}$$

has a minimum. In this equation, the average is taken both with respect to atmospheric aberrations and to realizations of noise in the images.

Taking the derivative of L_n with respect to H_{nu} at each spatial frequency v and setting the result to zero gives

$$\langle c_n \mathcal{E}_v \rangle = \sum_u H_{nu} \langle \mathcal{E}_u \mathcal{E}_v \rangle. \tag{7}$$

By making the definition

$$R_{uv} = \langle \mathcal{E}_u \mathcal{E}_v \rangle, \tag{8}$$

Eq. (7) can be seen to define a matrix equation determining the filters H_{nu}. We note that R_{uv} is the same for all aberrations n. This matrix can be inverted once and its inverse be used to give all filters H_{nu}.

We have calculated the matrix R_{uv} in several different ways, including the use of Monte Carlo simulations with granulation images and random wavefronts. A simpler approach is the following: For small aberrations, we can express $\mathcal{E}$ as a Taylor expansion

$$\mathcal{E} = \sum_{m=4}^{\infty} \frac{\partial \mathcal{E}}{\partial c_m} c_m \tag{9}$$

where $\partial\mathcal{E}/\partial c_m$ are calculated with aberration–free transfer functions. For Karhunen–Loeve polynomials, this is particularly simple, because $\langle c_n c_m \rangle$ is zero when $n \neq m$. This gives

$$\langle c_n^2 \rangle \frac{\partial \mathcal{E}_v}{\partial c_n} = \sum_{m=4}^{\infty} \langle c_m^2 \rangle \frac{\partial \mathcal{E}_u}{\partial c_m} \frac{\partial \mathcal{E}_v}{\partial c_m} H_{nu} + \text{noise bias} \tag{10}$$

which allows R_{uv} to be constructed from the weighted sums of the autocorrelation of the partial derivatives of the error function with respect to the aberration parameters. This should be a good approximation also for Zernike coefficients. By adding bias from random noise to the diagonal of the matrix R_{uv} the filters will be optimized both for noise and the expected RMS amplitudes of atmospheric aberration coefficients.

This method is particularly powerful with the object-independent error function proposed in this paper. We have established that this error metric requires only low resolution in Fourier space. Therefore, we can compress $\mathcal{E}_u$ by binning Fourier pixels. In the tests presented in this paper, we have compressed the original 128×128 pixels by 5 times to produce 21×21 pixel error functions. Similarly, each of the filters have only 21×21 pixels. This compression does not destroy any information about the aberrations, on the contrary, the compressed error functions in Fig. 1 show that further compression would be desirable in order to remove redundant information. The compression over 5×5 pixels reduces the number of multiplications needed to estimate the aberration coefficients by more than a factor 25. The total computational cost of the proposed method is thus completely dominated by the time required to take the two Fourier transforms of the focused and diversity sub-images, with very small additional cost even for estimating a large number of aberrations.

When compressing the error function in Fourier space, care should be taken with objects such as granulation, which have Fourier transforms similar to that of random noise. For such objects, direct compression of the error function is not desirable, because this may give residual noise spikes at Fourier frequencies where the object happens to have small amplitudes. Such noise spikes at large amplitudes can be seen in Fig. 1. Instead, we determine the ratio $C_u = D_{ku}/D_{0u}$ averaged in Fourier space by a least squares procedure, i.e. determining $\langle C_u \rangle$ such that $\sum_u |D_{ku} - \langle C_u \rangle D_{0u}|^2$, where the summation is over the binned Fourier pixels, has a minimum. This procedure also deals properly with Fourier frequencies at which either of D_{0u} or D_{ku} passes through zero.

In Fig. 2 are shown one set of filters H_{nu} (left) and the corresponding error function $\mathcal{E}_u$ for (small) Zernike aberrations 4–27. Because of the Hermitian properties of $\mathcal{E}_u$ and H_{nu}, the inner products involved in the estimates of the Zernike coefficients are such that the real and imaginary parts of the filter are multiplied with the corresponding parts of $\mathcal{E}$. The filters show that information about small aberrations with even radial power (4–6, 11–15, 22–28) is obtained from the real part of the error function whereas information about aberrations with odd radial power (7–10, 16–21) is obtained from the imaginary part of the error function. Although there are similarities between the filters and response for the same aberration, there are also large differences. This demonstrates that cross-talk from higher aberrations is a significant problem and that the filters are strongly influenced by the need to suppress this cross-talk.

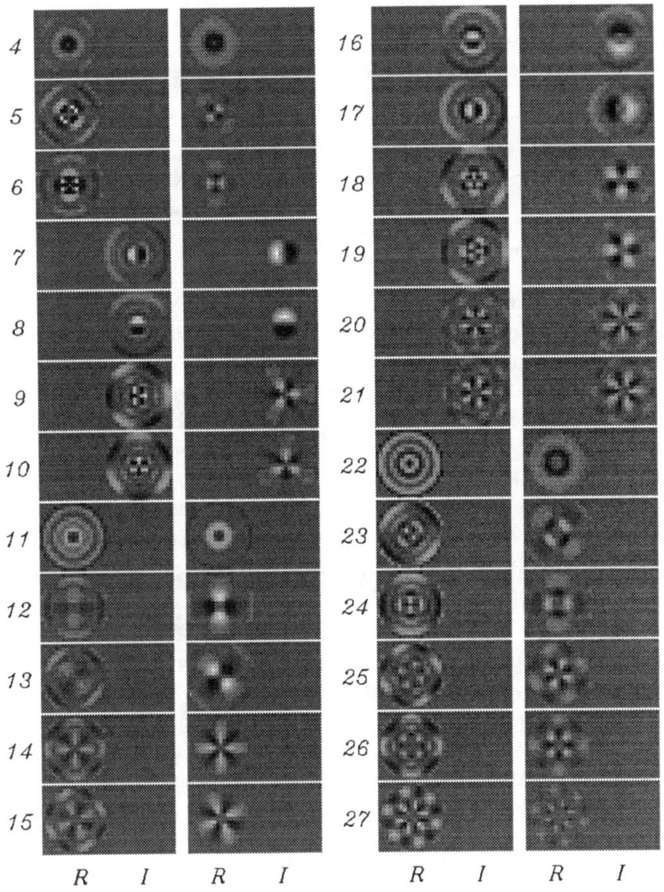

Figure 2. The real (R) and imaginary (I) part of some of the optimum filters (left) used for measuring Zernike aberrations 4–36 in the closed-loop simulations described in this paper. Also shown is the response of the error function (right) for a small amount of the same aberration. The large differences between the filters and the response demonstrate that cross-talk with other aberrations is important. The calculated filters minimize this cross-talk, taking into account the expected relative amplitudes of atmospheric aberrations.

It is straightforward to add filters that measure also the alignment between the two diversity images. This is modeled by adding wavefront tilt (Zernikes 2 and 3) to one of the two diversity images. This allows the wavefront sensor to be self calibrating. By measuring also alignment errors, the filters measuring

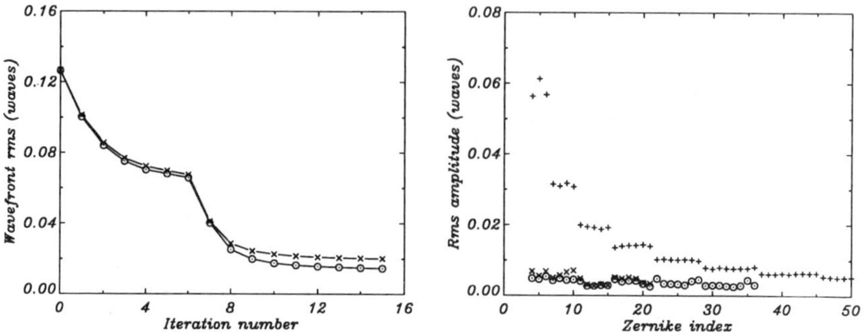

Figure 3. The convergence of the wavefront RMS (left) for the simulated closed-loop correction with input wavefronts containing Zernike 4–325 and estimated wavefronts for Zernike 4–21 (crosses) and 4–36 (circles). The wavefront RMS refers to Zernike 4–21. On the right, are shown the average RMS for each of the Zernike aberrations: The input wavefront (+), the residual wavefront of the converged solution with correction of Zernike 4–21 (x) and 4–36 (o).

Zernikes with odd radial degree (7–10,16–21) will be modified to suppress cross-talk from alignment errors and vice versa.

When adding noise bias to the diagonal of the matrix R_{uv}, the filters respond by giving less weight to the Fourier frequencies where noise dominates. Such filters are therefore fully optimized simultaneously for both noise and cross-talk.

With the filters shown in Fig. 2, we made tests to simulate closed-loop operation with an adaptive mirror. Random wavefronts containing aberrations up to Zernike 325 and scaled to the amplitudes expected for atmospheric seeing (Noll 1976) with $D/r_0 = 3$ were generated and "large" focused and defocused granulation images were calculated from these aberrations. We then extracted 128×128 pixel sub-images at random positions in the field and used the filters shown in Fig. 2 to estimate estimate the Zernike coefficients iteratively, creating new "large" images using the residual wavefront for each iteration. In total, we used 1000 realizations of random wavefronts. In Fig. 3 is shown the average residual wavefront RMS for Zernike 4–21 obtained from two separate experiments which corrected Zernike 4–21 (crosses) and 4–36 (circles) for fifteen iterations. During the first 6 iterations we corrected only Zernike 4–10 in both cases, using a different set of filters. It is evident from these plots that convergence is rapid and that the residual errors (0.02 respectively 0.015 waves RMS) of Zernikes 4–21 are well below what is required for well-corrected wavefronts. The wavefront RMS for uncorrected Zernike 22–325 are approximately 0.06 waves or 3–4 times greater than those of the corrected Zernikes, demonstrating the low cross-talk levels.

In the right-hand plots are shown the average RMS amplitudes of the input Zernike coefficients (pluses) and after correction of Zernike 4–21 (crosses) and Zernike 4–36 (circles).

For these simulations, approximately 7% of the cases did not converge or showed no improvement in the wavefronts, these cases are not included in the averages. We noticed that most non-converged cases correspond to input wavefronts having a negative value of c_4 such as to give opposite signs of defocus for the two focused-defocused images. This indicates that the chosen amount of defocus, corresponding to one wave phase error at the edge of the pupil, is too small in poorer seeing and that this is the primary reason why the simulated AO system failed to lock. It also indicates that using defocus as a means of adding PD to one of the images may not be ideal for measuring atmospheric seeing, since the atmospheric focus variations have, together with the two astigmatism terms, the largest amplitudes of atmospheric aberrations.

In order to understand the origin of the residual wavefront error, we made closed-loop simulations with truncated input wavefronts containing only the 21 first Zernikes. For these simulations, the residual wavefronts after fifteen iterations were approximately 0.008 waves RMS showing that there are no important effects of the sub-fielding and apodization processes. Our interpretation of this is the following: Sub-fielding and apodization is equivalent to a convolution in Fourier space, which degrades the information at each Fourier frequency. Expressed differently, it can be said that the sub-fielding and apodization processes reduces the effective "resolution" in Fourier space. Reducing resolution in Fourier space by binning pixels, as in the present method, is the appropriate way of extracting information from such data.

In an additional experiment, we used input wavefronts containing only Zernike 22 and higher aberrations and corrected the first 21 Zernikes. The residual wavefront error was then 0.02 waves which is due to a combination of contributions from cross-talk and sub-fielding. The previous experiment indicate that the residual wavefront errors of our simulations are dominated by cross-talk with higher-order uncorrected aberrations. This is consistent with the result that increasing the number of corrected Zernikes from 21 to 36 reduced the residual wavefront errors from 0.02 to 0.015 waves.

We have also made closed loop simulations with under-sampled 64×64 -pixel images, obtained by compressing the 128×128 -pixel images used in the experiment described above, which showed good performance.

These tests clearly demonstrate that the filters suppress cross-talk to low levels and that correction of many Zernike aberrations is possible. Increasing the number of aberrations does not lead to unstable behavior, as with the SVD method. The compression of the error function in the Fourier plane eliminates most of the noise in the data and filters optimized for noise will efficiently remove noise at high spatial frequencies. It is obvious from Fig. 1, that the error function in the image plane contains information only over a small area at the center. By applying a filter also in the image plane, final noise reduction is possible. This was implemented for the simulations presented here. Although we have not yet made simulations with noise, we feel confident that the proposed method allows noise to suppressed in a near optimum way, and will give significantly better wavefront estimates than with least-squares fits based on the error metric of Gonsalves & Chidlaw (1979).

4. Measurement of larger aberrations

The main problem of post-processing and adaptive optics control using phase-diversity is that the recorded intensities depend in a non-linear way on the aberrations. The approach used for closed-loop simulations was based on a linearization for zero aberrations and cannot be expected to work for large aberrations. Although we have not yet implemented methods for dealing with larger aberrations, we believe that fast such methods should be possible. The main reason for this optimism is that the proposed error metric requires calculations involving a very small number of Fourier pixels and that our simulations indicate that suppression of cross-talk is possible by proper filtering. The proposed error metric is in the form of observed data compared with a model of the wavefront. When combined with Eq. (1), this suggests the use of methods similar to those developed for measuring image shifts with cross-correlation techniques. This also suggests comparisons with a pre-calculated library of error functions as a basis for rapidly estimating larger aberrations. By proper filtering, it should be feasible to develop search strategies involving small sub-sets of aberration parameters, possibly using General Regression Neural Networks, such as proposed by Kendrick et al. (1994). The absolute difference algorithm, which is implemented as hardware instruction in e.g the new Compaq Alphas, can be used with a modified version of Eq. (1) for comparing such pre-calculated functions with error functions obtained from observed data. We are presently investigating methods for extending the lock-range to conditions corresponding to $D/r_0 = 6$.

5. Post-processing

For post-processing of data, the error function should drive the estimated transfer functions $\widehat{S}_{d0}$ and $\widehat{S}_{dk}$ to approximately equal the true corresponding transfer functions. In this case, it is not appropriate to use fixed "filters" which give accurate estimates only at small aberrations, corresponding to an AO mirror operating in closed-loop. We are presently implementing a post-processing method which uses the partial derivatives of the error function with respect to the aberration parameters to iteratively improve the solutions such that errors from noise and cross-talk are minimized. In this method, we calculate the partial derivatives using a "low-resolution" model of the pupil, which gives important reductions in computing time. This is possible in view of the conclusion made earlier, that the PD data do not contain any information at high resolution in the Fourier plane and therefore also not at the pupil plane. We expect that this method will lead to more accurate wavefront estimates and a faster code than that developed by Löfdahl & Scharmer (1994). Details on this will be published separately.

6. Conclusions

In this paper, we have demonstrated that the proposed object-independent PD metric offers several important advantages. In particular, this compresses the information about the aberrations in both the image and Fourier planes, such as to allow wavefront information to be extracted from a minimum of observables and computed parameters. Regularisation of the data can be made in both the Fourier and image planes to efficiently reduce the effects of noise. Gradients of the error function can be calculated without contamination by noise or sub-fielding effects from the data. The approach allows algorithms to be developed which should be insensitive to the choice of targets and fundamental limitations and possibilities of PD methods to be explored. In this context, we point out that the proposed error metric and the binning operation applied in the Fourier domain, suggest comparisons to be made with Shack–Hartmann wavefront sensing. The Shack–Hartmann wavefront sensor samples the pupil plane using a micro-lens array and is limited by aliasing effects from that sampling. The here proposed error metric leads to sampling in the Fourier plane using binned Fourier pixels which are implemented in software. The residual wavefront error from aliasing is for both methods given by the number of samples across the diameter corresponding to the pupil diameter. The disadvantage of the Shack–Hartmann wavefront sensor is usually its poor resolution across the pupil and strong cross-talk with higher-order aberrations. The advantage is that wavefront sensing can be made rapidly by searching first for image shifts of the individual sub-images, then using that information to reconstruct the wavefront. We believe that rapid search techniques can be developed for the proposed PD technique and that this may establish PD as the preferable technique for wavefront sensing also on larger telescopes.

We also conclude that knowledge about the statistical properties of atmospheric aberrations can and should be used to develop algorithms which minimize errors in the derived wavefronts, rather than residuals in the measured data, such as is the case with standard least-squares techniques. In particular, we emphasize the importance of optimizing PD wavefront sensing techniques also with respect to cross-talk with higher-order uncorrected aberrations.

Finally, we have applied the proposed error metric by developing a real-time PD technique which is fast enough to control an adaptive mirror with present low-cost work stations. Simulations indicate excellent performance for small aberrations. We expect that a combination of the methods discussed in this paper can be used for phase-diversity wavefront sensing with the AO system planned for the 1-meter Swedish solar telescope, to be installed on La Palma.

Acknowledgments. Mats Löfdahl is thanked for giving access to his subroutine library, for many interesting discussions and for help in producing this presentation.

References

Gonsalves, R. A. & Chidlaw, R., 1979, in Applications of Digital Image Processing III, Proc. SPIE, (ed. A. G. Tescher), pp. 32

Kendrick, R. L., Acton, D. S. & Duncan, A. L. 1994, AO, 33(27), 6533

Löfdahl, M. G. & Scharmer, G. B. 1994, A&AS, 107, 243

Paxman R. G., Schulz, T. J., & Fienup J. R., 1992, JOSA A, 9(7), 1072

Press, W.H., Vetterling, W.T., Teukolsky, S.A., & Flannery, B.P., 1992, Numerical Recipes in C, (2nd Edition)

High Resolution Solar Physics: Theory, Observations, and Techniques
ASP Conference Series, Vol. 183, 1999
T.R. Rimmele, K. S. Balasubramaniam, R.R. Radick, eds.

Optimum Apodization for Speckle Imaging of Extended Sources

C. U. Keller

National Solar Observatory, Tucson, AZ 85719, USA

Abstract. Speckle imaging of extended sources such as the Sun has to be performed on small segments to reduce the impact of anisoplanatism. Since speckle techniques typically work in the Fourier domain, these segments are apodized to minimize the influence of edges on the Fourier transformed quantities. While apodization functions have been studied to optimally estimate the amplitudes of the Fourier transform of the true object, the influence of apodization on the Fourier phase estimates has not been studied. Indeed, commonly used apodization functions distort the phase estimates considerably. Here I show how apodization distorts the phase estimates and present a new type of apodization functions that leaves Knox-Thompson phase estimates unaffected. This leads to a substantial improvement of the quality of speckle reconstructions of the solar surface, in particular under mediocre seeing conditions.

1. Introduction

The solar disk is much larger than the isoplanatic patch in the visible and near-infrared parts of the spectrum, even at the best sites. In most cases, the detector size is also larger than the isoplanatic patch. Therefore, speckle and phase-diversity reconstructions are normally performed on small segments with a size comparable to the size of the isoplanatic patch (typically a few seconds of arc in the visible). Afterwards, the reconstructed segments are mosaiced together to obtain the final reconstruction of the scene that may span many isoplanatic patches.

Most reconstruction techniques work in the Fourier domain, where the apodization of the segments is crucial to minimize artifacts due to the edges. In speckle imaging the object amplitudes and phases are determined separately. Apodization functions have been investigated in detail for optimum amplitude estimates. The most reliable amplitude estimate would be obtained if 100% of the segment would be apodized with a function that falls off smoothly from the center of the segment towards the edges. However, if the apodization is too strong, the reconstructions become unreliable (e.g. von der Lühe 1993). Fig.1 shows a reconstruction of granulation near disk center under moderate seeing conditions with a 20% Hamming window apodization. A high-frequency noise artifact is easily visible at the centers of the segments. The magnitude of these artifacts increases with increasing amounts of apodization. Speckle reconstructions using the Knox-Thompson approach have mostly been low-pass filtered to remove these artifacts. The prime suspect for these problems are the phase estimates. However, the influence of apodization on phase estimates has not been investigated (Fienup 1992, private communication).

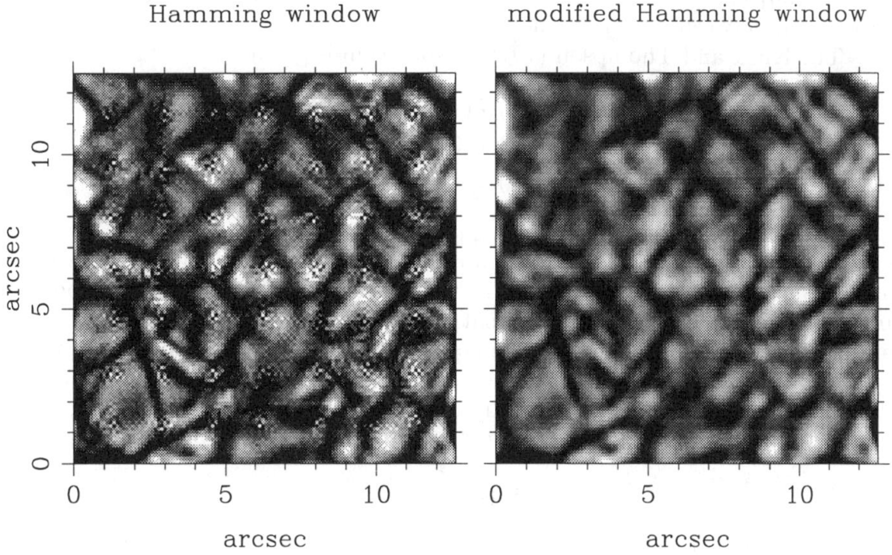

Figure 1. Knox-Thompson speckle imaging reconstruction of solar granulation near disk center. The data were obtained under moderate seeing conditions. A conventional Hamming window was used for apodization to generate the reconstruction on the left. The center of each segment shows a high-frequency noise artifact. The right side shows the reconstruction using a modified Hamming window as introduced in the present contribution, which completely removes the artifacts.

In the following, I will describe how apodization by common window functions distorts the phase estimates of Knox-Thompson speckle imaging. This insight leads to an algorithm to calculate apodization functions that do not distort the phase estimates. This modification has no net effect on the final reconstruction after mosaicing. The right side of Fig. 1 shows a reconstruction of the same data set using a modified Hamming window, which results in a reconstruction without any high-frequency artifacts.

2. Phase distortion

The formation of images through the turbulent atmosphere can be described in Fourier space by

$$I = OS + N = I_0 + N, \tag{1}$$

where I is the Fourier transform of the observed image, O is the true object as seen without atmospheric distortions, S is the optical transfer function, and N is the noise (photon statistics and detector noise). In the following, I assume that N is not correlated with I, which is a reasonable assumption for low-contrast objects such as the solar photosphere. This assumption does not hold for high-contrast objects where the Poisson distribution of the photon noise leads to a correlation between signal and noise.

The Knox and Thompson (1974) bispectrum is given by

$$\begin{aligned}\langle I(f)I^*(f-1)\rangle &= \langle I_0(f)I_0^*(f-\delta f)\rangle + \\ &\quad \langle I_0(f)N^*(f-\delta f)\rangle + \langle N(f)I_0^*(f-\delta f)\rangle + \\ &\quad \langle N(f)N^*(f-\delta f)\rangle,\end{aligned} \tag{2}$$

where * indicates conjugate complex quantities and angle brackets $\langle\ldots\rangle$ indicate ensemble averages over all possible states of the atmosphere. The ensemble average is usually approximated by the arithmetic average over many short-exposure images. Furthermore, we assume that signal and noise at the same frequency and noise at different frequencies are not correlated. Then, the influence of noise on the bispectrum vanishes on average.

However, this changes drastically if we add apodization. Apodization by a window function w can be described in real space by a multiplication

$$i = [o * s + n]w, \tag{3}$$

where $*$ is the convolution operator, and lower case letters represent the real-space counterparts of the capital letter Fourier transformed quantities. In Fourier space, the multiplication becomes a convolution of the form

$$I'(f) = [I_0(f) + N(f)] * W = I(f) * W. \tag{4}$$

For simplicity, we will assume that $\delta f = 1$, which is equivalent to the most simple version of the Knox-Thomson phase estimation algorithm. The following can easily be extended to the case where $\delta f \neq 1$, which is known as extended Knox-Thompson algorithm. Furthermore, for simplicity, I assume that W has only three Fourier components:

$$a \text{ at } f=-1,\ b=b^* \text{ at } f=0,\ a^* \text{ at } f=1, \tag{5}$$

where a and b are the Fourier coefficients of the window function. b is a real quantity while a is a complex number. The hermicity at $f=1$ and $f=-1$ comes from the fact that the window function is real.

Using these simplifications and assumptions, the equation of image formation with apodization becomes

$$I'(f) = I(f)b + I(f-1)a + I(f+1)a^*. \tag{6}$$

To show the main effect of apodization on the Knox-Thompson bispectrum, we realize that any reasonable window function has $\frac{|a|}{|b|} \ll 1$ because the window function should be close to unity over most of its support. I therefore neglect terms in $|a|^2$ in the following. Furthermore, the rapid fall-off of signal correlation with increasing frequency difference (e.g. von der Lühe 1988) leads me to disregard any terms with a frequency difference of more than δf. Again, we make the well-founded assumption that noise and signal at the same frequency are not correlated and that the noise at different Fourier frequencies is not correlated. The Knox-Thompson bispectrum then becomes

$$\begin{aligned}\langle I(f)I^*(f-1)\rangle &= |b|^2\langle I_0(f)I_0^*(f-1)\rangle + \\ &\quad a^*b\left\langle|I_0(f)|^2\right\rangle + ab^*\left\langle|I_0(f-1)|^2\right\rangle + \\ &\quad a^*b\left\langle|N(f)|^2\right\rangle + ab^*\left\langle|N(f-1)|^2\right\rangle.\end{aligned} \tag{7}$$

The first term is directly proportional to the "true" Knox-Thompson term $\langle I_0(f)I_0^*(f-1)\rangle$. Since only the phase of this term is needed for estimating the true phases of the object, any scaling with a real number has no influence. However, the four additional terms are complex numbers of magnitude $|a||b|$ times the power spectrum of the signal or the noise. Note that the phase of these terms is only determined by the window function itself, but not by the signal or the noise. Higher frequency Fourier components in the window function produce terms with larger frequency differences and therefore terms with even smaller magnitudes. The four additional terms identified above do not vanish on average and therefore change the phases of the bispectrum. It is easy to see that the relative magnitudes of the noise terms increase with decreasing signal such as due to seeing for a given noise level (recall that $I_0 = OS$).

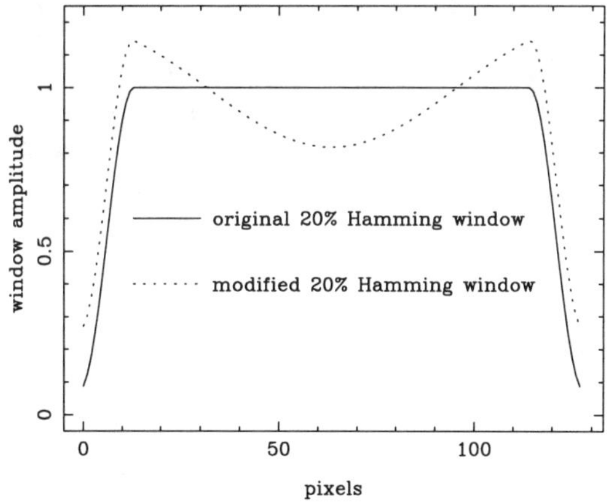

Figure 2. Comparison of regular and modified apodization functions. The modification consists in setting the amplitude of the first AC component to zero.

3. New apodization function

All undesired terms vanish for $a = 0$. Therefore, setting a to zero removes all terms, at least to first order. We can therefore modify existing apodization functions by removing their first AC component, i.e. Fourier-transforming the window function, setting the amplitudes at frequencies ± 1 to zero followed by an inverse Fourier transform. This leads to apodization functions with a pronounced "trough" in the middle and enhanced "ears" towards the edges. Fig.2 shows a regular Hamming window with the outer 10% of each edge (20% total) being apodized as well as the modified version with the first AC component removed.

As can be seen from Eq. (7), noise alone can produce artifacts. I therefore prepared an artificial time series which contained uniform noise only. The speckle reconstruction of this noise series with the original Hamming apodization function leads to strongly enhanced noise in the center. Figure 3 shows that these artifacts disappear when a modified apodization function is used.

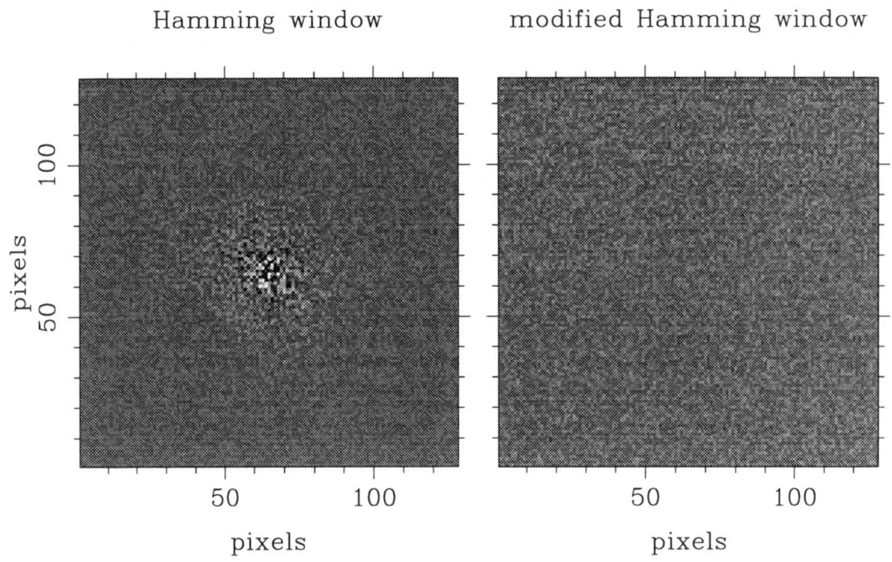

Figure 3. Speckle reconstruction of noise alone in one segment shows the disappearance of the artifacts in the center when a modified apodization function is used.

4. Mosaicing

The modification of the window function described above leads to a distortion of the true object due to the trough and ears of the window function. One would expect that this change would need to be removed. However, apodization windows cannot be simply taken out by dividing the reconstruction with the apodization function. This is due to the fact that the apodization mostly suppresses the edge effects, and a division by the apodization function after restoration would enhance those edge effects again.

However, appropriate segmentation and mosaicing can remove the effects of the trough and ears of the modified window function. The first AC component corresponds to a $cos(x)$ signal that is largest at the center of the segment (angle

0) and smallest at the edges of the segment (angles $\pm\pi$). The lack of this AC component leads to the trough and ears. Segmentation and mosaicing that overlaps segments by half the size of a segment (angular shift of π) removes the trough and ears since $cos(x) + cos(x + \pi) = 0$. Therefore, the mosaiced reconstruction is independent of the amplitude of the first AC component and the effect of the modification of the apodization function on the final, mosaiced reconstruction is removed. Fig.4 shows the overlapping apodization functions and the sum of the apodization functions.

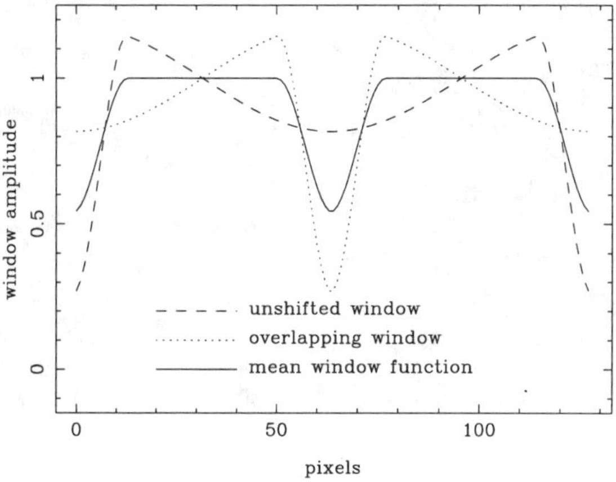

Figure 4. By overlapping segments by half their size, the modification of the apodization function has no effect on the final reconstruction. Indeed, the same mean window function would be obtained without the modification. The dips in the center and at the edges of the mean window function are removed by appropriately weighting the edges of the individual segments such that they do not contribute to the final reconstruction.

In the mosaicing process, the segment edges are given much less weight than the centers of the segments (e.g. von der Lühe 1993). The weighting must be done with care so that the modification of the apodization function still disappears in the final reconstruction. This removes the influence of the edges without having to divide by the apodization function.

5. Discussion

It has long been suspected that the apodization has a detrimental influence on the Knox-Thompson phase estimates. Here I have shown how the apodization influences the reconstruction of the object phases. The apodization introduces a correlation between adjacent frequencies, which leads to additional terms in the bispectrum that do not vanish on average. I have shown how any apodization function can be modified to make these additional terms vanish. With a careful choice of the segmentation and mosaicing parameters, the influence of the modification disappears after the mosaicing step.

These modified apodization functions have significantly improved Knox-Thompson speckle reconstructions under moderate seeing and/or high noise conditions and have led to a major improvement for speckle polarimetry (Keller & von der Lühe 1992). Fig.5 shows the reconstructed magnetograms of a very quiet part of the solar photosphere for 20% Hamming windows.

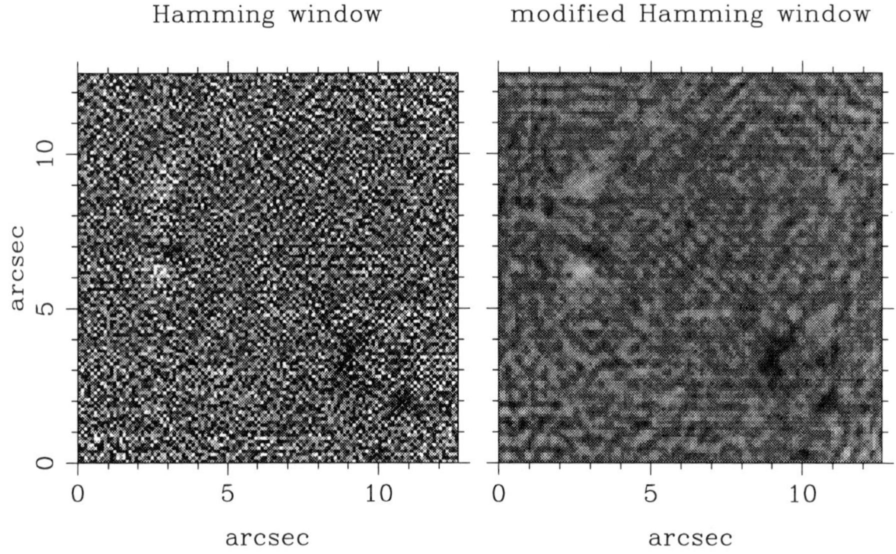

Figure 5. The noise level of a reconstructed magnetogram is reduced by a factor of 4 by using a modified apodization function. The maximum polarization signal in this quiet region near disk center is 2%.

This modification of existing apodization functions can easily be applied to the extended Knox-Thompson algorithm by setting the n-th AC component to zero for a Fourier frequency difference of n. Other reconstruction algorithms such as speckle masking might also take advantage of such modified apodization functions.

Acknowledgments. The National Solar Observatory is one of the National Optical Astronomy Observatories, which are operated by the Association of Universities for Research in Astronomy, Inc. (AURA) under cooperative agreement with the National Science Foundation.

References

Keller C.U., von der Lühe O. 1992, A&A 261, 321
Knox K.T., Thompson B.J. 1974, ApJ 193, L45
von der Lühe O. 1988, JOSA A5, 721
von der Lühe O. 1993, A&A 268, 374

High Resolution Solar Physics: Theory, Observations, and Techniques
ASP Conference Series, Vol. 183, 1999
T.R. Rimmele, K. S. Balasubramaniam, R.R. Radick, eds.

Spectral and Spatial High Resolution Observations of the Sun with Image Reconstruction and Deconvolution Methods

M. Koschinsky, J. Krieg, F. Kneer, C. Ritter, M. Wunnenberg

Universitäts-Sternwarte Göttingen, Geismarlandstraße 11, D-37083 Göttingen, Germany

Abstract. In order to produce observations of both high spectral and spatial resolution of the solar surface, we employ two different image reconstruction approaches based on speckle interferometry and the phase diversity method (PD), respectively. Both of them use the reconstruction of a broad-band image to determine the optical transfer function of the single raw images. These are then used to deconvolve simultaneously taken narrow-band images with FWHM of 0.003 – 0.004 nm. The observations are obtained with the Vacuum Tower Telescope (VTT) on Tenerife. We use a double Fabry-Perot interferometer for narrow-band imagery. At particular spatial frequencies the instantaneous optical transfer function can be very small or zero and the information is lost. To avoid this to a large extent, we use five pairs of broad-band images and their corresponding narrow-band images to reconstruct an image from the solar atmosphere with high spatial as well as spectral resolution.

1. Introduction

Observations of the Sun with both high spectral and high spatial resolution are needed for understanding the structure, dynamics and development of small scale features in the solar photosphere. To achieve this we combine narrow-band observations using Fabry-Perot interferometers with image reconstruction methods to overcome degradation of spatial resolution due to seeing and telescopic aberrations.
After a short presentation of the optical setup, two different ways to reconstruct narrow-band images are described, and first results are shown. In the last section future possibilities, provided by the powerful means of image reconstruction applied to narrow-band images, are discussed.

2. Optical Setup

The observations are obtained with the Vacuum Tower Telescope (VTT) at the Oberservatorio del Teide on Tenerife/Spain (Bendlin & Volkmer 1995, Bendlin, Volkmer, & Kneer 1992, Bendlin 1993). The image of the Sun in the prime focus of the telescope is transferred into an optical laboratory by simple transfer optics, consisting of two lenses L1 and L2 (see Fig. 1). An interference filter

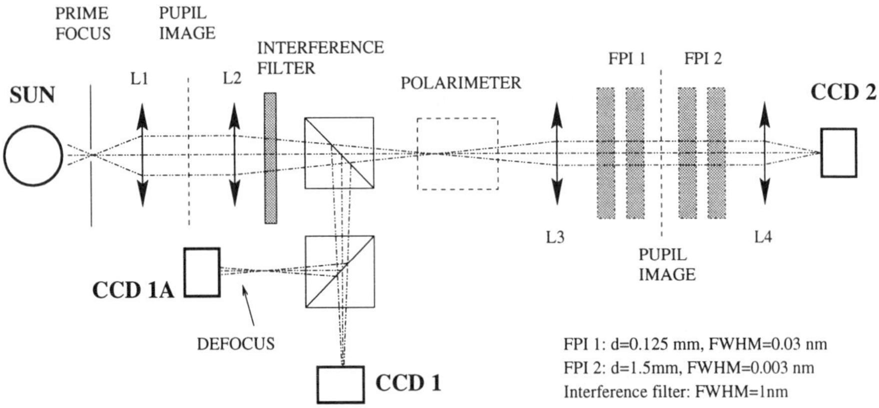

Figure 1. Optical setup at the Vacuum Tower Telescope (VTT) of the Observatorio del Teide/Tenerife

with a bandwidth of 0.3 – 1 nm is used to preselect the wavelength range. 95% of the light passes straight through a beam splitter and forms the narrow-band images at CCD 2. Further transfer optics consitst of the two lenses L3, L4. The Fabry-Perot interferometers are placed close to the image of the pupil formed between L3 and L4. The first FPI acts as a prefilter with a bandwidth of 0.03 – 0.04 nm, the second one limits the bandwidth down to 0.003 – 0.004 nm. To investigate magnetic features by Zeeman splitting of spectral lines, a polarimeter can optionally be placed near the secondary focus.
For the reconstruction of the narrow-band images we take exactly simultaneously broad-band or so called 'white light' images in another part of the optical setup. The remaining 5% of the light at the first beam splitter are again divided half by half and detected by two CCD cameras. One of them (CCD 1) is exactly in focus, the other one (CCD 1A) is defocussed by a known amount. The simultaneity of the exposures of all three cameras is needed in order to obtain the same optical perturbations for the three images due to the rapidly changing seeing.

3. Principle of the Observations

To obtain spectral information at each point in the twodimensional image plane the FPIs scan through a chosen spectral line, step by step by changing the spacings of the etalons. By taking an image triple at each wavelength position we obtain a threedimensional data cube at CCD 2 together with the according white light images at CCD 1 and 1A. The white light images are used to retrieve the information about the instantaneous optical transfer function of each single raw image. This information is applied to the narrow-band images in order to improve their spatial resolution. The major problem is that a satisfying reconstruction of just one image is not possible in most cases, because the instantaneous optical transfer function may be zero or very small at some spatial frequencies, so there is no information in that image at these frequencies. We overcome this problem by taking at least 5 images at each wavelength position, which improves the SNR as well. A typical scan contains 28 wavelength steps at CCD 2 with 5 images at each step, that is 140 images altogether at each CCD camera. To obtain time series these bursts can be repeated after about 100 seconds, which includes also the time to save the data.

4. Image Reconstruction

There are two somewhat different ways to deal with the data. The first one is based on the well known speckle interferometry and uses the information about the instantaneous optical transfer function in an implicite way. The second one is to calculate the optical transfer functions directly by means of the PD method and the deconvolution of each narrow-band image with the corresponding optical transfer function. So we assume the image degradation to be identical for the broad- and narrow-band channel.

4.1. The speckle method

A speckle reconstruction is calculated using all the broad-band images of one burst of CCD 1 (de Boer 1993, Keller & von der Lühe 1992). Once the 'true scene' is known, and let us adopt the speckle reconstruction $\hat{o}_{\mathrm{b}}$ to be the true scene in white light, we can determine the point spread functions and therefore the optical transfer functions S_l for each single image $i_{\mathrm{b},l}$ of CCD 1. According to our assumption they are the same for the corresponding narrow-band images $i_{\mathrm{n},l}$ at CCD 2. These optical transfer functions can now be used to deconvolve the narrow-band images without calculating them explicitely. In Fourier space, the relations read, within an isoplanatic patch,

$$I_{\mathrm{b},l} = \hat{O}_{\mathrm{b}} S_l \qquad \text{and} \qquad I_{\mathrm{n},l} = \hat{O}_{\mathrm{n}} S_l \,. \tag{1}$$

The $\hat{\ }$ denotes an estimate. Here the indices b and n are used to distinguish between the broad- and narrow-band channel, respectively, l refers to the 5 images taken at the same wavelength position.
Inserting the term for S_l from the broad-band image into the term for the narrow-band image, and summing over all available l, gives an estimate of the narrow-band object

$$\hat{O}_{\mathrm{n}} = H \frac{\sum_l I_{\mathrm{n},l} I^*_{\mathrm{b},l}}{\sum_l |I_{\mathrm{b},l}|^2} \hat{O}_{\mathrm{b}} \,. \tag{2}$$

H is a noise filter which prevents amplification of noise at spatial frequencies, where the SNR is too low (Brault & White 1971).

Preliminary results are shown in the next figures (cf. Krieg et al. 1998). Fig.2 shows 5 raw images from the broad-band channel at CCD 1, together with the speckle reconstruction, obtained from the whole burst of 140 broad-band images. Fig.3 and 4 show the corresponding 5 narrow-band raw images and the appertaining reconstructions, using the above formula. Fig.3 corresponds to the observations in the line wing of $\mathrm{NaD_2}$ 0.13 nm off the line center. The FWHM of these images is 0.02 nm. The great improvement in spatial resolution and contrast of the reconstruction can clearly be seen. Fig.4 shows a similar

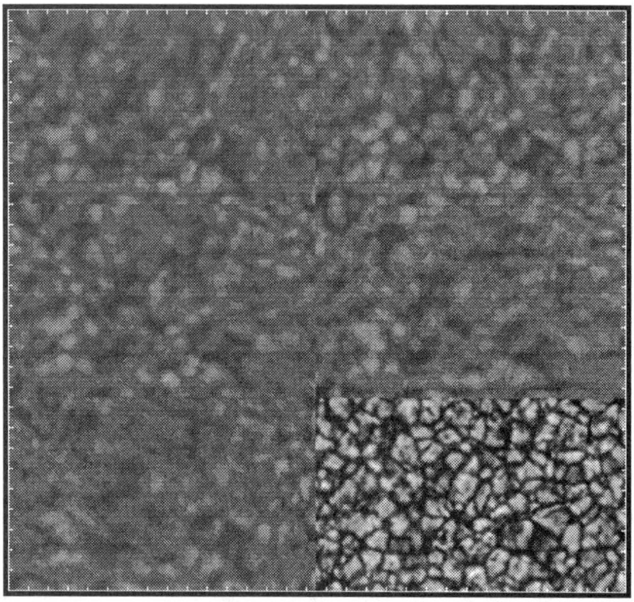

Figure 2. Five broad-band raw images (CCD 1) and the speckle reconstruction. Tickmarks are 1″ distance.

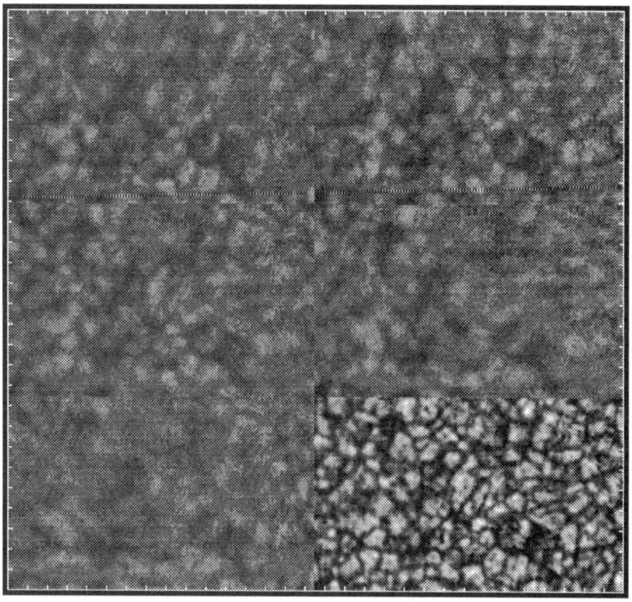

Figure 3. Five narrow-band raw images (CCD 2) and the reconstruction. Images are taken in the line wing of NaD_2, FWHM is 0.02 nm.

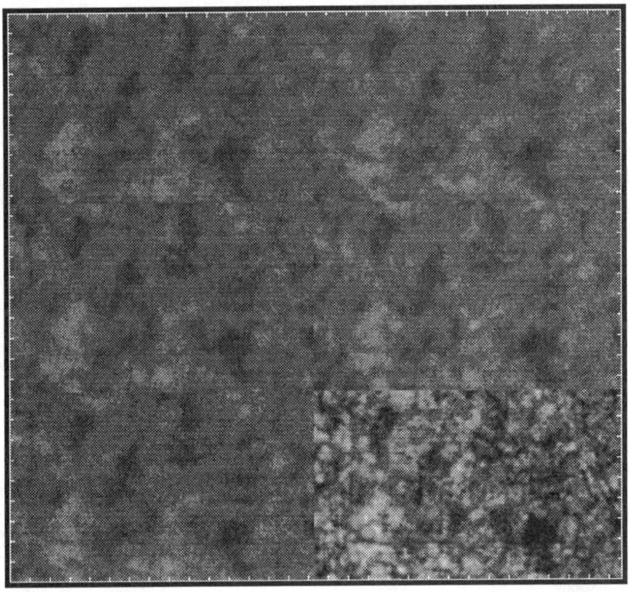

Figure 4. 5 narrow-band raw images (CCD 2) and the reconstruction. Images are taken in the line center of NaD_2, FWHM is 0.02 nm.

result in the line center of NaD_2. Due to the lower SNR there, the quality of the reconstruction is not as good as in the former case, but again an obvious improvement compared to the raw images can be seen.

4.2. The phase diversity method

The other method to calculate reconstructions of the narrow-band data is to use the PD technique. For each image of CCD 2 we have a focussed image of CCD 1 and a defocussed image of CCD 1A. These can be used to calculate the optical transfer function and therefore the point spread function directly for each image pair in white light, which we again suppose to be the same for the corresponding CCD 2 image. For this calculation we use a PD-IDL code from Scott Acton, based on the PhD-Thesis of Mats Löfdahl (Löfdahl 1996, Gonsalves 1982). An example is given in Figure 5. Raw image pairs from CCD 1 (focussed) and CCD 1A (defocussed) are shown, together with their corresponding point spread functions, derived from the PD calculations. The different image quality and the different shapes of the point spread functions due to the fast change of the seeing conditions from image to image can clearly be seen. Once the optical transfer functions S_l are known for each time l we are able to calculate reconstructions from each set of five images, following the standard procedure. Thus the estimated object is given by

$$\hat{O} = H \frac{\sum_l I_l \hat{S}_l^*}{\sum_l |\hat{S}_l|^2} \quad , \tag{3}$$

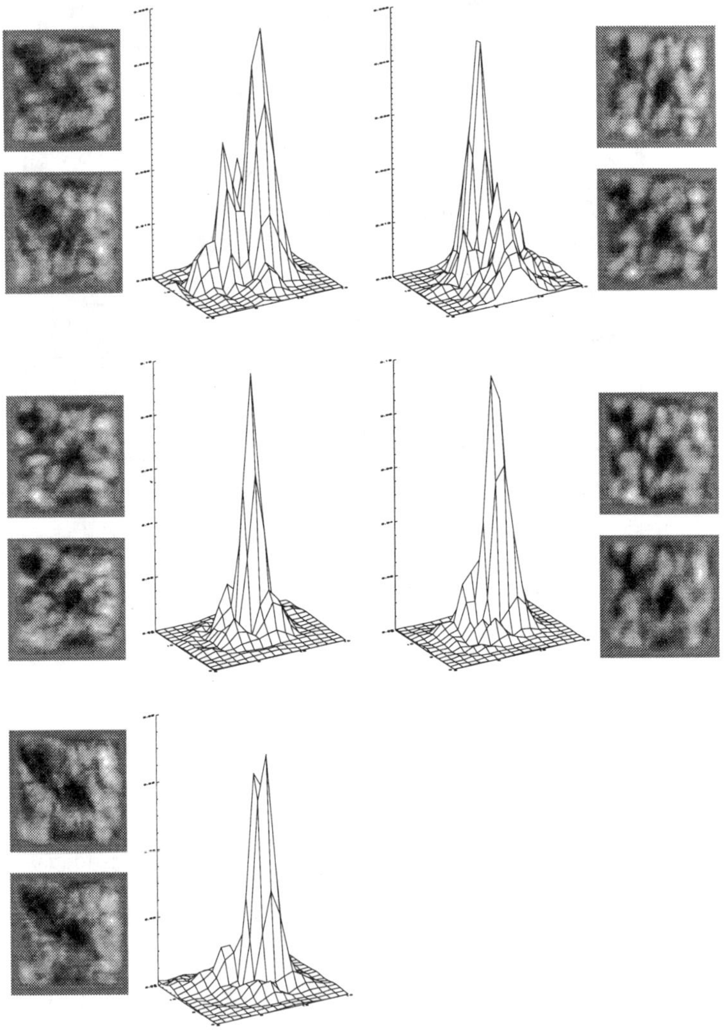

Figure 5. Five pairs of a focussed (upper) and a defocussed (lower) broad-band image at CCD 1 and CCD 1A, respectively, together with the corresponding point spread functions, derived from the PD calculations.

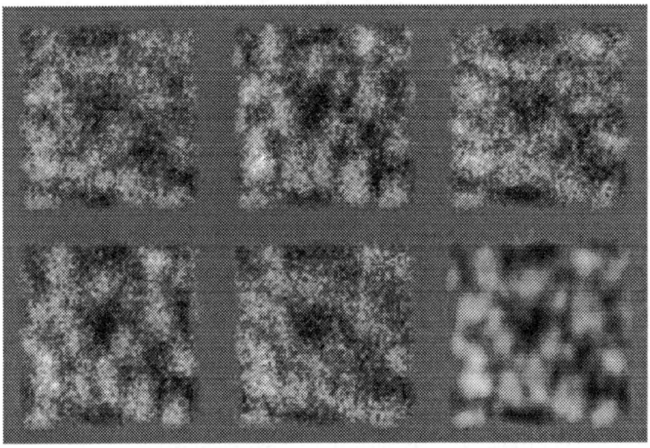

Figure 6. Five narrow-band raw images (CCD 2) and the reconstruction by applying the 5 Point spread functions from Fig.5. Image size is 7″. Images are taken in the line wing of Fe I 557.6 nm, FWHM is 0.003 nm.

where H is again a noise filter (Brault & White 1971). This indeed works both in the broad-band and also in the narrow-band case. The resulting narrow-band reconstruction applying the 5 point spread functions from Fig.5 to the corresponding raw images from CCD 2, together with these raw images, are shown in Fig.6. The images were taken in the line wing of Fe I 557.6 nm, FWHM is 0.003 nm. Again the great improvement of this kind of reconstruction in spatial resolution and contrast is obvious, compared to the raw images.

5. Conclusion and Outlook

Two dimensional imaging of the Sun with high spatial resolution has been possible already for some time. High spectral resolution *together* with high spatial resolution is becoming feasible. Here we have presented ways of image restoration that lead to a substantial increase of information on small-scale structures seen in narrow wavelength bands. Further evaluation of the data will then develop its full advantage over spectroscopy with conventional slit-spectrographs in studying small-scale processes in the solar atmosphere. The change of intensity patterns depending on the formation height of spectral lines or the dynamics and evolution of small-scale magnetic features could be two of many future applications, established by our new ways of data processing. Further improvements should be possible by refinements of the reconstruction methods and updates in detector techniques, that is faster readout and larger field of view. However, the smaller the bandwidth the larger is the photon noise. Thus, for high spatial, spectral, and temporal resolution large telescopes are needed.

References

Bendlin, C., Volkmer, R., Kneer, F., 1992, A&A 257, 817

Bendlin, C., 1993, PhD thesis, Göttingen University

Bendlin, C., Volkmer, R., 1995, A&AS 112, 371

Brault, J.W., White, R.O., 1971, A&A 13, 169

de Boer, C.R., 1993, PhD thesis, Göttingen University

Gonsalves, R.A., 1982, Optical Engeneering 21(5)

Keller, C., von der Lühe, O., 1992, A&A 261, 321

Krieg, J., Kneer, F., Koschinsky, M., Ritter, C., Starck, J.-L., 1998, Ground-based observation with high spatial and spectral resolution. In: Harris, R.H., Moreno Insertis, F., Priest, E. (eds.) A Crossroads for European Solar and Heliospheric Physics. ESA SP-417, p. 317

Löfdahl, M.G., 1996, PhD thesis, Stockholm University

High Resolution Solar Physics: Theory, Observations, and Techniques
ASP Conference Series, Vol. 183, 1999
T.R. Rimmele, K. S. Balasubramaniam, R.R. Radick, eds.

Jets and Eruptions in the Transition Region Observed with CDS, EIT and TRACE

P. Brekke, O. Kjeldseth-Moe

Institute of Theoretical Astrophysics, University of Oslo, Oslo, Norway

T. D. Tarbell

Lockheed-Martin Solar and Astrophysics Lab., 3251 Hanover Street, Palo-Alto, CA 94304

J. B. Gurman

NASA/Goddard Space Flight Center Greenbelt, MD 20771, USA

Abstract. Observations of active regions at the solar limb have revealed ejections of transition region material with unexpectedly high velocities. We present two types of ejections observed by CDS and EIT on SOHO and by TRACE, i.e. active region jets and active region eruptions. The jets are spike-like ejections of material and have been observed with CDS and EIT on SOHO. The plasma appears to follow large scale magnetic field lines to heights up to 170 Mm. The velocity perpendicular to the line of sight approaches 180 km s^{-1}. Active region eruptions consist of much more spatially extended ejections of plasma. One of these was observed on the west solar limb on 19 May 1998 by CDS, EIT and TRACE, and was probably associated with a C4.4 flare recorded by GOES. Doppler shifts corresponding to velocities of 300 km s^{-1} away from the observer, i.e. parallel to the solar surface, were observed by CDS at the leading edge (top) of the eruption while high cadence observations with TRACE show that the plasma velocity perpendicular to the line of sight, i.e. along the solar vertical, reached at least 200 km s^{-1}. Later the plasma apparently falls back down, partly along the same trajectories as it came up, but also along other paths. One may speculate that the earlier eruption, presented here, represents a "CME" that did not make it.

1. Introduction

Active regions have been observed with the EUV/UV instruments on the Solar and Heliospheric Observatory (SOHO) and recently also by the Transition Region and Coronal Explorer (TRACE). In particular observations of active regions on the solar limb have revealed that loops existing at transition region temperatures are highly dynamic in nature (e.g. Kjeldseth-Moe & Brekke 1998). Active region loops are difficult to distinguish on the disk because of their low contrast to the disk emission.

During some of the observations on the limb we see plasma ejected from the active regions. Spike-like jets of material have been observed with the Coronal Diagnostic Spectrometer (CDS) and the Extreme ultraviolet Imaging Telescope (EIT). Another type of eruption was observed by CDS, EIT and TRACE on 19 May 1998. In this latter case a spatially much more extended ejection of plasma on the west limb was observed, probably associated with a C4.4 flare registered with GOES. Below we will describe both types of eruptions.

2. Instruments and Observations

Spectral observations of the features presented here were obtained with CDS, while high cadence broad band images were taken with EIT and TRACE. Relevant instrument features are briefly described below.

CDS is a dual extreme ultraviolet (EUV) spectrometer covering the wavelength range 150–780 Å. The normal incidence spectrometer (NIS) gives stigmatic spectral images along a 240″ long slit in two broad wavelength regions (308–381 Å and 513–633 Å). The astigmatic grazing incidence spectrometer (GIS) covers four wavelength bands in the range 150–780 Å. The wavelength ranges of both NIS and GIS contain emission lines covering temperatures from the chromosphere to the hot corona at T ≈ 3 MK. Spectral and angular resolution of CDS-NIS is $\lambda/\delta\lambda \approx 3500 - 4500$ and $\approx 4''$, respectively. For more details of the instrument and its performance we refer to Harrison et al. (1995).

The CDS data used in this investigation were obtained with NIS using the 4″ slit. By rastering, i.e. moving the solar image perpendicular to the slit in steps of 4″, the spectrometer builds up images of solar features in a series of selected spectral lines emitted at different temperatures. Using 40 slit positions, each with an exposure time of 10 s, a 160″×240″ field of view was covered in 10 min. Each image retains detailed spectral information and can be used to give maps of total emission, velocities, line widths, plasma densities etc. Using this technique CDS can distinguish differences in the emitting plasma on a finer graded temperature scale than for example EIT.

EIT (see Delaboudinière et al. 1995)can produce images of the solar atmosphere with a resolution limited by the 2.59″ pixel size. The multi-layer coating of different quadrants of the mirrors are tuned to four wavelengths centered around spectral lines formed at 80,000 K (He II 304 Å), 1 MK (Fe IX/X 171 Å), 1.5 MK (Fe XII 195 Å), and 2.0-2.5 MK (Fe XV 284 Å).

TRACE (see Handy et al. 1998) is a Small Explorer Mission (SMEX) devoted to the study of the evolution and propagation of fine-scale magnetic fields and plasma structures throughout the solar atmosphere. The instrument consists of a telescope with a 30 cm primary mirror, normal incidence coatings for three EUV bands at 171, 195 and 284 Å, and interference filters for UV bands with maximum transmission at 1216, 1550, 1600, and 1700 Å, as well as a “white light”channel with actual contributions from 1700 Å to 10 000 Å. The 1550 and 1600 Å bands contain the C IV resonance lines at 1550 Å. This allows for a selection of temperature ranges from 6 000 K to 2.5 MK. The 1024×1024 pixel CCD camera has a field of view of 8.5′×8.5′ with a spatial resolution of 1 arcsec and exposure times ranging from 0.002 s to 260 s with a cadence as short as two seconds are available.

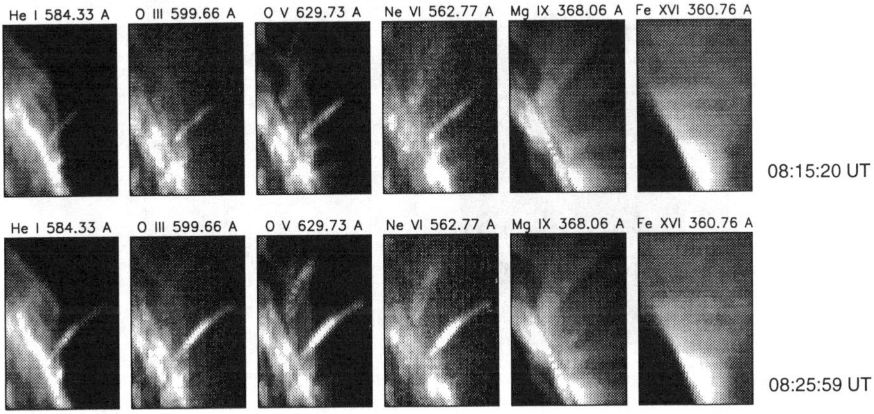

Figure 1. Raster images of the active region jet on 9 December 1997 observed with CDS. Image size is $160'' \times 240''$ and cadence is 10 minutes. The images are ordered from left to right with increasing temperature from He I at 1.6×10^4 K to Fe XVI at 2.7 MK.

3. Active Region Jets

Observations of spike-like jets of plasma were made with CDS and EIT on 9 December 1997. The plasma emerged from NOAA AR8113 located on the NW limb. Fig.1 shows raster images of the active region and the jet above the solar limb in 6 lines observed with CDS ordered from left to right with increasing temperature of formation. The lines include He I 584.3 Å (1.6×10^4 K), O III 599.6 Å (9.1×10^4), O V 629.7 Å (2.3×10^5), Ne VI 562.7 Å (4.3×10^5), Mg IX 368.1 Å (0.96 MK), Fe XVI 360.8 Å (2.7 MK).

The raster images demonstrate the clear difference between the transition region emission and the hotter coronal lines, both in the active region loops and in the jet. In the transition region up to 0.5 MK, including the He I line, loops are clearly outlined and the jet is prominent and changing rapidly with time. In the corona the emission is more fuzzy and the jet is not visible. The same general temperature dependence in the dynamical behavior, with a distinguishing difference at 0.5 MK, is also common for Doppler shifts registered in active region loops without ejections or jets.

The plasma at transition region temperatures seems to be ejected along a curved large scale magnetic structure. The 10 minute time cadence is not fast enough to follow the phenomenon, and this makes it difficult to estimate the perpendicular velocity, i.e. normal to the solar surface. However, the CDS raster images retain their spectral information and the line of sight velocities can be determined at each image element. In the O V raster image in Fig.2 (left panel) the jet appears to be moving with velocities of 45 km s^{-1} and 75 km s^{-1} towards the observer at the positions marked A and B, respectively,(see right panel). The highest velocity occurs in the upper region, B. This is consistent with a motion of the gas along a loop curving towards the observer. It seems likely that the actual flow velocities are much larger than the values measured from the Doppler shifts, since the curved trajectory probably has a considerable

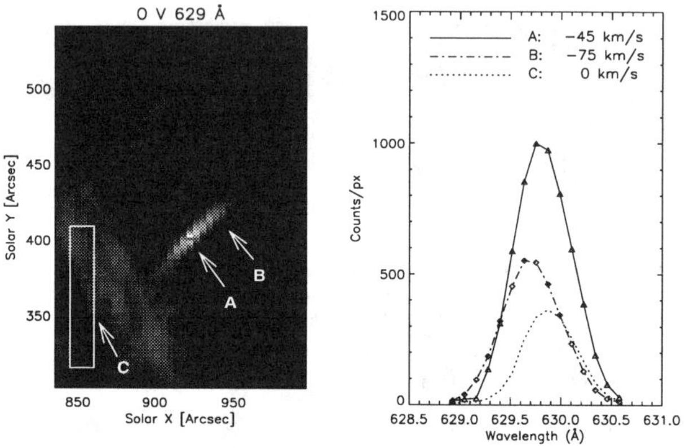

Figure 2. The left panel shows a monochromatic image of an active-region loop system above the west limb observed in O V 629 Å. The image field of view is 160″×240″. Line profiles in three spatial positions, A, B and C, have been marked and are displayed in the panel to the right. The area marked C represents the average disk with zero velocity.

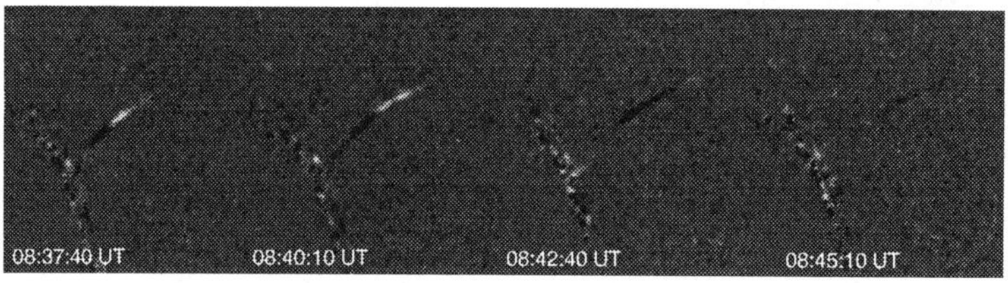

Figure 3. Four EIT images in the Fe XII 195 Å band showing the active region jet of 9 December 1997. The images are selected from a larger series with 30 s cadence.

component perpendicular to the line of sight. Region C represents the area on the disk defining the zero velocity value.

The jet was also co-observed with EIT running in a sub-field mode with a time cadence of 90 s and tuned to the Fe XII 195 Å band (T=1.5 MK). Selected exposures from the EIT observing sequence are displayed in Fig.3 The images show a running difference to better enhance the changes. From the EIT time sequence we have estimated that the ejected plasma is moving with a velocity of 180 km s^{-1} perpendicular to the line of sight. This is not inconsistent with the line of sight velocities derived from CDS spectral images described above. The plasma appears to follow large scale magnetic field lines to altitudes of 170 Mm.

4. Active Region Eruptions

The second type of ejection, the active region eruption, was observed during a Joint Observing Program (JOP) with CDS, EIT and TRACE, starting at 07:00

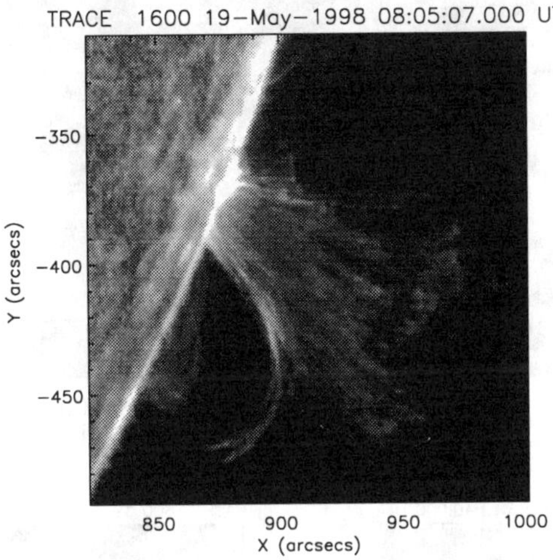

Figure 4. TRACE image in the 1600 Å band showing the detailed structure of the active region ejection from 19 May 1998 at 08:05:07 UT, 5 minutes after the eruption started.

UT on 19 May 1998. Active region NOAA AR8218 had just rotated around the west limb when a large eruption of plasma was observed, most probably related to a spiked, brief flare of type C4.4 registered by GOES. The flare started at 07:56 UT and ended at 08:04 UT. The TRACE instrument recorded images in the 1216 Å and 1600 Å channels as well as in the white light channel. The cadence between images in each channel was 30 s. Figure 4 shows the eruption observed in the 1600 Å band at 08:05:07 UT, 5 min after it started. Outside the solar limb it seems reasonable to assume that the C IV lines at 1550 Å (T $= 10^5$ K) are the only significant contributors to the image intensity, with only weak contributions from the continuum, and from neutral and singly ionized lines within the band. We note the abundant and inhomogeneous detail in the structure. The extent in height of the eruption at this time is 62 Mm, and later it reached a maximum of 120 Mm 12 min after start.

Fig.5 shows a series of TRACE images consisting of every fifth frame in the 1600 Å band sequence. The eruption throws plasma to a great altitude with plasma velocities along the solar vertical reaching 200 km s^{-1} as measured from the expansion of the leading edge of the eruption. The upward motion then seems to slow down and the plasma apparently falls back down, partly along the same trajectories as it came up, but also along other paths. A possible picture is of plasma being transported up one loop leg, with part of it going beyond the summit point before falling back to the solar surface. Similar structures and time development are observed in the 1216 Å band. Perhaps more surprisingly, the ejection is also observed in the white light channel, albeit at low altitudes during first 5-10 min of the eruption.

Figure 6 shows a selected frame of the eruption from EIT in the 195 Å band together with a simultaneous TRACE image. The EIT frame is a difference image from the previous exposure to enhance the erupting material. It is

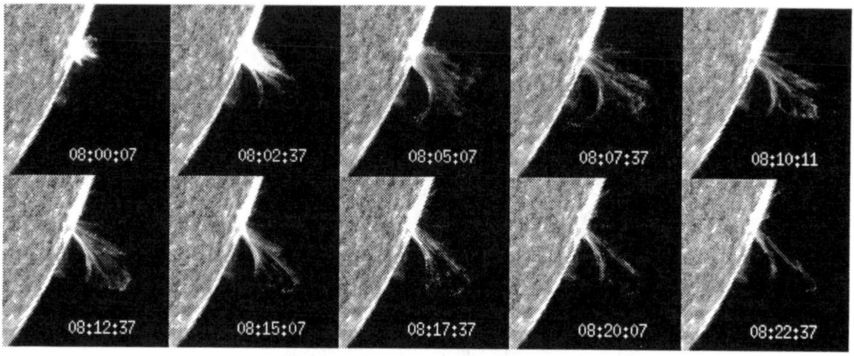

Figure 5. Series of every fifth TRACE frame in the 1600 Å band from the active region ejection on 19 May 1998, showing its development with time.

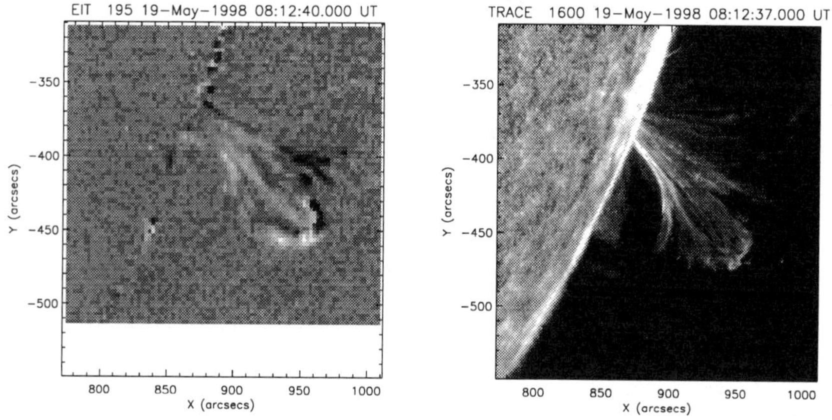

Figure 6. Simultaneous EIT and TRACE images, showing the eruption from the active region ejection on 19 May 1998.

interesting that the eruption also can be seen in the EIT images, emitted from a plasma at much higher temperatures than the 1600 Å band TRACE images.

Figure 7 shows two selected sets of CDS raster images made during the eruption. Again the size of the rasters is $160'' \times 240''$ and the cadence is 10 min. The lines are the same as in Figure 1 for the jet, i.e. at the same temperatures. The rasters in the two sets start at 08:00:54 UT and 08:22:21 UT respectively, i.e. a time difference between the sets of 22 min. The first set is taken at the time when the TRACE images showed the initial rise in the eruption, i.e. between 08:00 UT and 08:11 UT. The second set is taken at a time when material is falling back to to the solar surface. In the eruption phase the material appears to fill a large volume, revealed by the TRACE observations as a densely packed trunk of many finer structures. The material falls back through an increasingly simplified set of loops that are only partly the same as those emitting during the rise phase. The emitting loops at temperatures below 0.5 MK are similar in structure and time development, and may appear co-located to within the accuracy of the observation. This does not rule out different temperatures occurring in different

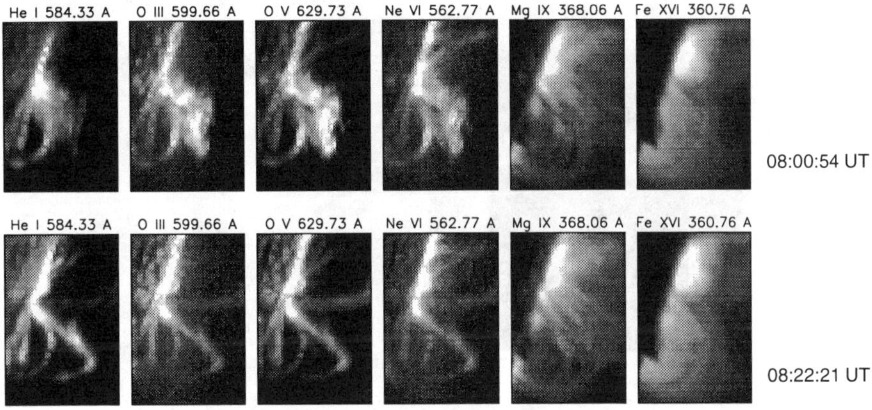

Figure 7. CDS raster images of the active region eruption 19 May 1998. Field of view, cadence and spectral lines (i.e. temperatures) for the images are the same as in Figure 1 above. The two sets of images are from the early (top) and late phases (bottom) of the eruption with starting times at 08:00:55 UT and 08:22:21 UT, respectively.

locations, but means that the loops may have fine structures, thin strands that are far smaller than the instrument resolution. In the hot plasma of 1 MK or more, we do not note the eruption. Again this agrees with the dynamic behavior of non-erupting loops described by Kjeldseth-Moe and Brekke (1998). In contrast the eruption is clearly visible in the EIT 195 Å band (see Figure 6). The difference between CDS (1 MK and 2.7 MK) and EIT (1.5 MK) images might be a "true" effect i.e. co-located solar plasmas emitting below 0.5 MK and at 1.5 MK, but not at a temperature of 1 MK. Alternatively the EIT bandpass may contain lines that are cool and sufficiently strong to cause structures very similar to the TRACE C IV eruption at 100 000 K. Both explanations would require a very special mix of hot an cool gas in the particular case of the eruption.

In contrast the eruption is clearly visible in the EIT 195 Å band (see Figure 6) which is emitted from a 1.5 MK plasma. The difference between CDS (1 MK and 2.7 MK) and EIT (1.5 MK) could either be a solar effect (eller kanskje bare REAL) or the EIT bandpass may be contaminated by some cooler emission. The latter should be possible to confirm by co-observing with EIT and CDS/GIS which includes the Fe XII 195 Å line.

Figure 8 shows in the left panel the CDS image in O V 620 Å together with the observed line profiles (in the right panel) from tree locations marked A, B and C. Location C represents the disk region defining the reference wavelength. We notice how the observed Doppler shifts correspond to material moving away from the observer with velocities of 120 km s^{-1} and 290 km s^{-1} in A and B, respectively. Location B is at the top of the expansion and may represent a purely horizontal velocity at the summit of a magnetic loop. In this case the 290 km s^{-1} velocity would represent the true flow speed in the loop. This, however, is not certain since the eruption continue to expand another 40 Mm before reaching its maximum altitude, i.e 120 Mm.

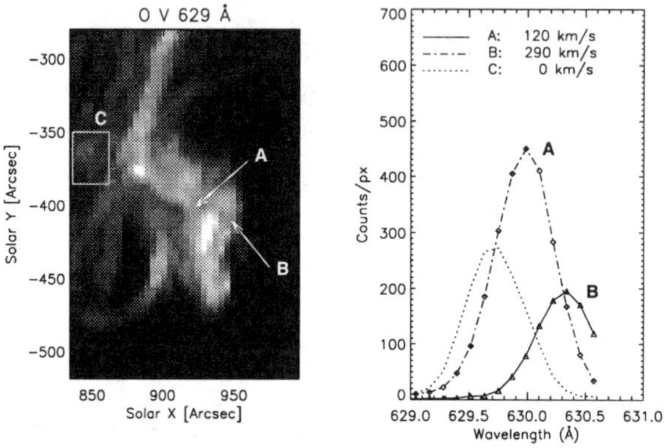

Figure 8. Monochromatic image of the eruption above the east limb observed in O v 629 Å. Line profiles in three spatial positions, A, B and C, have been marked and are displayed in the panel to the right. The area marked C represents the average disk with zero velocity.

5. Concluding Remarks

The two types of ejections, jets and active region eruptions, are remarkably similar in velocities and in temperature dependence of the dynamical behavior. Perhaps they are the same phenomenon on different geometrical scales. Secondly, we note that the centripetal force from the magnetic field applied to the gas flowing in the loops during the eruption is considerable, i.e tree times the solar gravity at the loop summit in Fig.8. Thus, the eruption may perhaps represent a "CME" that did not make it.

Acknowledgments. Thanks to Nils Brynildsen for careful reading of the manuscript. The TRACE mission is supported by NASA contract NAS5-38099 at the Lockheed Martin Solar & Astrophysics Laboratory. SOHO is a project of international cooperation between ESA and NASA.

References

Kjeldseth-Moe, O., & Brekke, P., 1998, Solar Phys. in press

Delaboudiniére, J.-P.,1995, Solar Phys., 162, 291.

Handy, B. N., Bruner, M. E., Tarbell, T. D., Title, A. M., Wolfson, C. J., Laforge, M. J., & Oliver, J. J., 1998, L. Golub, T. Barbee, Jr., Solar Phys. in preparation

Harrison 1995, Solar Phys. 162, 233.

High Resolution Solar Physics: Theory, Observations, and Techniques
ASP Conference Series, Vol. 183, 1999
T.R. Rimmele, K. S. Balasubramaniam, R.R. Radick, eds.

Coordinated Observations of Transition Region Dynamics using TRACE and the SVST

T. E. Berger, B. De Pontieu[1], C. J. Schrijver, A. M. Title

Lockheed-Martin Solar and Astrophysics Laboratory, 3251 Hanover St., B/252, O/H1-12, Palo Alto, Ca., 94304.

G. Scharmer

Stockholm Observatory, Swedish Royal Academy of Sciences, S-13336 Saltsjöbaden, Sweden

Abstract. We present coordinated observations from the Transition Region and Coronal Explorer (TRACE) satellite and the Swedish Vacuum Solar Telescope (SVST) on La Palma, Spain. The observations focus on AR 8227 on 30-May-1998 and the analysis of a novel extreme ultraviolet (EUV) emission pattern over plage regions seen in the Fe IX/X 171 Å bandpass of the TRACE instrument. The emission appears as a bright dynamic network with dark inclusions on scales of 2,000–3,000 km, confined to layers approximately 1,000-3,000 km thick at heights between about 2,000–4,000 km above the photosphere. The fact that the emission does not occur over all plage regions implies that it is not a low-temperature contamination of the TRACE bandpass. SVST Hα images establish that the dark inclusions correspond to jets of chromospheric material. Preliminary analyses suggest that we are imaging the upper transition region and its interaction with chromospheric structures for the first time.

1. Introduction

A primary goal of the TRACE mission is to observe the characteristics of the transition region between the 10^4 K chromosphere and the 10^6 K corona. Although much theoretical work has been devoted to the transition region (see the book by Mariska, 1992, for an excellent overview), high resolution observations of the solar atmosphere in this temperature regime have for the most part been limited to spectral data (Cook 1991) which do not provide the high temporal cadence images necessary to study the spatial structure of the region. The EIT telescope on SOHO (Delaboudiniere, Artzner, et al. 1995) images the same EUV wavelengths regions as TRACE, but with a full-disk resolution of only $5''.2$, which is inadequate for studying the interaction of chromospheric structures with the upper atmosphere.

One of the most striking features of TRACE EUV images of solar active regions is the occurrence of a fine-scale, low-lying, network over certain plage regions. Because of its resemblance to the ground-covering plant, this emission pattern has been termed "moss". It has been previously recognized in lower-

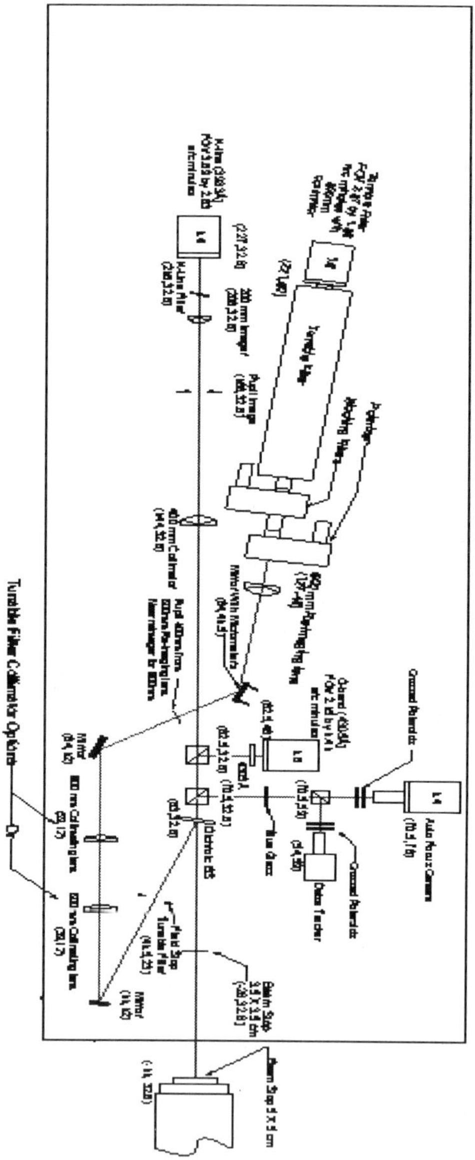

Figure 1. Optical layout of the SVST on 30-May-1998 during coordinated active region observations with TRACE. Courtesy of D. Torgerson, LMSAL.

resolution sounding rocket images (Peres, Reale, & Golub 1994) and correctly deduced as fundamentally distinct from large-scale active region EUV loops. It is also detectable, though not resolved, in EIT 171 Å images. With the improved spatial and temporal resolution of TRACE, we can now study the structure and dynamics of this novel emission for the first time.

On 30-May-1998 we observed AR 8227 in excellent seeing at the SVST over a period of approximately 10 hours. The observations were coordinated with the TRACE observing program, which took Fe IX/X 171 Å images on a roughly 30 second cadence and white-light images of the active region every two minutes. The resulting dataset, along with data from the SXT instrument on Yohkoh, and the MDI instrument on SOHO, has established the basic nature of this newly resolved feature of active regions in the EUV. We report here on the very high resolution data obtained at the SVST and its role in investigating the characteristics of the moss.

2. Observations

2.1. Instruments

TRACE is a Small Explorer (SMEX) satellite in a 600 km Sun-synchronous orbit (Handy et al. 1999). It consists of a 30 cm Cassegrain telescope with a four-quadrant multi-layer coating on the primary and secondary mirrors. Three of the quadrants image the Sun in EUV bandpasses near 171, 195, and 284 Å, respectively. The fourth quadrant acts in combination with a 12.5 cm LiF entrance window and dual focal plane filters to select several bandpasses from the H-Lyα 1216 Å band up to a wide-band visible light bandpass. All quadrants image the Sun onto the same 1024$\times$1024 pixel lumigen-coated CCD, with the wavelengths sequentially recorded in an operator-specified order. The full field-of-view (FOV) in any bandpass is a circle roughly 512 arcseconds in diameter. The EUV spatial resolution is pixel-limited to $0\farcs9986$ (J. P. Wüelser, TRACE sci-ops email, 1998). Table 1 lists the wavelengths and temperature responses of the various TRACE bandpasses.

The SVST is a 47.5 cm vacuum refractor located at 2400 m ASL on the island of La Palma, Spain (Scharmer, Brown, et al. 1985). On 30-May-1998 we observed AR 8227 using the three camera setup shown in Fig 1. Excellent seeing conditions prevailed for most of the day, with the best seeing occurring during the period from about 12:00-16:00 UT. Post-processing of the SVST data entailed flat-field/dark current correction, rigid alignment of each wavelength set to remove arcsecond-scale jitter and image rotation, and (in the case of the G-band and magnetogram sets only) destretching to remove geometric seeing distortions. In addition, the G-band and magnetogram datasets were space-time Fourier filtered with a 4 km/sec linear filter to attenuate p-mode brightness variations.

2.2. Dataset Alignment

Major effort was required to align and scale the various SVST and TRACE wavelength datasets to a common coordinate system since each camera at the SVST has its own plate scale and field-of-view (FOV), each of which is entirely differ-

Wavelength Å	Bandpass FWHM, Å	Observed	Temperature K
5000	broad	photosphere	4–6.4×10^3
1700	200	UV continuum	4–10×10^3
1600	275	C IV, Fe II, cont.	4–10×10^3
1550	20	C IV	60–250×10^3
1216	84	H I Lyα	10–30×10^3
171	6	Fe IX,X	0.16–2×10^6
195	7	Fe XII	0.5–2×10^6
284	11	Fe XV	1.2–4×10^6

Table 1. TRACE Bandpass Characteristics

ent from that of the TRACE instrument. We chose the SVST G-band dataset as the reference set since this was the highest resolution dataset in the group. Thus all datasets were rotated to have solar North up with a final plate scale linearly interpolated, if necessary, to the G-band value of 12 pixels/arcsecond.

Inter-wavelength alignment was accomplished in the following sequence: common features in the TRACE white light image nearest in time to the initial G-band image were used to determine the G-band/TRACE white-light alignment. The TRACE EUV dataset was then aligned to the G-band dataset using calibrated TRACE white-light/EUV offset values (R. Shine, TRACE sci-ops email, 1998). The Ca II K-line images were aligned to the G-band using common bright points associated with magnetic elements. The magnetograms were aligned to the G-band dataset using the RCP pre-magnetogram image set which also shows magnetic element bright points; the LCP images were subsequently "stretched" onto the RCP set before magnetogram creation. The Hα datasets were aligned to the G-band dataset by using magnetic element bright points visible in the Hα-700 mÅ filtergrams and then assuming that all other Hα wavelength offset images possessed the same relative alignment to the G-band images. Residual seeing distortions and alignment drifts in the SVST datasets limit the random frame-to-frame image alignment accuracy to about 1–2$''$, although subframes can be manually adjusted to align with better than 1$''$ accuracy across most of the FOV.

The final images in all datasets were formatted to 2028×2044 pixel 16-bit arrays; in those datasets with smaller FOVs (such as the Hα and magnetogram sets), the images were imbedded in the arrays with zero-valued borders in the blank areas. We used the G-band dataset as a temporal reference as well, creating nearest-neighbor interpolated symbolic links in the other datasets where necessary to arrive at a uniform number of files covering the period from 7:50

Dataset	Wavelength (Å)	Atmospheric Region	Cadence (sec)	FOV (arcmin)
G-band	4305	PS	20	2.82×2.82
Ca II K-line	3933	Low CS	15	2.76×2.14
	-700 mÅ	Low CS	120	
	-350 mÅ	Mid CS	120	
Hα	6563	CS	120	2.87×1.88
	+350 mÅ	Mid CS	120	
	+700 mÅ	Low CS	120	
Fe I	6302	PS Magnetic Field	120	2.87×1.88
TRACE Fe IX/X	171	Upper TR	30	8.5×8.5
TRACE White Light	5000	PS	120	

Table 2. Coordinated dataset list for AR 8227 on 30-May-1998. PS = photosphere, CS = chromosphere, TR = transition region. All datasets cover the period from 7:50 to 16:50 UT. FOV values are original values; final datasets are clipped or imbedded into G-band FOV.

to 16:50 UT. This resulted in 1540 images in all datasets with a total storage requirement of approximately 22 Gbytes. Table 2 lists the coordinated datasets and their properties.

3. Image Analysis

Fig.2 shows AR 8227 on 30-May-1998 14:40 UT (heliographic coordinates N26°.7 E10°.4) as seen in the TRACE Fe IX/X 171 Å 171 Å bandpass. The dotted box outlines the FOV of the aligned and scaled dataset used in all subsequent analyses. Moss regions are prominent throughout the central area of the image. In particular, Boxes A and C outline regions of strong moss emission. In contrast, Box B shows very low-level Fe IX/X 171 Å emission.

Fig.a–d show representative SVST images of AR 8227 corresponding to the dotted outline shown in Fig.2 The locations of Boxes B and C are outlined in

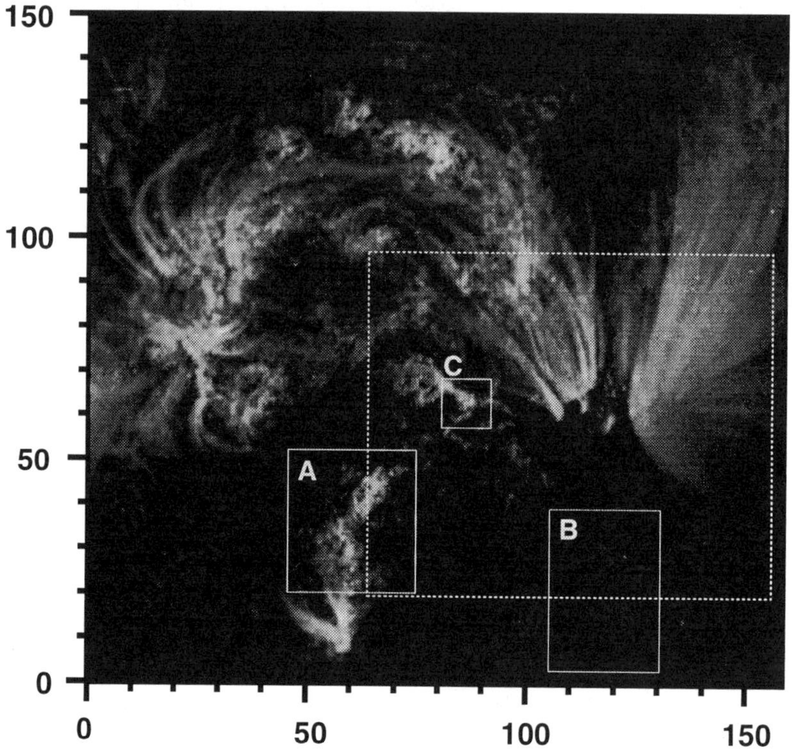

Figure 2. TRACE Fe IX/X 171 Å image of AR 8227 on 30-May-1998 14:40 UT. Heliographic position is N26°.7 E10°.4. The dotted box outlines the maximum field-of-view achieved in coordinated observations at the SVST. The active region consists of a unipolar spot group (just to the north of Box B) and a large trailing plage complex.

each figure. Comparison of TRACE images with SVST Fe I 6302 Å magnetograms and Ca II 3933 Å K-line images shows that moss regions in the SVST FOV occur exclusively over plage magnetic field regions. On 10,000–20,000 km scales, outlines of the moss regions follow the magnetic network patterns closely. However there are conspicuous areas of plage that lack moss: Box B in Figure 2 for example. This indicates that moss is not a ubiquitous feature of the magnetic network in active regions; the presence plage is a necessary but not sufficient condition for the occurrence of moss. This finding implies that the moss emission in the TRACE 171 Å bandpass is not from a low temperature contamination line such as the O VI 173 Å line formed at a temperature of about 300,000 K. Since all plage regions are visible in spectroheliograms in lines with formation

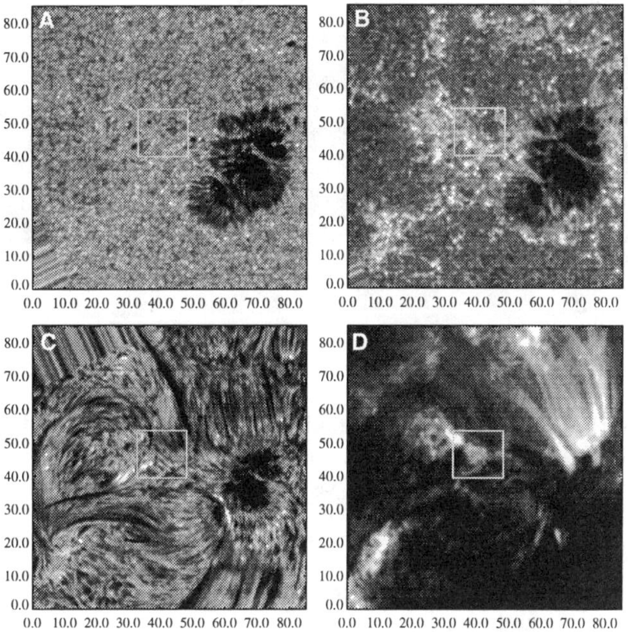

Figure 3. Representative images of AR 8227 obtained at the SVST on 30-May-1998 in coordination with TRACE. (a)G-band 4305 Å filtergram, (b)Ca II K-line 3933 Å filtergram, (c), Hα-350 mÅ filtergram and (d)Fe I 6302 Å magnetogram. All images have been aligned and scaled to the G-band image. Tickmarks are labelled in arcseconds. The box corresponds to Box C in Fig. 2.

temperatures up to about 400,000 K, a low-temperature contamination source for the moss would result in moss occuring over all plage regions.

Moss elements move and vary in brightness on time scales of one minute or less. The motions are random displacements of the bright EUV elements on scales of 1000 km. We see no systematic drifts or flows of the bright elements, nor do we detect any intrinsic "spinning" or vortex motions within these structures. In the large majority of TRACE active region observations to date, moss patterns appear relatively static on scales of plage regions (10^4 km). As with the Ca II K-line plage pattern, the large-scale moss pattern usually evolves slowly on time scales of 10–20 hours.

The rapid brightness variations seen in the moss are due to two independent mechanisms: intrinsic variability of the EUV irradiance in the bright elements and extinction of the line-of-sight EUV irradiance by relatively cool spicule-like "jets." Extinction by dark jets is most clearly visible in limb images. Intrinsic variablity of the EUV emission is difficult to quantify with the existing datasets

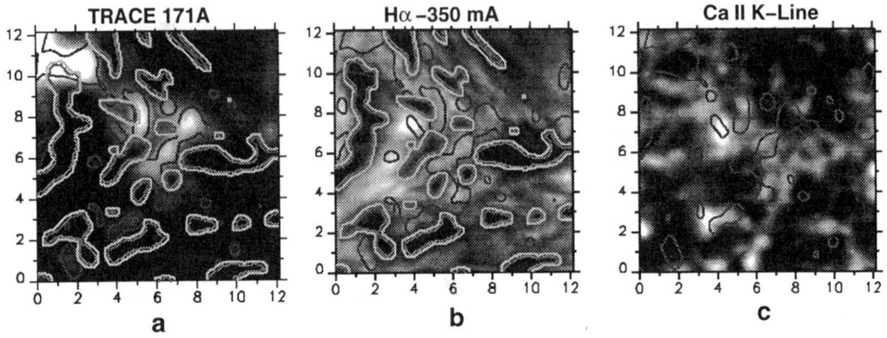

Figure 4. Comparison of TRACE 171 Å moss contours to contours outlining dark jets in simultaneous Hα-350 mÅ filtergrams. (a)TRACE 171 image of Box C in Fig. 1, (b)SVST Hα-350 mÅ filtergram of Box C, and (c)SVST Ca II K-line filtergram of Box C. All images taken within one minute of 14:38 UT.

due to temporal resolution limitations. However it is clear that the moss is a highly dynamic phenomenon.

On the disk, TRACE Fe IX/X 171 Å movies show clear cases of dark material both projecting horizontally over bright moss elements to obscure the emission as well as "pushing" the bright elements laterally. The dark material corresponds nearly exactly to dark absorbing features seen in simultaneous SVST images taken in the wings of the Hα 6563 Å line. Specifically, the dark inclusions in the moss correlate most closely to dark absorbing features in "summed images" created by adding nearly co-temporal Hα ±350 mÅ filtergrams. Fig.4 cxhibits this in a single snapshot: the bright moss contours defined in Panel A are complementary to the double contours that delineate dark features in the summed Hα image in Panel B.

Such red/blue shifts in Hα are consistent with upward-directed jets accompanied by backfalling material, as in classical spicules. However we note that these "jets" are on average somewhat smaller than classical spicules and are found over plage regions. Whatever the details of these chromospheric events, comparisons of TRACE and SVST Hα ±350 mÅ movies show that moss temporal variations are closely tied to their dynamics. However the occurrence of chromospheric jets is apparently not linked to moss regions since we see similar jets over all magnetic network in SVST movies, and particularly in the moss-less plage region of Box B of Figure 2.

4. Conclusions

Given the high-temperature and low altitude of the moss emission, we hypothesize that it is due to thermal conduction from overlying hot coronal loops. Such a mechanism was successfully modelled by Peres, Reale, & Golub (1994) assuming hydostatic loop models. Although the hydrostatic assumption is clearly dubious given the dynamic nature of the moss (and presumably any connected loops), thermal conduction is clearly a factor in upper transition region heating (Mariska 1992). We thus delineate between the bright moss EUV emission caused by thermal conduction and the dark inclusions caused by chromospheric jet extinction of the moss EUV emission.

If we associate the bright moss elements with the locations of coronal field lines along which thermal conduction heats the transition region, the misalignment with chromospheric flux elements magnetic field is intriguing. One interpretation is that large-scale coronal loops are formed from the interaction of many separate field-lines originating from a myriad of small-scale magnetic elements lower in the atmosphere. In this picture, coronal loops are not simple extensions of magnetic field lines rising from the photosphere into the corona, but are instead formed from a complex merging or tangling of the field lines caused by the constant relative motions of magnetic elements in response to convective flows in the photosphere. Downward thermal conduction leading to the EUV emission would then occur along coronal field lines that are not necessarily vertically aligned with the photospheric flux elements. A similar relationship of coronal field lines to photospheric flux tubes has been proposed in previous theoretical work (Spruit & Roberts 1983) and empirically modelled in recent G-band bright point work (Van Ballegooijen et al. 1998).

In summary, our data indicate that the TRACE Fe IX/X 171 Å "moss" emission is due to 10^6 K plasma in the plage footpoint regions of coronal loops. The emission is localized in relatively thin bright elements on 2–3×10^3 km scales typically about 3000 km above the photosphere. The characteristic appearance in high resolution EUV images is that of a dynamic low-lying network of bright elements interspersed with dark inclusions. The dark inclusions often correspond to jets of chromospheric plasma visible in simultaneous Hα images. Neither the bright EUV elements nor the dark jets align well with the locations of magnetic elements in the photosphere and lower chromosphere. These characteristics suggest that we are for the first time at least partially resolving the structure and dynamics of the upper transition region in the solar atmosphere.

Cook, J. W. 1991, in Mechanisms of Chromospheric and Coronal Heating, ed. P. Ulmschneider, E. R. Priest, & R. Rosner (Springer-Verlag, Berlin Heidelberg), 83

Delaboudiniere, J. P., Artzner, G. E.& Brunaud, J., et al. 1995, Solar Physis, 162, 291

Handy, B., Tarbell, & T., Wolfson, J. 1999, Solar Physics, in preparation

Mariska, J. T. 1992, The Solar Transition Region, Cambridge Astrophysics Series (Cambridge: Cambridge University Press)

Peres, G., Reale, F. & Golub, L. 1994, Astrophysical Journal, 422, 412

Scharmer, G., Brown, D. S., Peterson, L. & Reben, J. 1985, Applied Optics, 24, no. 16, 2558

Spruit, H. C., & Roberts, B. 1983, Nature, 304, 401

van Ballegooijen, A., Nisenson, P., Noyes, R. W., Löfdahl, M. G., Stein, R. F., Nordlund, Å. & Krishnakumar, V. 1998, Astrophysical Journal, in press

Group Discussion

Zirker : The moss-less region seems to have stronger magnetic fields and yet has no 3-5 MK loops, which are necessary for moss. Can you explain?

Berger : The mossy and moss-less regions do have comparable magnetic fields. The 3-5 MK loops are those main loops connecting the leader sunspot group to the trailing plage.

Bala : Could these structures be hot, low-lying loops brimming off the transition region surface?

Berger: If this were the case, one would expect the moss to be a ubiquitous feature of the magnetic network. Instead we see many regions of strong field network without moss. This fact argues against a "generic" source for the heating of the moss.

High Resolution Solar Physics: Theory, Observations, and Techniques
ASP Conference Series, Vol. 183, 1999
T.R. Rimmele, K. S. Balasubramaniam, R.R. Radick, eds.

Small-Scale Magnetic Reconnection in the Quiet Sun

Jongchul Chae

Big Bear Solar Observatory, New Jersey Institute of Technology
40386 North Shore Lane, Big Bear City, CA 92314-9672
chae@bbso.njit.edu

Abstract. Recent observations indicate that small-scale magnetic reconnection ubiquitously occurs in the quiet Sun as well as in active regions. Particularly interesting is the latest finding of close associations among (1) magnetic flux cancellation seen from photospheric magnetograms, (2) chromospheric upflow events seen in the Hα line and (3) transition region explosive events seen in UV lines. I present a brief review of this finding, and propose a schematic two-step magnetic reconnection model to explain it. The essence of the model is in the formation of magnetic islands as a result of slow magnetic reconnection occurring in the low atmosphere. Magnetic islands are ejected and accelerated upward by the so-called diamagnetic melon seed mechanism, and are eventually annihilated by overlying magnetic field lines through fast magnetic reconnection. I consider photospheric flux cancellation as a direct result of the first magnetic reconnection, and identify chromospheric upflow events with upward moving magnetic islands, and explosive events with the hotter material ejected from the second magnetic reconnection.

1. Introduction

A typical deep magnetogram shows that magnetic fields exist everywhere in the quiet Sun —i.e., both in supergranulation boundaries and in cells. Therefore it appears that the chromosphere and corona in the quiet Sun as well as in active regions is dominated by magnetic fields. Furthermore, the small-scale and mixed-polarity nature of the quiet Sun magnetic fields implies that small-scale magnetic reconnection should ubiquitously occur in the quiet Sun and may be important in the dynamics and energetics of the upper atmosphere. In the present paper, I give a brief review of some of recent observations which support small-scale magnetic reconnection in the quiet Sun, and propose a two-step magnetic reconnection model as an attempt to explain them in a self-consistent way.

2. UV Explosive Events, Flux Cancellation, and Hα Upflow Events

UV explosive events which were first discovered from the High Resolution Telescope and Spectrograph (HRTS) experiments (Brueckner & Bartoe 1983; Dere, Bartoe & Brueckner 1989) are small-scale dynamic features which can be barely

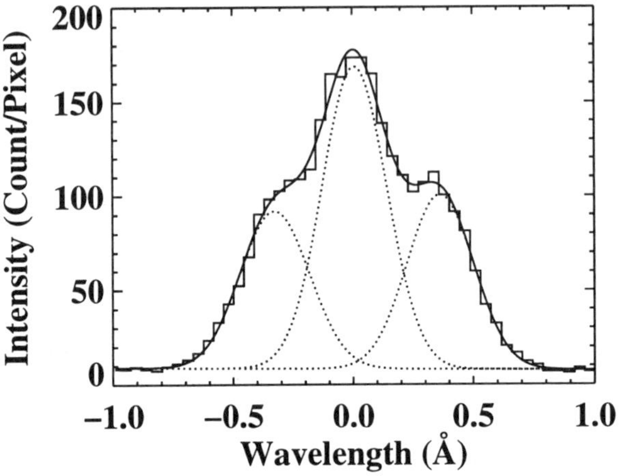

Figure 1. A typical explosive event seen in the Si IV $\lambda 1402$ line profile observed by SOHO/SUMER. The profile consists of three components, one stationary and two high velocity components (Paper II).

resolved at a spatial resolution of about $1''$. They ubiquitously occur in the quiet Sun as well as in active regions. Explosive events are characterized by very broad UV line profiles, and therefore have been considered to be a manifestation of small-scale magnetic reconnection (Porter & Dere 1991; Dere et al. 1991; Dere 1994). Recent observations performed by the Solar Ultraviolet Measurements of Emitted Radiation (SUMER) on board the Solar and Heliospheric Observatory (SOHO) confirm the magnetic reconnection origin of explosive events in a couple of ways. Firstly, the structure of the explosive events evolves in the manner predicted by theoretical models of magnetic reconnection (Innes et al. 1997). Figure 1 shows a typical explosive event seen in the Si IV $\lambda 1402$ line profile. Note that the broad line profile is mainly due to the existence of the two high-velocity components: one blue-shifted and the other red-shifted. The velocities of these components are comparable to the local Alfvén speed, demonstrating the bi-directional jet nature of the explosive event. It was also found from SOHO/SUMER that explosive events preferentially occur at the regions with weak and mixed polarity fluxes, and the majority of explosive events occur as bursts during magnetic flux cancellation which usually lasts longer than 1 hour (Paper I). Since flux cancellation is usually considered to be a result of magnetic reconnection (e.g., Wang & Shi 1993), the observed association between flux cancellation and explosive events is further evidence of the magnetic reconnection origin of explosive events.

Chromospheric upflow events are short-lived events which display blueshifts in Hα which were first discovered by Wang et al. (1998). A typical upflow event seen in the Hα line is presented in Fig.2. As seen from the contrast profile, the event appears dark and highly visible at the wavelength Hα-0.6 Å , but is not easily visible at the line center and the red wing. This characteristic can be well explained by the existence of an upward-moving cloud which can either absorb

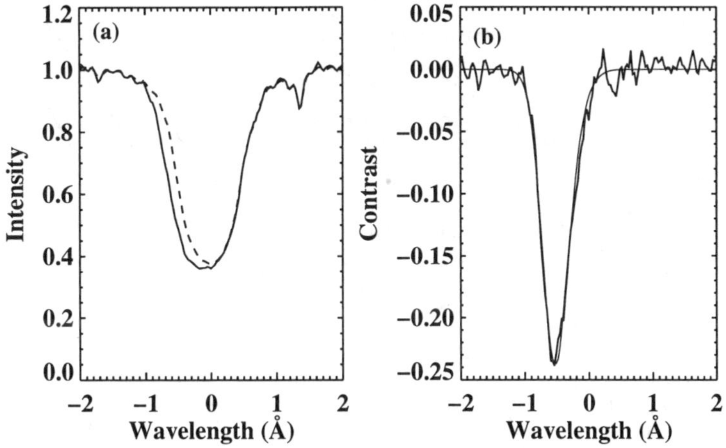

Figure 2. (a) Hα profiles of the average Sun (dashed curves) and a chromospheric upflow event (solid curve). (b) The Hα contrast profile (thick curve) and its cloud model fit (thin curve) (Paper II).

Table 1. Typical Parameters of Transition Region Explosive Events and Chromospheric Upflow Events (Ref. (1) Dere 1994; (2) Paper II).

Parameter	Explosive event	Upflow event
Size	1500 km (1)	1500 km (2)
Lifetime (FWHM)	60 s (1)	60 s (2)
Bulk motion	100 km s^{-1} (1)	20 km s^{-1} (2)
Nonthermal motion (ξ)	35 km s^{-1} (2)	$<$ 10 km s^{-1} (2)
Temperature	10^5 K (1)	10^4 K (2)
Electron density	7×10^{10} cm^{-3} (1)	3×10^{10} cm^{-3} (2)
Mass density	1×10^{-13} g cm^{-3} (1)	1×10^{-13} g cm^{-3} (2)
Global birthrate	600 s^{-1} (1)	?

or scatter the Hα light incident from below (Wang et al. 1998; Chae et al. 1998b, hereafter, Paper II). Hα upflow events appear physically distinct from spicules in that they have a round and compact shape, are visible only in the blue wing, display only blueshifts, and are found in both network (Wang et al. 1998) and intranetwork areas (Paper II). That upflow events are related with magnetic reconnection was inferred from the finding of the association between upflow events and converging magnetic dipoles (Wang et al. 98), and was supported by the later finding of the associations among upflow events, flux cancellation and explosive events (Paper II). It was found that chromospheric upflow events are similar to explosive events in size, lifetime, and mass density (see Table 1), and in that both the kinds are recurrent at the same place. But they have different temperatures and velocities so that they may represent different kinds of plasma ejection.

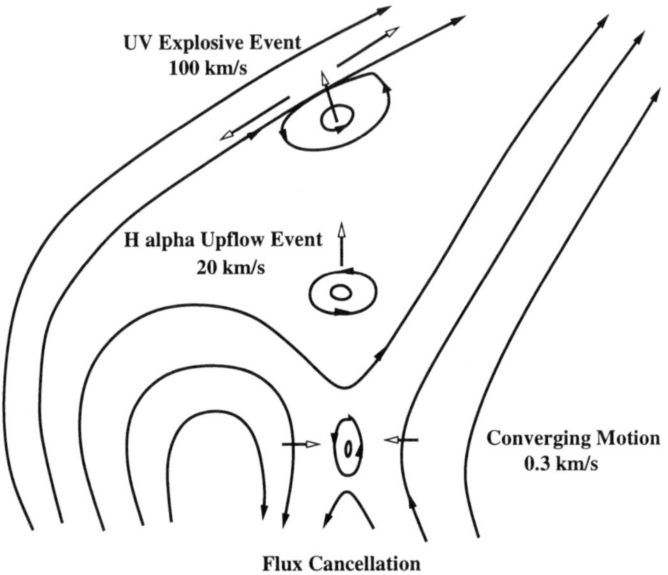

Figure 3. A schematic drawing of the two-step magnetic reconnection model.

What are the physical relationships among photospheric flux cancellation, Hα upflow events, and UV explosive events, which appear to be associated with one another? Answering this question requires a self-consistent picture of small-scale magnetic reconnection occurring in the quiet Sun, including the driving source of the magnetic reconnection. Paper II proposed that chromospheric upflow events represent the cool plasma material flowing into regions of magnetic diffusion whereas explosive events are a manifestation of the hot plasma material flowing out of the regions. This proposition can explain at least two important observational facts about explosive events: their lifetime and the fact that explosive events are not seen in lines cooler than 4×10^4 K. But the origin of the upflow events was not explained in Paper II. In the next section, I propose a two-step magnetic reconnection which attributes the upflow events to the dynamical development of magnetic islands formed by slow magnetic reconnection occurring in the low atmosphere.

3. A Two-Step Magnetic Reconnection Model

Our two-step magnetic reconnection model is schematically drawn in Fig.3 According to this picture, there are two kinds of magnetic reconnection: the first step magnetic reconnection occurring in the lower atmosphere and the second step magnetic reconnection occurring in the upper atmosphere. The first step magnetic reconnection is driven by the converging motion of two flux systems of opposite polarities like network-intranetwork and intranetwork-intranetwork collisions due to supergranular flow. The converging motion slowly develops a current sheet in the interface of two flux systems, which contains two Y-type neutral points and one O-type neutral point. The reconnected field lines above the upper Y point float upward and eventually form the overlying field lines in the upper atmosphere whereas the reconnected field lines below the lower Y point sink and eventually disappear from the photospheric level, which we identify with photospheric flux cancellation. On the other hand, the O-type neutral point forms a magnetic island that grows with flux pile-up due to the converging motion. After reaching a critical value of magnetic flux, the magnetic island may become dynamically unstable and begins to move either upward or downward. These processes in the low atmosphere proceed slowly and steadily, forming and ejecting magnetic islands continually.

The second step magnetic reconnection is driven by an upward moving magnetic island. As the magnetic island moves upward, it expands and is accelerated to have a size and a velocity large enough to be observable in the Hα line, which we identify with chromospheric upflow events. The magnetic island collides with the overlying field lines, forms current sheet in the interface, and eventually drives the second magnetic reconnection. During the magnetic reconnection, the cool material carried by the magnetic island is injected into the overlying field lines to be heated and accelerated, leading to explosive events. Since the magnetic island carries a finite amount of mass and magnetic flux, this process lasts for only a short period. A series of magnetic islands successively colliding with the overlying field lines cause this process to happen recurrently in the same region.

In this two-step reconnection model, the origin of chromospheric upflow events is attributed to the development of magnetic islands produced by the first magnetic reconnection in the lower atmosphere. For this idea to be physically plausible, we should be able to answer the following questions: 1) Can a magnetic island be formed during the first magnetic reconnection in the low atmosphere? 2) Is the magnetic island large enough to carry sufficient mass for an upflow event? 3) What accelerates the magnetic island to produce the upflow event? We address these questions below.

The structure of a magnetically non-neutral reconnection current sheet was studied in detail by Somov (1992) based on a two-dimensional steady flow assumption. One of the important results is that a slow magnetic reconnection has an O-type magnetic configuration at the central zone of the current sheet whereas a fast magnetic reconnection has a X-type magnetic field configuration which is consistent with the Petschek model. The speed of magnetic reconnection is characterized by the dimensionless parameter $\alpha = \frac{w}{h}\frac{v_{\rm A}}{v_{\rm in}}$, where w, h, v_{rmA}, and $v_{\rm in}$ are the thickness and the length (height) of the current sheet, the local Alfvén speed, and the inflow speed. Somov(1992) showed that if α is greater than a critical value α_c — i.e., if the magnetic reconnection is slow enough —, magnetic islands are formed inside the current sheet. He estimated α_c to be about 1.28. Recently, Oreshina & Somov (1998) found that α_c can be as low as 0.1 if radiative losses are taken into account. The thickness of the current sheet, w, is determined by the electrical conductivity, σ, and the inflow speed, and is roughly given by $w = \frac{c^2}{4\pi\sigma v_{\rm in}}$. Thus we see that the conditions of magnetic island formation are a small electrical conductivity and a small inflow speed if the length of the current sheet and the Alfvén speed are assumed to be constant.

The first magnetic reconnection occurring in the low atmosphere in our schematic model is likely to be a slow magnetic reconnection and thus form magnetic islands. First of all, it is driven by a small converging motion due to the supergranular flow, which is typically 0.3 km s^{-1} (e.g., Wang et al. 1996; Zhang et al. 1998). Secondly, there is observational evidence that there exist very cool regions in the low atmosphere, especially around the temperature minimum region (e.g., Ayres & Rabin, 1996), so that those regions should have a very low value of electrical conductivity, as asserted by Feldman (1993). The electrical conductivity at the temperature minimum can be as low as 1×10^{10} s^{-1} according to Wang's (1993) calculation, which is about 10^{-6} times the coronal value. The small inflow speed and electrical conductivity produces a fairly large estimate of the current sheet thickness,(2.4 km). The mass density at the temperature minimum is about 5×10^{-9} g cm^{-3} (e.g., Vernazza, Avrett, & Loeser 1981). The magnetic field strength at the same level of flux cancelling region may be smaller than 500 G which is known as a typical intrinsic field strength of the intranetwork magnetic element (Keller et al. 1994). Choosing 200 G as an estimate of the field strength at the temperature minimum, leads to an estimate of v_A,of 8 km s^{-1} at the same height. Thus the first magnetic reconnection must be slow and have the O-type magnetic field configuration if h is smaller than 50 km (no radiative loss) or 640 km (radiative loss). It appears that these values are physically plausible since they are roughly of the same order as an estimate of the intranetwork magnetic element diameter, 100-200 km (e.g., Wang et al. 1995). On the other hand, the second magnetic reconnection occurring in the upper atmosphere is not likely to be a slow magnetic reconnection. The electrical conductivity at 10^5 K is as high as 10^{14} s^{-1}, and the inflow speed is 20 km s^{-1} which is equal to the typical velocity of Hα upflow events. The thickness of the current sheet is then estimated to be 3.6×10^{-6} km. The Alfvén speed is about 100 km s^{-1}. Thus we see that the second magnetic reconnection could be slow if h is smaller than 1.3×10^{-5} km (no radiative loss) or 1.8×10^{-4} km (radiative loss). It appears that these values are physically unrealistic since they are extremely small compared with the size of the explosive events of 1500 km. This means that the second magnetic reconnection should be a fast magnetic reconnection that displays an X-point magnetic configuration.

Now let's examine whether the mass in a magnetic island formed during the first magnetic reconnection can account for the mass carried by an upflow event. The mass of an upflow event is estimated to be about $L^3\rho = 3 \times 10^{11}$ g if we use $L = 1500$ km and $\rho = 1 \times 10^{-13}$ g cm^{-3}. On the other hand, the mass of the magnetic island is of the order of $M = wh l\rho$ where l is the length of the current sheet in the direction of invariance. The length l can inferred from the flux loss rate, $l = \frac{d\Phi/dt}{v_{\rm in}B}$ and is estimated to be 100 km if we use $d\Phi/dt = 6 \times 10^{13}$ Mx s^{-1} (Paper I). Thus M is estimated to be in the range between 6×10^{10}–8×10^{11} g depending on the choice of h. This range is comparable to the mass of an upflow event of typically 3×10^{11} g.

It is quite likely that the so-called *diamagnetic melon seed* mechanism (e.g., Pneuman 1983) will eject and accelerate the magnetic island. The magnetic island is carrying a current whose direction is opposite to the directions of the currents that create the ambient quadrupolar magnetic fields so that the island must be expelled from the current sheet either upward or downward. Once the magnetic island is accelerated, its speed is soon as large as the Alfvén speed unless the effect of gravity is large. Thus a magnetic island ejected upward may have a velocity greater than the Alfvén speed at the temperature minimum,(8 km) s^{-1}, since the Alfvén speed usually increases with height. This will explain the typical observed velocity of upflow events, 20 km s^{-1}. That a magnetic island can be formed in the current sheet of a quadrupolar magnetic field geometry and is ejected at a local Alfvén speed was also demonstrated by numerical simulations (e.g., Karpen, Antiochos, & DeVore 1995).

4. Conclusion

I have proposed a two-step magnetic reconnection model to explain in a self-consistent way the three kinds of observed phenomena – photospheric flux cancellation, Hα upflow events, and UV explosive events which are closely associated with one another. This model chooses the temperature minimum region as a preferential height for the slow magnetic reconnection that is directly responsible for flux cancellation. In particular, it is demonstrated that this model can explain the physical origin of the mass and velocity of Hα upflow events. High spatial and high temporal resolution observations of flux cancelling features, which can cover simultaneously photosphere, temperature minimum region, chromosphere and transition region may be able to provide further evidence either for or against the proposed model.

Acknowledgments. I would like to thank Dr. Gwangson Choe for helpful discussions, and Martin Wooddard for carefully reading the manuscript. This work is partially supported by NSF under grant ATM-97-14796 and NASA under grant NAG5-4919 to BBSO.

References

Ayres, T. R., & Rabin, D., 1996, ApJ 460, 1042.

Brueckner, G. E., & Bartoe, J. F., 1983, ApJ 272, 329.

Chae, J., Lee, C.-Y., Wang, H., Goode, P. R., & Schühle, U., 1998a, ApJ 497, L109, Paper I.

Chae, J., Lee, C.-Y., Wang, H., Goode, P. R., & Schühle, U., 1998b, ApJ 504, L123, Paper II.

Dere, K. P., 1994, Adv. Space. Research, 14, 13.

Dere, K. P., Bartoe, J.-D. F., & Brueckner, G. E., 1989, Solar Phys. 123, 41.

Dere, K.P., et al. 1991, J. Geophys. Res. 96, 9399.

Feldman, U., 1993, ApJ 411, 896.

Innes, D. E., Inhester, B., Axford, W. I., & Wilhelm, K., 1997, Nature 386, 811.

Karpen, J. T., Antiochos, S. K., & DeVore, C. R., 1995, ApJ 450, 422.

Keller, C. U., Deubner, F.-L., Egger, U., Fleck, B., & Porel, H. P., 1994, A&A 286, 626.

Oreshina, A. V. & Somov, B. V., 1998 A&Ap 331, 1078.

Pneuman, G. W., 1983, ApJ 265, 468.

Porter, J. G. and Dere, K. P., 1991, ApJ 370, 775.

Somov, B. V., 1992, Physical Processes in Solar Flares, Kluwer Academic Publisher, Dordrecht, The Netherlands.

Wang, H., Tang, F., Zirin, H., Wang, J., 1996, Solar Phys. 165, 223.

Wang, H. et al. 1998, Solar Phys. 178, 493.

Wang, J. 1993, The magnetic and velocity fields of solar active regions: IAU Coll. 141, ASP. Conf. Ser. 46, (eds.) Zirin, H. et al., p. 465.

Wang, J. & Shi, Z. 1993, Solar Phys. 143, 119.

Wang, J., Wang, H., Tang, F., Lee, J.W., & Zirin, H. 1995, Solar Phys. 160, 277.

Vernazza, J.E., Avrett, E. H., & Loeser, R. 1981, ApJS 45, 635.

Zhang, J., Wang, J., Wang, H., & Zirin, H. 1998, A&Ap 335, 34.

Group Discussion

Jain : What is the time difference between the upflow event and the UV explosive event? Has this time-scale been used in your proposed model of slow and fast reconnection? Furthermore, every upflow due to cancellation of magnetic field is associated with an explosive event.

Chae : Both kinds of H-alpha upflow events and UV explosive events are short-lived events with lifetimes of about 1 minute. My model predicts a time difference of a few tenths of seconds between these two kinds of events. But our observations were not good enough to detect this time difference since their time resolution is 30 s. I have not studied the dynamics of slow and fast reconnection, so I can not tell how the relevant time scales can be used in my model. One thing I can say is that the observed velocity and the size of upflow events is consistent with the lifetime of upflow events and explosive events. As for the association between upflow events and explosive events, we performed a detail study of a couple of regions which display a cluster of explosive events on a cluster of H-alpha upflow events. We need more statistical studies to determine the degree of correlation between these two kinds of events.

Petrovay : What fraction of cancellation events is associated with UV explosive events?

Chae: We found the majority of explosive events are associated with flux cancellation(APJ. 497, L109), but we do not know whether all cancellation events are associated with explosive events.

High Resolution Solar Physics: Theory, Observations, and Techniques
ASP Conference Series, Vol. 183, 1999
T.R. Rimmele, K. S. Balasubramaniam, R.R. Radick, eds.

Internetwork Grains with TRACE

Robert J. Rutten

Sterrekundig Instituut, Utrecht, The Netherlands

Bart de Pontieu

Lockheed-Martin Solar and Astrophysics Lab, Palo Alto, USA

Bruce W. Lites

High Altitude Observatory, NCAR, Boulder, USA

Abstract. We demonstrate that TRACE data are well suited to studying internetwork grain patterns. We first show that wide-band Ca II K filtergrams register spatio-temporal internetwork morphology including the occurrence of K_{2V} grains, although the latter are a narrow-band phenomenon. We then show that the TRACE near-UV passbands image the same morphology. TRACE does so free of seeing deformations, over long times and large areas, and adds photospheric context as well.

1. Introduction

Ca II H_{2V} and K_{2V} grains thank their name to their appearance as spectrally narrow peaks just to the blue of line center in Ca II H & K spectrograms, without corresponding emission on the red side. This property distinguishes them from the network emission, which is also grainy in spatial morphology but spreads symmetrically over the full H & K cores and inner wings. Other differences are that K_{2V} grains appear exclusively in the internetwork regions, appear intermittently at intervals of a few minutes, and appear only a few times at a given location. Yet another difference is that the spectral formation mechanism of the K_{2V} grains is largely understood as a signature of weak acoustic shocks (Carlsson & Stein 1997; cf. reviews by Rutten & Uitenbroek 1991, Rutten 1994, 1995, Rutten et al. 1999), whereas the network emission mechanism, although undoubtedly betraying magnetic heating of some kind, is not identified.

2. Passband comparison

Studying the spatio-temporal patterns in which H_{2V} and K_{2V} grains appear is hard to achieve with a spectral passband as narrow as the peaks themselves, because that requires the spectral resolution of a spectrograph rather than a filtergraph and therefore brings a lack of spatio-temporal coverage. Fortunately, however, a much wider passband suffices. This is demonstrated by Fig. 1 which is constructed from a Ca II H spectrogram sequence taken with the Richard B.

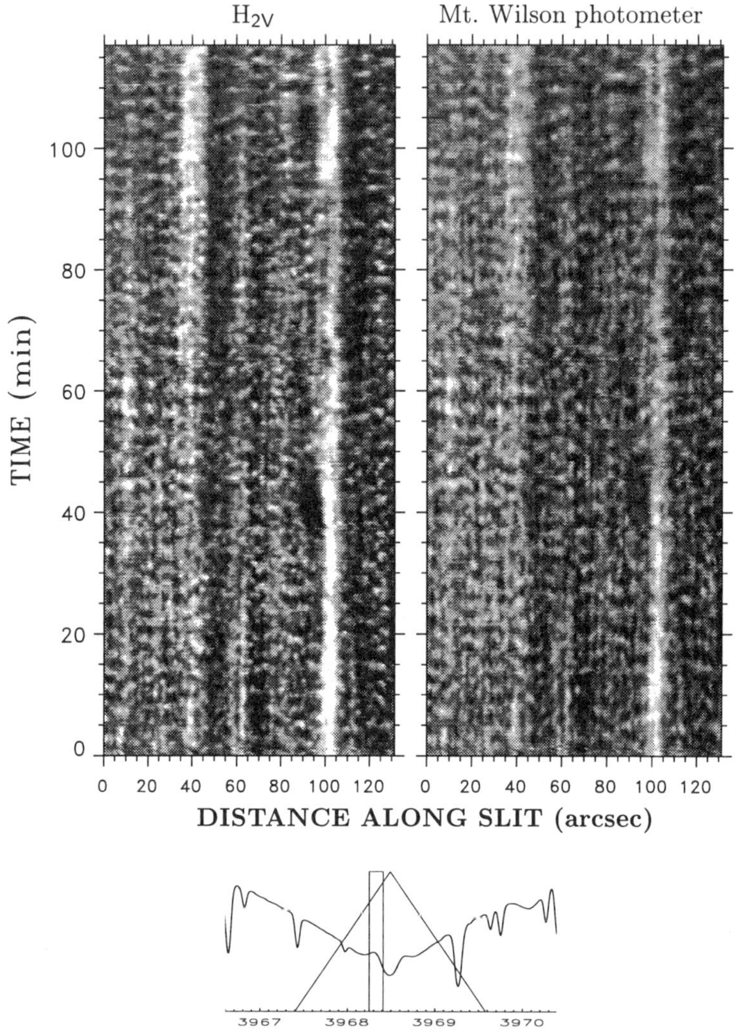

Figure 1. Comparison of internetwork grain signatures between different passbands from Ca II H spectrograms obtained with the horizontal spectrometer of the NSO/Sacramento Peak Dunn Telescope. The upper panels are "time slices" that display the spatio-temporal evolution of the intensity integrated over the passband defined in the lower graph. Left: integration over a narrow rectangular H_{2V} passband that selects only the H_{2V} peak blueward of the Ca II H line center. Right: integration over a triangular passband corresponding to the transmission profile of the Mt. Wilson stellar H & K photometer. Bottom: corresponding segment of the NSO solar flux atlas of Kurucz et al. (1984) downloaded from ftp://ftp.noao.edu/fts/fluxatl/ with the two passband profiles superimposed.

Dunn telescope at NSO/Sacramento Peak in 1991. The upper panels are $x - t$ diagrams that were obtained by spectral integration over the passbands shown in the lower graph. The lefthand panel is for a narrow band that isolates H_{2V} grains conform their spectral appearance. The righthand panel holds for the much wider triangular passband defined by the slits of the stellar H & K photometer at Mt. Wilson (Vaughan et al. 1978). Comparison of the two panels shows that the spatio-temporal patterns are virtually the same, even though the one at left is considerably sharper. The bright vertical stripes mark chromospheric network; in between is internetwork. Especially the rightmost internetwork area, with $x = 105 - 130$ arcsec, displays a stream of well-defined characteristically flashing H_{2V} grains. They appear less focused and less bright in the wide-band panel, but they are present there as well with very much the same spatio-temporal occurrence pattern. Only when these two panels are computer-blinked against each other does one notice a small phase shift, in the form of grains jumping slightly up and down with the narrow-band ones appearing slightly later than the wide-band ones.

The reason for the large similarity is that the H_{2V} grains appear as part of a more or less standard spectral evolution sequence in which bright "wing whiskers" (Beckers & Artzner 1974) develop and contract towards line center (Figs. 4–5 of Cram & Damé 1983, Fig. 2 of Rutten & Uitenbroek 1991, Fig. 3 of Hofmann et al. 1996, Figs. 16–17 of Carlsson & Stein 1997). They correspond to the upward propagating acoustic disturbances that become shocks slightly later and slightly higher up, and then produce peaks at the H_{2V} wavelength. The wider passband samples a larger contribution from these whiskers, but its high correspondence with the narrow-band panel shows that their spatio-temporal occurrence pattern is essentially identical to that of the grains proper.

The good agreement implies that, thanks to the presence of the whiskers, Ca II K filtergrams with passbands of a few Ångstroms sample K_{2V} grain patterning quite well — apart from the small phase difference and the loss of contrast. The contribution by Smaldone *et al.* in these proceedings demonstrates that the apparent grain density becomes smaller for wider passband at a specified contrast, but simple rescaling of wider-band filtergrams to the display dynamic range serves to map K_{2V} grain occurrence adequately. Examples of internetwork grains on high-resolution 3 Å Ca II K filtergrams are shown by Lites et al. (1999).

3. Filtergram comparison

Figure 2 shows a small segment of a quiet-Sun field observed by TRACE in its three near-UV passbands. The fourth panel (bottom) is a co-spatial and co-temporal Ca II K filtergram taken with the 3 Å filter at the Swedish Vacuum Solar Telescope (SVST) on La Palma. The TRACE images were rebinned to the SVST pixel resolution. They have lower angular resolution, but there is nevertheless a high degree of spatial agreement between the Ca II K SVST image and the TRACE images, in particular the 1700 Å one. Thus, at their resolution

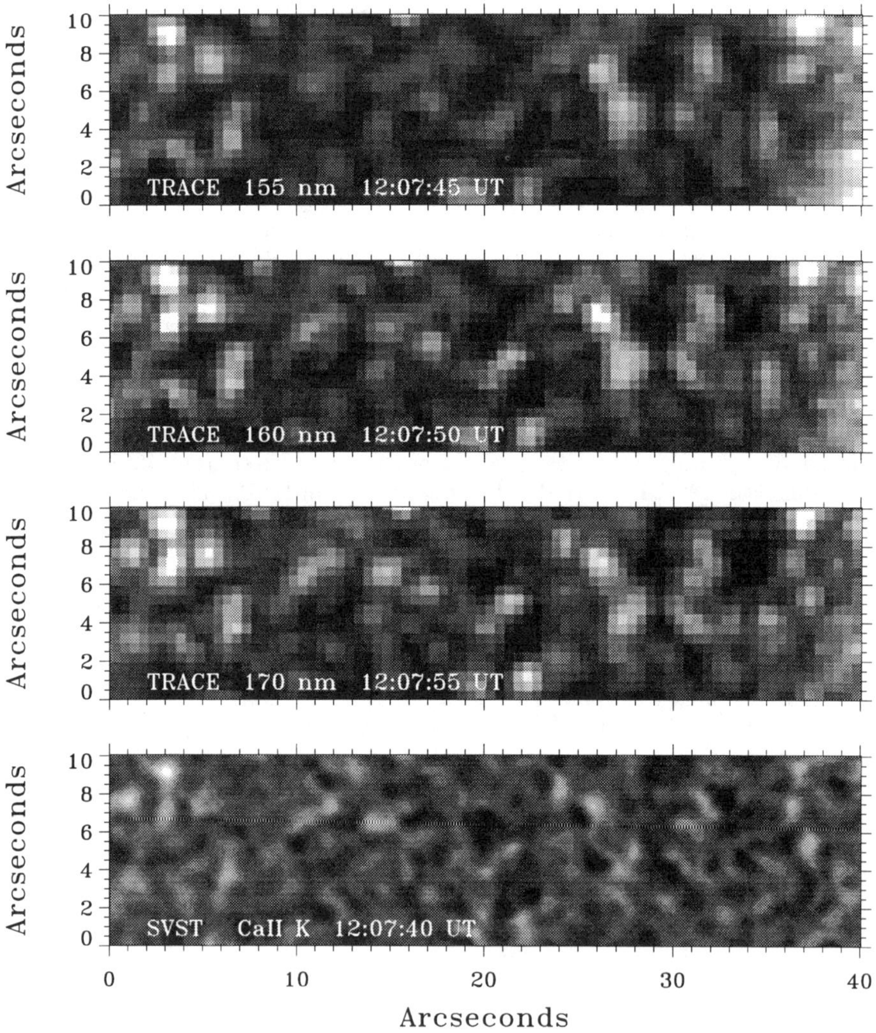

Figure 2. Enlargements of a small subfield covering a quiet solar region of filtergrams taken on May 30, 1998 at the specified times. Bottom panel: high-resolution Ca II K image taken with the SVST. Other panels: nearly simultaneous TRACE images taken in the three near-UV passbands as specified. The SVST and TRACE images were co-aligned using a larger field containing an active region.

these TRACE channels portray internetwork grain phenomena about as well as Ca II K filtergrams do[1].

4. TRACE as grain tracer

TRACE observing collects images without interruptions from night or clouds or bad seeing. There are no rubber-sheet pattern deformations as in ground-based observing. In addition, TRACE collects much larger spatial samples than can be achieved with spectrometry. Therefore, TRACE image sequences provide an excellent means to study spatio-temporal grain occurrence on time scales from tens of seconds to many hours. In addition, combined analysis of co-spatial and co-temporal white-light images from TRACE should serve to relate grain patterns to photospheric convective patterns down to granular scales, again without degradation from seeing, and permit improved studies along the lines of Hoekzema et al. (1998) and Hoekzema & Rutten (1998) for large data sets delivering high statistical significance.

In terms of Fourier characteristics, we expect to find intensity modulation power spectra similar to those derived from the Ca II infrared lines by Deubner & Fleck (1990) and from the 1 Å wide Ca II H-index by Lites et al. (1993a), *i.e.*, sampling the same layers that produce the whiskers in the inner wings of Ca II H & K. Power spectra from this upper-photospheric domain are characterized by high power at low frequency in both network and internetwork, a power dip around $f = 2$ mHz, and a power peak around $f = 4$ mHz that is higher and extends to higher frequencies for internetwork than for network. The low-frequency component may partially be due to gravity waves (cf. Lites *et al.* 1993a, 1993b).

The morphology in the 1600 Å and 1550 Å panels of Fig. 2 differs increasingly from the morphology in the 1700 Å and Ca II K panels. This may be due to the presence of signal with different nature from appreciably higher levels. The 1550 Å passband contains lines of C I and of C IV; in active regions, the latter contribution dominates whereas in the internetwork the C I lines may constitute most of the signal. However, the differences may also betray temporal retardation with respect to the lower layers sampled by 1700 Å and Ca II K. We have embarked on Fourier phase difference analysis to see to what extent this is the case.

Beckers J. M.,& Artzner G., 1974, Solar Phys. 37, 309
Carlsson M.,& Stein R. F., 1997, ApJ 481, 500
Cram L. E., Damé L., 1983, ApJ272, 355
Deubner F.-L.,& Fleck B., 1990, A&A 228, 506
Hoekzema N. M.,& Rutten R. J., 1998, A&A 329, 725
Hoekzema N. M., Rutten R. J., Brandt P. N.,& Shine R. A., 1998, A&A 329, 276
Hofmann J., Steffens S.,& Deubner F. L., 1996, A&A 308, 192

[1]We therefore feel that K_{2V} grains should be called "internetwork grains" more generally.

Kurucz R. L., Furenlid I., Brault J. W.,& Testerman L., 1984, Solar Flux Atlas from 296 to 1300 nm, NSO Atlas Nr. 1, National Solar Observatory, Sunspot, New Mexico

Lites B. W., Rutten R. J.,& Berger T. E., 1999, ApJ submitted

Lites B. W., Rutten R. J.,& Kalkofen W., 1993a, ApJ414, 345

Lites B. W., Rutten R. J.,& Kalkofen W., 1993b, in H. Zirin, G. Ai, H. Wang (eds.), The Magnetic and Velocity Fields of Solar Active Regions, Proc. IAU Coll. 141 (Beijing), Astron. Soc. Pac. Conf. Series, Vol. 46, p. 530

Rutten R. J., 1994, in M. Carlsson (ed.), Chromospheric Dynamics, Proc. Mini-workshop, Inst. Theor. Astrophys., Oslo, p. 25

Rutten R. J., 1995, in J. T. Hoeksema, V. Domingo, B. Fleck,& B. Battrick (eds.), Helioseismology, Proc. Fourth SOHO Workshop, ESA SP–376 Vol. 1, ESA Publ. Div., ESTEC, Noordwijk, p. 151

Rutten R. J., Lites B. W., Berger T. E.,& Shine R. A., 1999, in C. J. Butler, J. G. Doyle (eds.), Solar and Stellar Activity: Similarities and Differences, Procs. Armagh Workshop, Astron. Soc. Pac. Conf. Series, in press

Rutten R. J.,& Uitenbroek H., 1991, Solar Phys. 134, 15

Vaughan A. H., Preston G. W.,& Wilson O. C., 1978, PASP 90, 267

High Resolution Solar Physics: Theory, Observations, and Techniques
ASP Conference Series, Vol. 183, 1999
T.R. Rimmele, K. S. Balasubramaniam, R.R. Radick, eds.

Dynamical Nature of The Quiet Solar Outer Atmosphere

D. Banerjee, J.G. Doyle

Armagh Observatory, College Hill, Armagh BT61 9DG, N. Ireland

E. O'Shea

Dept. of Pure & Applied Phys., Queens University Belfast, Belfast BT7 1NN, N. Ireland

G.H.J. van den Oord

Sterrekundig Instituut, P.O. Box 80.000, 3508 TA Utrecht, The Netherlands

Abstract. We examine spectral properties of the network and internetwork chromosphere and transition region from the SUMER instrument on the *SOHO* spacecraft. Observations were obtained for a number of lines ranging in temperature from Log T=4.1 K to 5.5 K. The lower chromospheric line N I 1319Å show intensity oscillations at frequencies between 3 and 6 mHz. The C II 1335Å lines generally show oscillations in the 3 to 5 mHz range. For network regions, the intensity oscillations observed in N I are often accompanied with a blue-shifted line profile of 2–3 km s^{-1}. These oscillations in the network regions can be interpreted in terms of magnetoacoustic waves propagating upwards along thin magnetic flux tubes. However, for internetwork regions the simulations of Carlsson & Stein (1995, 1997) are not consistent with the observed lines. Perhaps this can also be explained in terms of some MHD effects, rather than in terms of upward propagating p-modes.

1. Introduction

In the last few years our understanding of the transition region and the chromosphere has changed dramatically with the realization that these are very dynamic and not static. Up till now only a limited number of analyses, based on *SOHO* data, have been published (see Doyle et al. 1999 for details). We present here an analysis of EUV lines formed in the vicinity of the transition region observed with SUMER onboard *SOHO* at disk center of the quiet Sun. An overview of typical lightcurves and power spectra of a large number of lines observed in different sequences with SUMER are presented.

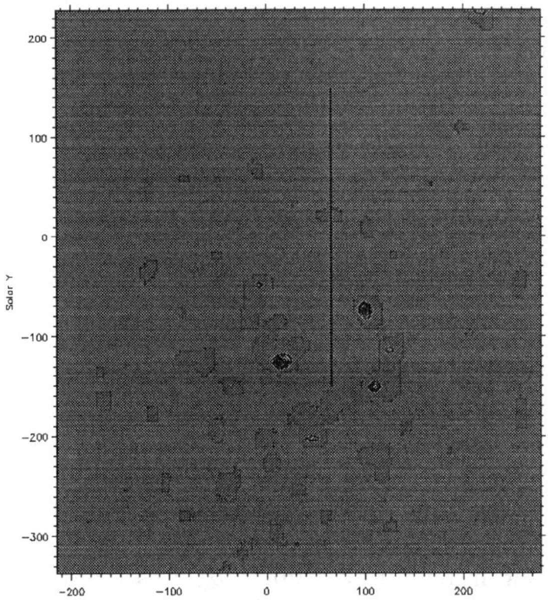

Figure 1. An enlarged view of the Mt. Wilson Observatory full disk magnetogram, taken in Na I 5896Å on 31 July 1996 at 20:55 UT. The SUMER slit location on 31 July 1996 at 22:47 UT is indicated. Solar rotation compensation has been used for proper superposition. The coordinate system is arc sec from disk center. The dark patches indicate the concentration of magnetic fields.

2. Observations

The Solar Ultraviolet Measurements of Emitted Radiation (SUMER) instrument is a stigmatic normal-incidence spectrograph operating in the range 400 – 1610 Å (Wilhelm et al., 1995) The observations reported here were obtained with the 1 × 300 arc sec slit (slit 2). The instrument was operated in a sit-and-stare mode with the SUMER standard rotation compensation switched off. For disk center pointing, this implies a new 1 arc sec region every 380 seconds (2.65 mHz). A log of the observations is given in Table 1. The overall field-of-view for one of the datasets is given in Fig. 1 where we show the SUMER slit position on a full disk magnetogram, taken in Na I 5896Å on 31 July 1996 at 20:55 UT. Due to the time difference between the N I and Na II images, solar rotation compensation has been used for proper superposition. We have also compared the slit position with several EIT images. Though the region observed by SUMER was a typical quiet Sun area, we do notice occasional concentrations of magnetic fields in the synoptic map (see Fig. 1) corresponding to the bright regions. Presumably the magnetic flux tubes are located in the chromospheric network and are responsible for these brightenings. The small patch in the central part of the slit in Fig. 1

(presumably the active network) corresponds to the brightening in region 190-202 (31 July) of Fig. 2. The morphology of network and internetwork has also been checked with Ca II images obtained from Big Bear Solar Observatory. Details of the Fourier analysis can be found in Doyle et al. (1999). Power

Table 1. Log of the datasets considered here. The observed lines in different sets are N I 1318.99 & C II 1334.53/1335.71 and O VI 1037.6 & C II 1037.2 Å. N is the number of data points.

Date	Pointing X,Y	Start UT	End UT	N	Lines observed
30 July '96	+54, 0	18:38	19:09	62	N I, C II
31 July '96	+54, 0	20:39	21:17	150	C II, O VI
31 July '96	+84, 0	22:47	23:49	124	N I, C II

spectra are obtained from the Fourier transforms of the auto-covariance function, multiplied by a window function to reduce the variance of the noise. Power spectra are normalized in such a way that the expected mean noise level equals 2. For power spectra we normally use a 99.9% confidence level. Note that when we refer to a 'normalized' lightcurve this implies a lightcurve with its mean subtracted and then divided by the square root of its variance. This results in normalized lightcurves which facilitate comparison of curves from lines with different emissivities.

3. Results

The spatial and temporal properties of the data are revealed in Fig. 2, which shows the total line intensity for the N I and the C II lines as a function of position along the slit and time. Fig. 2a,c indicate that there is more 'structure' in the emission of N I than in C II. The brightest features, which appear as almost horizontal streaks in Fig. 2 are related to network magnetic fields and observable in both N I and C II. Often these features are observed for the whole length of the observation but sometimes the in- or egress of a feature in the slit is observed. The typical dimensions along the slit are 5–8 arc sec. If circular in shape these features would be visible for 32–50 minutes,i.e. for the larger part of our observing runs. Fluctuations in the bright features are clearly visible and their appearance seems periodic. Due to the dominance of the network emission in Fig. 2a,c we used a standard contrast enhancement technique to bring out the weak features in Fig. 2 and to filter out the effect of strong long-lasting (network) features. The data were smoothed in the time direction with a running Gaussian of $\sigma = 15$ minutes. Then the original data were divided by the smoothed dataset. The results are shown in Fig. 2b,d. From the figure it is clear that the slowly varying network emission has been largely removed. The contrast enhanced intensity maps show a fairly uniform pattern of grain-like brightenings down to scales of a few arcseconds. This is especially found in the N I maps. In the C II maps the pattern is less regular and the individual grains are slightly larger. These grain brightenings have sizes of 2–4 arc sec along the

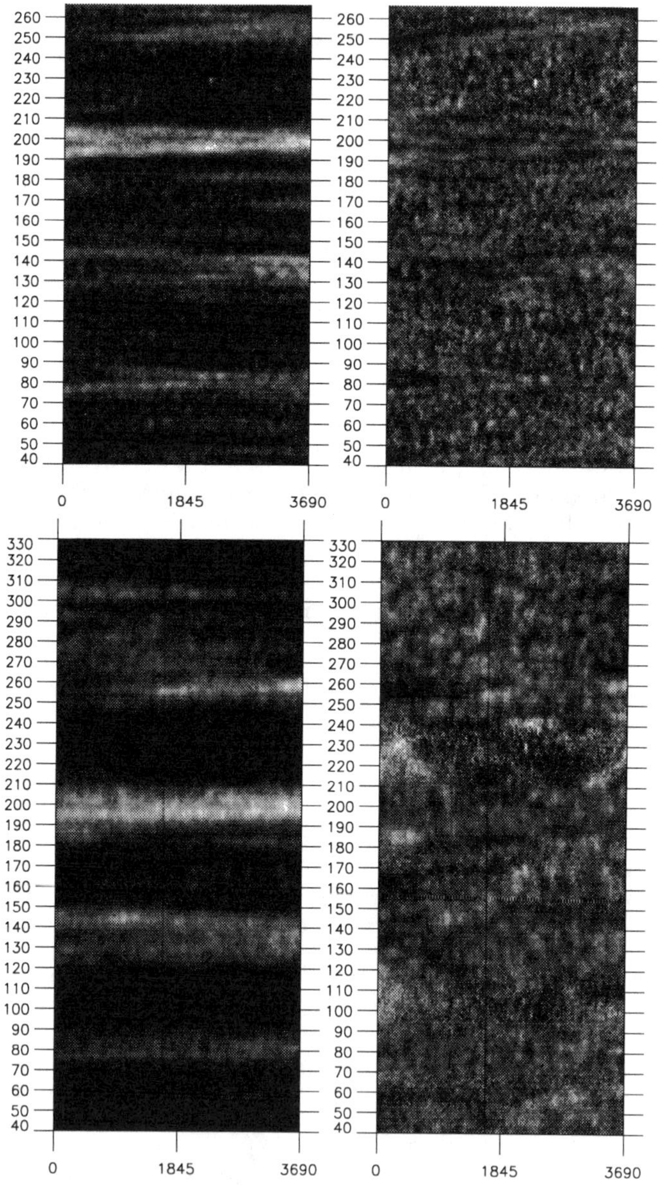

Figure 2. The line intensity as a function of position along the slit (y-axis) and time (x-axis) for the N I ([a,b] top) and the C II([c,d] bottom) line. The left panel shows the observed line intensity and the right panel shows the contrast enhanced intensity (31 July data).

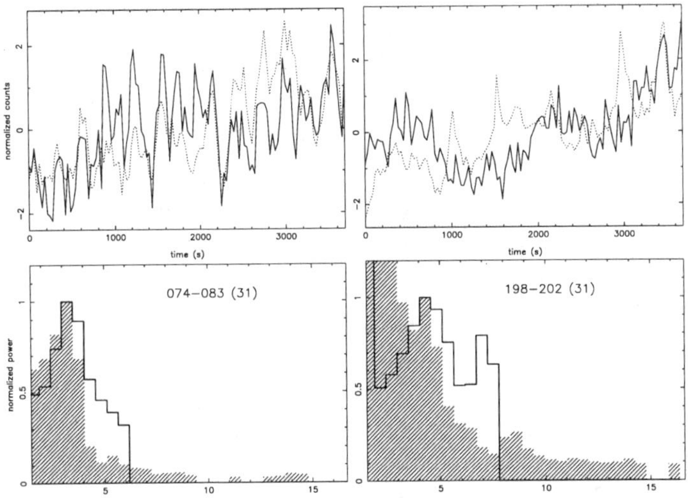

Figure 3. The top panel shows the normalized N I ($\sim$ 15,000K) (solid line) and C II ($\sim$ 32,000K) (dashed line) counts for typical network regions. The lower panel shows the normalized power spectra of N I (solid line) and C II (dashed area).

slit and last from less than one minute to a few minutes. The grains appear to be repetitive in time and, at a fixed position along the slit, show the same pattern of in- or egress as the bright network features discussed above. Over distances of a few arcseconds along the slit, the moments of maximum intensity in different 'streams' of these grains appear to be shifted by one to a few minutes. In Figs. 3 & 4, we show for two selected network and internetwork regions respectively, the normalized N I and C II lightcurves and their power spectra. Fig 5 shows the light curves and power spectra for simultaneous observation of O VI 1037.6 Å and C II 1037.2 Å , corresponding to another network region of dataset taken on 31st July 1996. The power spectra indicate that there is considerable power all along the slit so that the power is not confined to the brightest areas. Above 2 mHz, power is found between 3 and 6 mHz. For network regions (Figs. 3,5) the dominant power peaks coincides for both the lines (around 3.5 mHz). Our results indicate that for both network and internetwork regions the bulk of the power in the intensity power spectra is below 5 mHz and most likely in the range 3–$\lesssim$5 mHz.

4. Discussion

For internetwork regions we find a strong peak around 3 mHz and a weaker peak around 5.5 mHz for N I , but for C II the 5.5 mHz peak seems to be

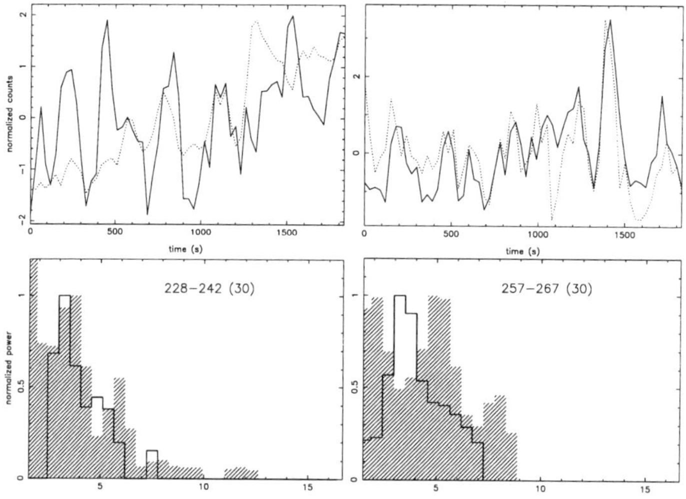

Figure 4. The top panel shows the normalized N I ($\sim 15,000$K) (solid line) and C II ($\sim 32,000$K) (dashed line) counts for typical internetwork regions. The lower panel shows the normalized power spectra of N I (solid line) and C II (dashed area).

absent. Steffens & Deubner (1999) have analysed velocity oscillations of 15 spectral lines observed with SUMER. For internetwork regions, they found that the chromospheric 3 min (5.5 mHz) oscillations increases with respect to the 5 min (3.3 mHz) oscillations with increasing formation height until a maximum is reached in lines formed at about 50,000 K. The absence of 5 min or 3 min peak in their power spectra, for lines formed at coronal heights lends supports to the assumption that the transition layer acts as a reflecting barrier for the internetwork waves. Ground based polarimetric data obtained recently by Lites, Rutten & Berger (1999) provide evidence that magnetic fields are very weak and dynamically insignificant in typical internetwork regions. Though the observation suggests that the gas dynamics changes radically in character, as the upward propagating shock waves interact with the overlaying magnetic fields, not much work has been done on the theoretical side for the interaction of shocks with magnetic canopy kind of structure. More recently, from the measurement of degree of polarization in the sodium doublet, Landi (1998) have ruled out the existence of turbulent magnetic fields and of horizontal, canopy-like fields stronger than ~ 0.01 G. The simulations by Carlsson & Stein (1995,1997) are reliable for heights below 1500 km for the quiet chromosphere and the non-magnetic internetwork. But the C II and O VI lines are formed higher in the atmosphere and can not be compared with existing simulations in the literature. Thus it is not at all surprising tha t we don't see any grain oscillation with 3 min period for higher chromospheric lines. Gouttebroze et al. (1998) also did not find any evidence of oscillations for lines formed at higher than $5x10^5$ K.

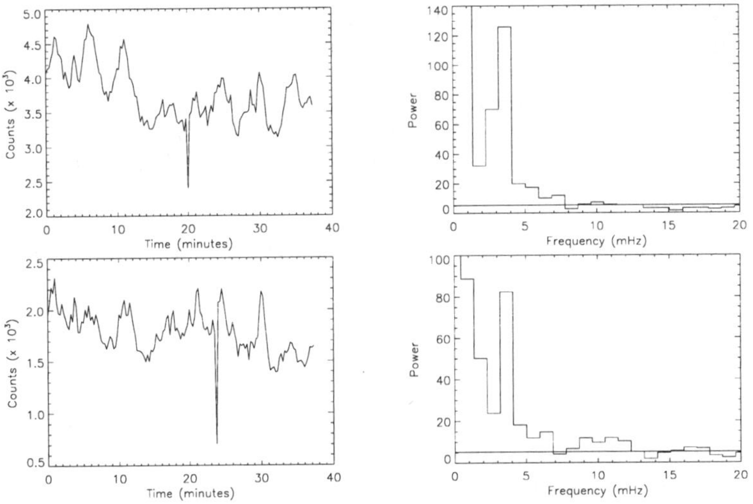

Figure 5. The light curve and power spectrum for O VI 1037.6Å ($\sim 300,000$ K)(top row) and C II 1037.2Å ($\sim 32,000$ K) (bottom row) in region 202-207 (network) of dataset taken on 31 July 1996. The horizontal line in the power spectra corresponds to the 99.9% confidence level.

They conclude that the waves in the frequency range 2.5 - 7 mHz are entirely reflected (or dissipated) by the chromosphere-corona transition region.

Simulations by Steiner et al. (1998), of magnetic flux sheets embedded in nonstationary, radiative convection have identified a number of dynamical phenomena. They found strong dynamic interaction between the flux sheet and the surrounding nonstationary convection. For example, occasional strong bending and rapid horizontal displacement of a magnetic flux sheet by the action of the external convective motion and the appearance of shock waves within the magnetic structures. These events possibly excite transverse MHD waves, which propagate along magnetic structures and transport mechanical energy in to higher layers of the atmosphere (Choudhuri, Dikpati, & Banerjee 1993).

Our network observations suggests that the lower chromospheric intensity oscillations come from small magnetic regions of at most few arc sec along the slit and lasts for 10–30 mins. These intensity oscillations could be related to the impulsive motions at the photospheric level. The clear presence of a blue-shift in the N I profile indicates an upward propagating wave. The observed 4–5 min network oscillations can be interpreted in terms of longitudinal magnetoacoustic sausage waves propagating upwards along thin magnetic flux tubes. Kink waves can be generated by random foot-point motions at the photospheric level. As they propagate within flux tubes, their amplitude grows exponentially with height and become non-linear, thereby undergoing a mode transformation becomes sausage waves, which are detected on the disk. The detailed analysis of the propagation of transverse magnetoacoustic waves in a magnetic flux tube will be presented in a forthcoming paper (Banerjee,Doyle, & O'Shea 1999). However we should point out that in a more extensive analysis, Doyle et al. (1999) have produced the power spectrum for individual pixels and then taken average over several pixels, in contrast to summing the pixels first and then producing the power spectrum (as we have done here). For the regions examined here, they show that the power at higher frequencies cancel, as expected if the oscillating regions are small in size. Although they find that individual regions (as chosen in this paper) show evidence of oscillations, the overall picture is somewhat confusing. On a statistical basis, there seems to be little difference between the bright (network) and dark (internetwork) regions. We hope to pursue this analysis in greater detail.

References

Banerjee, D., Doyle, J.G. & O'Shea, E. 1999, A&A(in preparation)

Carlsson, M. & Stein, R.F. 1995, ApJ, 440, L29

Carlsson, M. & Stein, R.F. 1997, ApJ, 481, 500

Choudhuri, A.R., Dikpati, M. & Banerjee, D. 1993, ApJ, 413, 811

Doyle, J.G., van den Oord, G.H.J., O'Shea, E. & Banerjee, D. 1999, A&A. (in press)

Gouttebroze, P., Vial, J.-C., Bocchialini, K., Lemaire, P. & Leibacher, J.W.: 1998, Solar Phys (in press)

Landi Degl'Innocenti, E. 1998, Nature, 392, 256

Lites, B.W., Rutten, R.J. & Berger, T. 1999 (in preparation)

Steffens, S. & Deubner, F.-L. 1999 (this volume)

Steiner, O., Grossmann-Doerth, U., Knölker, M. & Schüssler, M. 1998, 495, 468

Wilhelm, K., Curdt, W., Marsch, E., Schühle, U., Lemaire, P., Gabriel, A., Vial, J.-C., Grewing, M., Huber, M.C.E., Jordan, S.D., Poland, A.I., Thomas, R.J., Kúhne, M., Timothy, J.G., Hassler, D.M. & Siegmund, O.H.W. 1995, Solar Phys., 162, 189

Group Discussion

Zirker : Do you take into account the reflection at the temperature jump?

Banerjee : Yes, the transmission of power depends on the jump location. We also see a tunneling effect. We have considered the reflection at the transition layer. The energy transmission depends on the location of the transition layer and the temperature contrast. Due to the lower cut-off frequency of the upper layer as compared to the lower layer we observe a tunneling effect.

High Resolution Solar Physics: Theory, Observations, and Techniques
ASP Conference Series, Vol. 183, 1999
T.R. Rimmele, K. S. Balasubramaniam, R.R. Radick, eds.

Fine-Scale Structure of Coronal Loops and Consequences for Coronal Loop Interactions

E. Kh. Kaghashvili[1]

Harvard-Smithsonian Center for Astrophysics, 60 Garden Street, Cambridge, MA 02138, USA

R. N. Smartt

National Solar Observatory/Sacramento Peak, Sunspot, NM 88349, USA

Abstract. Visible coronal emission-line images reveal occasional coronal loop interaction (CLI) events, as evidenced by transient brightenings at the projected intersection of adjacent loops. The image cadence is one minute with an angular resolution of, at best, 1 arcsec. Based on details revealed in the observations, these events are generally interpreted as due to the localized transient coalescence of two loops. Soft X-ray images suggest that brightenings can occur also in single loops. However, at least in some of these cases image processing that improves the effective angular resolution reveals that two loops are involved. Nevertheless, it is found that a plausible scenario, based on a 'non-rotating plane', can allow for single-loop brightenings. This theoretical study is supported by some simple numerical simulations of particular mechanisms, and some rough estimates of the likely energies involved in such processes. Finally, the model of a single loop brightening due to energy deposition at the site of a non-rotating plane is related to the case of observed two-loop interactions.

1. Introduction

The emission corona is characterized by concentrated magnetic flux systems that are revealed as complex loop structures clearly evident in high-resolution images in the 530.3nm (Fe XIV) and 637.5nm (Fe X) coronal lines (Smartt, Dunn, & Fisher 1981), especially in the vicinity of the active latitudes. Such concentrations of separate magnetic flux systems (coronal loops), driven by photospheric motion, inevitably lead to occasional circumstances where a portion of one loop comes into close proximity with a portion of an adjacent loop to an extent that they interact, leading to observable brightening at the interaction site. These events can involve loops within an active region, or loops lying between active regions. Specifically, loop topologies leading to an interaction can, for example, occur where loops grow in height differentially, the relative orientations of adjacent loops change due to motion of their footpoints, or new loops appear. Alternatively, closely spaced loops forming an arcade might come into contact at their tops if the plane of one loop tilts relative to its neighbors or simply from the attractive forces arising from currents in adjacent parallel loops. The observations show clearly that brightenings commonly involve localized two-loop interaction. However, even with an angular resolution of no more than a few arcseconds, the loop geometry is often not clear for brightenings that occur low in the corona, where there are typically many loops in the line-of-sight, which allows for the possibility of single- or multiple-loop interactions in some cases. Here we make an attempt to find a connection between single- and multiple-loop interaction events by considering a simple model for a single-loop brightening and linking it with that possibly involved in a loop-loop interaction.

2. Morphology of Observed Coronal Loop Brightenings

Observation of coronal loop interactions were first discovered in visible emission-line data in the 1980's (Smartt & Zhang 1987). These kinds of observations (Smartt & Zhang 1992; Smartt, Zhang, & Smutko 1993) reveal that a brightening appears first in the 530.3nm ($T \simeq 2 \times 10^6$K) emission, reaching a maximum on average $\sim$ 15-20 minutes after the first evidence of the start of the event, the brightening then gradually fading on the same time-scale as the rise time. Following a maximum in this green-line emission, a similar brightening appears in 637.5nm ($T \simeq 10^6$K) emission, reaching a maximum on average nine minutes after that of the green line. The delay in brightening of this 637.5nm emission compared with that of the 530.3nm is interpreted as cooling of the plasma at the interaction site following the initial interaction. These events appear to be well-described by a process of localized loop coalescence involving partial magnetic reconnection (X-type) of two loops (Sakai & De Jager 1989; Sakai 1993). For typical events, it is estimated that $T_{max} \leq 5 \times 10^6 K$, $N_e \leq 6 \times 10^{10}\ cm^{-3}$ and a total energy, $E \approx 1 - 5 \times 10^{26}$ ergs (Airapetian & Smartt 1995). If the footpoints of interacting loops are twisted by photospheric motions, higher energy can be released by the magnetic pinch effect over a scale comparable with the loop radius (Sakai, 1993). Because of the above predicted high temperature produced in an interaction, it is expected that these events would also be evident at the initial stage in SXR and HXR observations, given sufficient energy involved in an interaction and adequate measurement sensitivity. While these events can occur high in the corona, it is interesting to observe that flare studies indicate that the interaction of current loops is an essential ingredient in the trigger of solar flare energy release (Machado et al. 1988; Nakajima 1993). The development of a major loop-loop brightening event observed in FeXIV(530.3nm) emission is shown in Figure 1 at a) the beginning, and b) the maximum stage, of the event. The x-point of interaction is at a height $\approx 4.7 \times 10^4$km.

Similiar enhancements have been observed in Yohkoh soft X-ray (SXR) data (Shimizu et al. 1994) and also reported in observations from CDS/SOHO (for example, Mathews & Harra-Murnion 1997). Some observations suggest that a single loop alone can brighten, although image processing to improve the effective angular resolution reveals that at least in some cases two loops are involved. Clear examples of two-loop interactions observed in Yohkoh events have been reported by several authors (Shimizu et al. 1992; De Jager et al. 1993), but single-loop brightenings have been commonly reported from Yohkoh

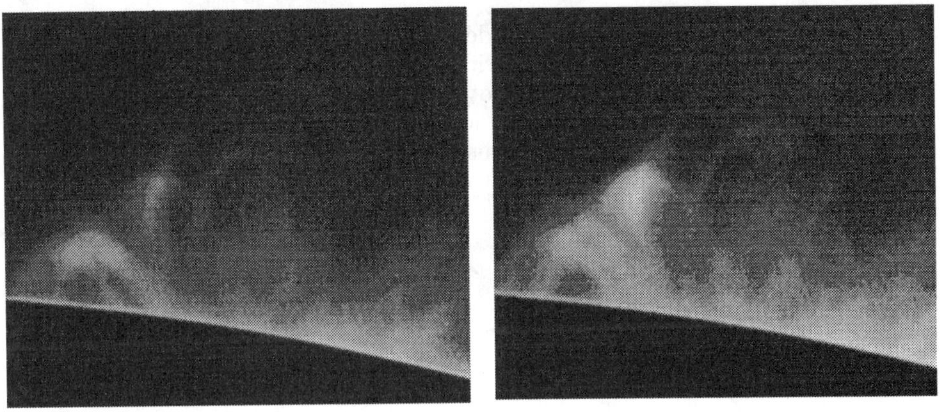

Figure 1. Coronal loop interaction event in 530.3nm emission (left) at the stage when it first becomes evident (17:47UT) and its maximum phase (right) fourteen minutes later (18:01UT)

data. We note that an apparent single, low-lying visible loop that brightened along most of its length has also been reported (Li, Zhang & Smartt 1994). Takakura et al. (1993) report transient brightenings taking place in different parts of single loops, with no clear evidence of a preferred place in a loop for these events to occur. These types of observational features are addressed below. For this study, useful statistics have been given by Shimuzu et al. (1992) who analyzed 144 transient brightening events using Yohkoh data. They grouped these data into three classes:

1. single-loop events (61 cases).
2. multiple-loop transient brightening events (69), and
3. point-like events (14).

These statistics thus indicate that single-loop events constitute a significant part of transient brightenings, in addition to multiple-loop interactions. It is not clear at this time how the mechanism proposed in this paper might explain point-like events. For this, higher-resolution observations, where details of the loop geometry are evident, are required.

3. Sheared Photospheric Motion: Existence of a Non-Rotating Plane

Random photospheric motion provides a source of stored magnetic potential energy within a loop. This can result from the rotation of one or both footpoints of a loop (Figure 2), producing axial rotation that propagates along the loop. It is noted from high-resolution magnetograms that at least some emergent flux sites display an axially rotational component (Pevtsov, Canfield, & Metcalf 1994). In this study, we consider the effect of some rotation of the loop footpoints that results in a non-rotating plane at some location along the loop, as indicated in Fig. 2. The location of the resultant 'non-rotating plane' in the magnetic flux tube depends on the relative rotation of the footpoints. If we assume almost the same averaged rotational momentum deposition to both loop feet (Figure 2), the non-rotating plane in this symmetric case is located at the top of the loop. Otherwise this plane is nearer to the footpoint with the smaller rotation component. Note that irregular motion of the footpoint would lead to a corresponding temporal variation of the position of this plane.

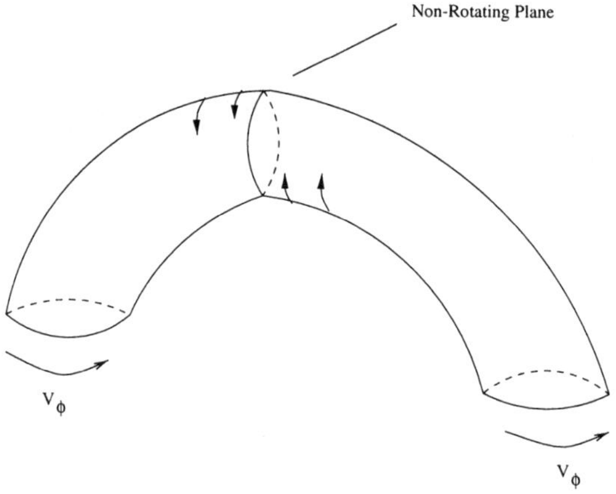

Figure 2. Single loop with rotating footpoints, indicating the non-rotating plane

4. Non-Linear Dynamics of the Non-Rotation Portion of a Loop

We now consider a simple model for a loop with some rotation imparted to the feet. The set of MHD equations are solved numerically to describe the initial non-linear behavior of the twisted magnetic flux tube; in particular, for the non-rotating portion of the loop, we have:

$$\frac{\partial \rho}{\partial t} + \nabla\cdot(\rho\vec{V}) = 0, \tag{1}$$

$$\rho\frac{\partial \vec{V}}{\partial t} + \rho\vec{V}\cdot\nabla\vec{V} = \frac{1}{4\pi}(\nabla\times\vec{B})\times\vec{B} - \nabla P + \underline{\rho\vec{g}} + \vec{F} + \underline{R_e^{-1}\Delta\vec{V}}, \tag{2}$$

$$\frac{\partial \vec{B}}{\partial t} = \nabla\times(\vec{V}\times\vec{B}) + \underline{R_m^{-1}\Delta\vec{B}}, \tag{3}$$

$$\frac{\partial P}{\partial t} + (\vec{V}\nabla)P + \gamma P\nabla\vec{V} = -\underline{\nabla\vec{q}} - \underline{L_r} + \underline{\frac{j^2}{\sigma}} + \underline{H}. \tag{4}$$

where ρ, $\vec{V}$, $\vec{B}$ and P are density, velocity, magnetic field and pressure, respectively; $\vec{g}$, R_e and R_m are gravity, and the viscous and magnetic Reynolds numbers, respectively.

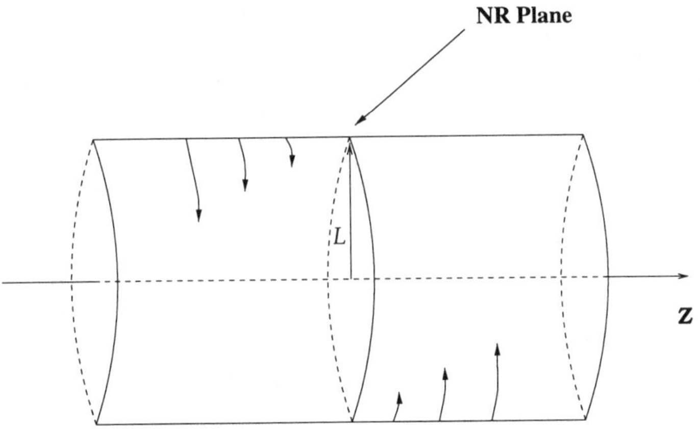

Figure 3. Local behavior of the non-rotating plane.

The right-hand terms in equation (4) characterize the energy losses of the system: the heat flux due to particle conduction, net radiation, the ohmic dissipation term and the term that represents all other heating sources (viscous dissipation rate, wave heating, etc.), respectively. In this simple simulation, in the initial state we ignore underlined terms based on the following kinds of arguments. The viscous and magnetic Reynolds numbers (R_e and R_m) are typically of the order of 10^8 or larger in the solar corona, and we therefore ignore the dissipation introduced by these terms in the initial state. Gravitation force is also ignored because of the large-scale size of its structuring in the corona. Here we refer to the paper by Lau (1995) where such a process in the total loop (considered anchored in the photosphere) was assumed to be gravity free, and investigate the non-linear dynamics of the area close to the non-rotating plane (the above terms play a negligible role in it). This results in the equations above becoming a closed set of a self-consistent system of equations. The steady-state hydrodynamical quantities are taken as:

$$\vec{V} = \left(\frac{\dot{a}(t)}{a(t)} \frac{r}{l}, \; v_\theta(t) \frac{z \cdot r}{l^2}, \; 0 \right) \qquad \vec{B} = \left(0, \; B_\theta(t) \frac{r}{l}, \; B_z(t) \right), \tag{5}$$

where $a(t)$ is the time-dependent parameter, l is the length of the loop portion involved in this process and z and r are the longitudinal and radial coordinates, respectively. The external force, $\vec{F}$ $(= f\frac{r \cdot z}{l^2})$, is assumed to be proportional to the z-component. In our simulation, we also ignore the thermal pressure term. Considering (5), we can derive a set of self-consistent equations:

$$\frac{d}{d\tau_a}\left(\frac{B_\theta}{B_0}\right) = \frac{1}{a^2}\left(\frac{V_\theta}{V_a}\right) - 2\,\frac{\dot{a}}{a}\left(\frac{B_\theta}{B_0}\right) \tag{6}$$

$$\frac{d}{d\tau_a}\left(\frac{da}{d\tau_a}\right) = -\,2\,a^3\left(\frac{B_\theta}{B_0}\right) \tag{7}$$

$$\frac{d}{d\tau_a}\left(\frac{V_\theta}{V_a}\right) = 2\,\frac{\dot{a}}{a}\left(\frac{V_\theta}{V_a}\right) + C\,a^2\,F(t), \tag{8}$$

where $B_0 = B_z(t)\Big|_{t=0}$ and $\tau_a = tv_a/l$ is the dimensionless Alfvén time-scale. Note that all other quantities can be expressed by the variables of the system:

$$\frac{B_z}{B_0} = \frac{1}{a^2}, \qquad \frac{\rho}{\rho_0} = \frac{1}{a^2}, \qquad \frac{T}{T_0} = \frac{1}{a^{2(\gamma-1)}}, \tag{9}$$

Here, we briefly describe the results of our numerical simulation, in agreement with the earlier analogous study by Lau (1995) and Zaqarashvili, Chargeishvili, & Sakai (1997). Figure 4a shows the temporal variation of the temperature of the system. It is seen that if the relative rotation is the same at both ends of the non-rotating plane, an instability (Lau 1995) threshold is reached in a few Alfvén time-scales. Note the appearance of a large temperature gradient that indicates the process where the plasma is heated very rapidly. This is consistent with the visible coronal emission line observations (Smartt, Zhang, & Smutko 1993), where the authors deduce that the interaction period of these events occurs typically in a very short time scale. Appearance of the large temperature gradient (Figure 4a) and an increase of the total current within the loop (Figure 4b) leads to the dissipation terms in equation (4) to be 'switched on', resulting in subsequent cooling of the plasma. This is also consistent with the visible emission-line data, which reveal plasma cooling continuing through the appearance of Hα (Airapetian & Smartt 1996).

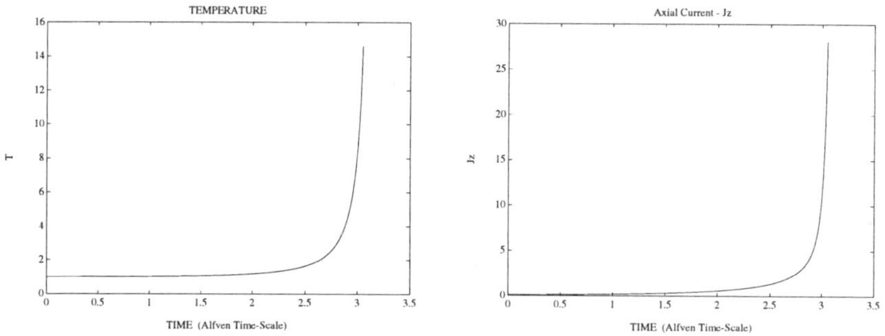

Figure 4. Time history of the temperature (a) and the total radial current along the z-axis (b)

We now consider the energy input due to this mechanism. The initial magnetic energy can be roughly estimated as $E_{mo} = r_0^2 l B_0^2/8$. For the coronal magnetic field, B_0=10 G (Lau 1995). Taking the cross-sectional radius, r_0=10^8m and $l = 8{\cdot}10^8$m gives $E_{mo}{\simeq}10^{26}$erg. The temporal variation of the magnetic energy that is several orders of magnitude more than the kinetic energy of the system, is given as: $E_m(t) = E_{mo}/a^2$ (Lau 1995). Our simulation shows that there is a sharp increase in the magnetic energy that is released by ohmic dissipation (Figure 4b) and heat transferred into the surrounding region, $E_h{\sim}10^{26}$erg. The numerical simulation also shows that after the rotational component formed according to (5), this consequently would lead to a fast build up of energy near the "non-rotating" plane. Such a dependence is valid when the relative momentum deposition to the two footpoints of the coronal loop are well correlated.

5. Coronal Loop Interaction: Relation to Single Loop Events

Now consider two loops that interact. In this case, we have two types of planes to consider in this loop system: 1) non-rotating planes determined by momentum deposition from the footpoints and, 2) the plane determined by a loop interaction where the magnetic field lines are tied (Figure 5). An absence of a real force to maintain a non-rotating plane at the same location along the loop will cause a non-rotating plane to move towards the region of the interaction. Consequently, one should expect that the single loop mechanism discussed above would operate there. Note that in this case it does not matter if the relative momentum deposition from both footpoints correlate, or not.

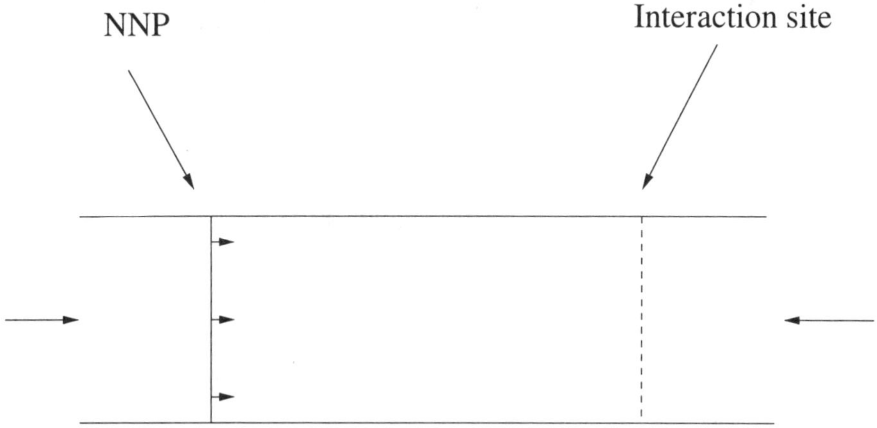

Figure 5. Physical picture of the Coronal Loop Interaction

6. Discussion

The key point is that rotation of coronal loop footpoints provides a mechanism of energy buildup at a localized region within a single loop. This is where there is a maximum energy deposition in the vicinity of the non-rotation plane in the case of a single loop. In the case of two interacting loops, non-rotating planes within the loops due to any twisting within the separate loops would migrate to the localized region of interaction, resulting in an increase in the level of energy release due to the magnetic pinch effect (Sakai 1993). A more detailed model of this work will be developed further. Finally, it is clear that observational tests of the possible validity of this type of model will require considerably improved angular resolution data over that currently available, together with the magnitude of the magnetic field, and, ideally, vector magnetic field information.

References

Airapetian, V.S., Smartt, R.N. 1995, ApJ., 445, 489

De Jager, C., Inda-Koida, M., Koida, S., & Sakai, J.-I. 1993, in Symposium on Current Loop Interaction in Solar Flares, ed. J.-I. Sakai (Toyama: Toyama University) 149

Lau Y-T. 1995, Phys. Plasmas, 2 (12), 4442

Li X-Q., Zhang Z., & Smartt R.N. 1994, A&A, 290, 963

Machado, M.E., Xiao, Y.C., Wu, S.T., Prokakis, Th., & Dialetis, D. 1988, ApJ, 326, 451

Mathews, S.A., & Harra-Murnion L.K. 1997, Solar Phys. 175, 541

Nakajima, H. 1993, in Current Loop Interaction in Solar Flares, ed. J-I. Sakai (Toyama: Toyama Univ.) 39

Pevtsov, A.A., Canfield R.C., & Metcalf T.R. 1994, ApJ, 425, L117

Pritchet, P.L., & Wu C.C. 1979, Phys. Fluids, 22, 2140

Sakai, J-I. 1993, in Current Loop Interaction in Solar Flares, ed. J-I. Sakai (Toyama: Toyama Univ.) 13

Shimuzu, T., Tsuneta, S., Acton, L.W., Lemen, J.R., & Uchida, Y. 1992, Publ. Astr. Soc. Japan, 44, L147

Shimizu, T., Acton, L.W., Tsuneta, S., Lemen., J.R., Ogawara, Y., & Uchida, Y. 1994, ApJ, 422, 906

Smartt, R.N., Dunn R.B., & Fisher R.R., 1981, SPIE, 288, 395

Smartt, R.N., & Zhang Z. 1987, in Theoretical Problems in High-Resolution Solar Physics, ed. G. Athay and D.S. Spicer (NASA Conf. Pub. 2483) 129.

Smartt, R.N. & Zhang Z. 1992, in Coronal Streamers, Coronal Loops, and Solar Wind Composition, ed. J. Mariska, V. Domingo, and A. Poland (ESA SP-348) 185

Smartt, R.N., Zhang Z., & Smutko M.F. 1993, Solar Phys., 148, 139

Smartt, R.N., Airapetian V.S., & Zhang Z. 1996, in Magnetodynamic Phenomena in the Solar Atmosphere - Prototypes of Stellar Magnetic Activity, ed. Y. Uchida, T. Kosugi, & H.S. Hudson (Dordrecht: Kluwer) 531

Spicer, D.S., & Švestka Z. 1983, Solar Phys., 87, 271

Švestka Z., & Poletto G. 1984, Adv. Space Res., 4, 287

Takakura, T., Inda, M., Makishima, K., Kosugi, T., Sakao, T., Masuda, S., Sakurai, T., & Ogawara, Y. 1993, Publ. Astr. Soc. Japan, 45, 737

Tsuneta S., Hara, H., Shimizu, T., Acton, L.W., Strong, K.T., Hudson, H.S., & Ogawara, Y. 1992, Publ. Astr. Soc. Japan, 44, L63

Zaqarashvili T., Chargeishvili B., & Sakai J. 1997, Astrophysics and Space Science, 246, 107

High Resolution Solar Physics: Theory, Observations, and Techniques
ASP Conference Series, Vol. 183, 1999
T.R. Rimmele, K. S. Balasubramaniam, R.R. Radick, eds.

Quiet Sun Bright Point Dynamics and Energetics as seen by SOHO and Yohkoh

P. T. Gallagher, F. P. Keenan

Department of Pure & Applied Physics, The Queen's University of Belfast, Belfast BT7 1NN, N. Ireland.

K. J. H. Phillips, P. Prés

Rutherford Appleton Laboratory, Chilton, Didcot, Oxfordshire OX11 0QX, U.K.

L. K. Harra-Murnion

Mullard Space Science Laboratory, Holmbury St Mary, Dorking, Surrey RH5 6HT, U.K.

Abstract.

Recent observations of EUV and X-ray bright points using both *SOHO* and *Yohkoh* suggest that stochastic mechanisms play an important role in heating the solar atmosphere at chromospheric and coronal levels. Using the Coronal Diagnostic Spectrometer (CDS) onboard *SOHO* we find rapid changes in both the velocity and intensity of lines formed on both the upper chromosphere and transition region demonstrating that these layers are extremely dynamic. Furthermore the transition region, and in particular the EUV network, appears to show larger amplitude variations than the chromosphere.

We also report on long duration bright points identified in the quiet Sun from *SOHO*/EIT Fe XII images which are found to have a time evolution from birth to decay which is extremely well correlated with the rise and fall of magnetic flux as determined from photospheric magnetograms made with the *SOHO*/MDI instrument. Radiative losses of the bright points are found to be much less than conductive losses but the sum of the two is comparable with the available energy of the associated magnetic field.

1. Introduction

Several instruments on the ESA/NASA *Solar and Heliospheric Observatory* (*SOHO*) are allowing us to view the quiet Sun atmosphere with spatial resolution and time coverage which is better than that obtained with any previous spacecraft. In particular, we are now able to understand the relation of extreme ultraviolet (EUV) emission to the photospheric magnetic field far better than before. As the EUV spectrum includes emission line emitted in a wide ($10^4 - 2\times10^6$K) temperature range, images from the EUV instruments on *SOHO* show structures (with ~ 2 arcsecond resolution) from the chromosphere to the corona.

This paper reports on a number of investigations we have been involved with in which the nature of the quiet-Sun atmosphere, including its dynamics (specifically short-lived, flare-like brightenings) and also its relationship with the photospheric magnetic.

2. Chromospheric and Transition Region Bright Point Observations

Observations of the EUV quiet-Sun network with the *SOHO* instruments including CDS show that it is in a continuous state of change, with large numbers of brightenings and motions. It is likely that this is due to small parcels of magnetic field emerging at cell centers and being transported by supergranular convection from the centers of cells to the network, where fields either add or cancel in some way. X-ray observations with the *Yohkoh* satellite show that tiny 'network flares' (Krucker *et al.* 1997) occur, while in the EUV 'blinkers' have been observed with the CDS instrument (Harrison 1997). These tiny flare-like features seem to be the quiet-Sun counterpart to those investigated within active regions by Shimizu (1995).

In work in progress, we are doing a systematic survey of them and their properties using CDS data - see Fig. 1 The recent period of solar inactivity (1996/1997) has been an ideal time to search for such activity in quiet areas of the Sun. We used a specially devised observing sequence called NTBRMDI, in this case consisting of a 4×4 arcmin^2 context image taken in He I (584 Å) and O V (630 Å) lines followed by 360 'single slit' images, the 4×240 arcsec^2 slit being used. The exposure time is 15 s, so for each run of the sequence there is around 2 hours' worth of observation in the single-slit mode. A further context image is then taken. While the single-slit images are formed, solar-feature 'tracking' (in which a feature is followed as the Sun's surface rotates) is disabled. Thus, features cross the slit at the very slow rate of 9 arcsec per hour. As indicated already, network features in transition-region lines are of order 15 arcsec wide, so a gradual rise and fall of intensity over the two-hour-long period of slit images can be expected as a network structure passes over the slit. The NTBRMDI sequence has been run on fourteen occasions between December 1996 and January 1998. Analysis of the enormous quantity of data has been a major task and is still being carried out.

Meanwhile, we can already state a few preliminary findings. First, the brightenings in the vicinity of the network structures are much more frequent than those in the cells. In a typical two-hour run of the NTBRMDI sequence, network brightenings are about a factor of two or three more intense than those in the cells, although their durations are not obviously different. The durations of brightenings average around three minutes though some last as long as 12 minutes. Brightenings that have durations longer than about 30 minutes begin to be indistinguishable from 'spatial' variations, i.e. the rise and fall of emission as network features pass across the slit (see Fig. 2).

Also from Fig. 1 it can be seen that network brightenings and in particular transition region network brightenings occur with a larger frequency and amplitude than in the chromosphere. These finding are not inconsistent with

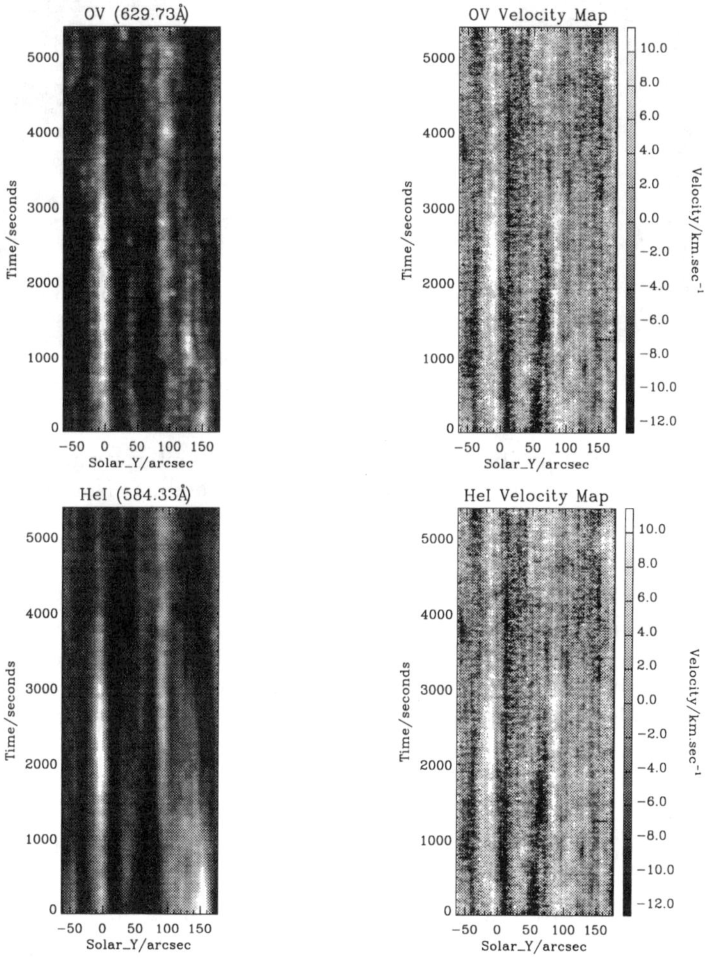

Figure 1. Line intensity and velocity 'Space-Time' plots for the transition region O V ion and the chromospheric He I ion. The abscissa gives position along the slit in arcseconds while the ordinate displays the time in seconds.

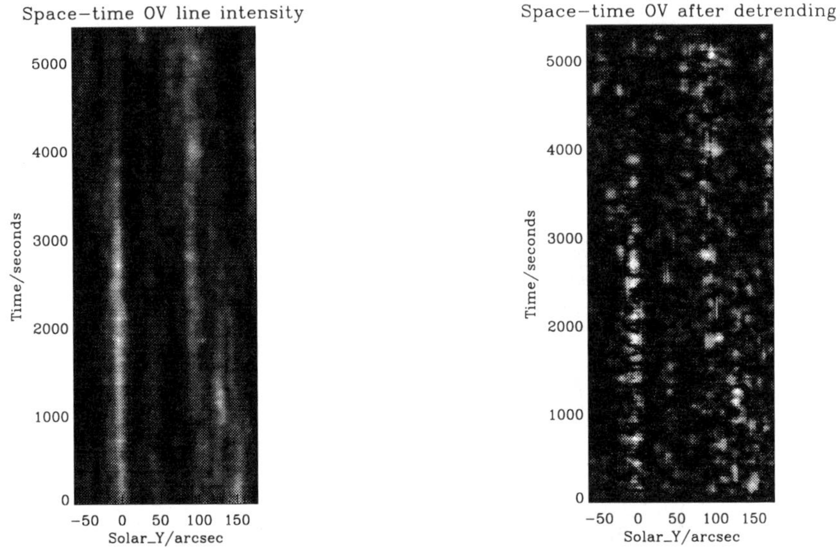

Figure 2. Slit position versus time for O V line intensity at left while the right hand image shows the data after detrending. Note the large number of bright points along the network.

the rebound-shock model of spicular material. In this theory it is assumed that cool chromospheric plasma is injected into the hotter upper atmosphere ($\sim 10^6$ K) where it then experiences intense heating. This hot plasma then rebounds from the transition region/coronal interface to form an extremely localized high velocity downflow resulting in large plasma emissivities and velocities.

3. Relation of Magnetic Field to Coronal Quiet-Sun Features

There was a period of very low activity in July and August 1997 when the *GOES* X-ray monitoring satellites registered a base level of A1 or even less. For a short period between July 31 and August 1, the SXT instrument on *Yohkoh* performed 10×10 arcmin^2 rasters in a quiet region near Sun centre supported by several *SOHO* instruments. Some CDS NTBRMDI context rasters were run for some of this period.

Our preliminary analysis of *SOHO* and *Yohkoh* data sets indicate very significant results which we outline here. Images from the Michelson Doppler Imager (MDI) on *SOHO* show, as has been reported elsewhere (Title *et al.* 1997), a profusion of photospheric magnetic monopoles which are in a continuous, almost 'Brownian', state of motion although the region observed is well away from any even minor centers of activity. At least three cases of magnetic events can be distinguished: merging of same-polarity monopoles, partial or complete cancellation of field when opposite-polarity monopoles collide, and appearance of magnetic flux as a bipole. On each of these occasions, coronal bright points,

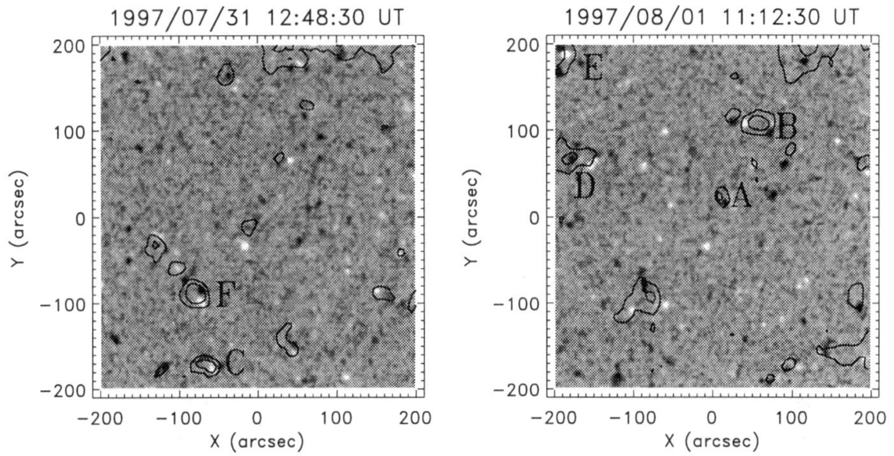

Figure 3. Magnetograms (grey-scale images) made with the MDI instrument compared with *Yohkoh* SXT emission (contours) for two times (July 31, 1997 at 20:48 U.T. and August 1, 1997 at 16:00 U.T.).

as observed by the SXT instrument and the Fe XII channel of the *SOHO* EIT instrument, are formed almost simultaneously. Fig. 2 shows EIT emission superimposed on MDI magnetograms at two times separated by almost a day. This channel views radiation in a bandpass centered on 195 Å which includes an intense Fe XII emission line (peak formation temperature 1.6×10^6K). Images from the Fe XII channel are made every 20 minutes, and so are very suitable for following the evolution of the several bright points which are formed.

We found first that there is an extremely close correlation between the flux of the Fe XII emission seen with EIT and the magnetic flux, as measured from the MDI data by the value of $\int \mid B_l \mid dA$ (B_l is the line-of-sight field strength) over areas (A) which include various bipolar regions as they form and disappear. This has been verified for at least five such regions in a field measuring 4×4 arcmin^2. There is a less tight correlation between the magnetic flux and the very weak and noisier SXT X-ray flux, as measured through the thin Al and Al/Mg filters (bandpass $\sim 3 - 18$ Å). The energy in the magnetic field, however, appears to be much more than the thermal energy of the hot plasma.

A second point is that several short-lived tiny X-ray bursts occur near the times of magnetic events. Thus, there was a small burst with sharp rise and slower decay (total duration 14 minutes) at 00:40 U.T. on August 1 simultaneously with a field cancellation event. A second slightly smaller event having a double peak and lasting only three minutes occurred in another region at 08:40 U.T. simultaneously with the emergence of a magnetic bipole. Each magnetic event was also associated with the formation of a bright point lasting several hours, though the two bursts were at different points in the bright point evolution – one in the early phase, the other late. Clearly these X-ray events are like the 'network flares' reported by Krucker *et al.* (1997). One can only speculate on the reason for flaring, but it might be, in the case of field cancellation

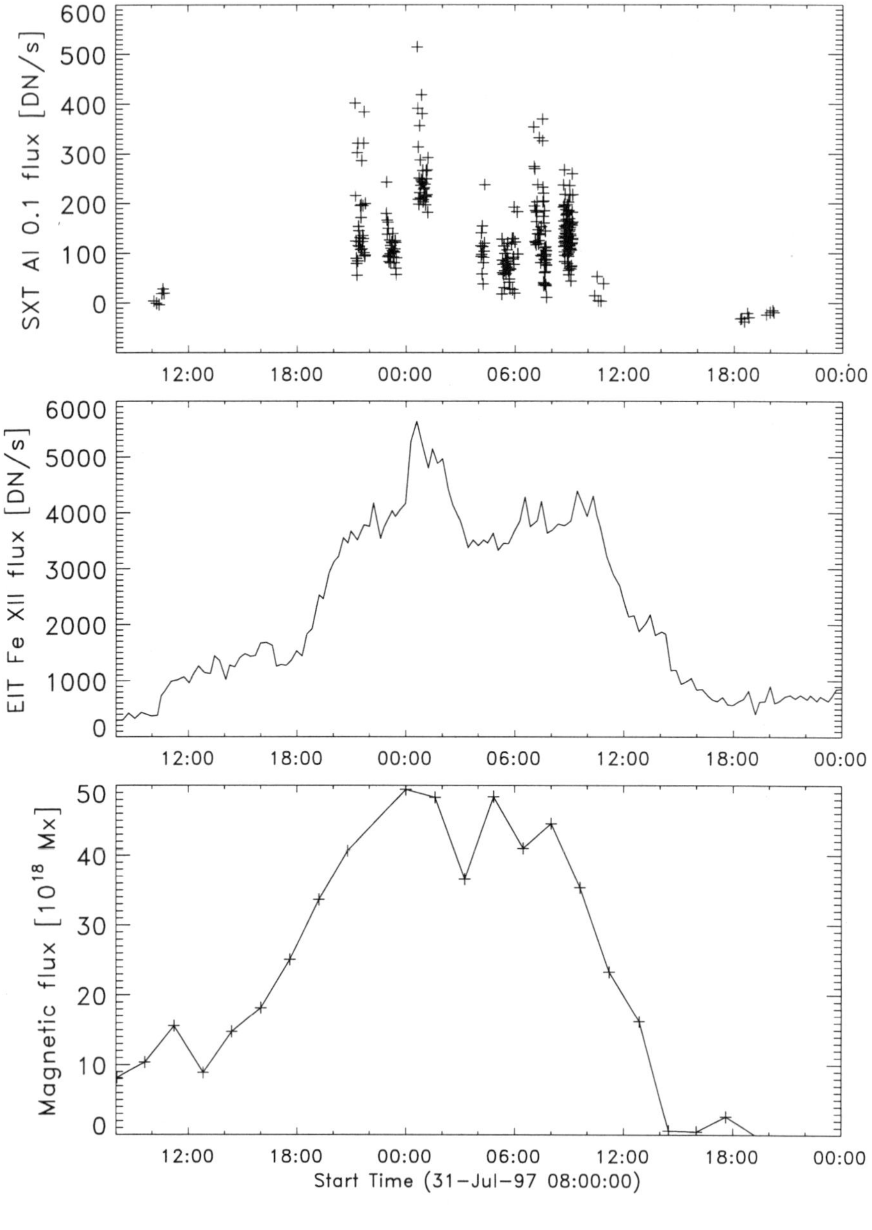

Figure 4. The SXT Al 0.1 (*top*) and EIT Fe XII (*middle*) light-curves of the coronal bright point A, together with (*bottom*) time variations in the local magnetic flux Φ from the MDI instrument. See Prés & Phillips, 1998.

events, that there are small areas of field concentration within the monopoles observed by MDI but beyond its spatial resolution that give rise to the flares.

These network flares can be truly described as microflares in that the emitted energy is of order 10^{25} erg (i.e. 10^{-6} of a large flare), as estimated from the SXT data. We are at present working on how important these flares are in the global energy budget of the solar corona.

4. Conclusion

Transient brightenings observed with CDS in He I and O lines occur largely in EUV network structures, and have durations of a few minutes. Coronal bright points (as seen with the EIT instrument in the Fe XII line) occur simultaneously with photospheric magnetic events with a very close relation between magnetic flux and the amount of Fe XII emission. Small X-ray flares often occur at such locations, apparently similar to the network flares previously noted.

References

Harrison, R. A. 1997., Solar Phys., 175, 467

Krucker, S., Benz, A. O., Bastian, T. S., and Acton, L. W. 1997. ApJ, 488, 499

Prés, P. and Phillips, K. J. H. 1998, ApJ, in press

Shimizu, T. 1995, PASJ, 47, 251

Title, A., Schrijver, C. J., Tarbell, T. D., Hagenaar, H. J., and van Ballegooijen, A. A. 1997. In Cool Stars, Stellar Systems, and the Sun, to be published

Group Discussion

Ayres : How did you calculate the conductive cooling?

*Gallagher :*This was calculated assuming classical Spitzer conductivity in which the energy is conducted down to the chromosphere along the magnetic field lines. The area and length of the loop was estimated from the FEXII and MDI data with the total conductive losses being given by: $E_c = (K \times T^{3.5} \times L^{-2}) \times A \times L$

Chae : Since you found large downflows in networks, you should be able to calculate the coronal energy loss by the enthalpy flow. According to my experience, in most cases, the enthalpy flow energy loss dominates over the conductive energy loss.

Gallagher : O.K.

High Resolution Solar Physics: Theory, Observations, and Techniques
ASP Conference Series, Vol. 183, 1999
T.R. Rimmele, K. S. Balasubramaniam, R.R. Radick, eds.

Effects of the Spectral and Spatial Resolution on the Detection of K2v Grains

L. A. Smaldone

Dipartimento di Scienze Fisiche, University of Naples "Federico II" Mostra d'Oltremare Pad. 19, I-80125 Napoli, Italy

A. Salvatore

Capodimonte Astronomical Observatory Via Moiariello, 16, I-80131 Napoli, Italy

Abstract. Using area scans of a quiet solar region, performed with the Horizontal Spectrograph at the NSO Vacuum Tower Telescope, we studied the statistical influence of the seeing conditions on the properties of K_{2v} grains. The spectral axis of our data has been then convolved with different instrumental profiles in order to simulate spectroheliograms and filtergrams of various spectral resolution. The resulting effect on the number of detected K_{2v} grains is impressive for the filtergrams.

1. Introduction

The core of the solar CaII K line profile shows typically a central doubly-peaked emission reversal, the K_{2v} and the K_{2r} toward the violet and the red wing respectively. Narrow band filtergrams and spectroheliograms of the solar quiet atmosphere, taken in the K_{2v}, show the calcium chromospheric network (boundaries of super-granulation cells) and intra-network small, bright elements : the K_{2v} grains.

In the seventies there are the first observations of the temporal evolution of the CaII K line profile in the K_{2v} grains and the attempts to relate them to the problem of the chromospheric temperature raising (Liu 1974). The oldest (visual) analyses of the spatial relationships between K_{2v} grains and small scale magnetic elements report a one-to-one correspondence (Sivaraman and Livingston 1982).

The new scientific interest about the K_{2v} grains is related to the L. Damè discovery, in 1984, that the intra-network 3 minute chromospheric oscillations are located in the K_{2v} grains (Rutten and Uitenbroek 1991, Von Uexküll and Kneer 1995, Hofmann, Steffens and Deubner 1996, Kalkofen 1996, Hoekzema and Rutten 1998, Wellstein et al. 1998). The relationships between K_{2v} grains and magnetic field have been, recently, reviewed using dynamical considerations (Steffens, Hofmann and Deubner 1996) or magnetic field proxy (Remling, Deubner and Steffens 1996) founding any correlation. Different physical phenomena, probably, are collected under the same name since *persistent* K_{2v} grains are well associated with magnetic elements (Brandt et al. 1992).

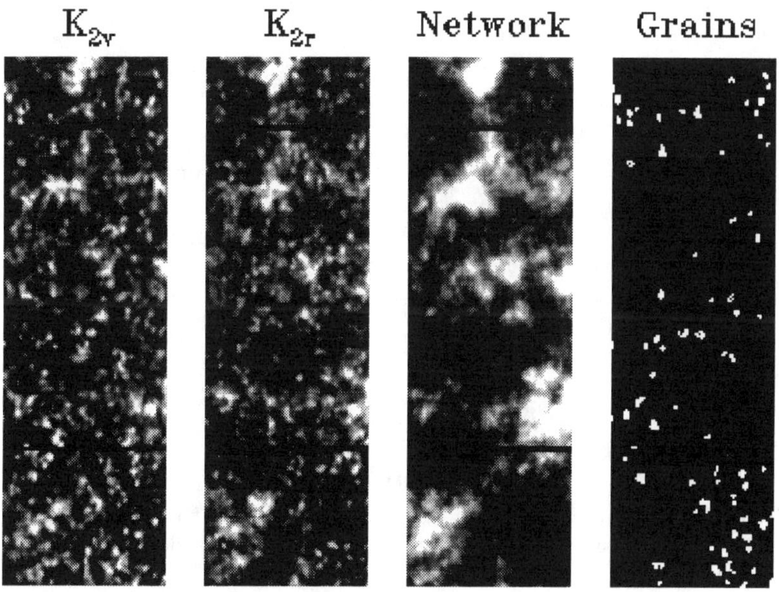

Figure 1. Spectroheliograms in K_{2v} and K_{2r} of a quiet area on disk center acquired on October, 16th 1996. We assume as network the bright pattern on the temporal mean, over 2 hours, of the weighted mean of K_{2v}, K_3 and K_{2r} spectroheliograms. The K_{2v} grains are found by thresholding the image of the K_{2v} *local* contrast.

Despite the renewed attention to these chromospheric structures, they are missing systematical studies not only of their spatial and dynamical properties but also of the best instrumental observing conditions: the only operative definition of the K_{2v} grains has been supplied by Kalkofen in 1996! Aim of the present work is to give an useful rationale for planning observations of K_{2v} grains.

2. Observations and data reduction

The observations have been performed with the Vacuum Tower Telescope (VTT) at the National Solar Observatory-Sacramento Peak on October, 16th and 18th 1996. After passing through a Correlation Tracker, the beam coming from the VTT fed the Horizontal Spectrograph (HSG). The spectral resolution was of the order of 720,000, the slit size $120'' \times .45''$ and the exposure time $4\,s$. The spectra, in the range 3928.25 Å- 3942.59 Å, were acquired on a TIJ CCD with an image scale of $0.243\ ''/pxl$. After the usual corrections for the dark-current and the table gain, the spectra were calibrated in wavelength and in relative intensity using the Liege spectral solar atlas.

Figure 1 shows the spectroheliograms in the K_{2v} and K_{2r} obtained from an area scan of October, 16th with a 128 mÅ spectral integration interval. The image size is $120'' \times 36''$.

The first step in the data analysis has been the detection of the chromospheric

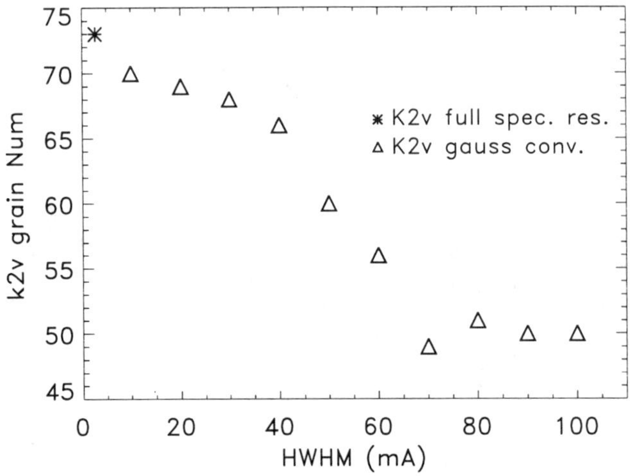

Figure 2. Detected K_{2v} grains versus the half-width at half maximum of the gaussian used to convolve the spectra in order to simulate different spectral resolution spectrographs.

network. As shown by the third panel of Fig. 1, the network is well pointed out by the the mean of the K_{2v}, K_3 and K_{2r} spectroheliograms averaged over the observation time , spanning about two hours. We assumed as intra-network the regions showing in this image a contrast less than 1.2 to respect to the mean intensity.
Due to the high spatial resolution of our data set, the intra-network is rich of structures and the Kalkofen K_{2v} grains definition criterion (areas brighter than 1.3 times the intra-network mean intensity) fails to detect some grains visually well evident. In our case a satisfactory detection is obtained referring to the mean *local* intensity, obtained replacing high gradient regions by the mean neighbour intensity. Consistently, we define as grains the intra-network areas brighter than 1.4 times the mean local intensity in K_{2v}.

The fourth panel of Fig. 1 shows the localisation of the grains present in the spectroheliogram of the first panel. Using the K_{2v} grains localisation masks for the area scans, it is easy to associate to each grain his properties in different spectral features.

3. Effects of the spectral resolution

We used the spectral axis of our data (14 Å centred on the CaII K) to simulate spectroheliograms and filtergrams of different spectral resolution. As the usual, we convolved our spectra with gaussian instrumental profiles for spectrographs and $sinc^2$ instrumental profiles for birefringent filters.
Figure 2 reports the number of grains found in spectroheliograms obtained convolving the spectral axis with gaussians of different width, the asterisk indicates the number of grains present in the original data. The 90% of grains are detected also using a resolving power of 70,000; a larger loss of K_{2v} grains, of the order of

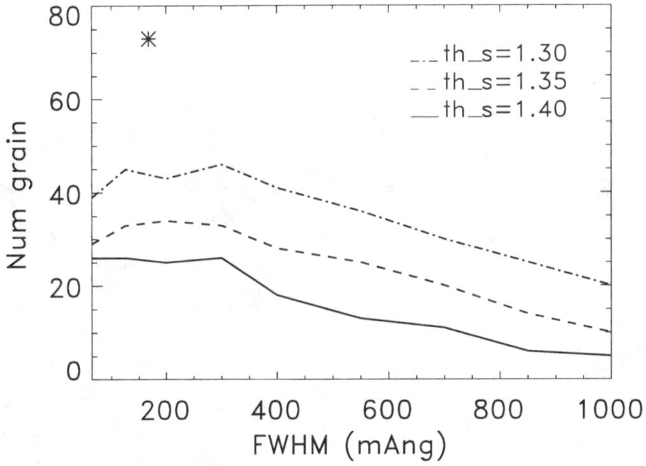

Figure 3. Detected K_{2v} grains versus the pass-band full-width at half maximum of the simulated filtergrams for 3 different detection thresholds in the local contrast. The asterisk indicates the number of grains present in the original data.

30%, happens when the width of the instrumental profile becomes of the same size of the spectral integration interval used to derive the spectroheliogram.

The loss of grains is really impressive in the case of birefringent filters: only 1/3 of the grains are detected observing with a 64 mÅ passband filter (Fig. 3). Also taking in to account the contrast reduction due to the filter passband, decreasing the detection threshold doesn't help too much.
The presence of the secondary lobes in the birefringent filters can also cause the appearance of spurious K_{2v} grains. Figure 4 shows the detected K_{2v} grains in the 125 and 550 mÅ passband simulated filtergrams. Some of them, not present in the original spectroheliogram, are due to K_3 bright points (diamonds), there are also grains of uncertain origin (triangles).

4. Effects of the spatial resolution

The typical K_{2v} grains size, of the order of 1.5″, is relatively large comparing with the typical seeing conditions in a good observing site, but the contrast reduction due to the seeing can have considerable influence on their detectivity.

During our observing area scans, the seeing spanned from .5″, limit imposed by our pixel size, to 1.7″. The width of the autocorrelation function, computed along the spectrograph slit direction at the farest wavelengths from the CaII K line core (the *pseudo-continuum*), can be a good index of the seeing conditions. In this way it is possible to associate to each spectrum of the area scan (each column of the spectroheliogram) its own seeing.

The contrast (ratio between the intensity and the mean local background intensity) is the grains parameter more sensitive to the seeing. The left panel

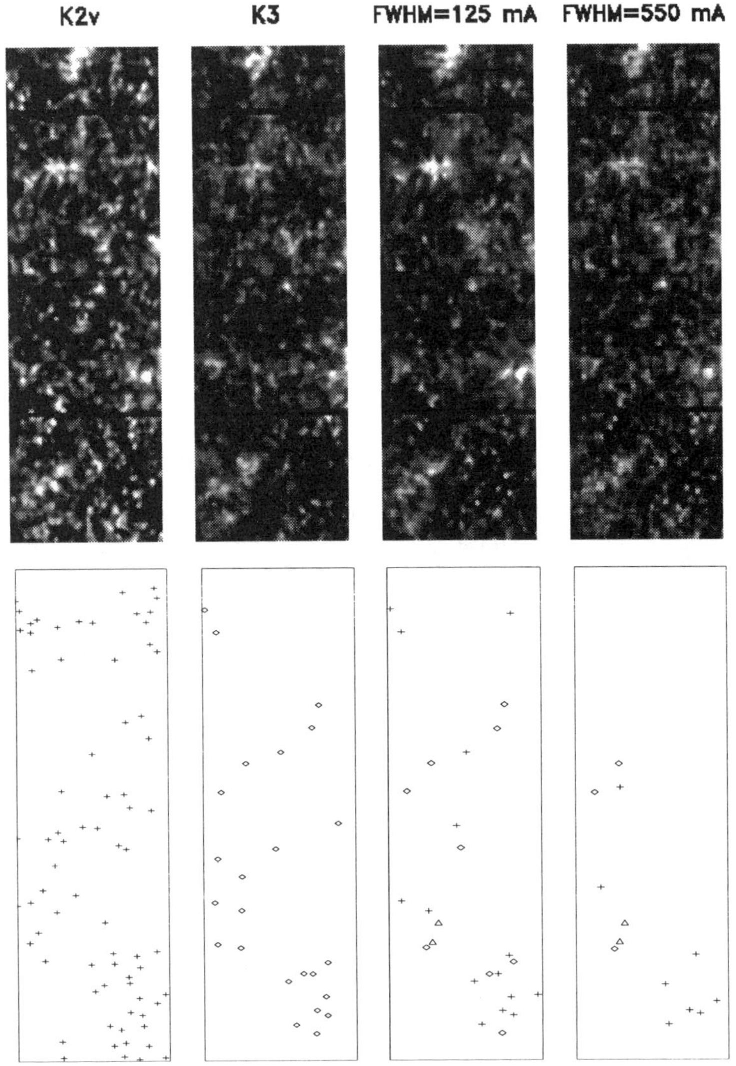

Figure 4. Top panels: spectroheliograms in K_{2v} and K_3 from the original data and simulated K_{2v} filtergrams with two different passbands. Bottom panels (from left): localisation of the original K_{2v} grains (crosses), K_3 bright features not present in the K_{2v} spectroheliogram (diamonds) and "K_{2v}" grains detected in the filtergrams. Some of the filtergrams K_{2v} grains are due to K_3 bright points; there are also grains of uncertain origin (triangles).

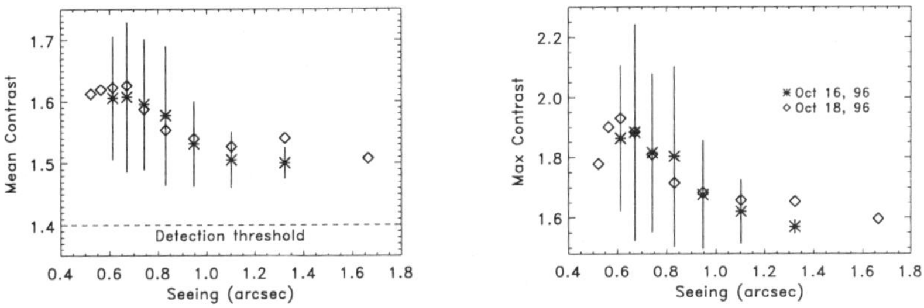

Figure 5. Left panel: seeing dependence of the mean contrast in the grain area. Right panel: behaviour of the maximum contrast in the grain with the seeing conditions.

of Fig. 5 shows the decrease of the mean contrast, averaged on the grain area, for the detected K_{2v} grains. The exponential decreasing function of the seeing is more evident in a parameter less biased by the thresholding detection procedure, the maximum contrast in the grain (Fig. 5, right panel). It spans from 1.9, with a variance of 0.3, during the best seeing condition to 1.6 at $1.4''$. Understanding the behaviour of the grains contrast with the seeing, it is easy to justify the seeing variations of different grains parameters.

The K_{2v} grains are slightly elongated structures. The mean ratio between the minor and the major axes is of the order of 0.7.
Figure 6 shows the statistical distribution of the grains *equivalent diameter*, i.e. the diameter of the circle having the same area, in different seeing conditions. The grain diameter spans roughly from $1''$ to $3''$, with a mean value of $1.63''$. As the seeing quality decreases, due to the contrast reduction effect, only the largest grains can be detected.

The numerical density of the K_{2v} grains is strongly depending from the seeing (Fig. 7). It is well described, in the seeing range $0.5''$-$1.4''$, by the following exponential law

$$n = 0.12e^{-2.7s}$$

where the numerical density, n, is the number of K_{2v} grains per arcsec^2 of intra-network area and the seeing, s, is in arcsec.

5. Discussion

In this work we described the preliminary results concerning the effects of observing conditions on the detection of the K_{2v} grains.
From the instrumental point of view: also a low resolution spectrograph should be preferred to narrow band birefringent filters. The contrast reduction and the spectral contamination, both due to the filters secondary lobes, override the advantage of the time resolution. My be it's time to design faster area scan systems.
Despite the relative large size of the K_{2v} grains, their detection is strongly related to the seeing conditions. The exponential dependence of the grains numerical

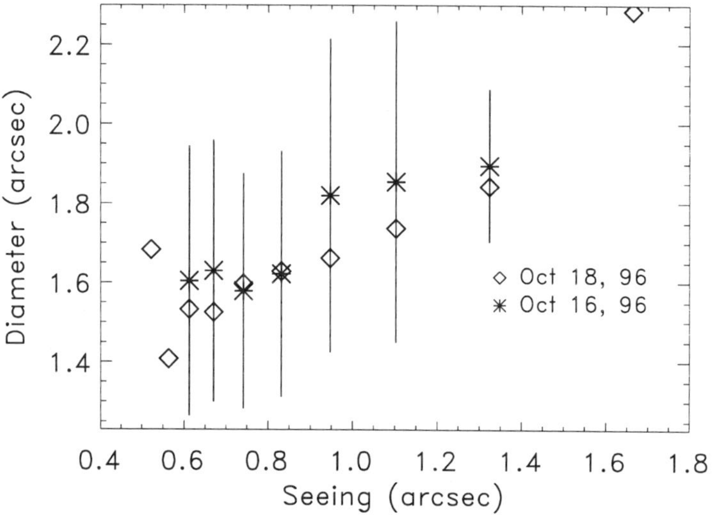

Figure 6. K_{2v} grains diameter statistics versus seeing conditions. The vertical bar indicates the diameter variance.

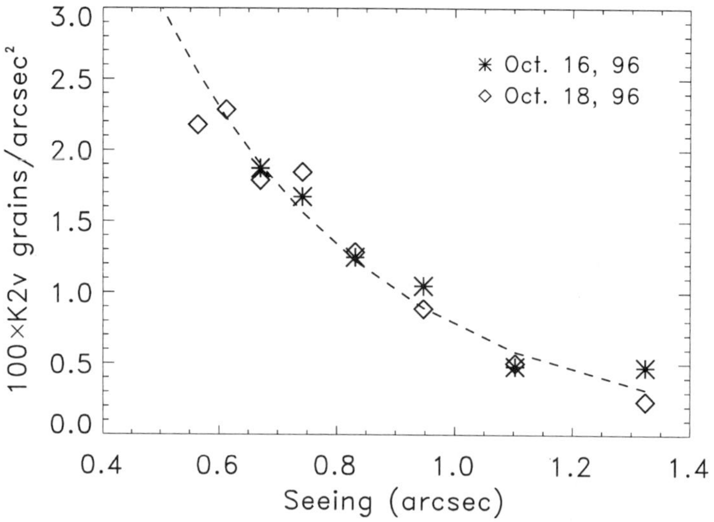

Figure 7. Detected grains (×100) per arcsec^2 of intra-network area in different seeing conditions.

density from the seeing suggest higher spatial resolution observations in order to determine their influence on the energetic balance of the chromosphere.

Acknowledgments. This work has been partially supported by the EOARD, WOS program No. 61708/98WW197, and by the MURST PRIN98 *Variabilità solare e relazioni Sole-Terra*. One of us (L.A.S) thinks is this the right place to express his gratitude to the London Office of the EOARD.

References

Kalkofen W. 1996, ApJ, 468, L69

Hofmann J., Steffens S., Deubner F.-L. 1996, A&A, 308, 192

Hoekzema N.-M., Rutten R.-J. 1998, A&A, 329, 725

Liu S.-Y. 1974, ApJ, 189, 354

Remling B., Deubner F.-L., Steffens S. 1996, A&A, 316, 196

Rutten R.-J., Uitenbroek H. 1991, Solar Physics, 134, 15

Sivaraman K.-R., Livingston W.-C. 1982, Solar Physics, 80, 227

Steffens S., Hofmann J., Deubner F.-L. 1996, A&A, 307, 288

Von Uexküll M., Kneer F. 1995, ApJ, 294, 252

Wellstein S., Kneer F., Von Uexküll M. 1998, A&A, 335, 323

High Resolution Solar Physics: Theory, Observations, and Techniques
ASP Conference Series, Vol. 183, 1999
T.R. Rimmele, K. S. Balasubramaniam, R.R. Radick, eds.

Oscillations and Heating in Chromospheric Fine Scale Structures

R. Kariyappa

Indian Institute of Astrophysics, Bangalore 560034, India

Abstract. We have analyzed a 35-min-long time sequence of spectra obtained simultaneously in CaII H-line, NaI D1 and D2 lines, and in strong and weak FeI lines in a quiet region at the center of the solar disc under high spatial, spectral, and temporal resolution using the VTT of NSO/Sac Peak. We derived the line profile at the sites of a large number of bright points and network elements, and extracted the various line profile parameters. The important results emerged from this analysis are: (i) The bright points and network elements exhibit 3-min and 5-7 min periodicity in their intensity oscillations respectively; (ii) The bright points are grouped into three classes according to their dynamical behavior, and it is surmised that the differences in their behavior are directly related to the inner network photospheric magnetic points to which they have been observed to bear a spatial correspondence; (iii) The period of intensity oscillations decreases outward from the low photosphere (5-min) to the middle chromosphere (3-min); (iv) The evidence for the existence of the constant period of oscillations in bright points with their brightness and the different period from network oscillations suggests that the mode of heating mechanism may be identical (by 3-min period waves) for all classes of bright points and it may be an entirely different mechanism (5-7 min period waves) in the case of network elements; (v) The physical process behind the cause for the bright points may be due to the interference of these 3-min period waves, and the 3-min bright points may be magnetic in origin.(vi) According to the current models, the energy transported by these 3-min period waves at the sites of the bright points can account for a substantial fraction of the radiative loss from the quiet chromosphere.

1. Introduction

One of the major problems of solar physics is to understand the way the chromosphere and the corona are heated and supported by the non-radiative energy from the layers below. Biermann (1948) and Schwarzchild (1948) have suggested that the outer atmosphere heated by acoustic waves generated in the hydrogen convection zone. In 1949, Schatzman has shown that the acoustic waves will be converted into shock waves and will dissipate the energy in the chromosphere and corona. Observational evidence has come from a famous total solar eclipse of February 25, 1952 with the flash spectra which led to the first models of the chromosphere based on excellent observations. Since then several theories attempted to explain the heating mechanism.

The most suited lines for the observations of middle and upper chromosphere in the visible region are MgI b1, H-alpha and H and K lines of CaII. The CaII H and K lines are resonance and collisionally controlled lines and are very sensitive to the variations in temperature and the magnetic field strength, therefore they are excellent indicators of the chromospheric structural changes related to solar magnetic activity. Using these lines, the chromospheric oscillations and heating at the sites of different chromospheric features can be studied in great detail. At high resolution CaII H and K spectroheliograms show the different chromospheric features, namely, the plages, network elements and intranetwork bright points. They are seen in emission and have a one-to-one correspondence with the underlying photospheric magnetic fields (Leighton, 1961, Skumanich, Smythe, & Frazier 1975; Sivaraman & Livingston 1982; Nindos & Zirin 1998). Thus the non-radiative heating seen as excess emission occurs only in regions associated with magnetic fields, irrespective of the size of the chromospheric structures. These features will contribute in the variations of total K- emission flux, in the heating of the chromosphere and in the variations of UV irradiance variability (Lean 1987; Foukal & Lean 1988; Kariyappa 1993; Kariyappa & Sivaraman, 1994; Kariyappa, Sivaraman, & Anandaram 1994; Kariyappa et al. 1995; Kariyappa 1994, 1996; Kariyappa & Pap 1996; Kariyappa & Kneer 1998; Kariyappa 1998).

Jensen and Orrall (1963) have shown observationally that the chromosphere oscillating with a period of 180-200 sec. Observations of high temporal, spatial, and long-time sequence of spectra of CaII H and K lines show that waves with a 3-min period transport, and dissipate a large amount of energy, and heat the chromosphere at the sites of the bright points (Punetha 1974; Liu 1975; Cram & Dame 1983). Although these observations provide an overall understanding of the oscillations and heating, some of the important problems remain unanswered.

2. A Brief Summary on Observations and Analysis

We made an attempt to solve some of the problems in using a 35-min time sequence of spectra obtained (Beckers et al. 1972) at the VTT of NSO/Sac Peak simultaneously in CaII H line, NaI D1 and D2 lines and large number of FeI lines at the center of the solar disc in a quiet region. The spatial resolution is about 1 arc sec or better and the time resolution is 12 sec. We derived the absolute line profiles for 30 bright points and 2 network locations and extracted the line profile parameters and plotted them as a function of time in order to obtain their light curves. More details on the reduction of the spectra are presented in Kariyappa et al.(1994); Kariyappa(1994,1996), and further data handling will also be described below, together with the results.

3. Results and Discussion

The bright points can be grouped into three classes depending on the brightness enhancement during their dynamical evolution. The differences in their behavior are directly linked with the inner network photospheric magnetic points to which they have been observed to bear a spatial correspondence. The light curves of the bright points give the impression that the "main pulse", which is the upward propagating disturbance carrying energy, throws the medium within the bright point into a resonant mode of oscillation that is seen as the follower pulses. The mechanical energy transported by the main pulses at the sites of the bright points over the entire visible solar surface can account for a substantial fraction of the radiative loss from the quiet chromosphere, according to the current models. The bright points can provide 50% of the energy required to heat the quiet chromosphere. This together with the contribution from the network can account for the energy requirement of the quiet chromosphere (Kariyappa, et al. 1994). The bright points are associated with 3-min periodicity, whereas the network elements exhibit 5-7 min periodicity in their intensity oscillations (Kariyappa 1994, See histogram and power spectra plots in Figures 2, 3, 5 and 6). We have analyzed the line profiles of NaI D1 and D2 , large number of strong and weak FeI lines together with the CaII H line obtained simultaneously. All these lines are forming at different heights in the solar atmosphere ranging from the low photosphere to the middle chromosphere. We examined the central intensities of these lines at the sites of the bright points and found that the period of intensity oscillations in the chromospheric bright points decreases outward from the low photosphere (5-min) to the middle chromosphere (3-min) (Kariyappa 1996, See Fig.6 and Table II).

We have derived the period of oscillations using power spectrum analysis and the peak intensity value of H2V for the main and follower pulses separately from the time series plots (Kariyappa et al. 1994, Kariyappa 1994) for all the 3 classes of bright points. In addition, we estimated the average peak intensity value of H2V of all the main and follower pulses for all the 3 classes of bright points. We have plotted in Fig. 1 the intensity of the main pulse and an average intensity of the main and follower pulses against the period of intensity oscillations, respectively.

We have examined the behavior of intensity oscillations at the sites of the bright points for the 35-min duration and we find that the amplitude of the intensity of the main and follower pulses of variety of bright points decay exponentially with time. We derived the decay rate from the exponential curves and plotted in Fig. 2, the decay rate versus the intensity of the main pulse and an average intensity of the main and follower pulses respectively. It is clearly seen from Fig.1 that the period of intensity oscillations seen at the sites of a variety of bright points in the interior of supergranulation cells is *independent* of their intensity enhancements.

In addition, the amplitudes of the main and follower pulses of a variety of bright points decay exponentially with time. Fig.2 clearly shows that the decay rate is also *constant* with their brightness. In other words, the wave period and the intensity amplitude are *uncorrelated* and the decay rate of the intensity is exponential and is *independent* of the intensity amplitude. We find evidence for a of constant period of the oscillations in bright points with their brightness and the different period from network oscillations. This suggests that the heating mechanism may be identical (by 3-min wave period) for all classes of bright points and it may be an entirely different mechanism (5-7 min wave period) in the case of network elements. This result will answer the question raised

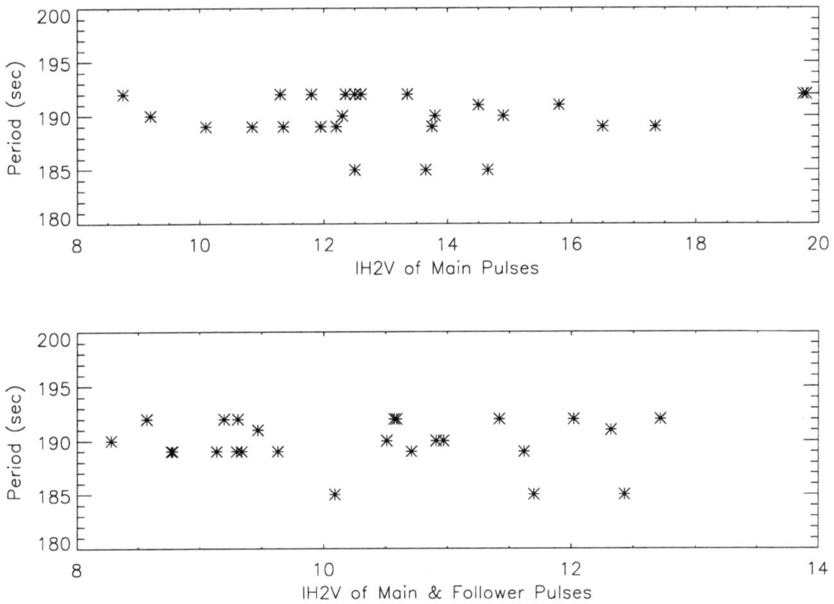

Figure 1. Variation of the period of intensity oscillations with the peak value of I_{H2V} of the main pulses (top) and the average peak value of I_{H2V} of the main and follower pulses (bottom) of all the classes of bright points. The correlation coefficient (r) is 0.12.

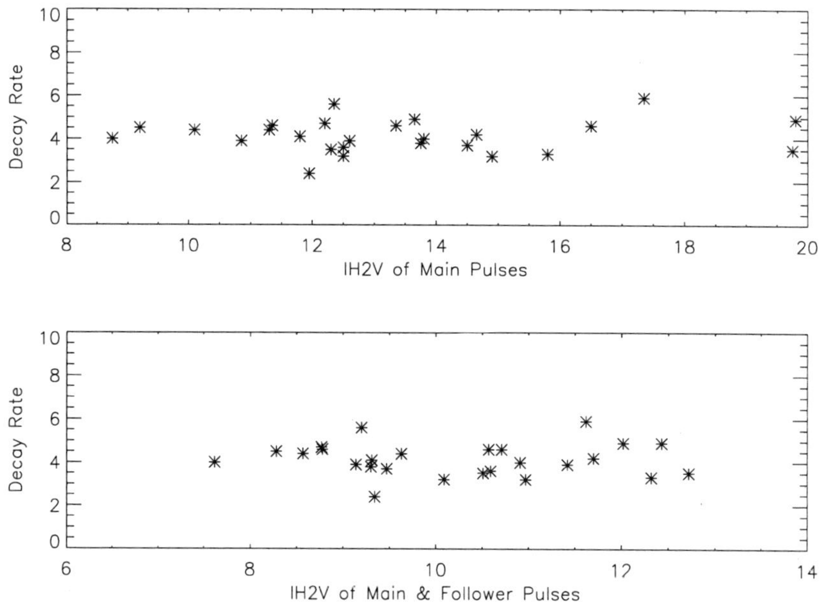

Figure 2. Variation of the decay rate in amplitudes of the pulses with the peak value of I_{H2V} of the main pulses (top) and the average peak value of I_{H2V} of the main and follower pulses (bottom) of all the classes of bright points. The correlation coefficient (r) is 0.05. The y-axis values will be multiple of 10^{-4} in % of the continuum per sec.

by Kalkofen (1989), whether the layers in the network elements and the bright points are heated in the same way. The physical process behind the cause for the bright points may be due to the interference of these 3-min period waves, and the 3-min bright points may be magnetic in origin.

4. Acknowledgments

I wish to thank Drs. J. M. Beckers, Raymond Smartt and Sivaraman for the data used in this study. I am grateful to Professor R. Cowsik, Director, Indian Institute of Astrophysics and the SOC Chair and members of the 19th NSO Summer Workshop for the financial support to attend the meeting.

References

Beckers, J.M. et al. 1972, Solar Phys. 25, 81

Biermann, l. 1948, Zs. F. Ap. 25, 161

Cram, L.E., & Dame, L. 1983. ApJ 272, 355

Foukal, P., & Lean, J. 1988, ApJ 328, 347

Jensen, E., & Orrall, F. Q. 1963, ApJ 138, 252

Kalkofen, W. 1989, ApJ 346, L37

Kariyappa, R. 1993, Ph.D. Thesis, Bangalore University, India

Kariyappa, R., Sivaraman, K. R., & Anandaram, M. N. 1994, Solar Phys. 151, 243

Kariyappa, R., & Sivaraman, K. R. 1994, Solar Phys. 152, 139

Kariyappa, R. 1994, Solar Phys. 154, 19

Kariyappa, R. 1996, Solar Phys. 165, 211

Kariyappa, R., & Pap, J. M. 1996, Solar Phys. 167, 115 (Paper I)

Kariyappa, R., et al. 1995 in J.T. Hoeksema, V. Domingo, B. Fleck, & B. Battrick (eds.), 4th SOHO Workshop on Helioseismology ESA SP-376, 2, 429

Kariyappa, R., & Kneer, F. 1998, under revision

Kariyappa, R. 1998, Sol. Phys. (Paper II), under revision

Lean, J. L. 1987, J. Geophys. Res. 92, 839

Leighton, R. b. 1961, in R. N. Thomas (ed.), 'Aerodynamic Phenomena in Stellar Atmosphere', *IAU Symp.* 12, 321

Liu, S.Y. 1974, ApJ 189, 359

Nindos, A. & Zirin, H. 1998, Solar Phys. 179, 253

Punetha, L.M. 1974, Bull. Astron. Inst. Czechoslovakia 25, 207

Schatzman, E. 1949, Ann. Astrophys. 12, 203

Schwarchild, M. 1948, ApJ 107, 1

Sivaraman, K.R., & Livingston, W.C. 1982, Solar Phys. 20, 227

Skumanich, A., Smythe, C., & Frazier, E.N. 1975, ApJ 200, 747

High Resolution Solar Physics: Theory, Observations, and Techniques
ASP Conference Series, Vol. 183, 1999
T.R. Rimmele, K. S. Balasubramaniam, R.R. Radick, eds.

$v - v$ Phase Relations in the Vicinity of the Transition Region Derived from SUMER Observations

S. Steffens, F. L. Deubner

Astronomisches Institut, Universität Würzburg, Am Hubland, D-97074, Würzburg, Germany

Abstract. We analyse velocity oscillations of 15 spectral lines observed with SUMER during three observational campaigns. The Doppler shifts of different lines were partly recorded simultaneously and spatially co-aligned. The $v - v$ phase between those lines is studied. We find that the chromospheric 3-minute oscillations increase with respect to the 5-minute oscillations with increasing formation height until a maximum is reached in lines formed at about 50 000 K. Above that height neither a 5-minute, nor a 3-minute peak can be distictly seen in the power spectra. The $v - v$ phase difference between all the doppler core displacements of the line pairs stay close to zero degrees at all resolved frequencies. Both power and phase spectra support the assumption of a standing wave field dominating the oscillation characteristics in the high chromosphere.

1. Introduction and Observations

The diagnostics of the chromosphere has suffered in the past from the small number of spectral lines, observable with ground-based telescopes. Chromospheric oscillations are still not well understood. On the one hand, the best simulations of the dynamics in this layer (Carlsson & Stein 1997, 1998) can reproduce, in 1-D, the shape of the prominent Ca II K-grains by modelling running waves. On the other hand, some observational facts like the existence of phase jumps (Staiger, Mattig, Schmeider, & Deubner 1984; Deubner, Waldschik, & Steffens 1996) and the $v - v$-phases of 0° between high chromospheric lines (e.g. Mein & Mein 1976; Mein 1977; Provost & Mein 1979; Lites 1994; Fleck et al. 1994; Bocialini & Koutchmy 1995; Hofmann, Deubner, & Fleck 1995), are not at all evident within these simulations.

We contribute here an analysis of EUV lines formed in the vicinity of the transition region, observed with SUMER onboard SOHO at the disk center of the quiet sun. The data set comprises 15 spectral lines observed during three observational runs in June and September 1996 and in May 1997. With total durations of the time series ranging from 5550 to 29070 seconds, a temporal resolution between 7.5 and 30 seconds was achieved. The usable slit lenght of the spectrograph variied between 112 and 320 arcsec with a spatial resolution of about 1 arcsec. A description of the SUMER instrument is given by Wilhelm et al. (1995).

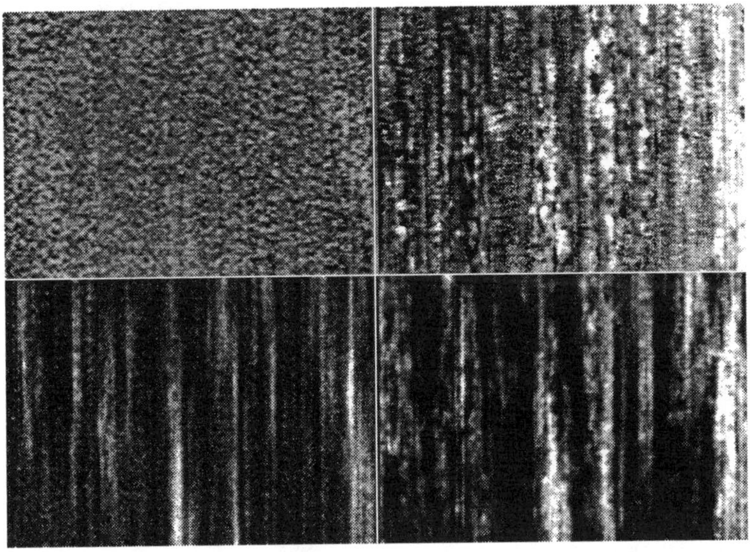

Figure 1. *xt*-diagrams of Si I 1256 (on the left) and O V 630 (on the right) simultaneously observed with SUMER in June 1996. These *xt*-diagrams show 78 minutes (vertical) and 290 arcsec (horizontal). The velocity (upper panel) ranges between ± 10 km/s. The intensity reaches $0.2 < \Delta I/I < 2.0$.

2. Results

The EUV-lines seem to split into two groups. One of them displays a regular wave field pattern with oscillations of a moderate amplitude; the other group – the lines of which originate higher up in the atmosphere – shows a quite eruptive behavior with irregular oscillations of higher amplitude concentrated towards the cell boundary of the chromospheric network. Fig. 1 shows Si I as an example of the regular type and O V as an example of the eruptive type. In general, the rms-values increase with formation height of the lines (Fig.2). With regard to the changed oscillatory pattern observed there, this fact can not be taken as evidence for a steepening of the 5-minute wave field in the higher layers. Shock waves seen in lines of the eruptive type seem not to be connected to the wave field in the lower layers. Corresponding to the different appearance of the *xt*-diagrams, the power spectra obtained with the two types of lines display a different behaviour: Spectral lines formed in relatively low atmospheric layers display a 5-minute-peak and a 3-minute-peak that gains amplitude with respect to the 5-minute peak with increasing formation height of the line. In He I 584 both peaks are about equally pronounced. This level marks the maximum of the 3-minute peak. Spectral lines formed above the He I line (i.e. at temperatures

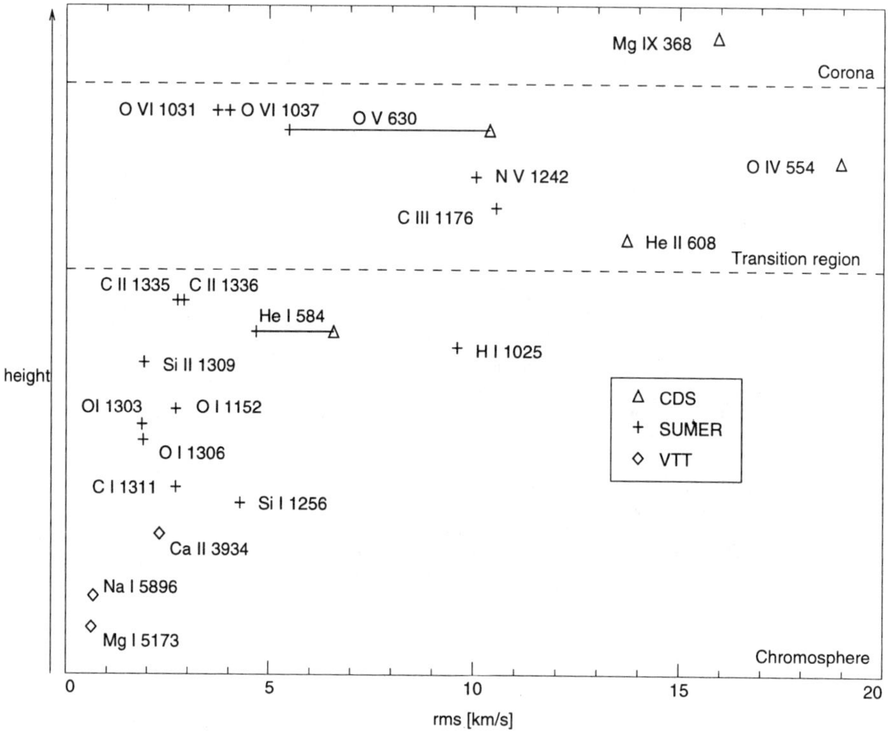

Figure 2. rms-values of the Doppler core displacements of all spectral lines observed in the 96/97 campaign. The rms-values of some lines observed with the VTT (Izaña) (diamonds) and CDS (triangles) are also shown. The lines are presented in rough order of formation height (heights of EUV lines after Wilhelm et al. (1995)).

higher than about 50 000 K) display power spectra in which oscillations at periods of some minutes can only weakly be seen in the velocity signal (Figs. 3 and 4), and not at all in the intensity (not shown here). The phase difference between the EUV lines observed with SUMER stays close to 0° (Figs. 5 and 6) and thus confirms earlier results from ground-based observations of spectral lines formed in the high chromosphere.

3. Discussion

Probably the highest layer in the solar atmosphere observable with ground based telescopes is the layer where the He 10830 line is formed. Fleck et al. (1994) report that the oscillations observed in this spectral line display a rms-value lower than those of lines formed in lower layers. In contradiction to this result, we find that the rms-values of the EUV lines formed in the high chromosphere and above generally continue increasing with formation height. The oscillations connected with these rms-values correspond to power spectra with peaks at 5

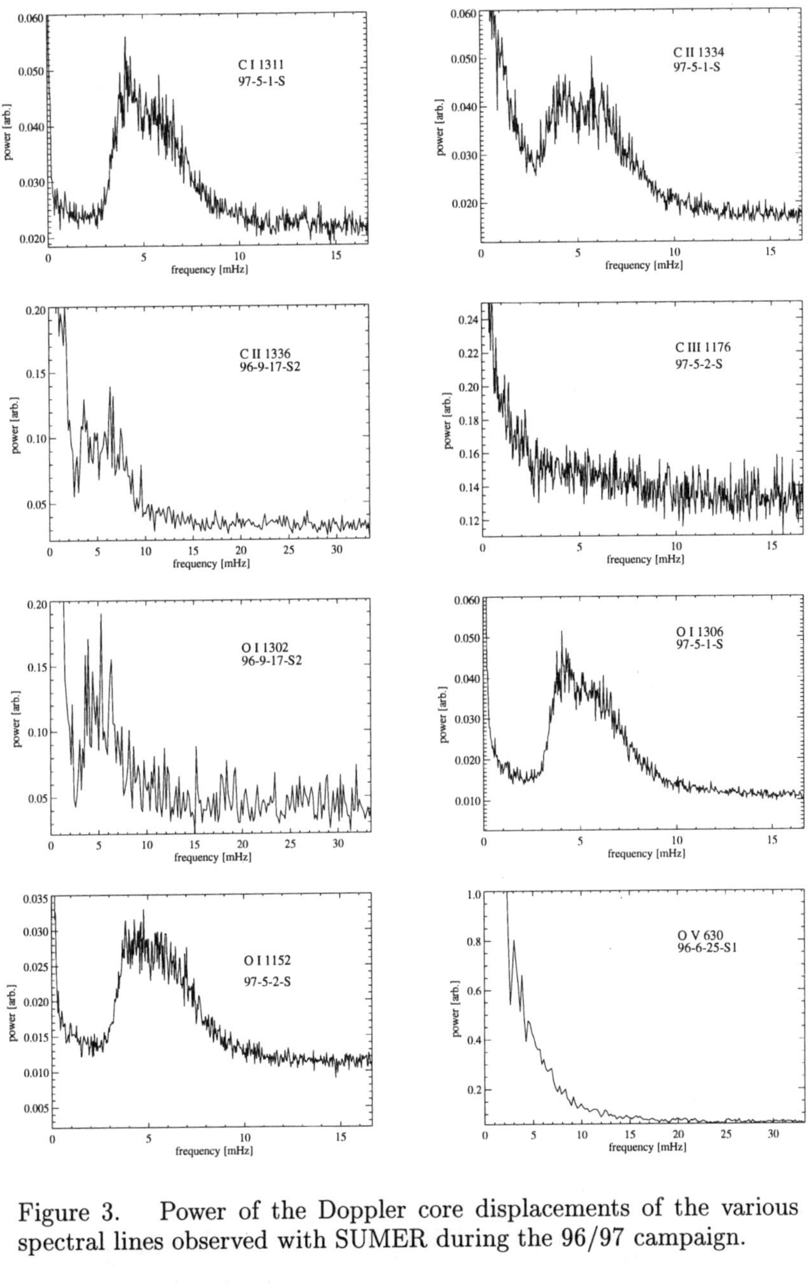

Figure 3. Power of the Doppler core displacements of the various spectral lines observed with SUMER during the 96/97 campaign.

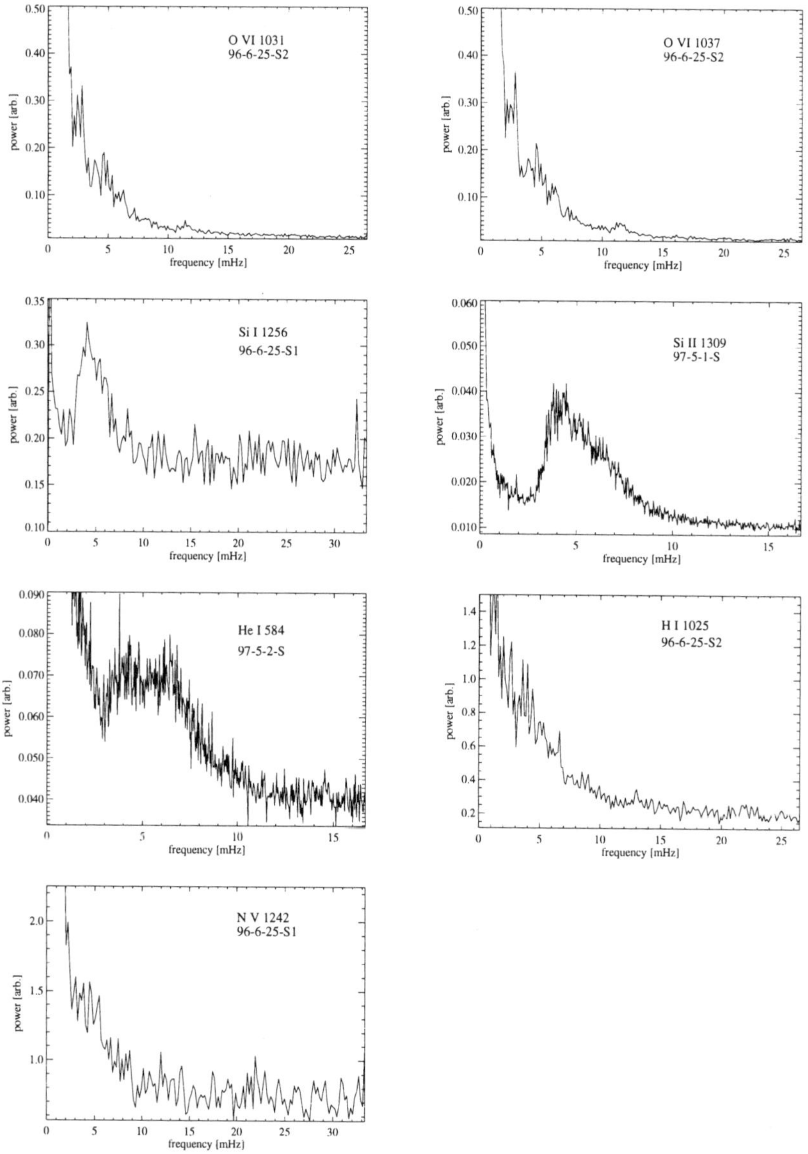

Figure 4. Power of the Doppler core displacements of the various spectral lines observed with SUMER during the 96/97 campaign.

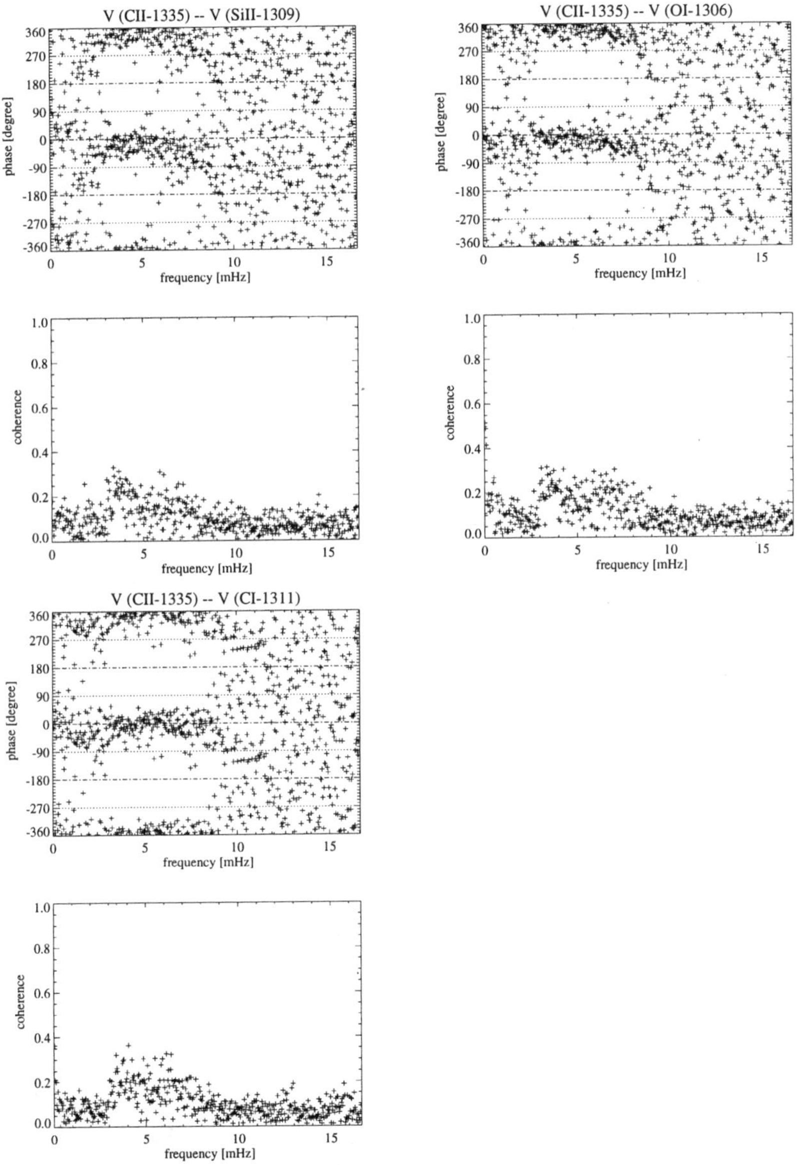

Figure 5. $v-v$ Phase differences between Doppler core displacement of C II 1335 and Si II 1309, O I 1306 and C I 1311 together with corresponding coherence spectra.

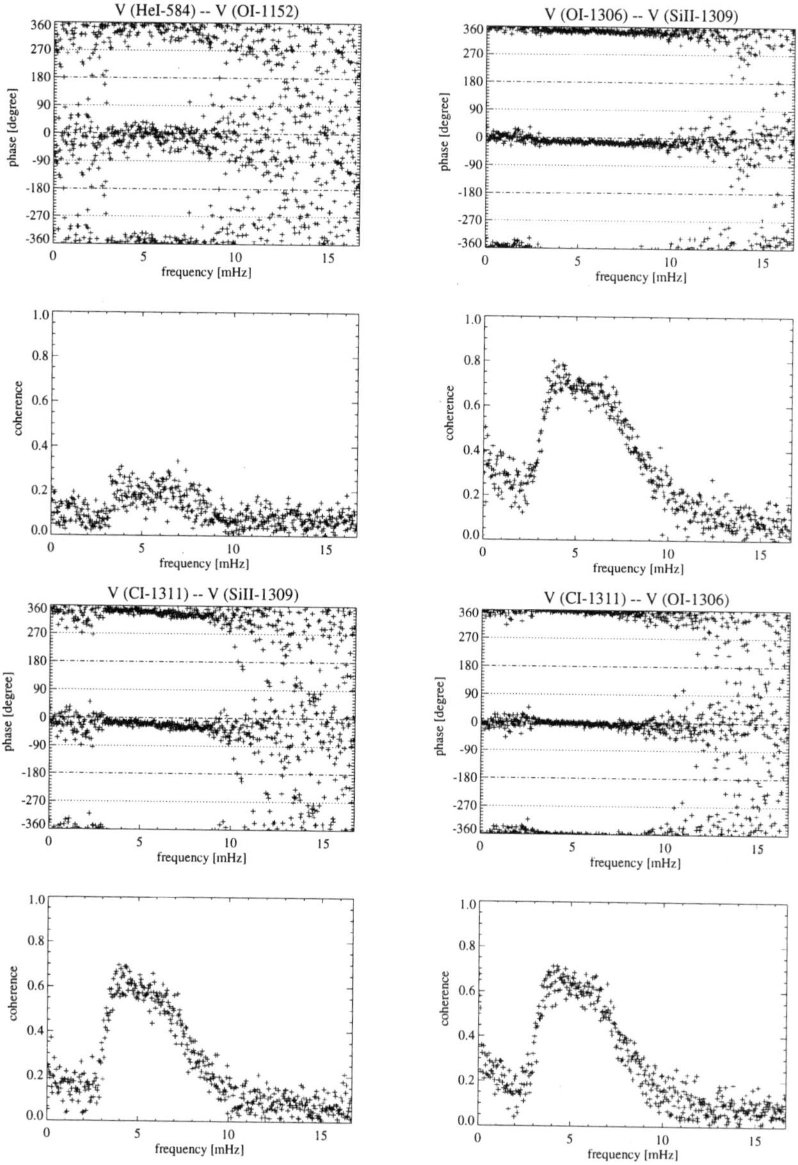

Figure 6. $v-v$ Phase differences between Doppler core displacement of He I 584 and O I 1152, O I 1306 and Si II 1309, C I 1311 and Si II 1309, C I 1311 and O I 1306 together with corresponding coherence spectra.

and 3 minutes. Especially the amplitudes of the 3-minute wavefield increase until the formation height of He 584 is reached. Above that height, we find still higher rms-values, but the fluctuations responsible for them are not related to a regular wave field anymore, but to irregular eruptive events. For all observed line-pairs the $v - v$ phase difference stays close to zero degrees. Mainly this result, but also the terminus of the regular 5- and 3-minute wave field in a certain layer, supports the assumption that the transition layer acts as a reflecting barrier for the internetwork waves. Further hints pointing in that directions are the phase jumps observed by Staiger et al. (1984) and Deubner et al. (1996) as well as the reflection ridge in the time-distance analysis of Jefferies et al. (1997). We conclude that the simulations of Carlsson & Stein (1997, 1998), though successfull in modelling one aspect of the chromosphere – the K-grains –, need to be revised with respect to their transmitting boundary condition in order to better reproduce the observational results. These seem to suggest a standing wave field dominating the higher layers of the chromosphere.

Acknowledgments. We gratefully acknowledge the help of the SOHO and SUMER teams and the financial support of the DFG.

References

Boccialini, K., Koutchmy, S. 1995, in: *4th SOHO Workshop on Helioseismology* (eds. J. Hoeksema, et al.), vol. 2, p. 499

Carlsson, M., Stein, R. 1997, ApJ 481, 500

Carlsson, M., Stein, R. 1998, in *IAU Symposium 185: New Eyes to See Inside The Sun and Stars* (eds. F.-L. Deubner, et al.), p. 435

Deubner, F.-L., Waldschik, T., Steffens, S. 1996, A&A 307, 936

Fleck, B., Deubner, F.-L., Hofmann, J., Steffens, S. 1994 in: *Proceedings of the Oslo mini-workshop on Chromospheric Dynamics, Inst. of Theor. Astrophys., Uni Oslo, Norway* (ed. M. Carlsson), p. 103

Hofmann, J., Deubner, F.-L., Fleck, B. 1995, in: *GONG 1994: Helio and Asteroseismology from the Earth and Space, ASP Conf. Ser.* (eds. R. Ulrich, et al.), vol. 76, p. 342

Jefferies, S., Osaki, Y., Shibahashi, H., Harvey, J., D'Silva, S., Duvall, T.J. 1997, ApJ 485, L49

Lites, B. 1994, in: *Proceedings of the Oslo mini-workshop on Chromospheric Dynamics, Inst. of Theor. Astrophys., Uni Oslo, Norway* (ed. M. Carlsson), p. 1

Mein, N. 1977, Solar Phys. 52, 283

Mein, N., Mein, P. 1976, Solar Phys. 49, 231

Provost, J., Mein, N. 1979, Solar Phys. 64, 43

Staiger, J., Mattig, W., Schmiede, B., Deubner, F.-L. 1984, Mem. S.A. It. 55, 147

Wilhelm, K., Curdt, W., Marsch, E., Sch"uhle, U., Lemair, P., Gabriel, A., Vial, J.C., Grewing, M., Huber, M., Jordan, S., Poland, A., Thomas, R., Kühne, M., Timothy, J., Hassler, D., Siegmund, O. 1995, Solar Phys. 162, 189

High Resolution Solar Physics: Theory, Observations, and Techniques
ASP Conference Series, Vol. 183, 1999
T.R. Rimmele, K. S. Balasubramaniam, R.R. Radick, eds.

Long Lasting Fine Structures in Type IV Radio Bursts and Coronal Effects: Spectrographic, Polarimetric, and Heliographic Observations

G. P. Chernov

IZMIRAN, 142092, Troitsk, Moscow region, Russia

M. Poquerusse and J.L. Bougeret

DESPA, Observatoire de Paris, Meudon, 92195, France

P. Zlobec

Trieste Astronomical Observatory, 34131, Trieste, Italy

Abstract. We present observations of spectra (IZMIRAN, ARTEMIS), polarization (Trieste AO), and spatial positions (Nançay Radioheliograph) of zebra-patterns in three long lasting type IV events (1989 03 12, 1990 06 05, and 1992 02 17) that are characterized by the absence of microwave bursts and big flares, and two other phenomena (1990 06 11 and 1994 10 25) formed by type II bursts and a rather short type IV continuum that are closely connected with $H\alpha$ flares and microwave bursts.

All those events were connected with active filaments. In particular the first one happened just during the 3- hour interval between two CMEs starting at different positions. As an example of more complex events, including an impulsive as well as a long lasting type IV burst, the event 1997 11 04 is considered, using Beijing and Culgoora dynamical spectra and interplanetary WIND/WAVES and Ulysses/URAP data. A general peculiarity of fine structures in the three long-lasting events is the long term evolution of zebra-patterns and pulsations (2-3 hours), that testifies repeated acceleration of fast particles in vertical current sheets during the restoration of magnetic arches after the escape of the CMEs. The magnetic configurations in the sites of impulsive events must be rather closed-to-oblique flare current sheets. Complex events can have two-step evolution: the first one due to a reconnection in flare low loops and the second one due to the reconnection high in the corona. A common observational fact in the zebra patterns of all events was discovered: between the source positions of zebra-lines in emission and those in absorption a displacement exists that is always about 30,000-70,000 km wide. This is a very important fact that must be considered in preparing a model of the radio emission mechanism.

1. Introduction

Generally it is considered that strong radio bursts of type II+IV include a big chromospheric $H\alpha$ flare, a shock wave and a long lasting stationary source (closed magnetic loop or trap) of radio emission that can present different types of fine structures (Krüger, 1979). The totality of the observations here considered made with the best solar instruments testifies a rather different scenario: the most long lasting events, showing fine structures of different type and duration in emission and in absorption (in particular the zebra-patterns in the dynamical spectrum) during some hours, are accompanied neither by big flares nor by microwave bursts. Additional data in SXR (Yohkoh/SXT data) help us to understand the magnetic structure in such events.

2. Observations

We present observations of spectra (IZMIRAN, ARTEMIS and Tremsdorf), polarization (Trieste AO), and spatial positions (Nançay Radioheliograph) of zebra-patterns in three long lasting type IV events (1989 03 12, 1990 06 05, and 1992 02 17), all developed in absence of microwave bursts and big flares, and in two shorter events (1990 06 11 and 1994 10 25), closely connected with $H\alpha$ flares and microwave bursts.

Some of the best European instruments were used to get more complete radio data, including the spectrographs:

1. *ARTEMIS* (Nançay, France) — digital recording in the range 100–500 MHz and time resolution of 0.01 sec ;

2. *IZMIRAN* (Troitsk, Russia) — analog recording in the range 180–270 MHz, with high frequency and time resolutions (0.2 MHz, 0.02 sec) in the domain 228–264 MHz; plus some fixed-frequency radiometers, including a 10-cm receiver.

The source positions and polarization were measured with the Nançay radioheliograph (NRH) at frequencies 164, 237, 327, and 407 MHz, with time resolution of 0.25 sec (Radioheliograph group 1989).

Circularly polarized left (L) and right (R) emission at higher time resolution were recorded by the radiopolarimeter at the observing station of the Trieste Astronomical Observatory (Italy) at the frequencies of 237, 327, 408 and 610 MHz with high time resolution (0.02 sec). We used also accessible SXR data (Yohkoh/SXT) to study the dynamical processes during the events considered.

For one event we used also dynamical spectra from Beijing AO and Culgoora station and interplanetary WIND/WAVES and Ulysses/URAP data.

3. Short Description of Events

3.1. 1990 06 05

The radio event occurred from 09:35 to 11:50 UT. The source was located at $N22^o E08^o$ near the small AR NOAA 6086. In this AR a faint subflare in $H\alpha$

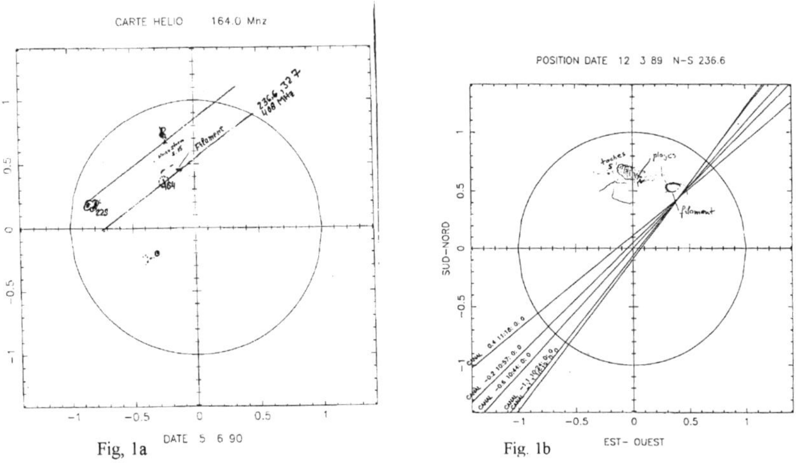

Fig, la Fig. lb

Figure 1. Positions of radio sources of fine structures, close (in projection) to active filaments (data of Nançay Radioheliograph; a: in the event 1990 06 05 at frequencies 164, 236.6, 327 and 408 MHz; b: in the event 1989 03 12, where a two-dimensional position was constructed using the crossing of one dimensional scans of NRH at 236.6 MHz during the event.

started about one hour after the radio brightening, i.e. from 10:49 to 11:03 UT. It therefore appears that the radio type IV event was related to a slowly evolving faint soft X-ray brightening (GOES class B8) with maximum at 11 UT (SGD 552, I), possibly after a filament activation and an eventual eruption. However the radio event was not initiated by a conspicuous flare (see more details in Chernov et al. 1994). The positions of radio sources at some frequencies are shown at Fig 1a.

3.2. 1989 03 12

The fine structures were observed with the IZMIRAN spectrographs between 09:00 and 12:00 UT. No impulsive phase was reported. The radio event started after the end of the 2B $H\alpha$ flare located in $N28^o E05^o$ (according to SGD N536I). The maximum of the $H\alpha$ flare (at 08:20) almost coincided with a small maximum M6.7 of the GOES-7 X-Ray detector (at 08:28) and with a weak maximimum of 313 SFU at about 08:30, as part of a multicomponent microwave burst at 10 cm (IZMIRAN). The strong continuum source was located near AR NOAA 5395 whereas the source of fine structures was far from it and located (in projection) above a dark filament (in $H\alpha$ images). We constructed a two-dimensional position of the source for Fig 1b, using the crossing of one dimensional scans of NRH at 236.6 MHz during the event. Two CMEs were observed during that day (according the SMM catalog of St. Cyr & Burkepile, 1990), one before the presence of our event above the east limb, and the other started during the lifetime of the event (11:49 UT) just above the position of the radio fine structures.

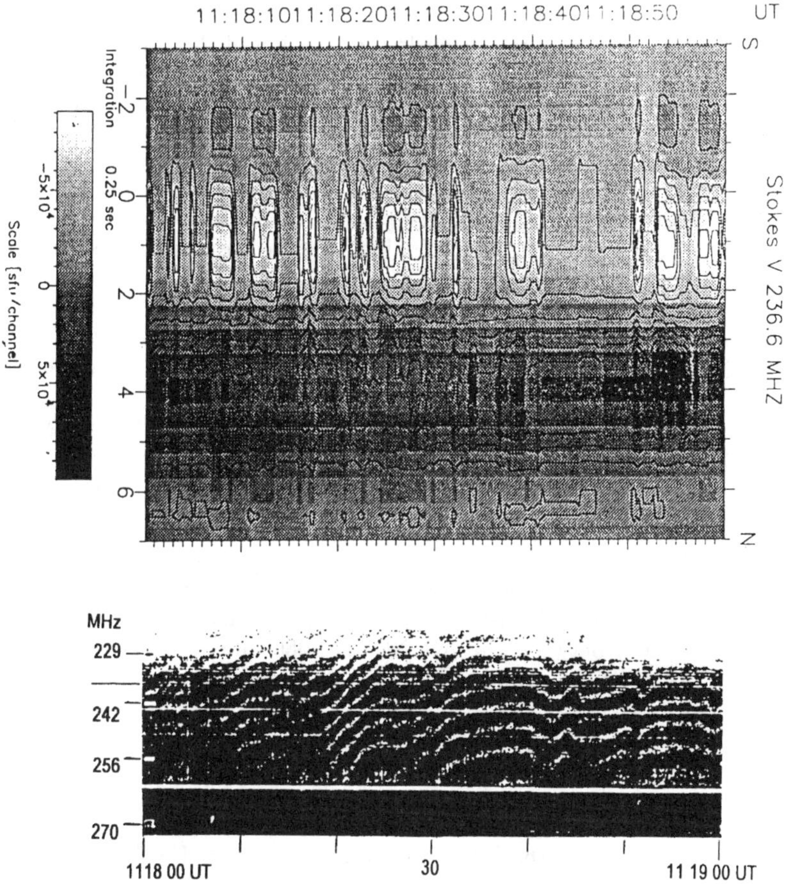

Fig. 2

Figure 2. Fig.2. Source positions and dynamic spectrum of zebra-pattern in the event 1989 03 12. The upper panel represents gray-scale plots of the one-dimensional (North-South) brightness distribution in polarization (Stokes V) measured by NRH at the frequency 236.6 MHz during one minute and the lower panel - the dynamic spectrum obtained with the IZMIRAN spectrograph. The vertical axis (channel number of NRH) give the coordinate along the terrestrial N-S direction (0 indicates the center of the solar disk, one channel have angular width of about 2.2' at 236.6 MHz).

Some details of the fine structure In Fig 2 the peculiarities of the fine structures are visible more clearly. All sources of zebra-lines are visible in white (R-polarized) at about the position of channel number -1; the strong continuum source is located at channel number +3. It appears in black (L- polarized) with numerous pulsating sources, chaotically shifted from the center of the continuum source predominantly directed towards the position of the zebra-patterns. The positions of the zebras in emission remained stable during each activity series, while the position of the zebras in absorption (remnant continuum emission, clear-gray at Stokes V plots or black lines between white lines in the dynamical spectrum) are always displaced with respect to the sources of the emission lines (in different directions). According to a more precise profile evaluation of the source positions, the averaged displacement value is about 0.5 channel, i.e. about $50 \cdot 10^8$ cm. This property remained distinctive during the whole event, i.e. about 3 hours.

However sources of zebra-patterns were not deconvolved completely by the Nançay radioheliograph (the source size is similar to the FWHM of the beam at 236,6 MHz), and therefore it is difficult to compare the source dimensions in emission and absorption. The great variations of the bandwidth of zebra-lines and the frequency separation between them during the event gives evidence for great dynamic changes of the magnetic structure in the radio source.

3.3. 1992 02 17

During this day a very intense and complex type IV event belonging to the type with weak $H\alpha$ and microwave activity was observed in the time interval 08-12 UT. A relationship should exist with the presence of active filaments. The radio source was located between the AR NOAA 7056 and 7058. Imaging observations by Yohkoh/SXT (Tsuneta 1991) show some variability in the brightest compact loop in NOAA 7058 during the orbit from 1045 UT to 1131 UT, while NOAA 7056 contains coronal structures on larger scales (Fig 3a). Three small H_α flares (importance SF C 4.1, 1N, and SF C 3.6, respectively) were reported during the first half of this event between 08 and 10 UT in the active region NOAA 7058 (SGD 576 II). They were accompanied by three small X-ray enhancements (at wavelengths 1–8 Å), recorded by the GOES-7 soft X-ray detector (SGD 571 I). Also an active dark filament appeared after 1110 UT (SGD 576 II) above NOAA 7056 (at S08 W39), near the radio source position. The event that produced a wealth of fine structures for about 3 hours was described in Chernov et al. (1998).

4. Short Impulsive Events

4.1. 1994 10 25

The radio event in the metric range included a type II burst and a short lived type IV burst (flare continuum) with fine structures (pulsations and zebra-pattern) between 10:00-10:09 UT. A small microwave burst of 49 SFU (type GRF) was observed during about one hour, while a longer lasting soft X- ray burst (importance C4.7) was detected by GOES-7 $1 - 8\mathring{A}$. The $H\alpha$ flare started 09:40 (maximum 10:04, end 12:36 UT) in NOAA AR 7792 (position S09, W12) (SGD

N 608-II). An active dark filament and an arch filament system were also observed above this AR.

This event was more active in the low corona. Some details of the radio type IV burst with different source positions outside the AR were discussed by Aurass et al. (1996). The data reveal some facts regarding the dynamical behavior of the magnetic reconnection in the flare current sheet. Repeated changes of the polarization sign were manifested during the event at 327 and 408 MHz. Manoharan et al. (1996) proposed that the flare triggered a large-scale magnetic reconnection. However the URAP experiment aboard Ulysses recorded at 10:10 UT (at about 1000 kHz) the radio emission of a type III burst caused by fast electrons accelerated at the onset time of the flare. This isolated type III burst was the strongest of five events observed by the Ulysses spectrograph during that day. Spectral ARTEMIS data in the range 100-500 MHz revealed negative frequency drift (typical for type III bursts) of numerous pulsations between 10:05:18-10:10:00 UT.

4.2. 1990 06 11

Different radio bursts were observed: the type II between 09:44-09:47 UT was crossed by a strong phenomenon formed by type III and type V bursts at 09:45:12-35, and fine structures (pulsations, zebra-patterns) were recorded in a slowly rising type IV continuum up to 11:13 UT. A strong microwave burst appeared presenting fast rise at 09:43 and slow fall up to 09:55 UT (maximum at 09:44 UT with 820 SFU). The $H\alpha$ flare in the AR (position N10, W22) was of type 2B (start at 09:34, maximum at 09:48 and end at 10:59). It presented high-speed dark filaments and some eruptive centers (SGD, 556-II).

5. Example of a Complex Event: 1997 11 04

More complex and more frequent events usually contain very strong impulsive and long-lasting post-eruptive phases. The metric type II burst usually follows in time just after the microwave burst. They have a continuation in interplanetary space including a CME and interplanetary type II and type III bursts. Such complex events had been described in numerous papers, but only within the past few years, with SOHO and Yohkoh data, have we the possibility to imagine the scenario of such events in the low corona and, with Ulysses and WIND/WAVE data, in the interplanetary space. As an example of such events we could present the complex type II+IV radio burst 1997 11 04 connected with large $H\alpha$ flare 2B, X2.1 in NOAA AR 8100 (05:54-05:59:-07:15 UT).

As the main peculiarity in this event we see an almost uninterrupted continuation of metric Culgoora type II bursts, connected with the blast flare shock wave, into the interplanetary (WIND/WAVES) part.

6. Discussion and Conclusions

Long lasting events As it can be seen in Figs 1a and 1b and Figs 3a and 3b, the radio sources providing fine structures were located outside the AR. The radio source of the event 1992 02 17 cannot be connected with a large

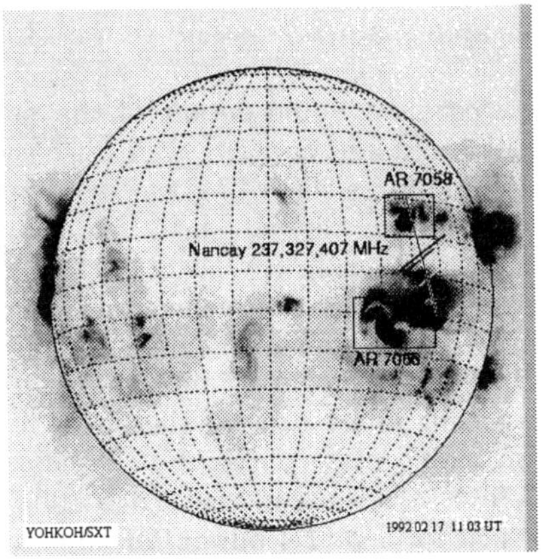

Fig. 3a

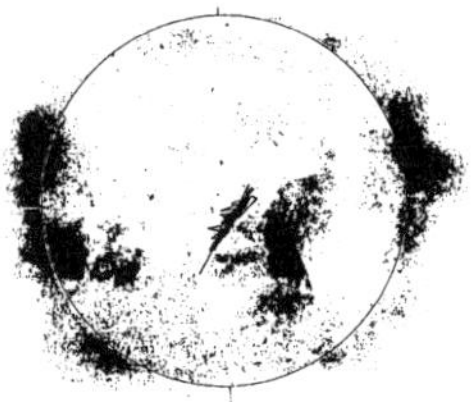

YOHKOH/SXT 1994 10 25

Fig. 3b

Figure 3. Fig.3. Projected radio positions measured with NRH overlaid on Yohkoh/SXT images (thin Al.1 filter); a: for the event 1992 02 17, crosses indicate half width of radio source (long bars at 237 MHz, short bars at 327 and 408 MHz); b: projected trajectories of zebra-lines emission (source centroid) at 164 MHz in the event 1994 10 25 (Klein & Aurass, private communication).

flare in a closed magnetic loop (magnetic trap), as it was located between ARs. Probably the source should be related to a large trans-equatorial loop structure connecting the ARs 7056 and 7058. The initial disturbance can be connected with the active filament, because during the event the polarization reversed (at 237 MHz) just at the same time as an active dark filament appeared near the southern AR, suggesting magnetic reconnection high in the corona (cf. Fig 1 in Steele and Priest, 1989). In this regard the long lasting energy release had taken place during numerous magnetic re-connections in the vertical current sheet high in the corona after a possible escape of a CME. The absence of microwave bursts in such events is explained by the dynamical processes driving fast particle acceleration that takes place high in the corona.

Similarly, the other two other long lasting events with fine structure (1990 06 05 and 1989 03 12) presented radio sources outside ARs connected with active filaments. We are confident that they are similarly disposed in vertical current sheets. Moreover in the event 1989 03 12 all fine structures had been observed, according to the catalog of St. Cyr & Burkepile, 1990, during about 3 hours between two CMEs. The continuum radio source was located above the AR, while the sources of the fine structures were outside the AR and above the active filament near the escape position of the second CME. That strengthens our idea that radio sources of fine structures are located in the vertical current sheets. The absence of strong type II bursts in all these events is a consequence of absence of a strong flare blast shock wave.

In the example of the complex event 1997 11 04 we observed really the propagation of the blast shock wave uninterruptedly into interplanetary space.

At two remote spacecraft we see different structures of low frequency bursts (below 1 MHz) that signify that WIND and Ulysses registered different bursts at the same time that may be of different origin. They are similar to SA (shock accelerated) bursts connected with CME.

Short lasting events Short lasting radio bursts with similar fine structures were closely connected with big flares, microwave bursts and strong blast shock waves. Their radio sources are probably located in the low corona where magnetic reconnections develop in flare loops. In such a case the magnetic reconnection is usually caused by a new emerging magnetic loop. The corresponding current sheet is relatively small and must be rather inclined tangently to the flare loop. The numerical simulation of similar current sheets are considered by Yokoyama & Shibata, 1996.

However, in the short event recorded during 1994 10 25, the source of fine structure was located outside the AR, as shown in the Fig 3b. Numerous short lines were formed on the Yohkoh/SXT image as projected trajectories of the pulsating zebra-stripes emission at 164 MHz. At the same time the source of the main continuum was located above the bright AR (in SXR) in the south-west direction. A more detailed analysis of Yohkoh/SXT images is given in Manoharan et al. (1996) and testifies some additional brightening outside the AR as a result of some re-connections of flare loops with neighboring loops. Thus, the remote position of the source of fine structures with respect to the continuum source in this event strengthens the notion that the impulsive energy release is in flare current sheets that might be slightly inclined.

The event recorded on 1990 06 11 presents clearly for the first time (at the frequency 236.6 MHz) a source displacement. The real source movement with the velocity of about 100 km/s (in projection) can be explained in connection with a CME. The lack of delay between the beginning of a type II burst and a microwave burst testifies a connection of the CME with a flare filament. The slow CME in this short event must be small in angular dimension and does not provoke a global reconstruction of the magnetic structure above AR. The simple behavior of the polarization of the radio emission should be related to a relatively stable magnetic configuration high in the corona above the AR.

Acknowlegements: G.C. is grateful for the support by CNRS and Paris Observatory (K.-L.Klein, J.-L.Bougeret, M.Poquerusse) that enabled him to work at Meudon, by the Russian Foundation of Basic Reserch, and by the Federal Program "Astronomy". The Yohkoh data were provided by the Yohkoh mission of ISAS, Japan, which was prepared and is operated by the international collaboration of Japanese, US and UK scientists under the support of ISAS, NASA and PPARC respectively.

References

Aurass, H., Klein, K.-L., & Martens, P.C.H. 1996, Proc. Yohkoh Conf. on Observations of Magnetic Reconnection in the Solar Atmosphere, ed. R.Bentley and J.T.Mariska v.111, 194-199.

Chernov, G.P., Klein, K.-L., Zlobec, P., & Aurass, H. 1994, Solar Phys., 155, 373.

Chernov, G.P., Markeev, A.K., Poquerusse, M., Bougeret, J.-L., Klein, K.-L., Mann, G., Aurass, H., & Aschwanden, M.J. 1998, Astron. Astrophys. 334, 314-324.

St. Cyr, O.C., & Burkepile, J.T. 1990, A catalogue of Mass Ejections observed by SMM Coronograph. Boulder.: NCAR

Krüger,A. 1979, Introduction to Solar Radio Astronomy and Radio Physics, D.Reidel Publ.Comp.

Manoharan, P.K., Van Driel-Gesztelyi, L., Pick, M., Demoulin, P. 1996, Astrophys. J. 468, L73-L76.

Radioheliograph group, 1989, Solar Phys. 120, 193.

Steele, C.D.C., and Priest, E.R. 1989, Solar Phys., 119, 157-195.

Tsuneta, S., et al. 1991, Solar Phys., 136, 37-67

Yokoyama, T., Shibata, K. 1996, Publ.Astron. Soc. Japan, 48, 353-376.

High Resolution Solar Physics: Theory, Observations, and Techniques
ASP Conference Series, Vol. 183, 1999
T.R. Rimmele, K. S. Balasubramaniam, R.R. Radick, eds.

Thermal Starting Plumes, Solar Granulation, and the Excitation of Solar Acoustic Oscillations

Mark Peter Rast

High Altitude Observatory, National Center for Atmospheric Research[1], *PO Box 3000, Boulder CO 80307 USA*

Abstract. Localized cooling events in the solar photosphere and consequent new downdraft formation are key ingredients in the dynamics of solar granulation, and granular evolution can be described as an advection-fragmentation process by considering the downflowing thermal starting-plume as a fundamental dynamical unit. Despite their dynamical importance, stability studies of compressible downflowing plumes suggest limited vertical coherence, implying that granulation may be a relatively shallow phenomenon. Nevertheless, granular downflows appear to play a significant role in the excitation of solar acoustic oscillations, and downflow plume initiation may be an important acoustic source.

1. Solar Granulation

Solar granulation is driven by strong radiative losses in the photosphere. The drop in radiative opacity there results in a rapid transition from nearly adiabatic convective heat transport to nearly unimpeded free streaming of radiation. The essential energy balance maintained by convection is that between advection of heat from below and radiative cooling from above. A simple balancing of the solar luminosity by vertical advection of enthalpy yields an estimated ascent velocity of about 2 km/s to sustain radiative losses (Nordlund 1985). Local reduction in vertical advection can lead to heat loss rates in excess of supply, localized cooling, and new downflow plume formation. It is this sensitivity to radiative loss which leads to granule fragmentation and governs granule evolution.

High spatial resolution observations of both intensity (e.g. de Boer et al. 1992, Hirzberger et al. 1997) and Doppler velocity (e.g. Nesis et al. 1992, 1993, 1997) indicate that regions of maximum granular upflow lie not in the centers of granules, but immediately adjacent to the intergranular downflow lanes. Additionally, many granules evolve by expansion, central darkening, and fragmentation along radially-directed lanes (e.g. Rösch 1960, Mehltretter 1978, Kitai & Kawaguchi 1979, Kawaguchi 1980, Namba 1986, Hirzberger et al. 1999). While the fragmentation process occurs particularly rapidly and vigorously during the phenomenon known as an 'exploding' granule, the scenario is common to much less vigorous fragmentations.

Compressible plume simulations (Rast 1998) suggest that both the observed granular flow topology and the process of granule fragmentation may be understood in terms of the properties of strongly driven compressible downflow plumes

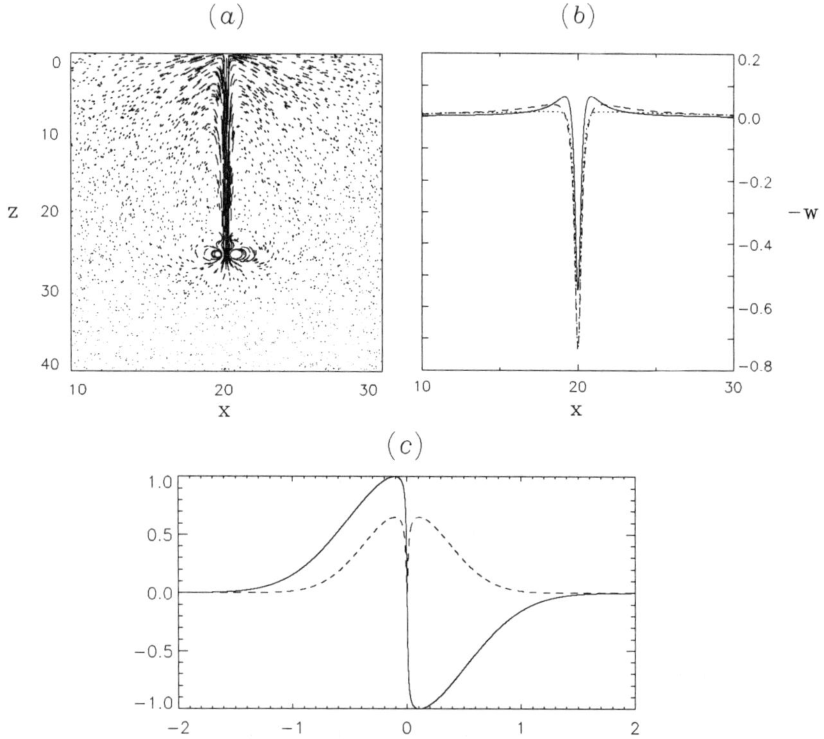

Figure 1. (*a*) The instantaneous vector flow field of a cool thermal plume buoyantly driven through an (initially) adiabatically stratified layer, and (*b*) horizontal cuts of the vertical velocity, $-w$ (upward plotted positive), as a function of horizontal position x at depths corresponding to 0.25, 1.0, and 2.5 density scale heights (units scaled as in text). (*c*) Analytic approximations to the horizontal (*solid*) and vertical (*dashed*) fluid velocities in the upper layers of the single plume simulation as a function of distance from plume center (units arbitrary).

(Rast 1995). Fig. 1a displays the instantaneous vector flow field of a plume descending through an initially adiabatically stratified layer. The plume emerges from the upper boundary layer due to an applied cool Gaussian temperature perturbation. Distances in the figure are scaled by the width of that perturbation and velocities are scaled by the isothermal sound speed at the top of the layer. Horizontal cross sections of vertical velocity through the plume at three depths, $N_\rho = 0.25$, $N_\rho = 1.0$, and $N_\rho = 2.5$, where N_ρ is the number of density scale-heights down from the top of the domain and corresponding to $z = 0.75$, $z = 3.80$, and $z = 17.2$ respectively, are plotted in Fig. 1b. At all depths, but particularly in the upper layers, the upflow velocity peaks immediately adjacent to the downflow site. The fluid in the neighborhood of the plume is accelerated

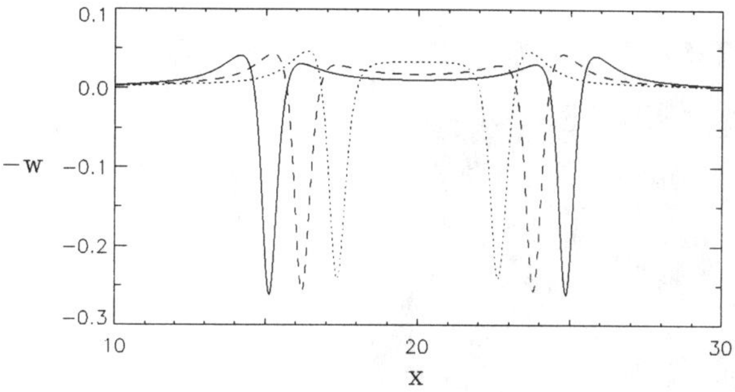

Figure 2. The vertical velocity, $-w$ (upward plotted positive), as a function of horizontal position x at a depth $z = 0.5$ for three realizations of increasing plume separation and equal plume strength.

upward and inward by local buoyancy forces and vertical pressure gradients induced in the vicinity of the downflow plume.

Given such a single-plume flow-structure, granule fragmentation can be interpreted as follows. Darkening and new downflow plume formation in the center of a granule results when the granule expands. The upflows are dynamically linked to the downflow lanes. Fig. 2 shows the vertical velocity at $z = 0.5$ in a domain similar to that of Fig. 1 but now through which two downflow plumes of equal strength are driven. Three realizations, differing only in the distance between the cool boundary perturbations applied, are shown. As the separation between the plumes increases, the vertical velocity between them decreases, leaving a central region of weak flow. As a granule expands the region of maximal upflow moves with the downflow and the central velocity weakens. The fluid in the center of an expanding granule is thus subject to radiative cooling, and ever diminishing vertical advection of heat to sustain those losses. When the central upflow velocity drops below about 2 km s^{-1} the gas cools until buoyancy forces become sufficient to initiate a new central plume. Neighboring flows then stagnate or reverse in response to pressure gradients induced by the new plume, and the cooling propagates outward from the granule center forming new intergranular lanes (Rast et al. 1993). Fragmentation occurs in those directions already predisposed to weak upflow, a predisposition determined by the distance to and strength of the intergranular downflow boundaries (Rast 1995). Granulation is thus seen as a surface phenomenon driven by radiative cooling in the photosphere. Its fundamental length scale has likely little to do with a cellular size determined by vertical stratification or the placement of ionization zones with depth, but is instead determined by the horizontal extent over which the upflows surrounding the strongly driven downflow plumes are able to advect sufficient heat vertically to maintain radiative loss.

This suggests a simplified model of solar granulation, one describing it as an advection-fragmentation process in the surface layers. Consider a collection of thermal plumes identified by position. Each of these induces in its vicinity vertical and horizontal fluid velocities of the form plotted in Fig. 1c. The fluid velocity at any point is given by the linear sum of that contributed by all the

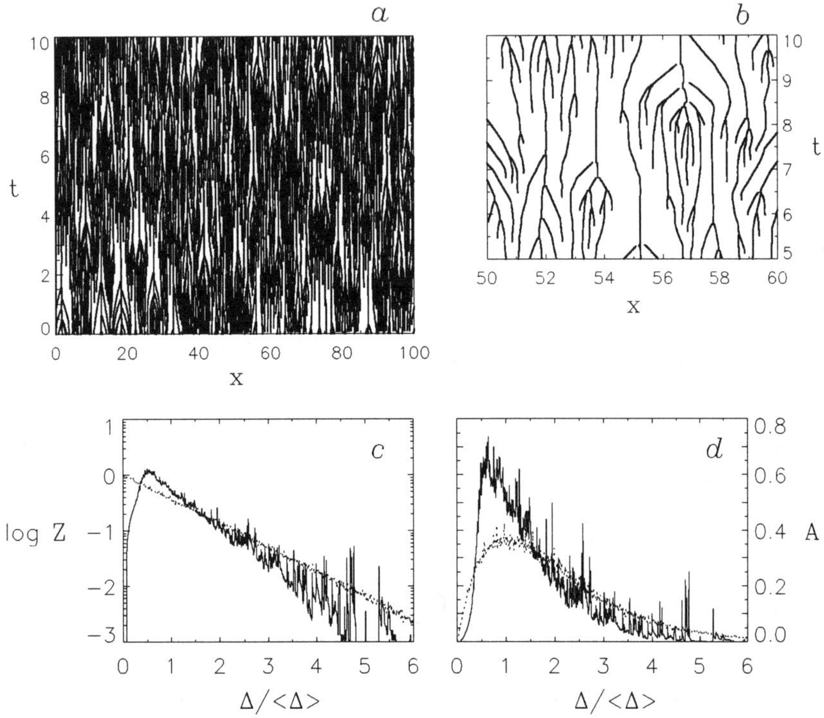

Figure 3. Simple advection-fragmentation model of granulation. In (*a*) downflow position x as a function of time t, with a subsection enlarged in (*b*). In (*c*) the logarithm of the probability distribution function Z of the nearest-neighbor plume separations in units of the mean separation $\Delta/<\Delta>$. In (*d*) the fractional "area" coverage by "granules" of each size as defined in text. Dotted curves in (*c*) and (*d*) plot the same quantities for randomly distributed downflow sites, and all distributions shown are the expectation values of many realizations.

plumes. Starting from a random initial distribution, the plume positions then evolve based on the following rules:

1. The velocity profiles shown in Fig. 1c are similar for all plumes, scaled only in height and width by the plume amplitude.
2. Plume positions are advected by the total horizontal flow at their position.
3. Merger of neighboring plumes occurs when their separation is less than the distance between the upflow maxima of either. Amplitudes are summed; the position of the merged pair is the amplitude-weighted average of the two.
4. New plumes are created when and where the total upflow velocity between plumes drops below a threshold value $w_{\rm min}$. New plume amplitudes are the average of their immediate neighbors.
5. Plume amplitudes decay toward $w_{\rm min}$ on a exponential time scale τ.

Fig. 3 illustrates the result of implementing these simple rules on a one-dimensional domain for parameter values $w_{\rm min} = 0.065$ and $\tau = 1.0$, where no attempt has been made to scale the parameters or the results to solar values. The results are qualitative at this time. In Fig. 3a plume positions are plotted along the horizontal axis with time running forward vertically. A section of the image is enlarged in Fig. 3b to show the evolution in more detail. At any one time the mean separation $<\Delta>$ between downflow plumes is about 0.5 units. A larger scale separation of stronger downflow sites also exists. They have a longer lifetime, persisting as local sites of flow convergence. It has been previously suggested that mesogranular scales may reflect the typical spacing of strong long-lived downdrafts (Rast 1995), and there is observational evidence that such downdrafts exist and may play an important role in the formation of magnetic flux tubes (Roudier et al. 1997). This highly simplified model suggests that such sites are maintained by the advection and merger of smaller plumes formed during granule fragmentation and so both granulation and mesogranulation reflect dynamics of starting plumes originating in the radiatively cooled boundary layer. The distribution of of "granular" scales produced by the model and their "areal" coverage are plotted in Fig. 3c and 3d, where the total "area" covered by granules of each size $A = Z\Delta/<\Delta>$. The size distribution has a peak at about one-half the mean size and decays exponentially to larger scales more steeply than a purely random distribution (dotted curves in Fig. 3c and 3d). In order to make a quantitative comparison of these distributions with observation it is important to extend this model to two dimensions. It will be interesting to see if such a simple model can in that case reproduce not only the size distribution of solar granulation but also the connectivity of the downflow lanes and the lineal fragmentation behavior.

2. Thermal Plume Stability

Given the importance of downflow plumes to the dynamics of solar granulation, one wonders to what depths below the solar surface they might penetrate. Two-dimensional compressible downflow starting plumes are subject to two modes of secondary instability: a symmetric pinch-like instability which detaches the plume head and a sinuous shear instability of the stem flow (Rast 1998). The first of these, illustrated by the second pair of images in Fig. 4, is unique to compressible downflowing thermal plumes and is not seen in incompressible plumes or compressible upflowing plumes. The downflow plume which emerges from the upper boundary layer first forms the familiar cap and stem structure seen in other settings (e.g. Moses et al. 1993). As the vorticity in the cap increases density fluctuations in the vortex core become negative. This is because the cap resists compression at the background adiabatic rate. With depth the integrated buoyancy force on the vortex cap is actually upward (an example of what Parker (1991) has called dynamically buoyancy). What keeps the cap propagating downward is the pressure gradient induced across it as it was spun up. The dynamical pressure increase above the cap also causes secondary head formation and the detachment of the leading cap. Consecutive pinches occur with each vortex cap weaker than the previous one and traveling downward through the layer slower than those ahead, the distance between them thus increasing (cf. Rast 1998). Note that the pinch instability is not a linear varicose mode of the stem flow. It is an intrinsically nonlinear process requiring large dynamical

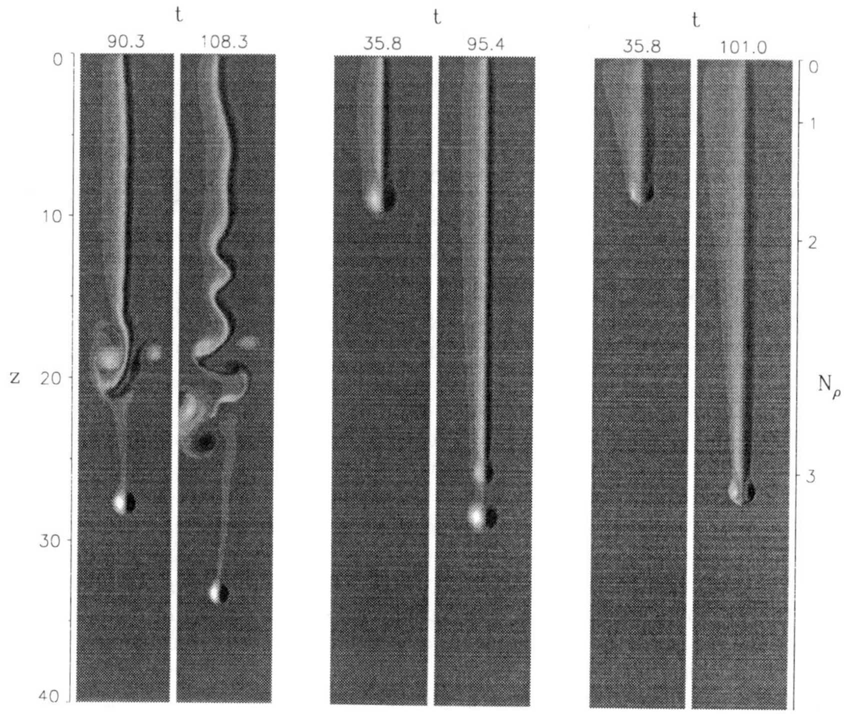

Figure 4. Horizontal vorticity snapshots from three simulations of starting plume flow. The first two image pairs illustrate stem-flow shear-instability and vortex cap detachment in two-dimensional flows. The last pair displays a cut through the center of a three-dimensional plume. In the first simulation the background was seeded with low amplitude random temperature fluctuations whereas in the the second two it is quiescent.

pressure fluctuations produced by Reynolds stresses. Whether such a pinch also occurs in three-dimensional plume flow is currently under investigation. The last pair of images in Fig. 4 displays preliminary results. While the pinch seems to be initiating, it may be somewhat delayed. Intricate structure developing within the cap makes it even more difficult to resolve the flow in three-dimensions than in two.

The second mode of instability realized in the two-dimensional simulations (not yet investigated in three-dimensions) is a linear shear-instability of the underlying stem flow. The first pair of images in Fig. 4 illustrates the result of seeding the domain with random initial low-amplitude temperature perturbations. These augment the onset of the stem shear-flow instability, which occurs first near the plume head and propagates upward along the stem disrupting it to a depth of about 1.5 – 2.0 density scale heights. The late time behavior of this two-dimensional plume is the quasi-periodic horizontal wagging of the remaining stem portion and the consequent shedding of vorticity into the domain below. Coherent penetration of the deeper layers does not reoccur after the initial stages, and the vortices shed travel along complex trajectories as they interact with those previously detached.

The importance of these secondary instabilities is in their effect on the global flow. For example, upflowing and downfowing compressible thermal plumes display significantly different dynamics (cf. Rast 1998). In an upflowing plume the pinch instability is never realized. Pressure fluctuations behind the head are weak due to the expansion of the rising vortex pair. When the stem flow becomes shear unstable the vortices shed expand and slow as they rise. Successive pairs travel faster than those ahead and ultimately penetrate and disrupt the previous ones as in the familiar leap-frogging of vortex rings seen in laboratory experiments (Yamada & Matsui 1978). Rather than the fluid moving out of the plume region, as it does in the compressible downflow plume, it is wrapped back in by these vortex interactions. The plume expands conically with height, reminiscent of incompressible turbulent plumes. Entrainment hypotheses and thermal plume scaling laws depend critically on this type of dynamics and the coherence it imposes on the plume as a whole. While such hypotheses may prove useful in describing compressible upflowing plumes in a stratified medium, as they have for incompressible plumes in a variety of settings (e.g. Turner 1986), they are likely inappropriate descriptions of compressible downflowing plume behavior. The downflowing plume in a stratified medium is fundamentally different. It sheds vorticity which leaves the plume region, descends through the medium, and remains compact due to compression with depth. Subsequent vortex interactions show little large-scale organization, and estimates of plume size, spacing, and strength at the base of the solar convection zone which depend on simple entrainment models are likely to be in error.

3. Acoustic Excitation

The initiation of downflow plumes in the solar photosphere may be an important acoustic source for both the excitation of the solar p-modes and the launching of upward propagating waves and consequent shock formation in the solar chromosphere (Skartlien 1998). The excitation process involves three plume-formation stages, radiative cooling, buoyant acceleration, and advective inflow, resulting consecutively in monopolar, dipolar, and quadrupolar emission (Rast 1999). The stages are not only spatial distinct but also temporally so. Pressure fluctuations are induced thermodynamically on the radiative cooling time-scale during the first stage and dynamically on the buoyancy and advective time-scales in the later two. The mechanism can be illustrated with the aid of a numerical experiment. A thermal starting plume is induced in an initially quiescent two dimensional atmosphere by the onset of a localized embedded heat loss. This is done by the inclusion of an internal energy loss term

$$Q \sim \left(1 + \tanh\left(\frac{t - t_0}{\sigma_1}\right)\right) \mathrm{e}^{-\sigma_2(\mathbf{x}-\mathbf{x}_0)^2} \tag{1}$$

in the energy equation, which introduces a Gaussianly distributed cooling at position $\mathbf{x} = \mathbf{x}_0$ for times $t > t_0$. Fig. 5a plots the temporal behavior of this

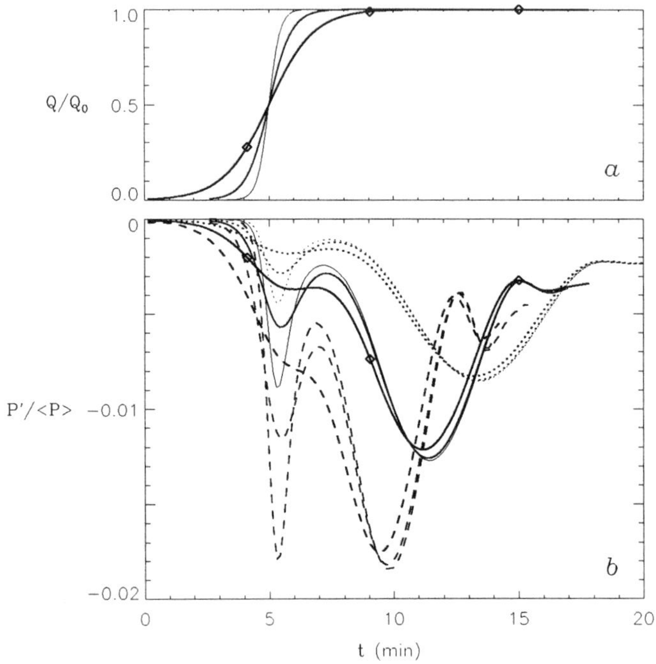

Figure 5. Normalized internal energy loss rate as a function of time (a) for onset widths of 1, 2, and 4 minutes (*thin*, *normal*, and *bold* line thicknesses respectively). Pressure perturbation response at the source location (b) for the three onset widths of (a) and three energy loss-rate amplitudes (in solar units equal to about 190, 380, and 760 erg/s/cm^3), plotted with *dotted*, *solid*, and *dashed* lines respectively.

cooling function for three onset widths, 1, 2, and 4, where time has been scaled to solar minutes. Three cooling amplitudes Q_0 were also considered, and the resulting pressure perturbations at the center of the source location are plotted for all nine combinations in Fig. 5b. Line-styles in that figure indicate cooling rate amplitudes, and within each line-style group line-thicknesses indicate temporal onset widths. Fig. 6 displays snapshots of both the instantaneous vector flow field and pressure perturbations in the vicinity of the cooling source. The times to which these correspond are indicated with diamonds in Fig. 5.

The three-stage excitation process is clearly apparent in the plots of Fig. 5 and 6. At the source location there are two stages of pressure drop followed by pressure recovery. The initial monopolar drop in pressure is the thermodynamic response of the fluid to the cooling. It leads to a rarefaction wave which propagates away from cooling site at the local sound-speed. The monopolar source term scales as $\partial Q/\partial t$ and thus the amplitude of the monopolar pressure fluctuation response in Fig. 5b depends on both the cooling amplitude and onset time scale. On the buoyancy time-scale a secondary dipolar component is induced by vertical fluid motions which dynamically compress the fluid below the initiating

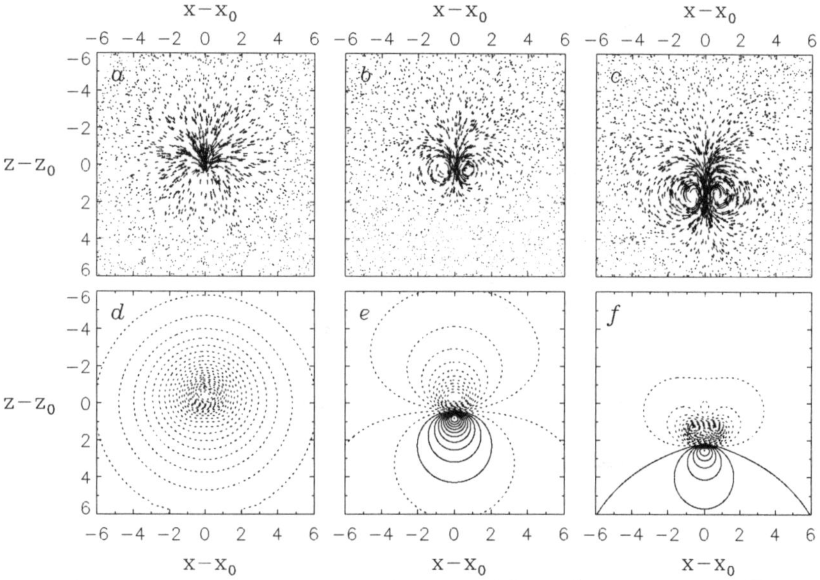

Figure 6. Instantaneous vector flow field (*top row*) and pressure perturbation contours (*bottom row*) at the times indicated by diamonds in Fig. 5. Cooling is centered at $x = x_0$ and $z = z_0$ with a spatial half-maximum width equal to one.

plume and evacuate it behind. The magnitude of this component depends on the integrated cooling and thus the dipole minima in Fig. 5b are relatively insensitive to onset time scale but reflect the amplitude Q_0. If the cooling onset time scale becomes comparable to the buoyancy time scale the monopolar and dipolar responses are mixed in the early stages, otherwise they are quite distinct. The third stage involves nonlinear pressure recovery by Reynolds stresses at the source location. Strong cooling produces a strong downflow, vigorous horizontal inflow behind the new plume, and rapid pressure recovery at the source location. Thus the total pressure pulse length in Fig. 5b depends inversely on the magnitude of Q_0. It is interesting to note that very similar time-dependence to that plotted in Fig. 5b can be seen in the continuum intensity at solar acoustic source locations (cf. Rimmele et al. 1995). The possible relationship of those observations to the source model presented here needs to be investigated.

One advantage of these simplified acoustic source simulations over those of more realistic turbulent convection is that in a large domain one can measure outward acoustic wave flux well away from the source location uncontaminated by compressible bulk fluid motions. This allows unambiguous determination of the efficiency of acoustic wave production, at least in the early stages. The acoustic efficiency of the source can be expressed as

$$\epsilon = \frac{2\pi r \int_{\frac{r}{c}}^{t+\frac{r}{c}} uP' \, dt}{\int_0^t \int Q \, d\mathbf{x} \, dt} \,, \tag{2}$$

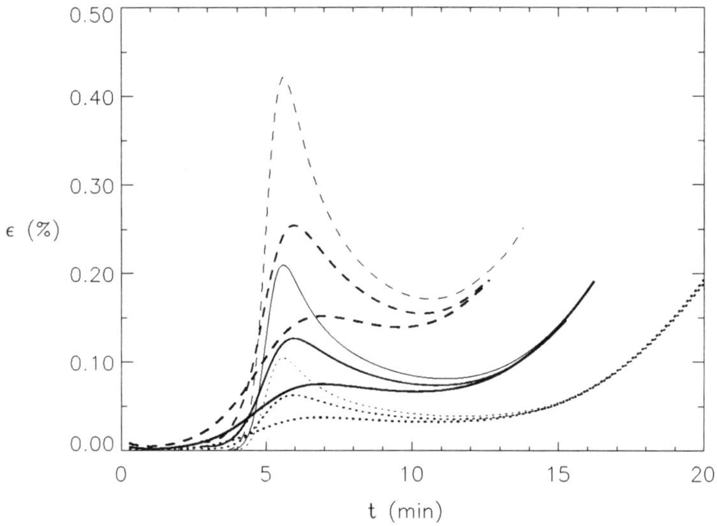

Figure 7. Efficiency of acoustic excitation by plume initiation (defined in text) as a function of time. Line-styles as in Fig. 5.

the ratio of the integrated acoustic flux at a distance r from the source to the total internal energy extracted by cooling at the source location. This ratio is plotted for all nine sources as a function of time in Fig. 7. That component of the source emitting the acoustic monopole (thermodynamic pressure drop due to cooling) is the most efficient. For a simple source in an unstratified medium the monopolar efficiency scales analytically as Q/σ_1. This is approximated by the simulations near the peak of monopolar emission ($t \approx 5.25$ in Fig. 7) despite an admixture of the dipolar component. Later in the time series the efficiency again increases due to the dipolar contribution which scales with the total integrated cooling, but bulk fluid motion likely contaminates the measured acoustic flux at those late times. In principle a large enough domain could allow unambiguous determination of the the dipolar and quadrupolar efficiencies if wave packet dispersion and dissipation were minimized, but in practice resolution of thermal plume motions at the source location make this difficult.

Finally, it is interesting to note that the excitation mechanism proposed here fits well with recent understanding of p-mode line-shapes. Solar acoustic power spectra exhibit non-Lorentzian line profiles, and observations of intensity and Doppler velocity show opposite asymmetry in the same region of the spectra (Duvall et al. 1993). The cause of asymmetry is generally attributed to a localized excitation source depth (e.g. Gabriel 1992, Duvall et al. 1993, Roxburgh & Vorontsov 1995, Abrams & Kumar 1996, Rast & Bogdan 1998), but while such a mechanism produces asymmetries which depend on both acoustical source type and depth, it does not produce, for a given source, oppositely signed asymmetries between different observed variables. It has recently been suggested (Roxburgh & Vorontsov 1997, Nigam et al. 1998) that the addition

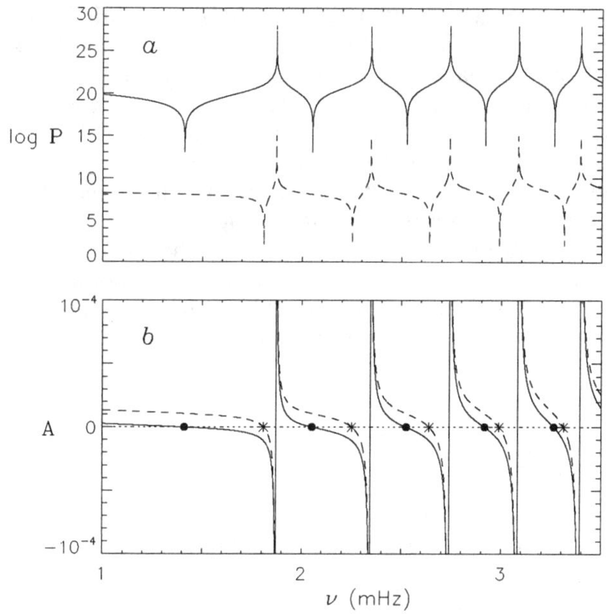

Figure 8. Power spectra of T'/T_0 from (Rast & Bogdan 1998) both with (dashed curve) and without (solid curve) an additive correlated noise component (*a*) and the corresponding amplitude spectra (*b*).

of a correlated noise component is needed. Note that what is meant by 'correlated noise' is noise with a particular phase relationship to the acoustic source, noise that is added to the amplitude spectrum before squaring rather that in quadrature.

To understand this point, consider the Doppler-velocity spectrum of the Sun to be given by its intrinsic spectral response to the acoustic sources, while the intensity signal combines the solar response to the source with direct observation of intensity fluctuations produced at the source location. This is consistent with the notion that the excitation source lies in the photosphere, is predominantly thermodynamic, and is thus directly associated with intensity rather than velocity fluctuations. For an impulsive intensity fluctuation with no temporal offset from the acoustic source, the correlated noise is a real constant. Even under such constraints its addition to the spectrum can alter the line-shapes (Schou 1998). This is illustrated by Fig. 8a which plots temperature-perturbation power spectra from Rast and Bogdan (1998) with and without such a correlated component. The two power-spectra exhibit oppositely signed line-asymmetry. In this simple case the Fourier amplitudes are purely real (undamped oscillations and intensity fluctuations cotemporal with the source) and are plotted for both spectra (with and without correlated noise) in Fig. 8b. Line-asymmetries are determined by spectral peak positions relative to the adjacent spectral minima (Roxburgh & Vorontsov 1995, Rast & Bogdan 1998). When a constant is added to the amplitude, the frequencies at which it equals zero shift (from the filled circles to the asterisks in Fig. 8*b*), and the power-spectrum line-shapes change.

What is significant is that to reproduce asymmetry observations the noise correlation introduced above must be of a particular sign, and that sign matches the excitation mechanism proposed in this paper. To match the observed intensity asymmetry, taking that of the velocity to be intrinsic (unchanged by a correlated noise component), the source must be either a negative pressure fluctuation associated with photospheric darkening or a positive pressure fluctuation associated with brightening. We expect both these combinations during new downflow-plume formation, one as a result of radiative cooling and the other due to subsequent inflow induced compressional heating, and suggest that it is the former that dominates the excitation process.

References

Abrams, D., & Kumar, P. 1996 ApJ, 472, 882

de Boer, C.R., Kneer, F., & Nesis, A. 1992, A&A, 257, L4

Duvall, T.L., Jr., Jefferies, S.M., Harvey, J.W., Osaki, Y., & Pomerantz, M.A. 1993, ApJ, 410, 829

Gabriel M. 1992, A&A, 265, 771

Hirzberger, J., Bonet, J.A., Vázquez, M., & Hanslmeier, A. 1999, ApJ, submitted

Hirzberger, J., Vázquez, M., Bonet, J.A., Hanslmeier, A., & Sobotka, M. 1997, ApJ, 480, 406

Kawaguchi, I. 1980, Sol. Phys., 65 207

Kitai, R., & Kawaguchi, I. 1979, Sol. Phys., 64, 3

Mehltretter, J.P. 1978, A&A, 62, 311

Moses, E., Zocchi, G., & Libchaber, A. 1993, J. Fluid Mech., 251, 581

Namba, O. 1986, A&A, 161, 31

Nesis, A., Hanslmeier, A., Hammer, R., Komm, R., Mattig, W., & Staiger, J. 1992, A&A, 253, 561

Nesis, A., Hanslmeier, A., Hammer, R., Komm, R., Mattig, W., & Staiger, J. 1993, A&A, 279, 599

Nesis, A., Hammer, R., Hanslmeier, A., Schleicher, H., Sigwarth, M., & Staiger, J. 1997, A&A, 326, 851

Nigam, R., Kosovichev, A.G., Scherrer, P.H., & Schou, J. 1998, ApJ, 495, L115

Nordlund, Å. 1985, Sol. Phys., 100, 209

Parker, E.N. 1991, ApJ, 380, 251

Rast, M.P. 1995, ApJ, 443, 863

Rast, M.P. 1998, J. Fluid Mech., 369, 125

Rast, M.P. 1999, ApJ, submitted

Rast, M.P., & Bogdan, T.J. 1998, ApJ, 496, 527

Rast, M.P., Nordlund, Å, Stein. R.F., & Toomre, J. 1993, ApJ, 408, L53

Rimmele, T.R., Goode, P.R., Harold, E., & Stebbins, R.T. 1995, ApJ, 444, L119

Rösch, J. 1960, in Aerodynamic Phenomena in Stellar Atmospheres, IAU Symp. 12, (ed. R.N. Thomas), Nicola Zanichelli, Bologna, 313

Roudier, Th., Malherbe, J.M., November, L., Vigneau, J., Coupinot, G., Lafon, M., & Muller, R. 1997, A&A, 320, 605

Roxburgh, I.W., & Vorontsov, S.V. 1995, MNRAS, 272, 850

Roxburgh, I.W., & Vorontsov, S.V. 1997, MNRAS, 292, L33

Schou, J. 1998, private communication (SOI Inversions Workshop)

Skartlien, R. 1998, PhD Thesis (University of Oslo)

Turner, J.S. 1986, J. Fluid Mech. 173, 431

Yamada, H., & Matsui, T. 1978, Phys. Fluids, 21, 292

Group Discussion

Rutten : Explain the usefulness of why and when acoustic events occur.

Rast : 1) Study of compressible convection-eg.efficiency of of acoustic energy production 2) True local helioseismology in vicinity of known sources 3) Connection between acoustic events and chromospheric shocks (comment after Goode talk)

High Resolution Solar Physics: Theory, Observations, and Techniques
ASP Conference Series, Vol. 183, 1999
T.R. Rimmele, K. S. Balasubramaniam, R.R. Radick, eds.

The Excitation of Solar Oscillations --Observations and Simulations

Philip R. Goode

Big Bear Solar Observatory, New Jersey Institute of Technology, Big Bear City, CA 92314

Louis H. Strous

Big Bear Solar Observatory, New Jersey Institute of Technology, Big Bear City, CA 92314
Lockheed-Martin Solar & Astrophysics Lab, Palo Alto, CA 94304

Thomas R. Rimmele

National Solar Observatory, Sunspot, NM 88349

R. F. Stein

Department of Physics and Astronomy, Michigan State University, East Lansing, MI 48824

Å. Nordlund

Theoretical Astrophysics Center and Astronomical Observatory, Dk-2100 Copenhagen, Denmark

Abstract. We compare observations of the excitation of solar oscillations with simulations. We pass synthetic line profile data through the same analysis scheme used for observed line profiles. We find in both cases that seismic events exciting the oscillations are largely confined to the dark intergranular lanes. The excitation is closely associated with the collapse of the intergranular lanes, rather than the continuous drumming of turbulent convection. The most significant difference in the results is the considerably longer lifetime of the observed events.

1. Introduction

It has been known for many years that solar oscillations are excited very close to the Sun's surface [Goldreich and Keeley 1977, Goldreich and Kumar 1989]. Rimmele, et al. (1995) observed that seismic events originate in the dark intergranular lanes as a part of a cooling and collapse of the solar surface. Goode, et al. (1998) showed that the seismic events feed power into the Sun's normal modes.

The observations are photospheric line profile data covering a two dimensional field of view. By measuring the velocity as a function of depth in the line, we are able to determine velocity and phase information as a function of altitude. This three dimensional velocity information coupled with the vertical phase gradient is the key to identifying the kind of seismic events we study. A seismic event is characterized by a sizeable vertical gradient, as though it were first travelling upwards. The last part of the event occurs a few minutes later. This part looks like a wave travelling back down in the same point in the field of view, as though it was reflected back down from above. This clear phase evolution, coupled with the spatial and temporal isolation of the events, enables us to distinguish the events. It is these seismic events that Rimmele, et al. found populating the dark lanes. They also found the power in the events was sufficient to drive the entire p-mode spectrum.

The synthetic line profile data comes from the large scale simulations of the velocity field in the outer layers of the Sun due to Stein and Nordlund (1998). These calculated "data" match the cadence of the observations, and are passed through exactly the same reduction scheme. The observations and data reduction are discussed in the next section.

Our purpose here is to compare and contrast seismic events from observations and simulations.

2. The Data and the Reduction of It

Our observations were made at the Vacuum Tower Telescope of the National Solar Observatory in Sunspot, New Mexico. The quiet Sun dataset discussed here is from Sept. 5, 1994. The data and the reduction of it are described in detail in Rimmele, et al. (1994, 1995), and references therein. One of the basic observational problems here is to distinguish the seismic event power from the dominant resonant mode power. To distinguish the two in our field of view ($60'' \times 60''$ patch of quiet Sun near disk center), Rimmele, et al. measured the velocity as a function of altitude in the photosphere for 65 min by observing the Doppler shift in the 543.4 nm Fe I absorption line.

We note that earlier spectrographic observations of the Fe I line, Stebbins and Goode (1987), but with a faster cadence and one dimensional field of view, the basic life cycle of the events was the same as ours, see Restaino, et al. (1993). Also, observations measuring other lines show the same basic results for the events, e.g., the well-known Fe I 5576 Å (g=0) line.

In the data reduction, we calculate $v(x, y, z, t)$, the four dimensional velocity cube. In the z-direction, ten depths in the 543.4 nm line yield velocities at ten altitudes spanning the photosphere. From this, we determine we determine the seismic flux which is defined as

$$u \propto \frac{V^2}{\omega}\frac{\Delta\phi}{\Delta z},$$

where V is the local velocity amplitude and $\frac{\Delta\phi}{\Delta z}$ is the vertical phase gradient and ω is the instantaneous oscillation frequency which is assumed to be constant. If the seismic flux were describing an acoustic travelling wave (above the acoustic cut-off), then the seismic flux would be identical to the ordinary mechanical flux. However, the seismic power is beneath the acoustic cut-off. Nonetheless,

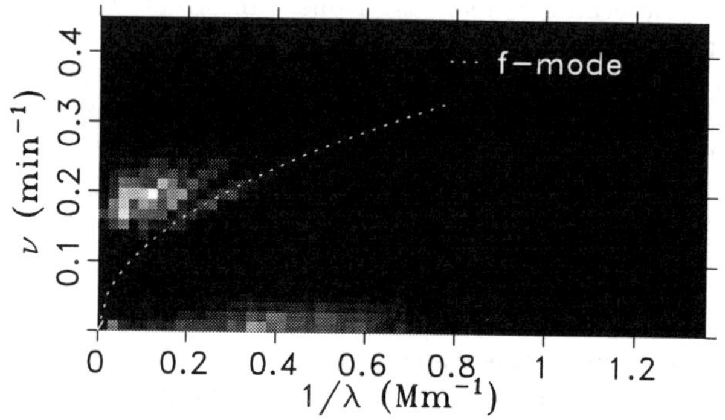

Figure 1. The k-ω diagram for the observational data. The dashed white curve represents the calculated f-mode ridge. Note that the sonic and subsonic regions are well separated. Note also that there is significant power in the region of the f-mode ridge.

the seismic flux is a useful measure of the events and enables us to distinguish the events from p-modes. After all, for p-modes $\frac{\Delta\phi}{\Delta z}$ is very close to zero. The k-ω diagram for the data is shown in Figure 1. The power is concentrated in two places−−the low frequency power arises from the convective flows. In fact, if we filter to pass only that power, the velocity field from near the continuum looks the same as that from a second CCD we employed to simultaneously measure the white light. The remaining power is in the p-mode region which is, of course, below the acoustic cut-off. This power consists of seismic events and p-modes, as Rimmele, et al. emphasized. A typical event is characterized by a 30° or more phase change through the photosphere at some time in its life. The lifecycle of events is described in detail in Rimmele et al. and Goode et al..

3. Seismic Events

In Figure 2, we show a contoured scatter plot of the seismic flux against the local convective velocity for the velocity cube. The events shown are the upper 10% of the events, and as such represent more that 2000 events as identified in Goode, et al. (1998). There is a linear relation between the velocity and seismic flux given by Strous, Goode and Rimmele (1998)

$$\frac{v_{\rm down}}{\rm m/s} = (-0.115 \pm 0.003) - (0.012 \pm 0.002)\frac{F_{\rm ac}}{100{\rm kW/m^2}} \pm 0.060,$$

where $F_{\rm ac}$ is the seismic flux and $v_{\rm down}$ is the local convective velocity. This relationship says that the greater the downgoing convective velocity, the greater the seismic flux. More significantly is the linear relation itself would be unexpected if turbulent convection itself drove the oscillations. In that case, one would expect the event power to go as a high power of the convective velocity,

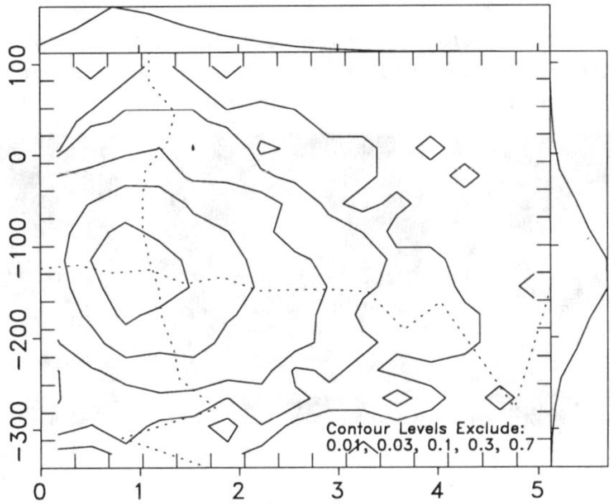

Figure 2. Extreme downward convective velocity vs. maximum seismic flux. A contour plot of the frequency distribution of the largest downward velocity (in m/s) versus the maximum seismic flux (in units of 100 kW/s) in all approximately 2000 seismic events. The contour levels indicate the fraction of data points that lies on the outside of the corresponding contour. For instance, the central contour excludes 70% of the data points. The small side panels show the associated one-dimensional frequency distributions.

Goldreich and Kumar (1990). We emphasize here that the observational data imply that the seismic events arise from cooling and collapse of the solar surface as a part of the convective overshooting, but not as the continuous drumming of the turbulent convection itself.

In Figure 3, we show the time evolution of the averaged seismic events shown in Figure 2. The events are pinned in time with t=0 s being the peak of the upgoing flux for each. The flux is shown in the white contours. The events are overlayed on the local granular field, shown in grayscale. The field is oriented with the local lane along the x-axis and the brighter feature on top. Clearly the events originate in the intergranular lanes. This is a robust feature surviving the averaging of more than 2000 events, Goode, et al. (1998). If we examine the components of the seismic flux as they evolve from the center of the mean event (see Figure 2 of Goode, et al.), we find that there is an excess V^2 power that is apparent out to beyond 2″ from the center of the event. On the other hand, the phase gradient shows ever diminishing event-like behavior (outgoing wave following by ingoing wave) which is absent beyond about 1″ from the event center. Thus, beyond 1″ the excess power shows essentially no vertical phase gradient, and, therefore, event power has changed form to look like normal mode power. In detail, this power looks like f-mode power as seen in Figure 1. Kumar

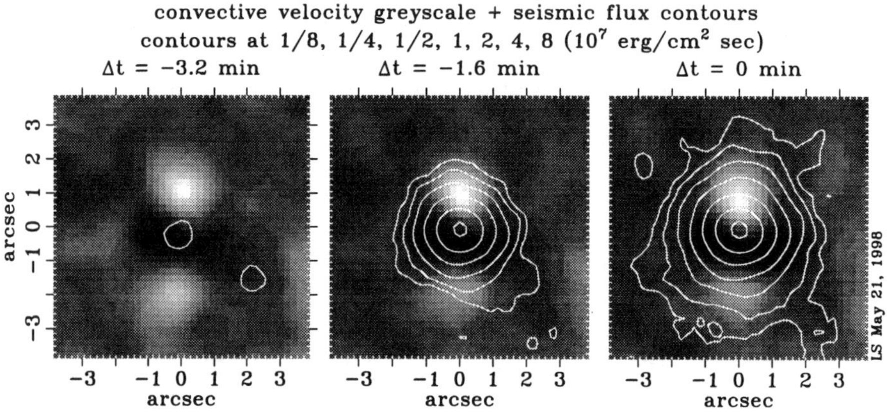

Figure 3. From observations: The superposed seismic flux, shown in white contours, is superposed on its local, averaged granulation altitude 150 km above the base of the photosphere. The contours are 0.25, 0.5, 1, 2, 4×10^7 ergs/cm^2/s. The time steps in the data collection are about 30 s, and at one time step before $T{=}-3.2$ min, the seismic flux is below 0.25×10^7 ergs/cm^2/s everywhere in the field of view.

(1993) has shown that one can expect seismic event power to be converted to p-mode power after a few subsurface skips. That is, their power would come to reside on the p-mode ridges of the k-ω diagram. However, the skip times and distances are so great that in a single skip a normal mode would be out of our field of view and its subsurface travel time would be longer than the temporal duration of our dataset. However, f-modes are, asymptotically, surface waves, so that their skip times and distances are small. Thus, these are the only normal modes we can expect to see resulting from the seismic events−−and we do see them. We also note that the V^2 power falls off as the distance squared, as one would anticipate. The averaged vertical phase gradient has a peak value of somewhat less than 3°. This value is so small because it arises from the averaging of many events. In fact this value is significant well above a 10σ level. Strous, Goode and Rimmele looked for seismic ripples beyond 4" from the center of the mean events. They found nothing that was inconsistent with noise.

In Figure 4, we show a scatter plot of the analyzed synthetic line profile data. This plot is meant to be compared to Figure 2. We see that the greater the downgoing convective velocity, the greater the local seismic event. Further, the relationship is linear, as in Figure 2.

In Figure 5, we see the k-ω diagrams from the synthetic line profile data. The plot on the right hand side of the figure is the k-ω diagram which has been degraded to account for the effects of atmospheric turbulence. As in Figure 1, the power is confined to two regions of the diagram. When we filter either of these diagrams to pass only the low frequency or convective power, we obtain

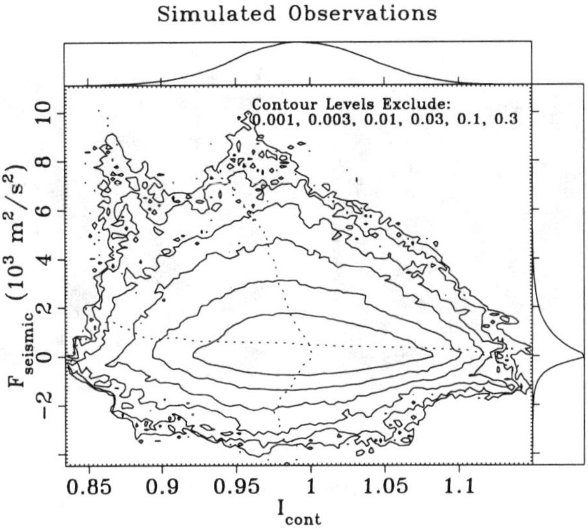

Figure 4. From simulations: The superposed seismic flux, shown in white contours, is superposed on its local, averaged granulation altitude 150 km above the base of the photosphere. The contours are 0.25, 0.5, 1, 2, 4×10^7 ergs/cm^2/s. The time steps in the data collection are about 30 s, and at one time step before T=-3.2 min, the seismic flux is below 0.25×10^7 ergs/cm^2/s everywhere in the field of view.

the granular field. When we pass the power for the higher frequency region, we have the p-mode power and the event power discussed here. Of course, we use the vertical phase gradient information to delineate between the latter two kinds of power. We note that there is also power with vertical phase gradients in the convective region. Much of this power is associated with granular evolution. In particular, if the edge of a granule moves across some point in the field of view, its horizontal motion, which is differential with altitude, may cause a signal which resembles a vertically propagating wave. Could there be seismic event power in the the convective region of the power spectrum? Well, if the spectrum of seismic events were to extend from the p-mode region to the convection region in the k-ω diagram, then we would expect event power in the convection region. However, there is no such connection between the regions.

In Figure 6, we show the simulated granulation and seismic flux at a particular, but typical instant—–both the results from the full velocity signal and that degraded in our effort to mimic the effects of seeing. The degraded signal looks quite like that from the observational data. It is clear that the seismic power lies in the intergranular lanes, just as was determined from the observations. Further, the events from the simulations do typically have the characteristic evolution of an upgoing event followed by a downgoing event. However, the downgoing flux in the latter stages of the events seems weak. This weakness is probably due to the fact that the outer boundary condition in the model is that all waves reaching the temperature minimum propagate out of the box, while we explain the more substantial downgoing flux in our observations as a partial

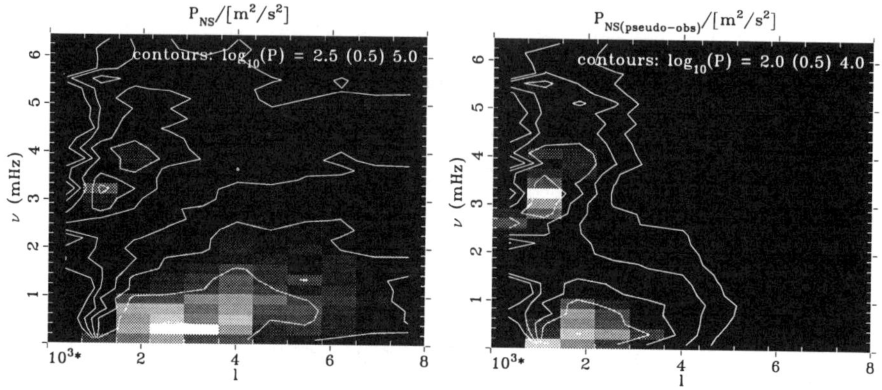

Figure 5. The k-ω diagrams for the simulations. The one on the right has the velocity signal degraded to account for seeing.

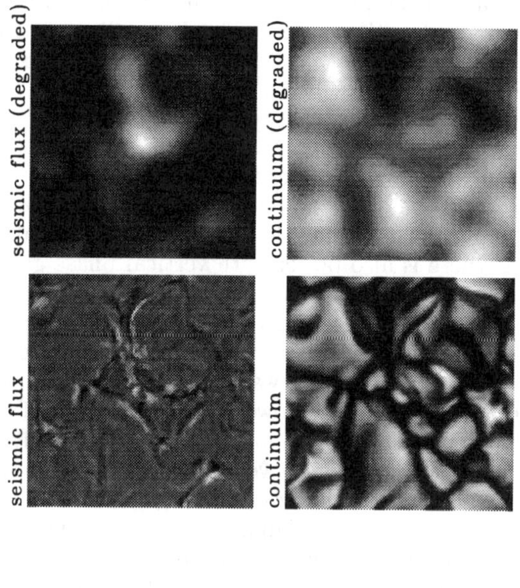

Figure 6. The left hand panels are a snapshot of the seismic flux and Doppler velocities for the full resolution simulations. Both the flux and granulation are shown in gray scale with bright meaning upgoing and dark meaning downgoing. On the right is the same snapshot in time, but with the velocity signal being degraded by seeing effects.

reflection of the preceding upgoing flux. The full resolution image reveals the role of small, but not necessarily weak granules. The small bright granules in the center of the field of view are at the heart of the strongest seismic flux in the image. In the events, these small granules collapse and bring down part of the neighboring granule. We also see this in particular examples of the observed seismic events. However, there are real differences between the observed events and those from the simulations. Most striking is the fact that the real events last 2-3 times longer than the simulated ones. This fact and other significant results of the comparison are currently under study, Strous, et al. (1998).

We are investigating the seismic potential of the events. However, we know that the probing will be very local. That is, no seismic signal is detected beyond about 3-4$''$ away from the center of an event. We believe the seismic probing would be most effective in probing nearby granules. The seismology will be aided by our ability to compare observed and simulated probings. We also will study the power of the events to probe small, weak magnetic features like pores.

P.R.G. and L.H.S. were supported in part by NSF-ATM-97-14796 and NAG5-4919. T.R.R. is supported by the National Solar Observatory which is operated by AURA under cooperative agreement with the National Science Foundation. R.F.S. was supported by NAG-1695, NAG5-4031 and NSF-AST-95-21785. Å.N. was supported in part by the Danish Research Foundation.

References

Goldreich, P. & Keeley, D.A., 1977, ApJ, 212, 243.

Goldreich, P. & Kumar, P., 1988, ApJ, 326, 462.

Goldreich, P. & Kumar, P., 1990, ApJ, 363, 694.

Goode, P.R., Strous, L.H., Rimmele, T.R.,& Stebbins, R.T., 1998, ApJ, 495, L27.

Kumar, P., 1993, in *GONG 1992: Seismic Investigation Of The Sun And Stars*, ed. T.M. Brown, A.S.P. Conf. Ser. 42, 15.

Restaino, S.R., Stebbins, R.T. & Goode, P.R., 1993, ApJ, 408, L57.

Rimmele, T.R., 1994, A&A, 290, 972.

Rimmele, T.R., Goode, P.R., Harold, E. & Stebbins, R.T., 1995, ApJ, 444, L119.

Stebbins, R.T. & Goode, P.R., 1987, Solar Phys., 110, 237.

Stein, R.F. & Nordlund, Å, 1998, ApJ, 499, 914.

Strous, L.H., Goode, P.R. & Rimmele, T.R., 1998, in preparation.

Strous, L.H., Goode, P.R., Rimmele, & T.R., Nordlund, Å. & Stein, R.F., 1998, in preparation.

Group Discussion

von der Luhe : The center of the plot showing average seismic flux appears to be off-set from the center of the lanes. Is this significant?

Goode : Yes, it turns out that hte seismic events are typically associated with the collapse of a small granule between larger granules. The event also brings down part of the nearing granule. When we average, the smaller granules are washed out and one sees the event closer to the brighter granule which was brighter in part because of the proximity of the small granule.

Rast : Is it possible that you might mistake focusing of p-mode energy by horizontal temperature gradients into the downflow lanes for acoustic sources? Comment: Connection of acoustic events with chromospheric shocks may also be an interesting possibility. *Goode :* Such gradients should generate considerable power at "high" frequencies, but we see no seismic power above the acoustic cut-off.

High Resolution Solar Physics: Theory, Observations, and Techniques
ASP Conference Series, Vol. 183, 1999
T.R. Rimmele, K. S. Balasubramaniam, R.R. Radick, eds.

On the Relation Between Solar Oscillations and Granulation

Michael Kiefer
Kiepenheuer-Institut für Sonnenphysik, Schöneckstraße 6, D-79104 Freiburg, Germany

Horst Balthasar
Astrophysikalisches Institut Potsdam, Telegrafenberg, D-14473 Potsdam, Germany

Abstract. We investigate the behavior of solar oscillations in the lower photosphere by means of a 2 hrs time series of FPI-filtergrams taken at the Fe I 7090 absorption line. We examine relations between granular structure and oscillation amplitudes by using filters for distinct wave types and by comparing rms wave-amplitudes at locations of different brightness in the granulation. We find that all types of intensity oscillations show a clear tendency to prefer the darkest locations in the granulation. The behavior of evanescent waves is dominated by the f-mode. Velocity oscillations show little tendency to prefer the dark granular locations. In all oscillation data-sets a preference for dark locations is most clearly visible in the regime of running sound waves. Our findings support theoretical predictions that oscillations should be excited preferentially at the locations of intergranular downflows and that diffraction of waves takes place due to temperature and velocity inhomogeneities induced by the convection in the subsurface layers and in the atmosphere.

1. Introduction

In recent years much attention has been paid to the question whether the convection in the outermost layers of the sun's convection zone, visible as granulation, exerts an influence on the excitation and propagation of waves in the sun.

There are essentially two theoretical approaches connected to the subject. The first one is the model of localized wave excitation (Brown 1991) in connection with the role of developing downdrafts as source of acoustic noise (e.g. Rast 1997). The second one is based on the diffraction of sound waves propagating through inhomogeneous media (Zhugzhda & Stix 1994, Stix & Zhugzhda 1996).

Observational efforts taken so far to investigate this topic were based on data of Doppler velocity in several heights in the photosphere (Rimmele et al. 1995), non-simultaneous intensity and velocity data (Espagnet et al. 1996), and intensity data (Hoekzema et al. 1998). Usually some filtering of the data was performed prior to the different analyses.

We use a data set of simultaneously obtained intensity and line of sight Doppler-velocity data to examine relations between solar oscillations and the granulation structure. We apply filters in the $\boldsymbol{k}$-ν-space to extract distinct wave types, i.e. p-modes, f-mode, and running sound waves, from the initial time series. Then we compare the rms-amplitudes of the oscillation time series for one wave type at locations of different brightness in the granulation.

2. Observation and Data Reduction

The data were taken on July 6, 1996 at the German VTT at the Observatorio del Teide, Tenerife. A field of $76'' \times 55''$ of magnetically quiet sun at disc center was observed for 119 min with the 2D-spectrometer of the Göttingen University (Bendlin & Volkmer 1995). The pixel size corresponds to $0\overset{''}{.}2 \times 0\overset{''}{.}2$. The initial data consist of two strictly simultaneous series of 160 temporal equidistant blocks of 40 images each. The first series consists of white-light images and the second one of filtergrams with wavelengths around the Fe I 7090 absorption line. The recording of the 40 images took 11 s, the time from block to block was 44.8 s. Dark and gain table corrections were performed and image motion and distortion were removed by correlation analysis and by a destretching algorithm. The corrections reduced the image size to $65'' \times 43''$.

For the subsequent analysis five time series were extracted. The first one is composed of 160 averaged images of the blocks of 40 white-light images. The second one consists of the best image of each block of 40 white-light images (that series therefore is not perfectly equidistant in time). The three other series were obtained from the filtergrams. Series three contains the images of the line-core intensity. In series four the Doppler velocity computed at 20% line depression is used, while the fifth series consists of the images of the Doppler velocity from the line core. We ascribe a photospheric height of approx. 100 km to the line-wing data and of approx. 250 km to the line-core data.

3. Data Processing

3.1. Filtering

If one looks for the behavior of waves it is advisable to take into account the different physical and propagation properties of p-modes, f-mode, and running waves. This is most conveniently done by filtering the initial series with appropriate filters. This then allows to examine the behavior of each wave type separately. All five time series were transformed into the Fourier domain, the $\boldsymbol{k}$-ν-space. Here $\boldsymbol{k}$ denotes the horizontal wavenumber vector, $\boldsymbol{k} = (k_x, k_y)$, and ν stands for the frequency. Power-, phase-, and coherence spectra were calculated from the transforms. Judging from the distribution of these quantities in a k-ν-plane (while assuming isotropic conditions on the solar surface) we defined filters for several wave types, namely p-modes, f-mode, and running waves. The former two are subdivided into two frequency ranges each, while the latter are divided into three frequency ranges. Further we defined a region which is supposed to represent the granulation. The filters for the running waves are such that the waves are restricted to the wavenumber range 0 Mm $< k <$ 2 Mm and to the three frequency intervals given by the ν-values $\{4.5, 5.63, 6.75, 7.88\}$ (in mHz). The shapes of the filters in the k-ν-plane can be found in Fig. 1 (super-

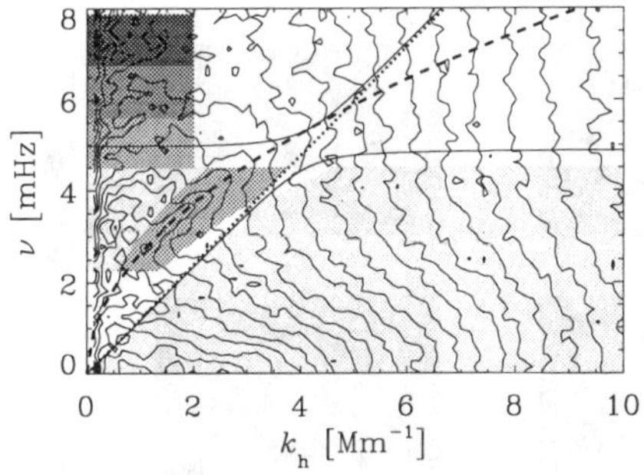

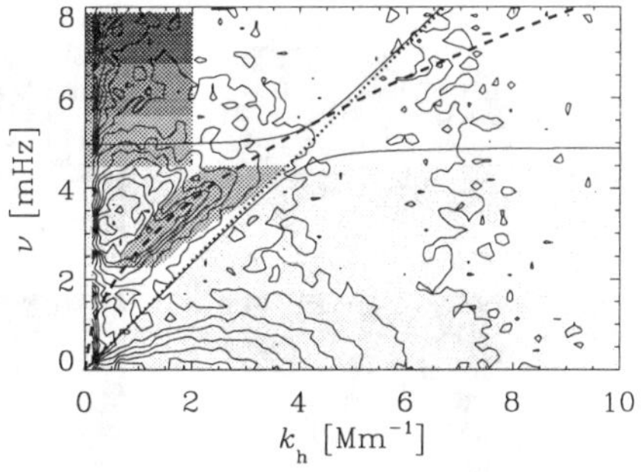

Figure 1. Powerspectra of intensity oscillations (left) and velocity oscillations (right). Contour lines are separated by a factor of $\sqrt{10}$. Dotted line: Lamb-mode with $\nu = \frac{1}{2\pi} c_s k_h$, heavy broken line: dispersion relation for surface waves (f-mode), $\nu = \frac{1}{2\pi}\sqrt{g k_h}$ ($c_s = 7.5$ km/s, $g = 274$ m/s^2). Below $\nu = 4.5$ mHz light shading outlines the areas of the filters for p-modes (left) and granulation (right) and medium shading shows the region of the f-mode filter. Above, the three filter regions of running waves are outlined.

imposed on contour plots of power spectra). The detailed definitions of the filter boundaries for p-modes, f-mode, and granulation filters are listed in Tab. 1. of Kiefer & Balthasar (1998). We shall call the combination of p-mode and f-mode waves evanescent waves henceforth.

All five time series were filtered in the Fourier domain and transformed back into the space-time domain again. The series of best white-light images was only filtered to get a "pure" granulation series, while the other four series were filtered with the diverse oscillation filters. The apodization in space and time of the time series applied prior to the Fourier transformation led to a final image size of $59'' \times 39''$ and a duration of 107 min of the series.

3.2. Further analysis

To obtain a measure for the relation between granulation and oscillation we used the following method, described in Kiefer & Balthasar (1998) in more detail. Each image of the granulation series was subdivided into ten brightness classes. In the first class, [0, 0.1], there are the ten percent darkest pixels, in the second class, [0.1, 0.2], there are the pixels for which 10% of all pixels are darker and 80% of all pixels are brighter, and so on. This method gives ten time series of pixel patterns out of the "pure" granulation series. The pixel patterns (ten per image) then were used to extract the corresponding amplitude values of the several oscillation time series. The amplitudes of the oscillations on the locations given by the pixel pattern were squared and summed over the whole series. The resulting single value per oscillation type (defined by the used filter) and per brightness/pixel class then was related to the rms-value of the amplitude of the whole time series of that oscillation type. This value is the area-related ratio between the rms value in certain brightness classes and the total rms value in the corresponding series. Shifting the series of patterns in time with respect

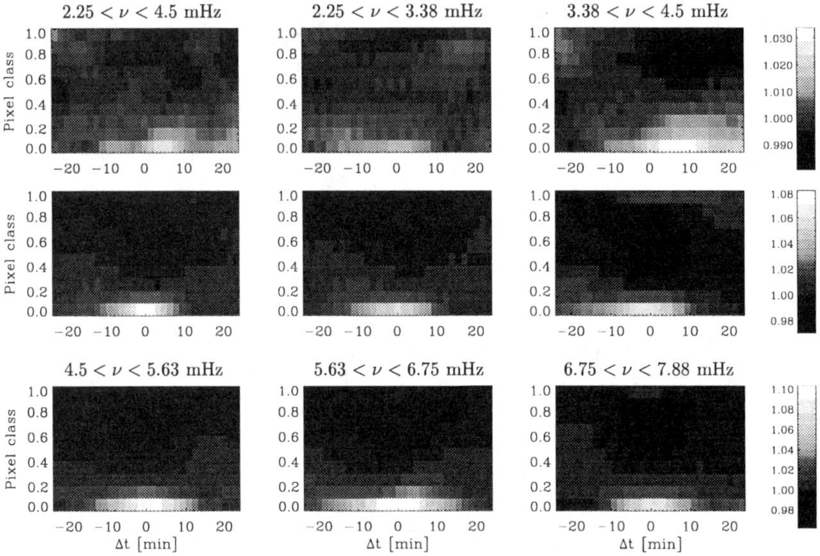

Figure 2. Grayscale plots of the ratio of rms oscillation amplitudes in intensity/pixel classes to the total rms amplitude for three types of oscillations in the white-light series. First row: p-modes, second row: f-mode, third row: running waves. On the top the frequency ranges of the applied filters are given. The abscissa gives the time delay between oscillation and pixel-pattern time series. The ordinate displays ten intensity/pixel classes. Each row is scaled individually in the range indicated by the bar to the right of it.

to the oscillation images yields information about a delay in the interaction of oscillations and granulation structure.

4. Results

Fig.1 shows the power in the k-ν-plane for the white-light series from averaged images (left frame) and for the series of Doppler velocities from the line core (right frame). In the intensity oscillations there is not much power located in the region of the p-modes. An almost equal amount of power can be found in the f-mode (for more details see Kiefer & Balthasar 1998, especially Tab. 2). No clear gap exists between the subsonic and supersonic regime. A quite different picture arises from the distribution of power in the velocity oscillations, where there is a pronounced gap between the subsonic and supersonic regime. Further, besides the still clearly discernible f-mode there is much power located in the p-mode region, some ridges are clearly visible.

In Figs. 2–5 grayscale plots of the ratio of rms amplitudes in intensity classes to the total rms amplitude are shown for all time series and for the three types of

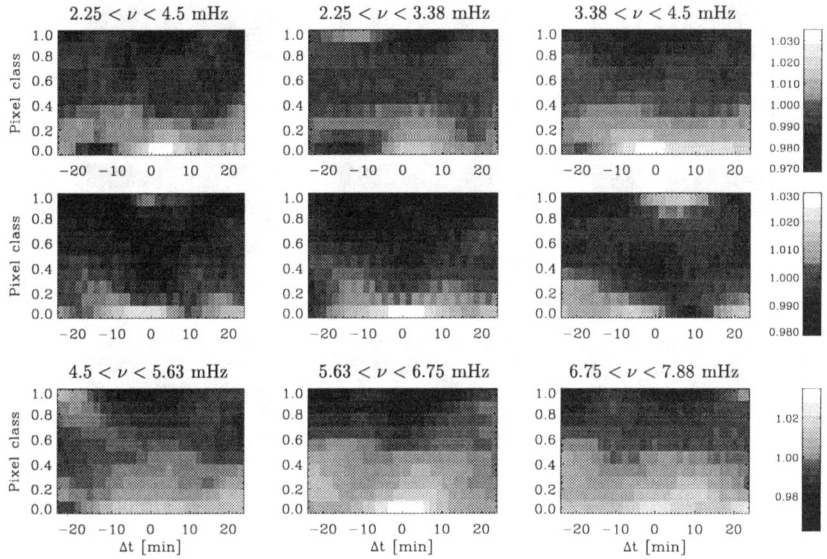

Figure 3. Same as Fig. 2 but for the oscillations in the line-core intensity in approx. 250 km height in the photosphere.

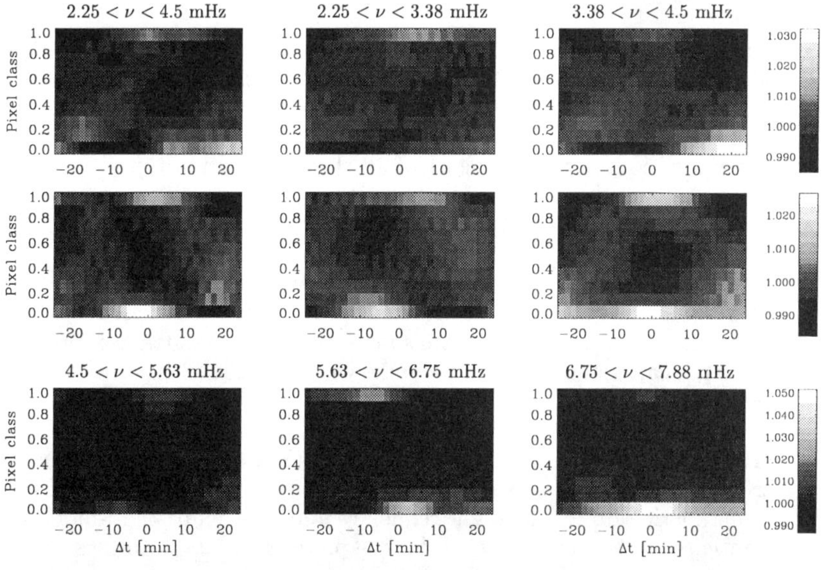

Figure 4. Same as Fig. 2 but for the oscillations in the Doppler velocity from the line wing in approx. 100 km height in the photosphere.

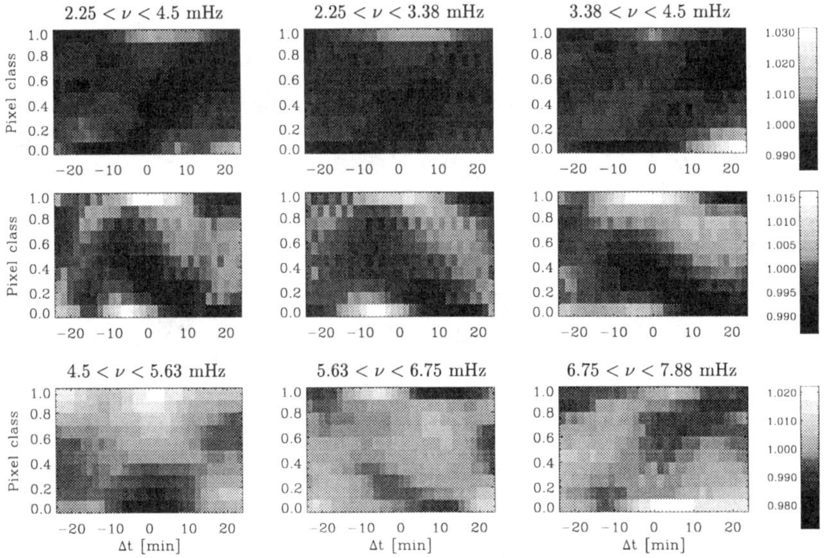

Figure 5. Same as Fig. 2 but for the oscillations in the Doppler velocity from the line core in approx. 250 km height in the photosphere.

oscillations, p-modes (first row), f-mode (second row), and running waves (third row). From the figures we find the following:

- **White-light series (Fig. 2):** The behavior of evanescent waves, p-modes, and especially the f-mode is discussed elsewhere (Kiefer & Balthasar, 1998). The running waves obviously prefer the darkest locations in the granulation with little or no time delay. For running waves the preference of dark locations is the strongest of all three oscillation types.

- **Core-intensity series (Fig. 3):** In general the appearance of the plots is fuzzier compared to the those of the white-light series. Still there is a slight tendency for p-modes and the f-mode to prefer the darkest granulation features. The f-mode does not feature a stronger effect than the p-modes any more. The running waves still show quite a clear tendency of preference for dark locations, however on a somewhat lower level now, compared to the white-light oscillations.

- **Series of Doppler velocity at 100 km (Fig. 4):** A preference of the darkest regions with a time delay of more than fifteen minutes seems to be characteristic of only the high frequency p-modes, while the f-mode tends to anticipate the emergence of dark structures by 2–5 minutes and also shows a preference for the very brightest granulation features. Running waves above 5.63 mHz clearly prefer the very darkest locations with a 1–2 minutes time delay. Like in the white-light and core-intensity data this is the strongest effect of all three oscillation types.

- **Series of Doppler velocity at 250 km (Fig. 5):** Apart from being much fuzzier in appearance there is no striking difference compared to the series of Doppler velocity at approx. 100 km in *p*-modes and the f-mode. The values are somewhat lower now. For running waves of low frequency there is a clear avoidance of dark and an equally clear preference for bright regions. In the high frequency running waves the level of preference of darkest locations is much smaller than in the series of Doppler velocity at approx. 100 km. The time delay in the case of the high frequency running waves is 4–5 minutes.

5. Discussion and Conclusions

The preference of the dark intergranular locations with a short time delay shown by the evanescent waves in white-light is compatible with the findings of Espagnet, Muller, Roudier, & Mein (1996). Further a part of the results of Hoekzema et al. (1998) with respect to the behavior of the running waves in white-light oscillations is confirmed by our findings. The behavior of the evanescent waves, especially of the f-mode, in white-light oscillations, is discussed elsewhere (Kiefer & Balthasar 1998).

We see indications for located wave sources and for wave diffraction in our data set. Our findings therefore support the two theoretical approaches cited in the introduction:

- **Located wave sources:** All time series show a preference for dark or even darkest locations in the granulation by the running waves, especially by the high frequency ones. This is indicative of upward running wave trains from sources located in the intergranular space. A strong support for this interpretation is that in the velocity data the time delay is longer in the data from 250 km height than in the data from 100 km. The physical picture of a transient sound source below the visible surface is fully compatible with our findings. From our data we can not distinguish whether the stronger preference of the darkest locations by the running waves of the highest frequency in the lower layer compared to the higher one is due to wave diffraction or whether it is an effect of the wave-source being located at finite depth below the solar surface (Brown 1991, Rast 1997).

- **Wave diffraction:** There is a common feature in the differences between data, either intensities or Doppler velocities, from different heights in the photosphere. In both cases the oscillations in the upper layer show diminished tendencies of the preference for dark locations. The low frequency running waves even show a clear preference for bright locations in the velocity data from 250 km height. Altogether this supports the picture of wave diffraction in the top levels of the convection zone and in the atmosphere due to temperature and velocity inhomogeneities (Zhugzhda & Stix 1994 and Stix & Zhugzhda 1996).

Further analyses, possibly with a longer time series of better quality, especially with more stable seeing conditions, are needed to confirm and extend our interpretations and conclusions. An application of the different analysis methods of Rimmele et al. (1995), Espagnet et al. (1996), Hoekzema et al. (1998), and Kiefer & Balthasar (1998) on one single high quality data set therefore seems to be a highly desirable task to clarify the abilities and sensitivities of the different methods.

Yet it is not clear whether the behavior of the f-mode in white-light intensity oscillations is due to advection processes and wave diffraction or due to opacity/optical depth changes in the strongly inhomogeneous topmost layers of the solar convection zone and overshoot zone. Opacity/optical depth effects can be excluded as an explanation for the findings in running waves because the behavior is similar for intensity and velocity data. Theoretical examination of this problem and modelling of the behavior of surface waves propagating in an inhomogeneous medium seems to be a necessary step for a better understanding and interpretation of the observational facts.

Acknowledgments. We thank W. Schmidt and M. Stix (KIS) and Y. Zhugzhda (IZMIRAN/AIP) for helpful discussions. The VTT on Tenerife is operated by the Kiepenheuer-Institut für Sonnenphysik (Freiburg) in the Observatorio del Teide of the Instituto de Astrofísica de Canarias. The Göttingen 2D-spectrometer was supported by the *Deutsche Forschungsgemeinschaft.*

References

Bendlin, C., Volkmer, R., 1995, A&AS 112, 371

Brown, T.M., 1991, ApJ 371, 396

Espagnet, O., Muller, R., Roudier, T., Mein P.,

Hoekzema, N.M., Brandt, P.N., Rutten, R.J., 1998, A&A 333, 322

Kiefer, M., Balthasar, H., 1998, A&A 335, L73

Rast, M.P., 1997, in: SCORe'96: Solar Convection and Oscillations and their Relationship, eds. F.P. Pijpers, J. Christensen-Dalsgaard, C.S. Rosenthal, Kluwer, Dordrecht, (1997), p.135

Rimmele, T.R., Goode, P.R., Harold, E., Stebbins, R.T., 1995, ApJ 444, L119

Stix, M., Zhugzhda, Y.D., 1996, in: Cool Stars, Stellar Systems, and the Sun, 9th Cambridge Workshop, PASP Conf. Ser. 109, ed. R. Pallavicini, p.163

Zhugzhda, Y.D., Stix, M., 1994, A&A 291, 310

High Resolution Solar Physics: Theory, Observations, and Techniques
ASP Conference Series, Vol. 183, 1999
T.R. Rimmele, K. S. Balasubramaniam, R.R. Radick, eds.

On the Relation of Photospheric Oscillations to Meso-Scale Flows

N. M. Hoekzema and P. N. Brandt

Kiepenheuer-Institut für Sonnenphysik, D-79104 Freiburg, Germany

Abstract. We use a high quality five-hour sequence of solar images taken at the Swedish Solar Vacuum Telescope, La Palma, to study relations of meso-scale divergence with oscillation amplitudes in the solar photosphere. Meso-scale flows are determined by local correlation tracking of the granulation. Statistical evaluation of the co-location probability of regions with more than twice the average Fourier amplitude and regions with high or low meso-scale divergence shows that high amplitude regions are preferentially located in meso-convergent regions and avoid meso-divergent regions.

1. Introduction

This paper builds on a series of papers (Hoekzema et al. 1998a, 1998b, 1998c) employing the correspondence factor C to explore local relations between fine structure and wave amplitudes in the quiet solar photosphere on the basis of image sequences obtained with the Swedish Vacuum Solar Telescope (SVST) on La Palma. The first paper gave evidence for the occurrence of localized wave sources and wave diffraction near the solar surface, while the second presented time delay alignment charts which suggest that localized sources have meso-scale patterning. The third paper addressed long-lived features in the solar granulation, so-called "intergranular holes", and quantified co-alignment of granular morphology with regions of high wave amplitudes.

Here we present results of a search for relations between Fourier amplitudes and meso-scale flow patterns: i.e., meso-scale convergent regions and meso-scale divergent regions. Much work has been done on acoustic emission from granular pistons (e.g., Goode et al. 1992, Rimmele et al. 1995), but the impact of larger-scale patterning such as meso-scale flow fields and supergranulation has received little attention. Whereas there is not a very great difference between the flow velocities associated with granulation, meso-scale flows, and supergranulation respectively — all are of the order of 1/2 to several $\mathrm{km\,s^{-1}}$ — it is worthwhile to compare their impacts on acoustic wave amplitudes.

The approach here follows that of the preceding three papers. We use similar reduction and measurement procedures, and apply our Fourier analysis technique on brightness histories (per pixel) of almost 90 min duration to produce Fourier amplitude maps for spatial and temporal correlation with the concurrent photospheric meso-scale divergence maps. We measure the amount of spatio-temporal alignment at varying time delays between different features through the statistical correspondence factor C that was introduced in the first paper of this series (Hoekzema et al. 1998a, henceforth called Paper I). It is used to quantify the co-spatiality of high oscillatory amplitudes with meso-scale morphology as measured by its divergence.

2. Data and analysis procedures

2.1. Observational data

The present analysis is based on a sequence of white light granulation images ($\lambda\,468 \pm 5\,\mathrm{nm}$) obtained on 5 June 1993 with the SVST on La Palma. The observing and intitial reduction procedures have been described in detail by Simon et al. (1994). From this sequence we analyze here 899 images, of 460 $\times$ 460 pixels covering a field of 41.7 Mm $\times$ 41.7 Mm (0.125 arcsec/pixel). The images used here had been aligned, destretched, and time interpolated to 21.05 s cadence.

2.2. Fourier mapping

Three non-overlapping 89.6-min segments were selected from the image sequence, each containing 256 images. Analogously to the procedure described in Paper I, they were apodized to an effective duration of about 60 min to produce pixel-by-pixel Fourier amplitude maps at different periodicities. This duration is short enough to resolve meso-scale pattern evolution, while long enough to resolve waves modes with periods as long as 20 to 30 minutes. In the analysis below we concentrate on regions of large amplitudes, i.e., locations with Fourier amplitudes exceeding twice the map average. They cover roughly 5% of the field. Fig. 1a (left) shows an example of an 8 min amplitude map (middle) together with the intensity averaged over 1 h (top) as well as the divergence of the granular flow pattern (bottom); here we show a somewhat larger field, i.e., (60 Mm $\times$ 64 Mm) than was used for the final correspondence analysis. The right column (Fig. 1b) exhibits an enlarged section of the lower left sub-field of Fig. 1a. A visual inspection of this figure shows hints of the co-spatiality of high power in the 8-min oscillation band and negative divergence, which can be quantified by the correspondence factor C (cf. Fig. 2a–c).

2.3. Correspondence factor

Spatial alignments are measured with the correspondence factor C, a statistical tool which quantifies the cospatiality of two pixel classes between two images or maps. It was introduced in Sect. 3 of Paper I and is defined as $C = f_{AB}/f_B$, where $f_{AB} = N_{AB}/N_A$ is the fraction of pixels of type A in one map that also belong to type B in another map and $f_B = N_B/N$ is the filling factor of pixels of type B. C quantifies cospatiality as a random-draw likelihood. $C = 1$ suggests that A and B are independent phenomena while values $C > 1$ imply preferential colocation and $C < 1$ implies spatial avoidance.

2.4. Divergence maps

Local correlation tracking (LCT) with a 1.36 arcsec FWHM window was used to derive granular proper motions from pairs of images 21 sec apart. The cor-

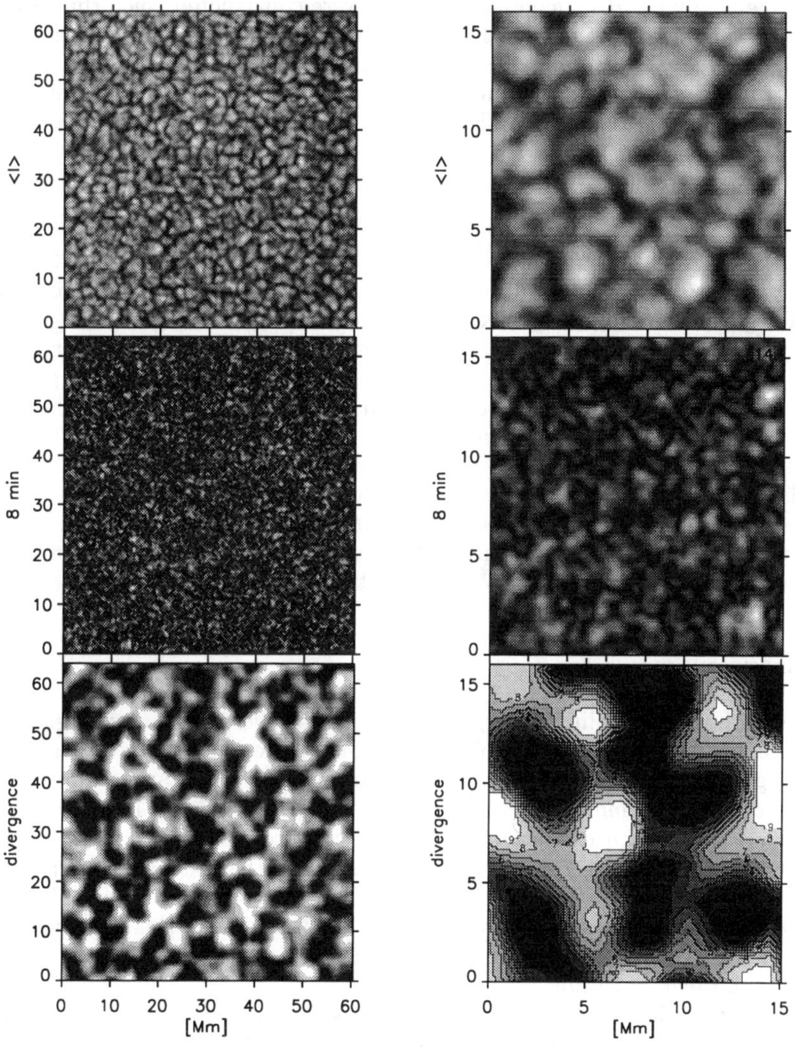

Figure 1. a) Left column: granulation pattern averaged over 1 h (top); amplitude of the oscillation at periods around 8 minutes (middle); divergence of the granular proper motion determined by LCT with positive divergence brighter than average (bottom). b) Right column: same quantities as in a) shown for the lower left sub-field of a).

responding flow fields were averaged over 1 h and the divergence of the flows was computed; in most cases it ranged between $\pm 7 \cdot 10^{-3}$ s^{-1}. From these averaged divergence maps 10 segmented sub-maps were constructed, such that each sub-map contained 10% of the pixels, binned between minimum and maximum divergence. Three such 1 h average sub-maps for the lowest divergence bin, three for an intermediate as well as the highest bin were then submitted to a test for co-location with the corresponding Fourier amplitude maps as described above.

3. Results

In Fig. 2a–c we present time delay charts of the spatial correspondence C between pixels with Fourier amplitudes exceeding twice the average value and pixels of regions with high negative divergence (= high convergence, Fig. 2a, top), high positive divergence (Fig. 2c, bottom), or with divergence near zero (Fig. 2b, middle). As an example, from Fig. 2a we see that 10% of the pixels with the *highest convergence* have a chance that is 10–30% larger than average to have *large Fourier amplitudes for most periods between 2.5 and approx. 20 min*, at that same moment and at the same location. The panel (Fig. 2b, middle) for pixels with divergence values near zero shows up almost blank and gives an indication of the noise. In all panels the distance between two contour lines is about $2\,\sigma$. These effects persist significantly for up to one hour or one hour and a half (Δt=-1 to -1.5 h) after the determination of the divergence.

4. Conclusion

Regions with Fourier amplitudes larger than twice the average show strong preference to coincide spatially with meso-convergent regions and to avoid meso-scale divergent regions. The effects demonstrated in Fig. 2 are rather similar to those observed for normal granulation in Fig. 7 of the paper by Hoekzema et al. (1998c). In that Figure intergranular lanes (convergent regions) show much larger correspondence with regions of more than twice the average Fourier amplitudes than the regions of divergence (granules). For periods below 3–4 min the meso-scale effects are smaller by roughly 25% than that of the granulation, and contrary to the granulation they are strongly significant as well for wave periods in the range 5–15 min.

Whether the effect that we find here is caused by the direct influence of the spatial velocity and temperature structure of the atmospheric layers involved or rather by changes of the granular structure on meso-granular scales (as found e.g. by Brandt et al. 1991) is a question left for future investigations.

Granulation and meso-scale flows display similar horizontal velocities and impact on Fourier amplitudes. As the flow velocities associated with the supergranulation also are of the same order, it would be interesting to analyze their relation with Fourier amplitudes.

Acknowledgments. We thank G. Scharmer (Stockholm) for the opportunity to observe at the SVST. The observations were performed by one of us (PNB) together with G. Simon (Sunspot). We are indebted to W. Wang (Stockholm) and R. Kever (La Palma) for observational support, and to M. Shand of

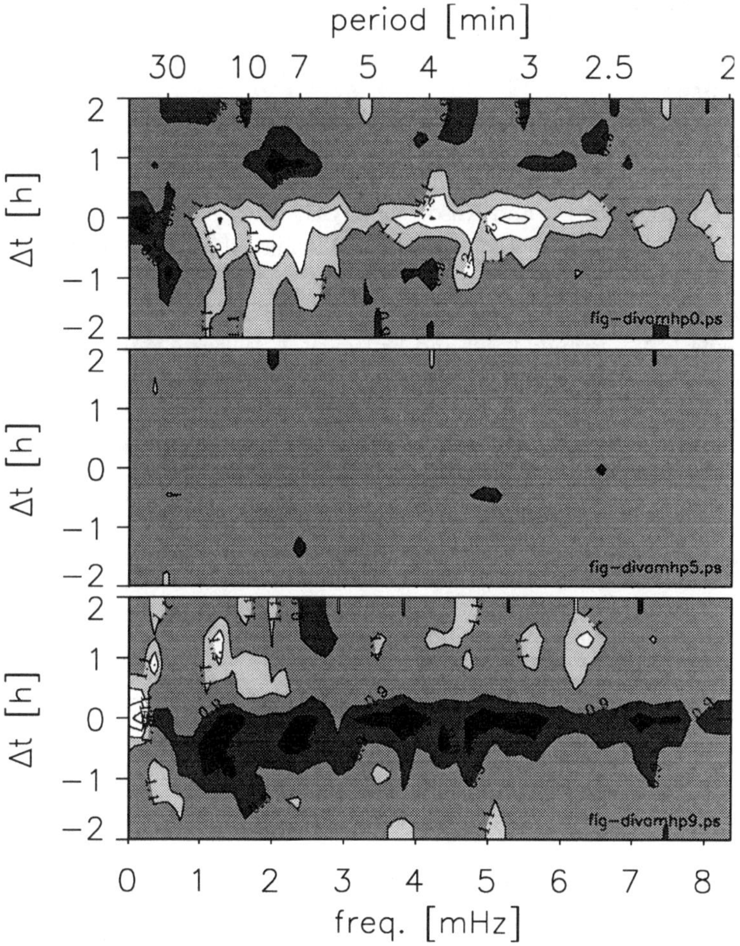

Figure 2. a) Top: contour plot of the spatial correspondence C between high oscillatory amplitudes (exceeding twice the average) and the regions of strongest negative divergence (convergence); abscissa is oscillation frequency in mHz, ordinate is time lag in h. b) Middle: same as a) for divergence values near zero. c) Bottom: same for regions of strongest positive divergence.

DEC (Paris) for financial and technical assistance with the data acquisition system. R. Shine (Palo Alto) and W. Mühlman (Graz) were involved in the pre-reduction of the data; their help is gratefully acknowledged. The SVST is operated by the Swedish Royal Academy of Sciences at the Spanish Observatorio del Roque de los Muchachos.

References

Brandt P.N., Ferguson S., Scharmer G.B., Shine R.A., Tarbell T.D., Title A.M., and Topka K. 1991, A&A 241, 219

Goode P.R., Gough D., and Kosovichev A.G. 1992, ApJ 387, 707

Hoekzema N.M., Rutten R.J., Brandt P.N., and Shine R.A. 1998a, A&A 329, 276

Hoekzema N.M. and Rutten R.J. 1998b, A&A 329, 725

Hoekzema N.M., Brandt P.N., and Rutten R.J. 1998c, A&A 333, 322

Rimmele T.R., Goode P.R., Harold E., and Stebbins R.T. 1995, ApJ 444, L119

Simon G.W., Brandt P.N., November L.J., Scharmer G.B., and Shine R.B. 1994, in *Solar Surface Magnetism*, (eds.) R.J. Rutten and C.J. Schrijver, Kluwer Ac. Publ., 261

Group Discussion

Goode : In our observations, we see no role for exploding granules (≈ 1 /correlation between seismic events and mesogranulation.

High Resolution Solar Physics: Theory, Observations, and Techniques
ASP Conference Series, Vol. 183, 1999
T.R. Rimmele, K. S. Balasubramaniam, R.R. Radick, eds.

Time Series of Solar Photospheric Spectrograms Bisector Analysis

A. Hanslmeier

Institut für Astronomie, Univ. Graz, Austria

J. A. Bonet and M. Vazquez

Instituto de Astrofisica de Canarias,, La Laguna, Tenerife, Spain

Abstract. Time series of spectrograms near solar disk center, obtained with the VTT telescope at the observatorio del Teide were analyzed. A 47 min time series was used to study the influence of oscillations on the behavior and variation of several line parameters, calculated from the line profiles.

1. Introduction

The convective overshoot of plasma flows from the solar convection zone into the photosphere produces to the well known phenomenon of solar granulation. Overviews about the properties of this phenomenon were given by Bray, Loughhead, & Durrant 1984, and Spruit, Nordlund, & Title 1990.

To study the evolution of the solar granulation time series of spectrograms or white light images are needed. However, it is very difficult to obtain long lasting high quality time series because of atmospheric seeing, image and telescopic motions etc. Since we want to study convective motions, oscillatory motions must be separated from convective overshooting. This is possible in the case of time series by applying a filter in the spatial-temporal domain (Title et al.1986). Furthermore, the use of a correlation tracker developed at the IAC (Instituto de Astrofisica de Canarias, La Laguna, Tenerife) guaranteed that always the same physical point on the sun was observed.

Parameters derived from such observations are needed for HD simulations. These can be divided into 2-D simulations (less accurate, however better spatial resolution, e.g. Nordlund (1976), Nelson (1978) (all are incompressible) and Steffen, Ludwig & Kruss 1989, Gadun (1995) (compressible)) and 3-D simulations (Nordlund (1982) (anelastic approximation in cont. equation), Stein and Nordlund (1989), Cattaneo, Hurlburt, & Toomre 1989, (fully compressible)).

2. Observations

We used a 1k× 1k CCD to take solar photospheric line spectrograms at the 70 cm VTT (Schröter, Soltau, & Wiehr 1985. The data consists of a 47 min time series the time interval between two successive spectrograms was about 20 s.

The entrance slit of the spectrograph covered a quiet region (this was checked by comparing with a $H-\alpha$ slit jaw), the spectrograph entrance slit width was set to 100μ. The following lines were observed:

Table 1. Important parameters for the lines analyzed (Moore, Minnaert, & Houtgast 1966. 1966)

Line	λ[Å]	W_λ[mÅ]	*height*[km]
line I	6494.994	165	500
line III	6494.476	69	200

The pixel resolution was 0.11 arcsec in the spatial direction, the exposure time 1.5s. The image was stabilized using a correlation tracker (Bonet et al. 1994). Our data set consists of 501 points in the spatial direction and 141 points in the time direction.

Flat fields where obtained by drifting the telescope in the same spectral region. Transparency changes in the earth's atmosphere were compensated by normalization of each spectrogram to its mean continuum intensity.

The line shifts were obtained from the profiles by a) finding the minimum using a 4th order polynomial fitting, b) reference to the positions of the terrestrial line.

Finally a 5 min oscillation filtering was applied (Title et al. 1989) using $\omega = 5k$. Such a filtering seems to be not correct from the mathematical point of view, since one needs a 2-D time series, which is not the case here. Our procedure was tested using a time series of 2-D images of solar granulation and applying the filter a) to the original 2-D time series and b) to successive 1-D rows of the time series and then reconstructing the image again. We found no remarkable difference between these two images (procedure a, b) .

3. Results

If we average all bisectors over space and time and apply this process to the unfiltered and filtered data, we find no difference between the filtered and unfiltered mean bisector. This can be used as an indication that the process of oscillation filtering worked properly, since when averaging over time and space these oscillations should cancel. A similar result was obtained for the deeper originating line however, due its deeper formation height smaller asymmetries were observed.

Plotting the whole set of continuum fluctuations as a function of time and spatial coordinate we can see that there are regions where the continuum intensity was lower than the average (dark zones) and regions where the continuum intensity was higher than the average (bright zones).

In order to see if there is any difference between the influence of oscillations for bisectors arising mainly from bright granular areas and those arising mainly

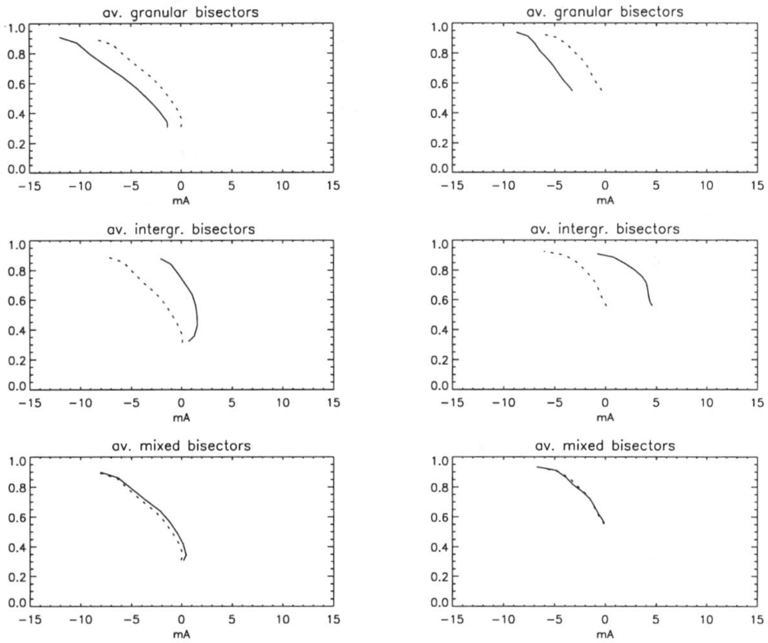

Figure 1. Bisectors of line I (left) and line III (right) averaged according to their evolution. a) mainly granular, b) mainly intergranular, c) mixed.

from dark intergranular lanes during the time series, we depicted all bisectors arising from line profiles with a continuum intensity $I > \bar{I}$, $\bar{I}$ denoting the averaged continuum intensity, from the data set and call them granular bisectors and consequently all bisectors belonging to areas where $I < \bar{I}$ as intergranular bisectors. The results can be seen in Fig. 1 for the higher originating line and Fig. 2 for the deeper forming line. In these plots we give also the results for the filtered data (dashed). It is clearly seen that the filtering has practically no influence on the "granular" bisectors, however, in the case of intergranular bisectors the asymmetries turn from red values (Fig. 1b, 2b full lines) for the unfiltered case to blue values for filtered data (Fig. 1b, 2b dashed lines). Thus the filtering process seems to influence only on the intergranular bisectors. In Figs. 1c, 2c we give also the so called mixed bisectors arising from zones where both higher and lower continuum intensities occurred.

In the following two figures we follow the evolution of a typical granulum and a typical intergranulum - again according to the above mentioned definition that in the case of a granulum the continuum intensity of the line profile must be greater than the average continuum intensity, $I > \bar{I}$ and in the case of the intergranulum $I < \bar{I}$.In order to show the variation of the individual bisectors, the bisectors were shifted by 2mÅ with respect to each other and numbered (only the even numbers are given here).

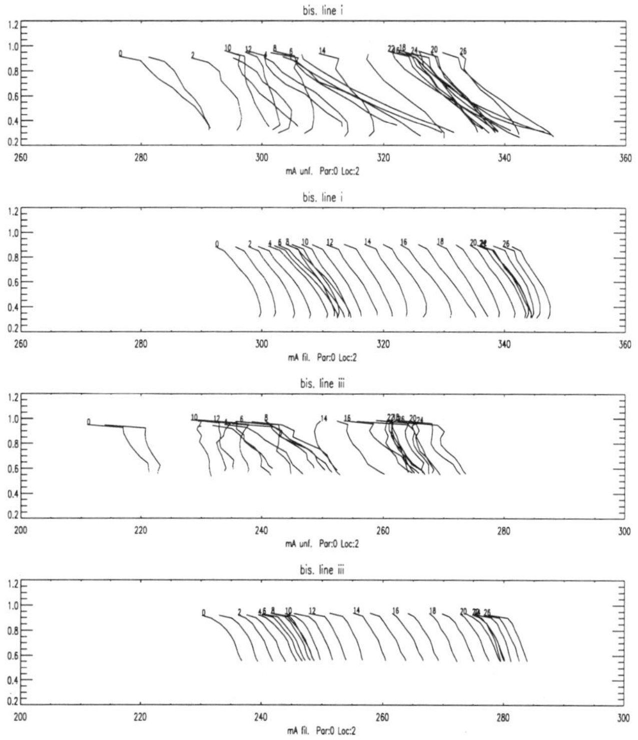

Figure 2. "Granular" bisector variations along a selected range from δI fluctuations; a) line I, b) line I filtered, c) line III, d) line III filtered.

In Fig. 3 we give the bisector evolution for a typical granular evolution; as it can be seen these bisectors are from areas with enhanced I values and have large blue asymmetries. Comparing the filtered bisectors with the unfiltered values the only difference is that in the cases for the filtered bisectors the arrangement seems to be better ordered and the bisector footpoints are stronger affected than the bisector endpoints. A similar trend but weaker can be observed for bisectors belonging to the weaker line III.

In Fig. 4 the bisector evolution is shown for a time interval during which the I values are relatively small, that means we follow the bisector evolution over a dark region (intergranulum). For the unfiltered data, in both cases, line I and line III one clearly gets what can be called " intergranular bisectors" with relatively large red asymmetry values. However, one should note that after filtering these red asymmetries completely vanish and turn into a blue asymmetry again. By carefully comparing the plots for the unfiltered and the filtered bisectors it seems that one can obtain the unfiltered bisectors from the filtered ones by just shifting the endpoints to the read and the footpoints toward the blue.

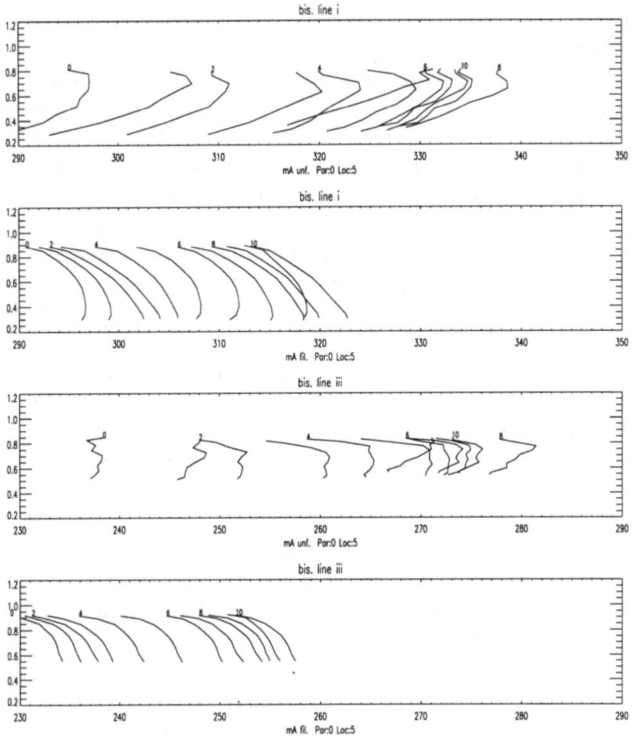

Figure 3. "Intergranular" Bisector variations along a selected range from δI fluctuations; a) line I, b) line I filtered, c) line III, d) line III filtered.

Considering the absolute shifts we note that here the whole sample of filtered bisectors is shifted toward the blue with regard to the unfiltered bisectors. In the case of the example of bisectors given by Fig. 5 we notice a slight blueshift of the sample of filtered bisectors with regard to the sample of unfiltered bisectors. This trends are even more pronounced in the samples of the bisectors belonging to line III.

4. Discussion

Two important conclusions can be drawn from this paper: a) it is possible to use 1-D time series for appropriate filtering for oscillations and b) the influence of the filtering greatly affects line asymmetries of lines originating in the intergranular lanes whereas for lines that originate in the granulum the influence is very small. This can be regarded as a hint that oscillations are mostly concentrated in the intergranulum, which was by different authors (e.g. see Goode et al. 1999 or Rast, 1999).

The influence of convection on the p-modes was studied by many different groups (e.g. Goldreich and Kumar, 1988) because it was found that convective eddies may be the agent of oscillations. Bogdan (1989) gave a description of the interaction of solar acoustic oscillations with convection from a theoretical point of view and pointed out the interaction between sound and vorticity.

In the paper of Bonet et al. (1991) asymmetries and shifts of the solar KI $\lambda 7699 \mathring{A}$ line were studied. They found that a line profile calculated by granulation models (using the code of Nelson,1978) does not match the observational results and studied the influence of 8min period gravity waves on the line bisectors. Introducing these oscillations into the calculation of the profiles they found decreasing bisector asymmetries.

Rimmele et al. (1995) and Goode et al. (1998) report on seismic events that are associated with the excitation of solar oscillations. They found that these events are located in the dark intergranular lanes and that the seismic events are preceeded by a further darkening of an already dark lane; on the temporal end of the seismic event a still further darkening was also found.

In a theoretical study given by Rast (1999) the onset of instabilities in downflowing intergranular areas are discussed. Such downflowing regions evolve when a local heat reduction in vertical flows leads to net energy losses and two kinds of instability occur: varicose and sinuous. The first one is compressive and is caused by dynamic pressure fluctuations behind a plume leading to secondary head formation. The second one results from shear instability of the stem flow. Such formed pressure fluctuations are the cause of solar acoustic excitation.

Our results are in good agreement with these findings because we also found that the oscillations are concentrated in the intergranulum.

Acknowledgments. A. Hanslmeier acknowledges the financial support of this project by the Austrian Fonds zur Förderung der wissenschaftlichen Forschung. All authors acknowledge the financial support from the Acciones Integradas.

References

Bogdan, T.J., 1989, ApJ 339, 1132

Bonet et al. 1994, JOSO Annual Report, p. 73

Bonet J.A., Márquez I., Vázquez M., Wöhl H., 1991, A& A 244, 492

Bray, R.J., Loughhead, R.E., Durrant, C.J., 1984, The Solar Granulation, Chapman & Hall, London, 2nd ed.

Cattaneo, F., Hurlburt, N.E., Toomre, J., 1989, in: Rutten R.J. and Severino G. (eds.), Solar and Stellar Granulation, Kluwer, Dordrecht, p. 415

Gadun, A.S., 1995, kinemat. i Fiz. nebesn. tel, 11, 3,54

Goldreich, P., Kumar, P., 1988, ApJ 326, 462

Goode, P.R., Strous, L.H., Rimmele, Th. R., Stebbins, R.T., 1998, ApJ 495, L27-L30

Goode, P.R., Strous, L.H., Nordlund, A. P., Stein, R.F., Rimmele, T.R.,1999, this volume

Moore, C.E., Minnaert, M.G.J., Houtgast, J., 1966, Rowland tables, 2nd rev.

Neslon G.D., 1978, ApJ 238, 659-666

Nordlund, A.,1976, A&A, 50, 23-39

Nordlund, A.,1982, A&A, 107, 1-10

Rimmele, Th. R., Goode, P.R., Harold, E., Stebbins, R.T., 1995, ApJ 444, L119

Rast, M.P., 1997, in Solar Convection and Oscillations and their Relationship, eds. F.P. Pijpers et al., 135, Kluwer Acad. Publ.

Rast, M., 1999, this volume

Spruit, H.,C., Nordlund, A., Title, A.M., Ann. Rev. Astron. Astrophys. 1990, 28, 263-301

Schröter, E.H., Soltau, D., Wiehr, E., 1985, Vistas Astron. 28, 519

Steffen, M., Ludwig, H.G., Krüss, A., 1989, A&A 213, 371-382

Stein, R.F., Nordlund, A., 1989, Ap.J. Lett. 342, L95-98

Title, A.M., Tarbell, T.D., Simon, G.W., and the SOUP Team, 1986, Adv. Space Res. 6(8), 253-262

High Resolution Solar Physics: Theory, Observations, and Techniques
ASP Conference Series, Vol. 183, 1999
T.R. Rimmele, K. S. Balasubramaniam, R.R. Radick, eds.

Imaging Spectroscopy of CO Lines Compared with Three-Dimensional Radiative Transfer

Han Uitenbroek
Harvard-Smithsonian Center for Astrophysics,
60 Garden Street, Cambridge, MA, 02138, USA

Abstract. Spectroscopic imaging observations of the CO fundamental vibration-rotation transitions at 4.6 μm, obtained at the Kitt Peak McMath-Pierce facility, show that the dynamics of both the solar granulation and, to a lesser extent, the five-minute oscillations play an important role in CO line formation. Spectro-heliograms made in the core of strong CO lines display an inverted granular contrast with dark areas corresponding to granule centers, and a bright network corresponding to the inter-granular lanes. This observation is confirmed by multi-dimensional radiative transfer modeling of CO line formation in a solar convection simulation snapshot. In addition, such transfer calculations allow us to interpret the observed pattern, and to predict that the darkest CO line cores at disk center are associated with the strong adiabatic expansion and cooling that occurs over granule centers when warm up-flowing material runs into the steep density gradient of the stable layer above the photosphere. Unfortunately, current granulation simulations do not extend to high enough layers in the atmosphere to model formation of CO lines close to the solar limb where they exhibit their anomalous temperature behavior.

1. Introduction

The low temperatures of 3700 K measured in the cores of strong vibration-rotation lines of molecular CO near the solar limb, first discovered by Noyes & Hall (1972) and later confirmed by Ayres & Testerman (1981), do not agree with temperatures measured in almost any other spectral diagnostics that forms at a similar altitude in the solar atmosphere. Intensities measured in, for instance, the UV continuum, in the millimeter continuum, and in strong lines like Ca II H and K indicate that minimum temperatures could be as high as 4400 K. However, the interpretation of observed intensities in terms of atmospheric temperatures is strongly model dependent, and the above statement merely indicates that no single one-dimensional static plane-parallel model can be found that accommodates both the spatially averaged spectra of the infrared CO lines, and the other diagnostics at the same time. This prompted Ayres and co-workers (Ayres 1981, Ayres, Testerman, & Brault 1986) to suggest that a solution might be found in a spatially differentiated atmosphere where the strong cooling provided by the CO molecule itself might produce a thermal bifurcation between non-magnetic inter-network on one hand and magnetic network elements on the other hand. In the network elements additional mechanical heating would destroy CO so that without its cooling the atmosphere would locally relax to a higher temperature.

To determine how CO absorption is distributed over the solar atmosphere and, eventually, to get a handle on the low temperatures observed near the limb, spatially and temporally resolved observations are required. These became possible for the first time with the installation of the Near Infrared Magnetograph (NIM1) at the McMath-Pierce facility at the Kitt Peak National Solar Observatory. Early observations with this instrument reported by Uitenbroek, Noyes, & Rabin (1994) showed that the *p*-mode oscillations play an important role in modulating CO absorption. These observations, however, were obtained under moderate seeing conditions. In this paper I present additional observations obtained under significantly better conditions, and I compare these observations with multi-dimensional radiative transfer calculations to improve their interpretation.

2. CO observations

Until 1994 CO observations at the McMath-Pierce (which is in principle the best telescope for observations at 4.6 μm since it is a large diameter reflector) were done mostly with the FTS which offered good spectral resolution, but limited spatial resolution as determined by the size of the entrance window. In addition, time resolved observations were difficult if spectral resolution was to be maintained because of the time it took to sample the spectrum in Fourier space. This situation was greatly improved when the Near Infrared Magnetograph (NIM1) instrument built around a 256×256 InSb detector (Amber Engineering) became available. This instrument has since made it possible to obtain time resolved long-slit spectra of the fundamental CO lines to study the temporal and spatial distribution of CO in the solar atmosphere. Even at these long mid-infrared wavelengths the 1.5 meter primary of the McMath-Pierce provides a reasonable theoretical resolution limit of $0\farcs8$. Cylindrical minification optics provided a field of view along the slit of approximately $100''$ over a spatial sampling of $0\farcs38$/pixel. The theoretical spectral resolution of the 13 meter vertical spectrograph with IR grating was 90 000, the sampling in the direction of the dispersion was 0.01cm^{-1}/pixel.

Series of long-slit spectra were combined to make time series of one slit position or, alternatively, to make a spatial scan by stepping the slit over the solar surface. These data cubes could then be used to construct time-slices or spectro-heliograms, respectively, of a particular function of wavelength, as for instance the intensity in the core of a line, or its Doppler shift. Fig.1 shows an example of three cotemporal time-slices obtained in May 1994 under good seeing conditions, approaching the theoretical resolution limit. The slices represent, from left to right, the intensity in the core of the strong 3-2 R14 line (converted to brightness temperature), the Doppler shift of that line, and its broadening. All three of these quantities were determined by fitting Gaussian profiles to the 3-2 R14 line for each position along the slit for each of the 512 spectra that make up the time series. A number of interesting features are apparent. The slit crossed two bright network elements which appear as the bright traces at x = 32 and 64 arcsec. Although there is no signature of the network in the

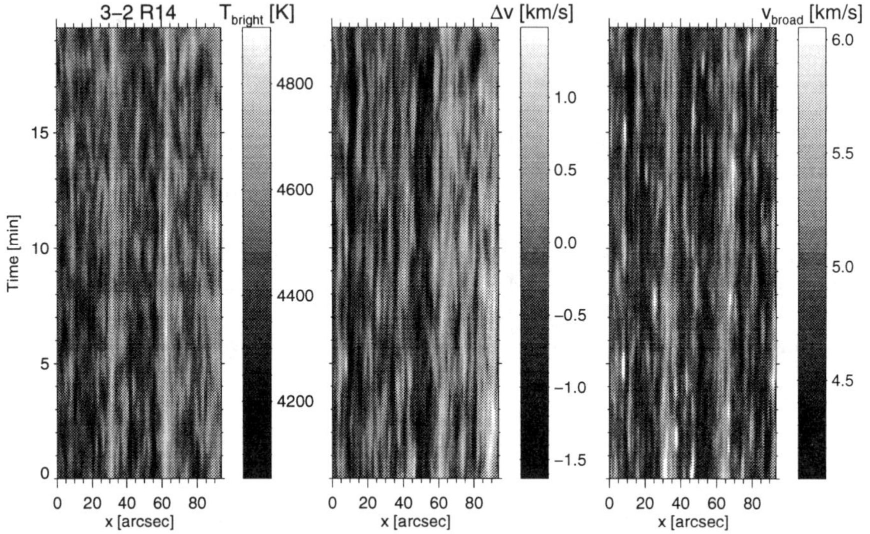

Figure 1. Time slices of the 3-2 R14 line core brightness (left panel), Doppler shift (middle panel), and line broadening. A total of 512 spectra was used for the time series at an average cadence of 2.3 sec.

Doppler signal (middle panel), it *is* visible in the line broadening (right panel), over even a wider area than in the intensity slice. There appear to be no areas in the inter-network regions that are persistently cooler than others: apart from the network all locations along the slit show similar bright and dark episodes. Most of these brightenings appear in a pattern of bright and dark streaks of several minutes duration, reminiscent of granulation patterns. The darkest (and presumably coldest) elements appear as part of this pattern. However, from the time-slices it is not clear whether they correspond to inter-granular lanes or to the interiors of "exploding" granules for instance. On top of this granular pattern the influence of the five-minute oscillation can be seen in the presence of bright and dark bands that appear at various angles in the intensity and Doppler signal (especially clear examples can be found at the right hand side of the left and middle panels of Fig.1

A better way to judge where the coldest elements in the CO line cores (at disk center) are located with respect to the granulation pattern is to look at a spectro-heliogram. Fig.2 shows a heliogram constructed in the core-brightness temperature of the 3-2 R14 line. It was obtained in the same region as the time-slices displayed in Fig.1 Both the heliogram and the time-slices were obtained at a viewing angle corresponding to $\mu = \cos(\theta) = 0.5$. The two bright patches of network crossed by the slit are easily recognized. It is clear from the heliogram that not the regular granulation pattern, but rather an inverted pattern of bright lanes with dark interiors is visible in the core of the 3-2 R14 line. This is reminiscent of the effect in the wings of the Ca II H and K lines described by Evans & Catalano (1972) where the granular contrast reverses at about 0.3 nm

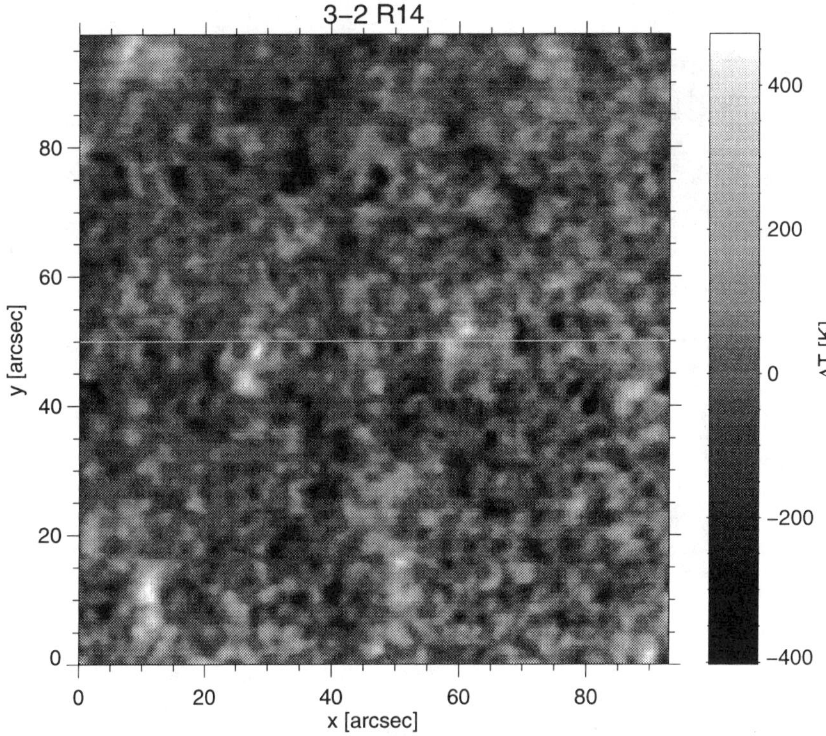

Figure 2. Observed spectro-heliogram in the core of the CO 3-2 R14 line. The horizontal line at $y = 50$ arcsec indicates the slit position used for the time-slices displayed in Fig. 1. The heliogram was obtained just before the start of the time series.

from line center when going from the wings of the line towards the core. This inverted contrast pattern indicates that the coldest elements visible at disk center in the cores of strong CO lines are associated with the centers of granules, rather than with the inter-granular lanes.

3. CO radiative transfer

Although the reversed contrast pattern of the granulation observed in the heliogram of the line-core brightness of strong CO lines like the 3-2 R14 line seems clear, it is observed close to the spatial resolution limit. In order to confirm the pattern, and to investigate what the reason behind the reversal is, multi-dimensional radiative transfer calculations of CO line formation in a snapshot of a hydrodynamic solar granulation simulation by Stein & Nordlund (1989) were performed. The temperature stratification and pattern of convective flows in a two-dimensional cut through such a simulation snapshot (which is inherently

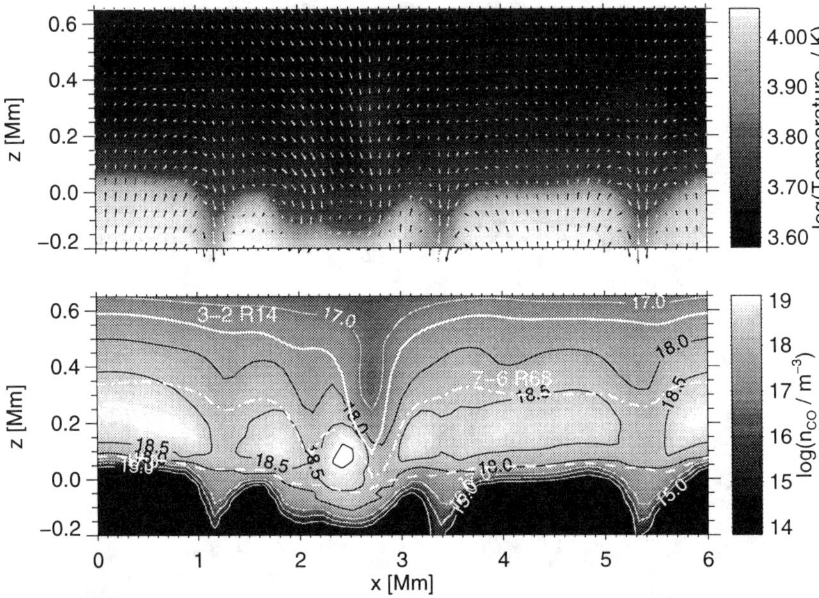

Figure 3. Upper panel shows two-dimensional cut through granulation snapshot. The false color image shows temperature distribution, while the arrows outline the convective flows. The bottom panels show the CO concentration in the cut. Note the enhanced concentration of CO due to the adiabatic cooling over the granules, and the reduced concentration over the lanes where the return flow of material is heated by compression. The run of optical depth unity in the strong 3-2 R14 line (thick solid curve), medium strength 7-6 R68 line (thick dot-dashed curve), and 4.6 μm continuum (thick dashed curve), respectively, is also drawn.

three-dimensional) are shown in the upper panel of Fig.3. Material rises through the broad upflow regions corresponding to bright granules in optical continuum images. When this upflowing material runs into the steep density gradient of the stable layer above the photosphere it has to expand rapidly to conserve mass flux causing it to cool, while the return flow is heated by compression when this lateral flow decelerates over the inter-granular lanes.

In such a convective model several radiative transfer effects play a role in the line formation process that do not occur in a one-dimensional spatially averaged model. Of these listed below only the first is a truly multi-dimensional transfer effect, the others are important because of the non-linearity of radiative transfer. Due to this non-linear reaction of the emitted spectrum to the variations of vertical temperature and density gradients over the surface, the horizontally averaged spectrum of the inhomogeneous atmosphere will in general not be reproduced by the spectrum of a one-dimensional atmosphere with the same (horizontally) averaged thermodynamic properties, even if lateral radiative transfer were to be neglected in the former. Three transfer effects that are important for the discussion here are:

1. In an exponentially stratified atmosphere radiation travels mostly vertically. Only elements smaller than a few density scale heights are affected by horizontal exchange of radiation (see Jones 1986, and Uitenbroek 1998).
2. Due to the strong temperature sensitivity of the H^- opacity (which increases with temperature when more electrons become available) hot material is masked, resulting in a more homogeneous radiation field compared to what would be expected on the basis of temperature differences at a fixed geometrical height (see Avrett 1990, Uitenbroek 1998).
3. In a convective atmosphere the adiabatic cooling of overshooting granules and the compression heating of the return flow results in a steep temperature gradient over hot granules and a shallow one over the inter-granular lanes, and eventually gives rise to a reversal of contrast in stronger lines.

In the transfer calculations presented here it was assumed that the CO lines had LTE source functions. However, because of the presence of electron scattering opacity, amounting to about 1% at the top of the atmosphere in the current models, the radiation field had to be lambda-iterated to get a consistent Non-LTE multi-dimensional transfer solution. The chemical balance between molecules of CO, H^-, H_2, H_2^+, CH, CN, and H_2O and their constituent atoms was solved assuming equilibrium according to Saha. Horizontal boundary (inherent to the original hydrodynamical simulation) conditions were used to solve radiative transfer along rays at an angle with the normal direction of the atmosphere. The effects of Doppler shifts due to the granular flows were taken into account in calculating the emergent line profiles.

The concentration of CO in a two-dimensional cut parallel to the $x - z$ plane through the simulation snapshot is shown in the bottom panel of Fig. 3. This figure also plots, as a function of horizontal position, the curves of optical depth unity for the cores of the strong 3-2 R14 and the medium strength 7-6 R68 lines, and the 4.6 μm continuum. Since the continuum opacity is almost exclusively due to H^- free-free processes, its optical depth unity curve more or less follows the contours of temperature resulting in a low-contrast emerging continuum intensity (see effect no. 3 above.) The opacity at the wavelengths of the CO line cores is the sum of the line and background opacity and is affected both by the CO concentration, the H^- free-free opacity, and to a much lesser extent (at least at disk center) by the electron scattering at the top of the model. From the optical depth unity curve of the 3-2 R14 line it is clear that this strong line forms too high in the granulation model to be modeled realistically. This is because the hydrodynamic simulation has its proper upper boundary at 500 km above the photosphere. The model is extended to 650 km with a fiducial layer to allow this boundary to be open. At disk center the weaker 7-6 R68 line is not affected by the boundary as it forms deep enough in the atmosphere, however, calculations at the limb would be less reliable as they sample the atmosphere at higher altitudes.

Fig.4 shows the emergent disk center intensity in the core of the medium strength 7-6 R68 line from a full three-dimensional transfer calculation. It clearly

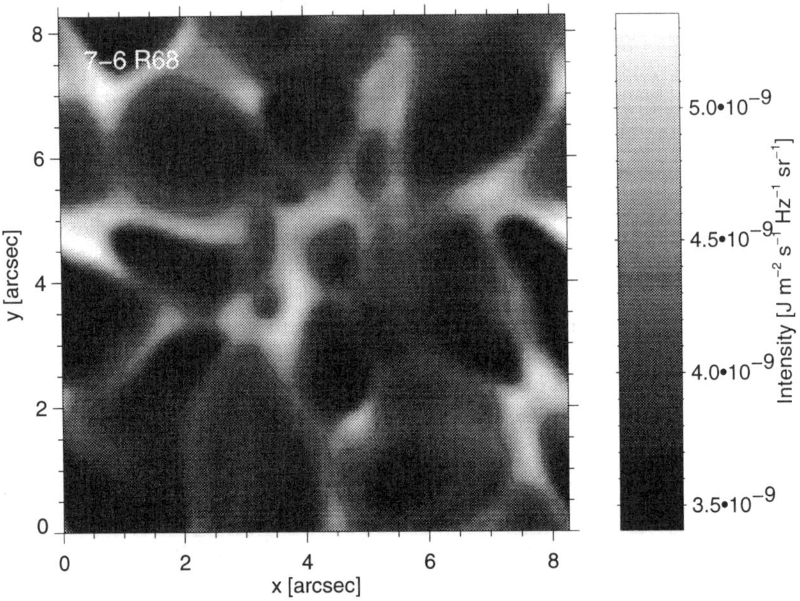

Figure 4. Calculated emergent intensity in the core of the medium strength 7-6 R68 line.

confirms the observed pattern of the reversed granular contrast. Moreover, we can understand how this reversal takes place by looking at the height of formation of the line as delineated by the location of optical depth unity (thick dot-dashed curve in the bottom panel of Fig.3 Over the hot granules the strong horizontal expansion of the flow causes the atmosphere to be cooled. This cooling increases the amount of CO in these regions (see the display of CO concentration in the bottom panel of Fig. 3) pushing line formation outward to cooler temperatures which map directly into lower emergent intensities due to the LTE source function of the CO lines. Over the inter-granular lanes the compression heating of the return flow destroys CO allowing radiation from deeper and hotter layers to escape in the line resulting in higher emergent intensities.

4. Conclusions

Combining spectroscopic imaging observations of the CO fundamental vibration-rotation transitions at 4.6 μm with theoretical multi-dimensional transfer calculations we now have a better understanding of the location of the coldest material in these lines resides when observed at disk center. "Thermal bifurcation" of the CO lines appears to be mainly introduced by convective flows rather than by the distinction between magnetic and non-magnetic areas, although the magnetic network appears consistently bright in the CO cores. CO line-core intensities and shifts show similar temporal variations everywhere in the quiet-Sun and no persistently cool areas seem to exist. Heliograms obtained in the cores of CO lines show an *inverted* granular contrast with the darkest areas corresponding to granule centers, and the inter-granular lanes appearing bright. Multi-dimensional transfer calculations exhibit the same pattern and make clear that the darkest CO line cores correspond to the adiabatically cooled areas due to the rapid horizontal expansion over strong granules. Unfortunately, current hydrodynamic simulations do not extend to heights great enough to study the behavior of CO lines towards the limb where they exhibit their anomalous brightness temperatures.

Acknowledgments. It is a pleasure to thank Doug Rabin and Dave Jaksha of the National Solar Observatory at Kitt Peak for developing the NIM1 instrument and for their support during the observations. This work was supported by grants NAGW-2545 and NAG5-3106 from NASA.

References

Avrett, E. H. 1990, in IAU symp. 138, Solar Photosphere: Structure, Convection, and Magnetic Fields, ed. J.-O. Stenflo, Kluwer, Dordrecht, p. 3

Ayres, T. R. 1981, ApJ 244, 1064

Ayres, T. R., & Testerman, L. 1981, ApJ 245, 1124

Ayres, T. R., Testerman, L., & Brault, J. W. 1986, ApJ 304, 542

Evans, J. W., & Catalano, C. P. 1972, Solar Phys. 27, 299

Jones, H. P. 1986, in Deinzer, W., Knölker, M., & Voigt, H. H. (eds.), *Small Scale Magnetic Flux Concentrations in the Solar Photosphere*, Abhandl. Akad. Wiss. Göttingen, Math.-Phys. Klasse Dritte Folge Nr. 38, p. 127

Noyes, R. W., & Hall, D. N. B. 1972, BAAS 4, 389

Stein, R. F., & Nordlund, Å. 1989, ApJ 342, L95

Uitenbroek, H., Noyes, R. W., & Rabin, D. M. 1994, ApJ 432, L67

Uitenbroek, H 1998, ApJ 498, 427

High Resolution Solar Physics: Theory, Observations, and Techniques
ASP Conference Series, Vol. 183, 1999
T.R. Rimmele, K. S. Balasubramaniam, R.R. Radick, eds.

High Resolution Information on Granulation from Full Disk Fe Line Asymmetry

William Livingston

National Solar Observatory, NOAO, Tucson, AZ 85726

Lloyd Wallace

Kitt Peak National Observatory, NOAO, Tucson, AZ 85726

You-ran Huang

Dept of Astronomy, Nanjing University, Nanjing, China

Elena Moise

Astronomical Institute, 75212 Bucharest, Romania

Abstract. Full disk, Sun-as-a-star, measurements of Fraunhofer line bisectors show that the amplitude of line asymmetry varies over the solar activity cycle but NOT in phase with it. Assuming the presence of surface magnetism inhibits granular convection, one can expect that bisector amplitude would vary inversely with activity. New FTS measurements taken with the McMath-Pierce image 1 meter out of focus compare plage vs quiet sun areas and verify this assumption. On average the plage bisector amplitude is reduced. Based on this one would expect bisector amplitude to track full disk total magnetic flux. Instead, as of July 1998, the solar cycle based on bisector amplitude has just reached solar minimum. According to most solar indices, solar minimum was attained approximately in 1996. Why the difference? High resolution quiet Sun granulation pictures that display filigree generally show that the latter are confined to intergranule lanes. In plage, where the fields are more crowded, the separation between granules and magnetic bright points is not so clear. Maybe this is why we get a near plage-type bisector. We conjecture that active region magnetism interacts only slightly with the convective process and that there is another regime of magnetism (the 'weak fields') that drive the full disk line asymmetry signal.

1. Introduction

There are approximately 5×10^6 granules on the whole solar disk at any time (Muller 1995). The convective motions of these granules will impart an asymmetry to a moderate strength Franhofer line. Fe lines are inherently symmetric because there is only one dominant isotopic component ($Fe^{56} = 91.8\%$), and they show no hyperfine structure. This asymmetry assumes a 'C' shape for the following reasons (Dravins, Lindegren, & Nordlund 1981; Bray, Loughhead, & Durrant 1984; Gray 1988). The core of the line lies mostly above the granulation layer and so is undisplaced. As we go to lesser line depth the brightness-velocity correlation results in an increasing blue shift from rising bright granules. As we approach the continuum in the line wing we begin to sense the red shift from the descending gas of the intergranule lanes in preference to the blue component of the granule center, for reasons of changing opacity, and the line bisector swings redward. The resulting line asymmetry is very stable because of the large number of granules being sampled. Harvey (1985) estimates the full disk granular noise is approximately 0.7 m/s. This stability is illustrated in Fig.1 where we compare the bisectors of a number of selected (unblended) Fe lines.

2. Observations

Our time series of full disk bisector observations began 20 May 1980 and has continued at a frequency of 2-4 times a year. The spectra are observed with the 1 meter FTS set to a resolving power of 500000. Thanks to the excellent initial advice of J.Brault in terms of choice of resolution and avoidance of vignetting from internal and telescope sources the configuration has remained constant.

Each observation represents 30 minutes of FTS scans covering 5000 to 6300Å. Reduction is by the BDECOMP program, a version of Decomp with special subroutines written by C.Mahaffey. Y.Huang carried out a lengthly appraisal of Fe lines in this spectral window and selected 17 as being nearly blend-free.

3. Results

The results of this long-term program are basically summarized in Fig.2, which is a plot of bisector amplitude as a function of time. Calculations of Spiegel and Weiss (1980) showed that convective motions tend to be inhibited in the presence of a magnetic field. Resolved disk measurements of plage areas compared to nearby quiet Sun apparently confirm that this applies to the photosphere (Livingston 1980; Cavallini, Ceppatelli, & Righini 1985; Brandt and Solanki 1990). We made some more observations this summer with the same result, Fi g.3.

4. FTS Errors

Recently we have discovered that our FTS bisector archives contain some errors. When we examine Fig.2 we notice that, especially after 1992, there are occasions when bisector strength changes abruptly. This does not make sense physically because surface magnetism cannot change globally in a short time.

So we must question the instrument. A good diagnostic of FTS performance is the asymmetry of the sharp telluric oxygen lines at 6303Å, which we expect to have a bisector that is constant with time at a given air mass. In fact we find some variability in oxygen line asymmetry. This variability has been traced to a lack of co-alignment between the FTS and the telescope. The FTS has always

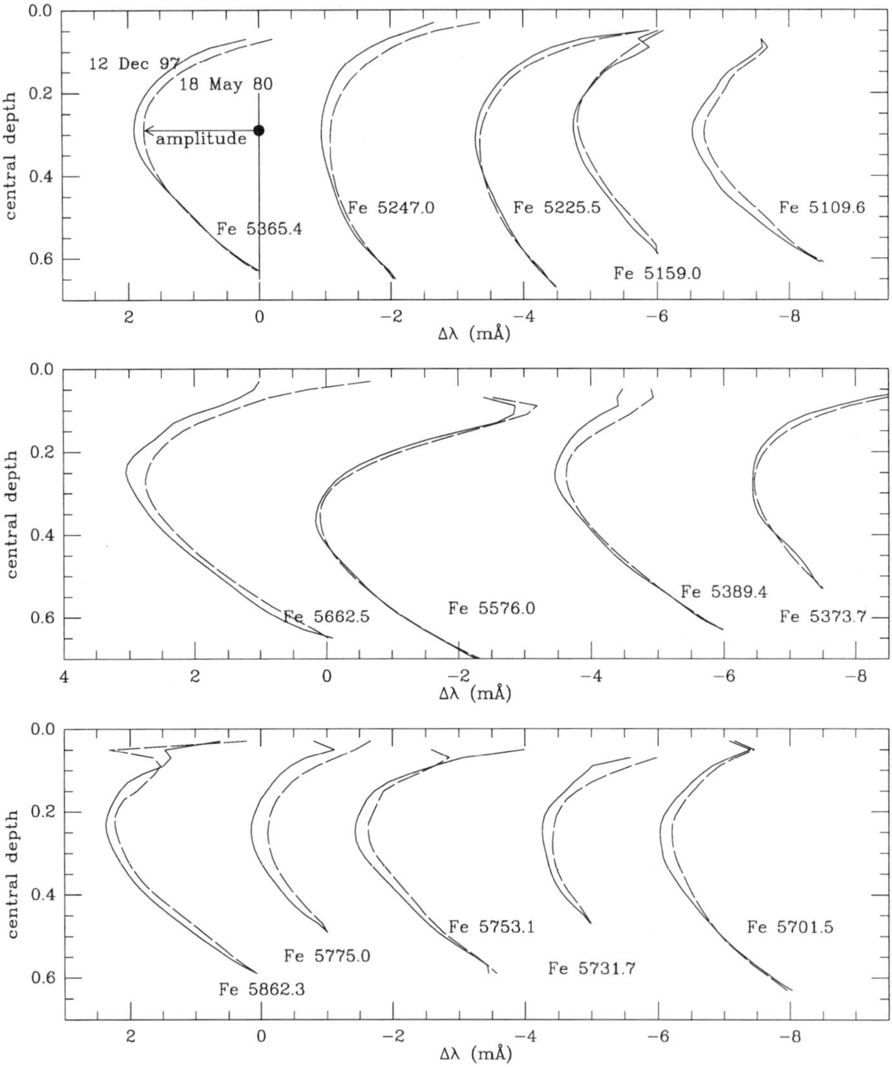

Figure 1. Bisectors for representative Fe lines selected to be free of solar and telluric blends. Results are given for two different epochs corresponding to the beginning and recent dates. It happens that 1980 was near solar maximum while the FTS thinks 1997 is a minimum year. How we define line asymmetry amplitude is shown. Note that 1 mÅ corresponds to 60 m/s.

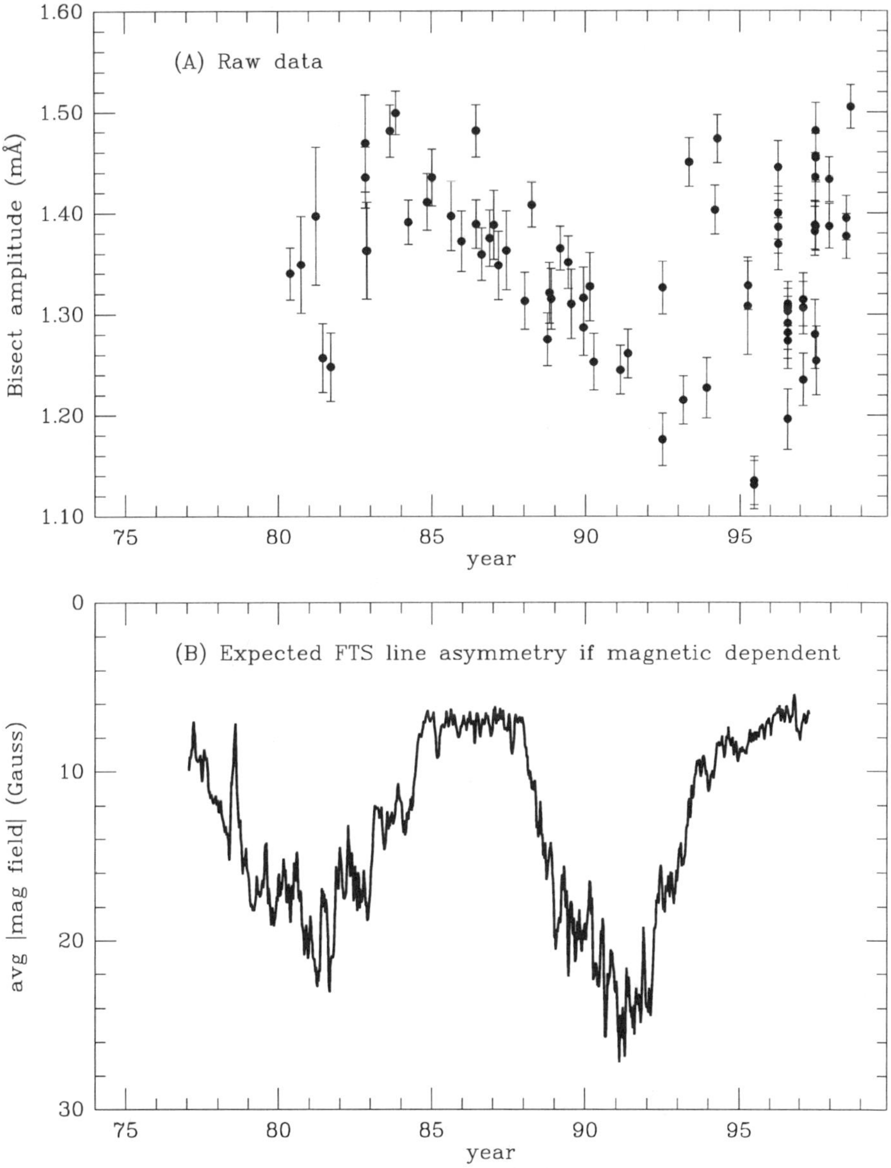

Figure 2. Top, full disk bisector amplitude for all dates. The length of the error bars reflect the number of observations that contribute to the day average. Bottom panel, smoothed total magnetic field strength from the Kitt Peak vacuum telescope synoptic program (J. Harvey). Note that the magnetic data are inverted are inverted for comparison. If granular convection varied because of the presence of solar cycle magnetism, one would expect similar records in the two panels.

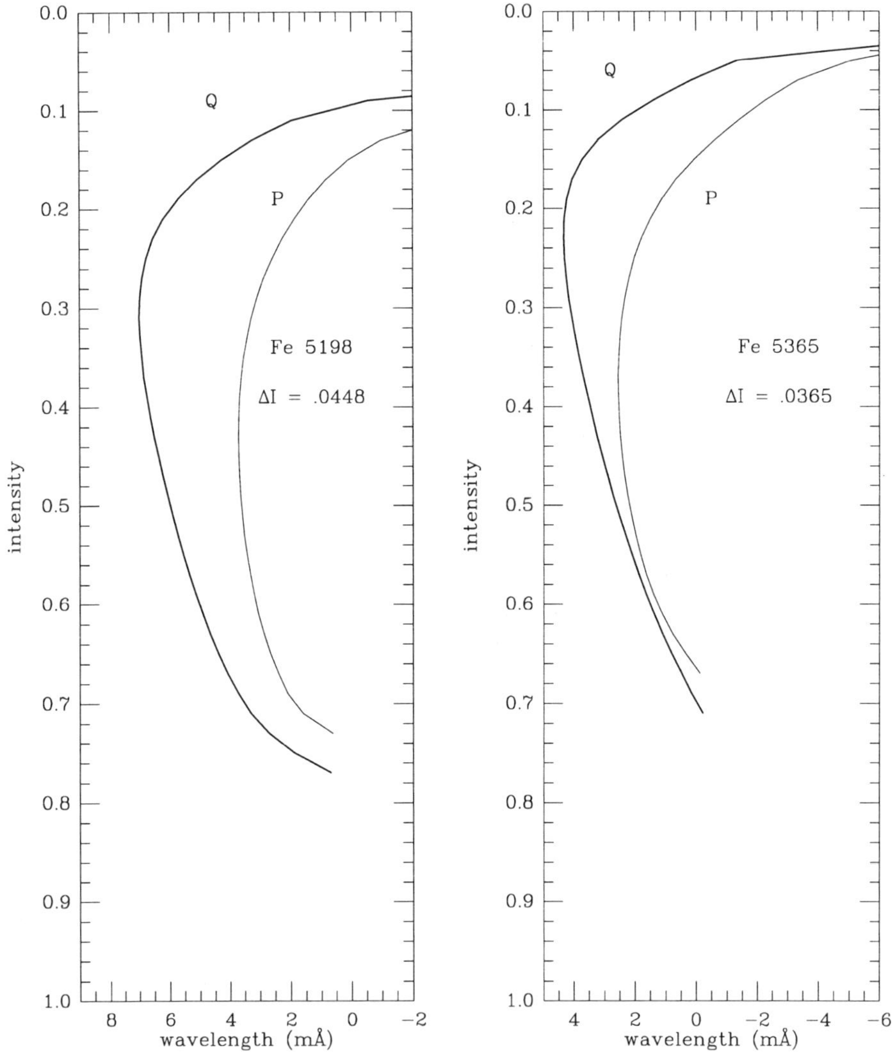

Figure 3. Observations this summer demonstrate how bisector amplitude is attenuated and the shape altered systematically a small amount between plage and quiet regions. By throwing the image out of focus by 1 meter we average over about 1 arc min of the solar disk. Note that point of maximum amplitude is always nearer the continuum for the quiet area. Here ΔI is the change of line depth.

been internally collimated before observing. But an additional requirement is co-alignment with the telescope. This has not been the practice and represents an oversight. Michael Dulick, FTS Scientist, believes that it may be possible to correct the archival FTS data set by using the oxygen lines as a diagnostic. But this is yet to be done.

5. The Magnetic Connection

Based on the observed reduction of bisector amplitude in plage we can expect that a time series plot of full disk bisector amplitude would correlate with activity indices and, in particular, with the total magnetic field. As seen in Fig.2 this seems not to be the case. (Note that FTS errors as discussed above cause an increase in line asymmetry. If we mentally ignore the sprinkle of points after 1992 that have high values, the temporal behavior is dominated by the lower bounds of the data.) The lowest recent points occur in 1995, whereas solar maximum (where one expects a minimum bisector amplitude) was in 1990-92. About a 2-3 year delay is indicated in bisector response.

It has been noticed that the horizontal motions of granules are reduced by a factor of three in plages (Simon et al. 1988; Spruit, Nordlund, & Title 1990). Whether this means the that granular convection is affected has not been determined.

High resolution observations in the G-band (Berger et al 1995) show that the bright points, or filigree, are mainly confined to intergranule lanes and suggest a lack of penetration of the fields into the granules. Another facet of the high resolution picture is the absence of fields in quiet regions. The average granule is thus unaffected by active region solar cycle magnetism.

There is another realm of surface magnetism not represented above, that has only recently began to be explored. This is the 'intranetwork' or 'weak' magnetism (Livingston & Harvey 1975; Martin 1988; Keller et al 1994; Lin 1995; Lin, this workshop; Zhang et al 1998; Meunier, Solanki, & Livingston 1998). By weak we mean flux; the field strength has yet to be determined. Any temporal variation in this magnetism is not known. We propose that it is these fields that interact with granulation to alter convection and induce line asymmetry. Perhaps the term 'granular field' is then appropriate here.

6. The Stellar Connection

David Gray has made a long series of measurements of line bisectors for the G8 dwarf ξ Boo and has found that convection changes lag behind the magnetic cycle (H and K) by 2.1 years, or comparable to what we determine for the Sun (Gray et al 1996). With less certainty he believes this phase shift decreases for late spectral type stars becoming zero at K0 (Gray 1994).

Another consequence of bisector variability and its phase shift has to do with planetary detection. Although we have not been able to measure it, there will be a wavelength shift that accompanies bisector change (see for example Dravins 1999). This is the quantity measured by Deming and Plymate (1994). To estimate the time variation of this velocity biasing effect, one should not expect it to be in phase with the star's magnetic cycle.

7. Conclusions

Despite some problems with the FTS due to alignment errors it is clear that the full disk line asymmetry changes with time. This change does not appear to be directly related to solar cycle magnetism. We conjecture that the asymmetry driver is a variation in the poorly understood weak magnetism.

References

Brandt, P.N., & Solanki, S.K. 1990, A&AS, 231, 221

Berger, T.E., Schrijver, C.J., Shine, R.A., Tarbell, T.D., Title, A.M., & Scharmer, G. 1995, ApJ, 454, 531

Bray, R.J., Loughhead, R.E., & Durrant, C.J. 1984, The Solar Granulation, Cambridge Univ. Press

Cavallini, F., Ceppatelli, G., & Righini, A. 1985, A&A, 143, 116

Deming, D., & Plymate, C. 1994, ApJ, 426, 382

Dravins, D., Lindegren, L., & Nordlund, Å. 1981, A&A, 96, 345

Dravins, D. 1999, in Proc. IAU Colloquium 170 on Precise Stellar Radial Velocities, Hearnshaw, J.B. and Scarfe, eds., in press

Gray, D. 1988, in Lectures on spectral-line analysis: F, G, and K stars, The Publisher

Gray, D.F. 1994, PASP, 106, 145

Gray, D.F., Baliunas, S.L., Lockwood, G.W., & Skiff, B.A. 1996, ApJ, 465, 945

Harvey, J. 1985, in Proc. ESA Workshop on Future Missions in Solar, Heliospheric, and Space Plasma Physics, ESA SP-235, p.199

Lin, H. 1995, ApJ, 446, 421

Livingston, W., 1980 Nature, 297, 208

Livingston, W., & Harvey, J.W. 1975, BAAS, 7, 346

Keller, C.U., Deubner, F.-L., Egger, U., & Povel, H.P. 1994, A&A, 286, 626

Martin, S.F. 1988, Solar Phys., 117, 243

Muller, R. 1995, private communication

Meunier, N., Solanki, S.K., & Livingston, W.C. 1998, A&A, 331, 771

Simon, G.W, Title, A.M., Topka, K.P., Tarbell, T.D., Shine, R.A., Ferguson, S.H., Zirin, H. and The SOUP Team 1988, ApJ, 327, 964

Spiegel, E.A., & Weiss, N.O. 1980, Nature, 287, 616

Spruit, H.C., Nordlund, Å., & Title, A.M. 1990, ARA&A, 28, 263

Zhang, J., Lin, G., Wang, J., Wang, H., & Zirin, H. 1998, Solar Phys., 178, 245

High Resolution Solar Physics: Theory, Observations, and Techniques
ASP Conference Series, Vol. 183, 1999
T.R. Rimmele, K. S. Balasubramaniam, R.R. Radick, eds.

Quiet-Sun Variability with the Solar Cycle

R. Kariyappa

Indian Institute of Astrophysics, Bangalore 560034, India

Abstract. In evaluating the contribution of various chromospheric features to total CaII K-emission flux and UV irradiance variability, the quiet-Sun is also one of the essential and important component. We have segregated the disc center intensity in a quiet-network and intranetwork regions, considered as a quiet-Sun component, from a digitized CaII K-spectroheliograms of 1992 and 1980, observed at the National Solar Observatory at Sacramento Peak. It is shown, *for the first time in the K-line*, that the longer-term variation of the quiet-Sun during the solar activity cycle is in phase with the full-disk K-emission flux (the spatial K index), and UV irradiance measured in MgII h and k lines. Thus, it indicates that the quiet-Sun will contribute in the variations of UV irradiance, and its contribution should be taken into account in UV irradiance models.

Using a standard FFT time series analysis, our results show, *for the first time*, the 51-day periodicity in the quiet-Sun variability. This evidence suggests that the 51-day period may be related to a new emerging weak magnetic fields, and the emission flux disappearance continues in the quiet-Sun regions at a slower rate. In addition, the variability of the chromospheric quiet-Sun suggest that it is very dynamic, *not quiet*, and is crucial for understanding of the chromosphere itself.

1. Introduction

Determining the contribution of the various chromospheric features to changes in UV irradiance is one of the main principal concerns for improving the current UV irradiance models and hence for both climatic and solar physics studies, since the solar radiative output is the main driver of the physical processes within the Earth's atmosphere. Thus the monitoring and study of the changes in UV irradiance are extremely important. The long-term irradiance variations are attributed to the changing emission of bright magnetic elements (Foukal & Lean 1988), and the short-term irradiance variations are directly associated with active regions as they evolve and move across the solar disk (Lean 1987). The current irradiance models are based on full-disk surrogates, therefore they cannot provide adequate information on the physical mechanisms of solar irradiance variability, detailed analysis of spatially resolved data is essential. The quantitative estimation of contribution of individual chromospheric features to UV irradiance variability is still an unexplained problem in solar physics.

The CaII H and K resonance lines are good indicators for identifying regions of chromospheric activity and for structural changes related to solar magnetic activity. A two-dimensional image of the Sun (spectroheliogram) in the CaII H or K line under high spatial resolution shows that the three agencies responsible for CaII emission are: (i) plages, (ii) the network elements, and (iii) the intranetwork elements. The changes in the network and intranetwork elements related to solar activity are less understood, especially because of the lack of systematic and quantitative measurements of these chromospheric features. Therefore, detailed analysis of the chromospheric fine structure is crucial for understanding the underlying physical mechanism of solar irradiance variability and to determine the contribution of the above-listed structures to the changes in UV irradiance (Sofia, Oster, & Schatten 1982; Lean et al. 1982; Kariyappa & Sivaraman 1994; Kariyappa & Pap 1996, henceforth Paper I) , and to incorporate the results into the irradiance models. It should be noted that one of the largest problems in analyzing the images of the Sun is the selection of the quiet-Sun regions which are used to normalize the final values of the full-disk indices. The contribution of quiet-Sun variability to the changes in solar UV irradiance is unresolved and its variability as function of solar magnetic activity is an unexplained problem in solar physics.

In our previous Paper I, we have presented the results on a new method of analysis of CaII K-spectroheliograms, and the variation of chromospheric features for the time interval of 1992 and 1980. Our aim in this paper is to separate and examine the variability and importance of the quiet-Sun with the solar cycle.

2. Observations

In our earlier Paper I, we have discussed the observational details of CaII K spectroheliograms of National Solar Observatory at Sacramento Peak (NSO/Sac Peak). We have selected and digitized about 1500 daily spectroheliograms for a 6-year-long time interval, when each year represents a different phase of the solar cycle, namely 1980 (maximum of SC21), 1985 (minimum of SC21), 1987 (the beginning of the ascending phase of SC22), 1988 and 1989 (ascending phase and maximum of SC22), and 1992 (descending phase of SC22). We have described the reduction procedure in great detail in our previous Paper I.

In the present paper, we have used 424 images for the time interval of 1980 and 1992 and extracted the disc center intensity in a very quiet regions covering the area of 36 arcsec x 36 arcsec on the Sun. Further data handling will also be described below, together with the results.

3. Results and Discussions

The solar chromosphere is highly structured by the emission features; the variability of these chromospheric emission features is the primary modulator of the solar UV irradiance on time scales of hours to years. To better understand the physical process and nature of the solar UV irradiance variations, it is important to know the spatial and temporal character and evolution of the chromospheric emission features including the background chromosphere and the quiet-Sun.

In order to understand and clarify the variability of quiet-Sun, we have extracted a quiet-Sun component at the center of the solar disc from 424 CaII

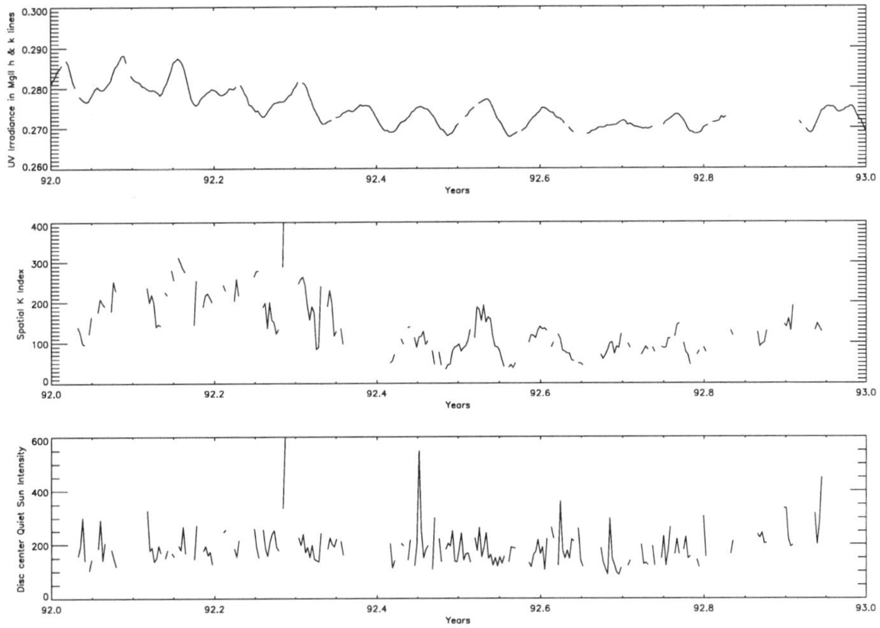

Figure 1. Time series of (from bottom) quiet-Sun intensity at the center of the solar disc, the spatial K index, and MgII c/w ratio measured with NOAA9/SBUV2 satellite for the period 1992.

K spectroheliograms of 1992 and 1980, after correcting for the photographic background emulsion. We have averaged the pixel intensity values of an area: 36 arcsec by 36 arcsec at the center of the disc in a very quiet-network and intranetwork regions from all the images. In Fig. 1 we present the time series of the quiet-Sun intensity, the spatial K index (the full-disk intensity, Paper I), and MgII c/w ratio of NOAA9/SBUV2 instrument for the year 1992. It can be seen from Fig. 1 that the quiet-Sun component will vary in general with the spatial K index and MgII c/w ratio, but may not show a peak-to-peak correlation in variation. The general solar cycle trend can be clearly seen in the quiet-Sun variability. In Fig. 2, similarly we have presented the time series of the quiet-Sun intensity, the spatial K index, and MgII c/w ratio of Nimbus7/SBUV1 instrument for the year 1980. The similar solar cycle trend in the quiet-Sun variability can also be seen for the year 1980. The longer-term variation of the quiet-Sun intensity during the solar activity cycle is nearly in phase with that of the spatial K index and UV irradiance in MgII h and k lines, suggesting that emission flux disappearance continues in the quiet-Sun regions, though it occurs at a slower rate than is observed in active regions. Harvey (1992, 1994) presented the total magnetic flux variation in active and quiet regions, and she suggested that the magnetic fields in the quiet-Sun are a important aspect of the total magnetic flux on the Sun, particularly during the minimum phases of the cycle when activity is minimal. In addition, she has listed the possible contributions

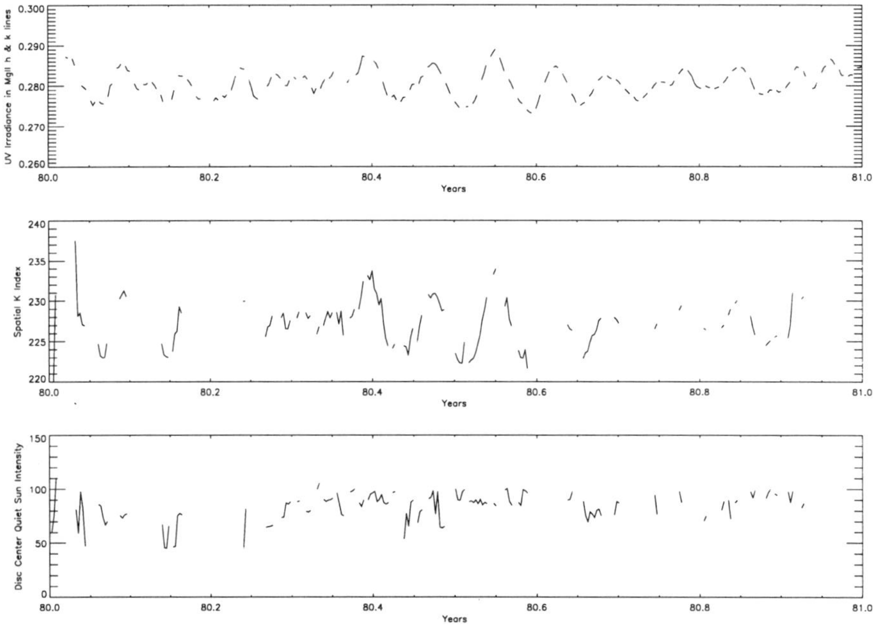

Figure 2. Time series of (from bottom) quiet-Sun intensity at the center of the solar disc, the spatial K index, and MgII c/w ratio measured with Nimbus7/SBUV1 satellite for the period 1980.

to the magnetic fields in the quiet-Sun are: (i) the dispersing active region fields from previous old cycle or cycles and the incoming new solar cycle; (ii) ephemeral regions; and (iii) an ever present background field. There are evidences for the contribution of intranetwork regions to K-emission flux and UV irradiance variability (Kariyappa, Sivaraman, & Anandaram 1994; Kariyappa 1994, 1996, 1998; Paper I), and Martin (1988) suggested that the intranetwork fields will be continuously produced and contributing to total magnetic flux. Thus, the emission flux in the quiet-Sun varies in phase with the solar activity but by a smaller factor than that in spatial K index, and UV irradiance. We suggest that the quiet-Sun will contribute quite significantly to the total K-emission flux and UV irradiance variability observed in MgII h and k lines, and show a solar cycle variation.

The study of periodicities in solar observational data has long been a high interest. The existence of a long-term cycle about 11-year, and short-term 27-day periods have been well established in numerous solar indices and in active regions. A search for periodicities in quiet regions would have fundamental significance for our understanding of solar magnetic activity, and the dynamics of the chromosphere. We have done a power spectrum analysis on the time series of quiet-Sun data of 1992 and 1980 separately. The power spectrum plots (not presented here) show a strong peak around 51-day period. This evidence suggests that the 51-day period may be related to a new emerging weak magnetic fields, and the emission flux disappearance continues in the quiet-Sun regions at a slower rate. The 51-day period is also found in other indices: total irradiance, and projected areas of active regions (Frohlich & Pap 1989, Pap, Tobiska, & Bouwer 1990), and in the occurrence rate of major flares (Bai 1994).

4. Conclusions

We have used the digitized CaII K spectroheliograms (Paper I) observed at the National Solar Observatory / Sacramento Peak for the period 1992 and 1980 to extract the quiet-Sun component at the center of the solar disc in a very quiet-network and intranetwork regions, and compared with the spatial K index (the full-disk intensity) and the UV irradiance measured in MgII h and k lines. It is shown, *for the first time in K-line*, that the quiet-Sun component will vary in correspondence with the solar magnetic activity. The longer-term variation of the quiet-Sun is in phase with the full-disk K-emission flux (the spatial K index), and UV irradiance. These results suggest that the variations in quiet-Sun will contribute in the variations of UV irradiance and its contribution should be taken into account in UV irradiance models.

The results of a standard FFT time series analysis show, *for the first time*, the 51-day periodicity in the quiet-Sun variability. The existence of the 51-day periodicity in the quiet-Sun variability suggests that it may be related to a new emerging weak magnetic fields, and the emission flux disappearance continues in the quiet-Sun regions at a slower rate. The variability of the chromospheric quiet-Sun suggest that it is very dynamic, *not quiet*, and is crucial for understanding and clarifying of the chromosphere itself.

5. Acknowledgments

I wish to thank Dr. J. M. Beckers for allowing me to use the CaII K spectroheliograms at the National Solar Observatory at Sacramento Peak. The strong encouragement and support provided by Dr. Judit M. Pap, and a valuable suggestions and discussions with Dr. J. R. Kuhn are highly appreciated. I am very thankful to the staff of the NSO/Sac Peak for their help and warm hospitality while digitizing the spectroheliograms. The Nimbus7/SBUV1 and NOAA9/SBUV2 MgII c/w ratio data was provided by Drs G. Rottman and L. Puga. The part of the work was carried out while I was at NASA's Jet Propulsion Laboratory, California Institute of Technology, Pasadena, CA, USA as NRC Resident Research Associate. The help provided by Dr. A. Vigouroux in the power spectrum analysis is highly appreciated. I am grateful to Professor R. Cowsik, Director, Indian Institute of Astrophysics and the SOC Chair and members of the 19th NSO Summer Workshop for the financial support to attend the meeting.

References

Bai, T. 1994, Solar Phys. 150, 385

Foukal, P., & Lean, J. 1988, ApJ 328, 347

Frohlich, C., & Pap, J. 1989, A & A 220, 272

Harvey, K. L. 1992, in R. F. Donnelly (ed.), The Solar Electromagnetic Radiation Study for Solar Cycle 22, p113

Harvey, K. L. 1994, in J. M. Pap, C. Frohlich, H. S. Hudson & S. K. Solanki (eds.), The Sun as a Variable Star, Cambridge Univ. Press, p217

Kariyappa, R., Sivaraman, K. R., & Anandaram, M. N. 1994, Solar Phys. 151, 243

Kariyappa, R., & Sivaraman, K. R. 1994, Solar Phys. 152, 139

Kariyappa, R. 1994, Solar Phys. 154, 19

Kariyappa, R. 1996, Solar Phys. 165, 211

Kariyappa, R., & Pap, J. M. 1996, Solar Phys. 167, 115 (Paper I)

Kariyappa, R. 1998, Solar Phys. under revision

Lean, J. L., White, O. R., Livingston, W. C., Heath, D. F., Donnelly, R. F. & Skumanich, A. 1982, J. Geophys. Res. 87, 10307

Lean, J. L. 1987, J. Geophys. Res. 92, 839

Martin, S.F. 1988, Solar Phys. 117, 243

Pap, J., Tobiska, K. W., & Bouwer, D. S. 1990, Solar Phys. 129, 165

Sofia, S. L., Oster, L., & Schatten, K. H. 1982, Solar Phys. 80, 87

High Resolution Solar Physics: Theory, Observations, and Techniques
ASP Conference Series, Vol. 183, 1999
T.R. Rimmele, K. S. Balasubramaniam, R.R. Radick, eds.

Time Evolution of Solar Granulation

J. Hirzberger, A. Hanslmeier

Institut für Astronomie, Universitätsplatz 5, A-8010 Graz, Austria

J.A. Bonet, M. Vázquez

Instituto de Astrofísica de Canarias, E-38200 La Laguna, Tenerife, Spain

Abstract. An 80 min time series of highly resolved white light images has been used for studying the characteristics of the evolution of solar granulation. For this purpose a sample of 2643 granules has been tracked automatically and another sample of 481 granules has been tracked manually. The histograms of granular lifetimes show for both samples a good agreement with exponential laws. The granular areas as well as the intensities increase linearly with lifetime but a change of the slopes at a lifetime of 12 min can be found for both parameters. For the manually tracked granules the dependence of the lifetime and the evolution of granular areas and intensities on the birth and dying mechanisms has been analyzed. We find remarkable differences in the behavior of small and large granules that gives some justification to classify granules into two populations defining two different physical regimes of convective motions on the Sun.

1. Introduction

The time evolution of solar granulation has been studied repeatedly in the past, but mainly with the aim of determining a mean or typical lifetime of the granules. The results cover a range between 5 min (Title et al., 1989) and 16 min (Mehltretter, 1978), depending on the tracking methods and the definitions of the birth and the death of a granule. Most of the morphological studies of solar granulation (e.g. LaBonte, Simon, & Dunn, 1975 or Mehltretter, 1978) show that granules can only be born by fragmentation of pre-existing ones and die either by fragmentation, merging with adjacent granules or dissolution to the diffuse intergranular background. Other studies (e.g. Dialetis et al., 1986) allow also a birth by merging and a spontaneous origination from the background.

According to Title et al. (1989) the area of the granules increases linearly with lifetime until about $T = 6$ min. For longer living granules they have found a flattening of this distribution and almost no variation for the longest living granules. Kawaguchi (1980) has also found a linear increase of the granular areas with increasing lifetime but only for granules dissolving to the background. For fragmenting granules he did not find this behavior.

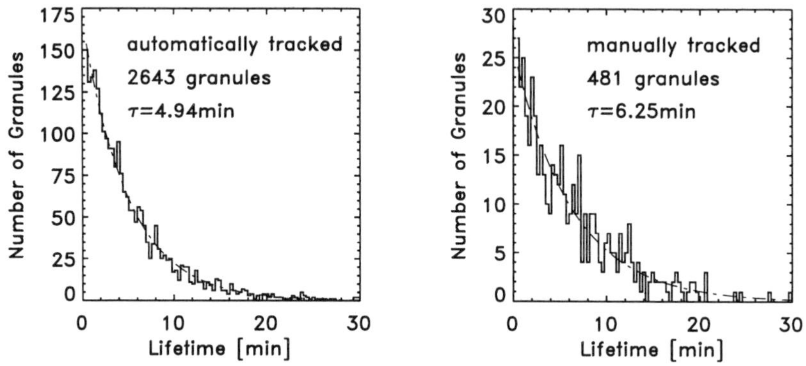

Figure 1. Lifetime distributions for automatically tracked (a) and manually tracked (b) granules.

2. Observations and Image Processing Techniques

The data material used for this study was taken at the Swedish Vacuum Solar Telescope (SVST) in La Palma and consists of a spatially highly resolved time series of white light images ($\lambda = 5257\,\text{Å} \pm 29\,\text{Å}$). The image size is $30'' \times 30''$ and the mean time interval between two images is 18.9 s. The images have been corrected for rotation, global shifts and the instrumental profile. Additionally, a subsonic filter has been applied to remove acoustic waves. Furthermore, an automatic granule finding algorithm has been developed and a sample of 100780 granules has been extracted from the time series. (For a detailed description of the image processing techniques see Hirzberger et al., 1997.)

The tracking of the granules has been performed firstly, using an automatic tracking algorithm (see Hirzberger et al., 1998) and secondly, manually i.e. by visual inspection. Hence, we have produced two samples of granular life-histories: the first one (automatically tracked) contains 2643 and the second one (manually tracked) contains 481 granular life-histories.

3. Results

3.1. Lifetime distribution

The distributions of the lifetimes of granules in both samples are displayed in Fig. 1. Both curves show good agreement with exponential laws. The decay times (which are equal to mean lifetimes in exponential distributions) are $\tau = 4.94\,\text{min}$ ($\sigma_\tau = 2.24 \times 10^{-2}$ min) for the automatically tracked granules and $\tau = 6.25\,\text{min}$ ($\sigma_\tau = 1.62 \times 10^{-1}$ min) for the manually tracked granules, respectively. These values are in the lower limit of those reported in the literature and are a consequence of our restrictive definitions of the birth and the death of a granule rather than due to the excellent quality of the used time series or possible errors produced by the automatic tracking algorithm.

3.2. Areas and intensities

To estimate the dependence of the granular areas and the intensities on the lifetime T we have calculated a twofold average: firstly, mean values of both parameters for each granular life-history were calculated and afterwards all the mean values in each bin of the lifetime histograms (see Fig. 1) were averaged. The results are plotted in Fig.2. It can be seen that the entire lifetime range can be divided in into two parts: $[0 < T \leq 12\,\text{min}]$ and $[12\,\text{min} < T < 30\,\text{min}]$. In both sub-ranges the areas as well as the intensities increase linearly with lifetime but with sizeable smaller slopes in the the second sub-range forming kinks in the distributions at $T = 12\,\text{min}$. Similar results (for granular intensities) have been found by Ploner, Solanki & Gadun (1998) from model calculations but with a kink at $T = 8\,\text{min}$ instead of $T = 12\,\text{min}$ as we do. Due to the much smaller number of granular histories in the sample of manually tracked granules, the corresponding results show higher fluctuations but a very similar trend can be recognized. (To show this more clearly we have produced linear fits to the results of the automatically tracked granules and overplotted these fits to the results of the manually tracked granules.)

3.3. Birth and death mechanisms

For the manually tracked granules it was possible to ascertain the birth and the death mechanism of each granule. We allow three birth mechanisms: fragmentation (F), merging (M) and spontaneous origination from the diffuse background (D) as well as three death mechanisms: again fragmentation (f), merging (m) and dissolution to the background (d). Combining each birth with each death mechanism we get nine families of granules. The numbers N of granules in each family and the corresponding mean lifetimes $\bar{T}$ are shown in Table 1. Only four of the nine families (F – f, F – d, F – m and M – f) contain enough granules which makes a statistical analysis possible. The numbers of granules in the other five families are very small, leading to an assumption that these are misidentifications and the corresponding combinations do not occur.

birth	death	N	N [%]	$\bar{T}\,(\pm\sigma)$ [min]
F	f	99	20.6	9.23 (±5.92)
F	d	72	14.9	6.07 (±5.62)
F	m	235	48.9	5.45 (±4.38)
D	f	0	0.0	-
D	d	3	0.6	6.83 (±4.47)
D	m	7	1.5	2.16 (±1.50)
M	f	51	10.6	4.49 (±4.72)
M	d	2	0.4	3.31 (±1.56)
M	m	12	2.5	6.17 (±3.26)
all		481	100	6.19 (±5.16)

Table 1. Division of manually tracked granules with respect to their birth and death mechanisms

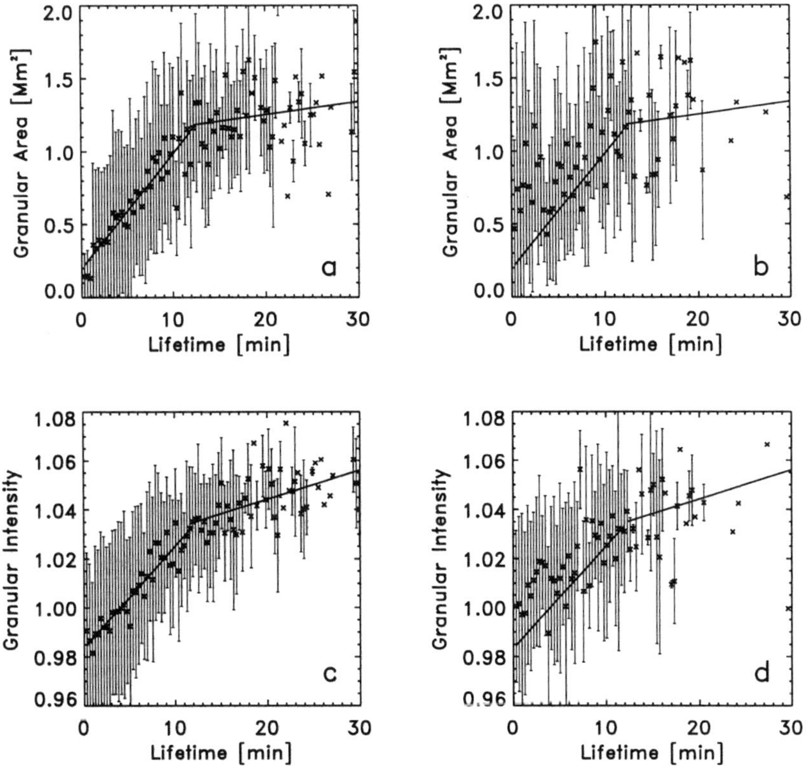

Figure 2. Averaged granular areas (panels a and b) and intensities (panels c and d) in dependence on the lifetime T. The left panels show the results of the automatical tracking, the right panels those of the manual tracking. The error bars denote standard deviations of the second averaging process (see text) and the straight lines denote linear fits in the ranges $[0 < T \leq 12\,\mathrm{min}]$ and $[12\,\mathrm{min} < T < 30\,\mathrm{min}]$ to the results of the automatic tracking.

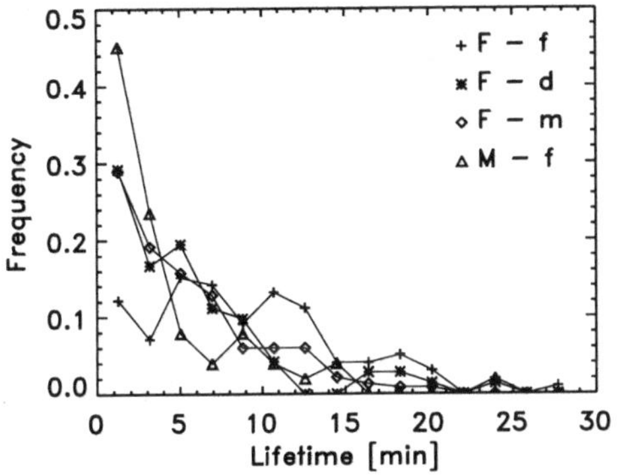

Figure 3. Normalized histograms for granular life-histories in the families: (F – f), (F – d), (F – m) and (M – f).

The mean lifetime is highest in the family (F – f) and lowest in the family (M – f) although the standard deviations are very high in all cases. Therefore, we have calculated normalized histograms for the four mostly populated families (Fig. 3). According to this figure the number of granules in family (F – f) is almost equal for lifetimes $T < 13\,\mathrm{min}$ and decreases slowly for longer lifetimes. The histogram for family (M – f) has a complete contrary behavior: it shows a peak for very short lifetimes ($T \approx 2\,\mathrm{min}$) and drops sharply towards longer lifetimes. From this behavior it could be guessed that the granules in this family are only transitory connections of two or more stable elements. The histograms for the families (F – d) and (F – m) lie somewhere between these two extremes and show good agreement with exponential laws.

3.4. Averaged evolutions

To find out a common behavior of the granules in a certain family, we have divided each of the four mostly populated families into four lifetime ranges and averaged all the granular evolutional histories in each of the 16 resulting family-lifetime bins (for a detailed description of the averaging procedure see Hirzberger et al., 1998). The width of these ranges are: $[1.0\,\mathrm{min} < T < 2.2\,\mathrm{min}]$ (I), $[2.5\,\mathrm{min} < T < 5.0\,\mathrm{min}]$ (II), $[5.4\,\mathrm{min} < T < 11.0\,\mathrm{min}]$ (III) and $[11.3\,\mathrm{min} < T < 21.7\,\mathrm{min}]$ (IV). The resulting mean evolutions of granular areas and intensities in the 16 family-lifetime bins are plotted in Fig. 4 and 5, respectively.

For family (F – f) it can be seen that the initial areas of all the granules are around $1\,\mathrm{Mm}^2$ and a linear growth with time can be detected. The growth is the steeper the longer the granules live. For families (F – d) and (F – m) the evolution is contrary. All the granules are born with areas smaller than $0.8\,\mathrm{Mm}^2$ and decrease the steeper the longer they live. For family (M – f) no common behavior can be detected, i.e., the assumption that those granules are transitory connections is confirmed.

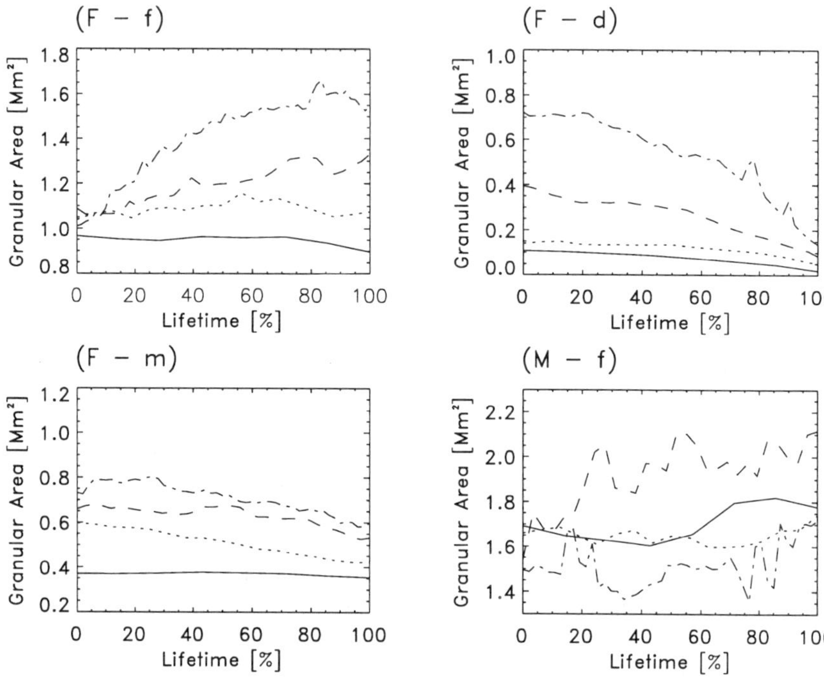

Figure 4. Averaged evolutional curves of granular areas in the 16 family-lifetime bins. Solid curves: lifetime range I, dotted curves: lifetime range II, dashed curves: lifetime range III and dash-dotted curves: lifetime range IV.

The granular intensities (see Fig. 5) are in all families the higher the longer the granules live. Granules in families (F – f), (F – m) and (M – f) are born rather faintly, become brighter in the first halve of their lives and become again fainter in the second halve of their lives, i.e., the intensity curves show one maximum (exceptions are in all families the very short living granules which do not change their brightness at all). For the granules in the family (F – d) the intensities run parallel to the areas, i.e., they decrease monotonically and become monotonically fainter too.

4. Conclusions

After the discussion of Table 1 and Fig. 4 we can conclude that granules can only be born by fragmentation of pre-existing ones, whereas they can die either by fragmentation, merging or dissolution to the background. Granules that are born by merging are only transitory and rather unstable connections of two or more granules. Furthermore, granules which are born by fragmentation can be divided into two populations: if they are larger than 0.8 Mm^2 they will increase and fragment again, if they are smaller than this value they will either

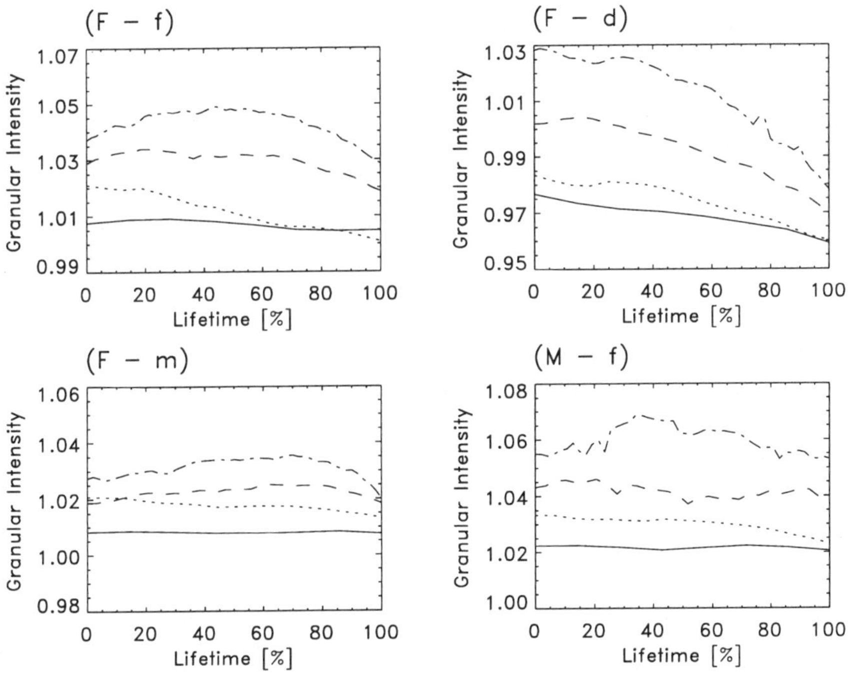

Figure 5. Averaged evolutional curves of granular intensities in the 16 range-family bins. The line codes are the same as in Fig. 4.

dissolve or decrease and become absorbed by an adjacent one (merge). A similar division has already been made by Hirzberger et al. (1997) regarding the fractal dimension and the area-intensity distribution of granules. In that division the value $d_G = 1\farcs39$ appeared as a critical diameter separating two populations of granules. This value coincides exactly with a granular area of $0.8\,\mathrm{Mm}^2$ which separates the two populations in the present work. Further hints that granules can be divided into two populations are the kinks in the lifetime-area and the lifetime-intensity distributions shown in Fig. 2.

A physical explanation for this behavior can be found in the turbulent cascade of the convective cells in the outer convection zone. According to Espagnet et al. (1993) this turbulent cascade should be non-dissipative if structures larger than 1000 km ($1\farcs38$) are observed. For smaller structures heat conduction becomes more and more important and a change in the hydrodynamic properties of the eddies produced by the turbulent cascade should be found. Applying this theory to our measurements we find again a very good agreement with the numbers separating the two populations of granules found in the present work.

Acknowledgments. J.H. and A.H. acknowledge gratefully the partial finance of this project by the Österreichische Nationalbank (Jubiläumsfondsprojekt 4993), the Austrian Academy of Sciences and the European Office of Aerospace Research and Development (EOARD).

References

Dialetis D., Macris C., Prokakis T., Sarris E.: 1986, A&A, 168, 330

Espagnet O., Muller R., Roudier T., Mein N.: 1993, A&A, 271, 589

Hirzberger J., Vázquez M., Bonet J.A., Hanslmeier A., Sobotka M.: 1997, ApJ, 480, 406

Hirzberger J., Bonet J.A., Vázquez M., Hanslmeier A.: 1998, (submitted to ApJ)

Kawaguchi I.:1980, Solar Phys. 65, 207

LaBonte B.J., Simon G.W., Dunn R.B.: 1975, BAAS, 7, 366

Mehltretter J.P.: 1978, A&A, 62, 311

Ploner S.R.O., Solanki S.K. Gadun A.S.: 1998, Space Sci. Rev. (in press)

Title A.M., Tarbell T.D., Topka K.P., Ferguson S.H., Shine R.A., the SOUP Team: 1989, ApJ, 366, 475

Group Discussion

Berger: What year was the data taken and what post-processing was performed?

Hirzberger: In 1993; flat-fields and dark currents, a Wiener filter for the telescope profile and an acoustic filter with a cut-off phase velocity of 5km/s-no destretching.

High Resolution Solar Physics: Theory, Observations, and Techniques
ASP Conference Series, Vol. 183, 1999
T.R. Rimmele, K. S. Balasubramaniam, R.R. Radick, eds.

Implementation of a Filter for the Restoration of Solar Granulation Images

M. Sánchez Cuberes, J. A. Bonet, M. Vázquez

Instituto de Astrofísica de Canarias, E-38200 La Laguna, Tenerife, Spain

A. D. Wittmann

Universitäts-Sternwarte, D-37083 Göttingen, Germany

Abstract. Statistical analysis of brighness fluctuations in solar granulation at different positions on the solar disk is seen as a powerful technique to model the variation of the temperature with depth in the photosphere. Nevertheless, prior to any measurement, an unavoidable task is to correct the observed images for the degradation induced by the terrestrial atmosphere and the telescope. For this purpose we describe here a restoration filter based on estimates of the PSF from the lunar limb in solar eclipse observations. Independently of the procedure for estimating the PSF, the central point of our presentation is the description of a novel method for constructing a noise filter for images of solar granulation.

1. Introduction

Several diagnostic tools are routinely used to model the variation with the optical depth, τ, or the geometrical height, z, of the physical variables related to photospheric structures. Thus, the analysis of the center-to-limb variation of different spectral lines allows $\Delta T(z)$ and $v(z)$ to be determined. The study of the dependence of the continuum intensity fluctuations, ΔI, with wavelength is another way to infer the temperature as a function of z. $\Delta T(z)$ can also be derived from measurements of ΔI at different heliocentric positions $\mu = cos\theta$. We are concerned here with the latter technique; namely, determinations of power spectra and normalized contrast, $\Delta I_{\rm rms}$, of the intensity fluctuations in the quiet granulation as a function of μ.

Granulation contrast is severely attenuated by the degradation caused by atmospheric turbulence and telescope instrumental profile and aberrations. Several procedures have been developed to compensate for these effects (see Bonet, 1998 for a recent review). Partial solar eclipses offer an opportunity to get an estimate of the point spread function (PSF) representing the degradation produced by the telescope and the atmosphere. In this paper we will concentrate on the implementation of a filter based on an estimate of the PSF from partial solar eclipse observations, to restore images of granulation at different positions on the solar disk. Special emphasis will be put on the procedure for the construction of the noise filter to be applied prior to the restoration itself. The discussion on the determinations of power spectra and contrasts, as well as other statistical considerations is postponed to a forthcoming paper.

2. Data Material

The material analyzed here was obtained at the Swedish Vacuum Solar Telescope (SVST, Observatorio del Roque de los Muchachos, La Palma; see Scharmer et al. (1985) during the partial solar eclipse of 1994 May 10.

A set of CCD images showing quiet granulation fields at different μ-positions on the solar disk, partially covered by the Moon's limb, was recorded at $\lambda 6708 \pm 20$ Å with an exposure time of 13 ms and an image scale of $0.''124$/pixel. From this set only the best images were selected for pre-treatment (flat-fielding and dark current subtraction) and further processing.

3. Image Reconstruction

3.1. Presentation of the problem

The image of an incoherent object $i_o(x,y)$, formed by an optical system, will be recorded as

$$i(x,y) = i_o(x,y) * s(x,y) + n(x,y)\,, \tag{1}$$

where $*$ represents spatial convolution, $n(x,y)$ the noise assumed uncorrelated with the signal, and $s(x,y)$ the PSF characterizing the optical system composed in the present case by the atmosphere and the telescope.

The recovery of the object (i.e. the "true" intensity distribution in solar granulation) from the observed image, can be easier performed in the Fourier transform domain since there convolutions turn into products. Thus the suitable formula to do that is

$$\hat{i}_o(x,y) = \mathcal{F}^{-1}\left[\frac{I(u_x,u_y)\cdot\Phi(u_x,u_y)}{S(u_x,u_y)}\right], \tag{2}$$

where $\mathcal{F}^{-1}$ stands for the inverse Fourier transform; $I(u_x,u_y)$ and $S(u_x,u_y)$ are the Fourier transfroms of $i(x,y)$ and $s(x,y)$, respectively; $S(u_x,u_y)$ is also called optical transfer function (OTF) of the imaging system. $\hat{i}_o(x,y)$ is an estimate of the real object. $\Phi(u_x,u_y)$ in (2) require that, prior to performing the image restoration, i.e. the quotient $I(u_x,u_y)/S(u_x,u_y)$, noise filtering in the recorded image is essential, otherwise the noise contribution would be amplified and have a disastrous effect, mainly on the high frequency range where the signal-to-noise ratio is poorer.

The lunar limb crossing the solar disk has been taken as a reference object to obtain an estimate of the OTF assuming isotropic degradation (see e.g. Bonet et al. 1995). We shall denote this function by $M(\nu)$, where $\nu = u_x^2 + u_y^2$. For this purpose we have measured the mean photometric profile of the transition Moon–solar photosphere all along the lunar limb included in each particular frame and compared the result with the "known true photometric profile", i.e.

the Heaviside step function. The average of the above-mentioned profiles implies that isoplanatism is assumed.

In short, in our image treatment the following approximations are implicit: a) uncorrelated signal and noise, b) isoplanatism, and c) isotropic degradation. The first assumption is based on the fact that we expect the noise to be predominantly photon noise and that the solar granulation is a low-contrast structure. Isoplanatism could be partially justified by the uniform quality observed in our selected frames. Finally, we are compelled to assume isotropic degradation, i.e. azimuthal symmetry in $s(x, y)$ because our reference object (the lunar limb) is substantially a 1D structure. We are aware of the weight of the last two assumptions. They are essentially equivalent to estimating a sort of average degradation on each frame. Given the excellent quality of our images, this would mean, to our understanding, that the restoration process should not introduce substantial bias in the statistical analysis that we intend to do taking advantage of a singular phenomenon offered by nature (viz., the eclipse) and of the low computational cost of the method as compared with other more recent reconstruction techniques.

3.2. Noise filtering

Noise filtering is performed by weighting the spectral components of the image at various frequencies in some optimum way. Thus, our filter $\Phi(u_x, u_y)$ used in (2) is based on the so-called *optimum filter* described by Brault & White (1971), and is derived from the observed images themselves. This filter is constructed by imposing the reasonable condition that, in the measuring domain, the deviation of the image smeared by the PSF and not affected by noise (i.e. $i_o * s$), from the image which results from the filtering, is a minimum in the root-mean-square error sense.

Assuming the noise to be uncorrelated with the signal, this optimum filter can be formulated as

$$\Phi(u_x, u_y) = \frac{|I(u_x, u_y)|^2 - |N(u_x, u_y)|^2}{|I(u_x, u_y)|^2}, \tag{3}$$

where $|I(u_x, u_y)|^2$ stands for the power spectrum of the observed image, and $|N(u_x, u_y)|^2$ for that of the noise. In the practical realization of $\Phi(u_x, u_y)$ it is usually adequate to replace both the signal and the noise power spectra by smooth fitted models.

In the present study the noise has been modeled as a constant power ("white") noise. White noise is typical for the statistical fluctuations in photon flux, which are the most prominent noise source in solar images taken with a good CCD camera. However the construction of a smooth model that fits the observed power spectrum well is much trickier. An accurate fit is increasingly important as the power level of the signal approaches the power level of noise in the high spatial frequency range.

In the paper by Brault & White (1971) are discussed simple functional forms to be fitted to the observed power spectra of 1D signals (spectral line profiles), depending of their peculiarities. In image restoration, the problem is much more troublesome. To the difficulty of dealing with 2D data is added the fact that the shape of the power spectra significantly changes depending on the nature of the structures in the considered image, and consequently the choice of suitable functional forms to be fitted in each particular case is a very difficult task.

What saves us is the nature of the solar granulation. Brightness fluctuations due to granulation at the center of the solar disk can be considered as a statistically isotropic phenomenon and hence, the corresponding power spectrum is expected to be azimuthally symmetric. Away from the disk center, granular structures become progressively narrower in the direction perpendicular to the solar limb as one moves toward the limb. This effect produces power spectra closely approximating a bell with elliptical horizontal sections, which simplifies the fitting process. Since the shape (the ellipticity of the sections) of the power spectrum changes with the position in the solar disk, the noise filter has to be specific for each μ-position and this compelled us to deal with small boxes for constructing the filter as well as for performing the restoration process. The size of the working boxes has been fixed to 100×100 pixels ($12''.4\times12''.4$).

Several attempts to fit 2D splines and Gaussian elliptical bells to the power spectra have failed, on the one hand due to the oscillations obtained in the spline fit (very dependent on the specific node positions), and on the other hand because the power bells have elliptical sections with variable ellipticity and orientation at different power levels so that they do not match Gaussian bells well. Consequently, we have chosen the fitting procedure described below.

To obtain the power spectrum of the observed signal at a given μ-position on the solar disk, an average of power spectra is performed—the resulting mean power spectrum will hereafter be denoted by PS. To this end, several boxes placed within a range $\mu \pm \delta\mu$ ($\delta\mu$=0.05 for $\mu \geq 0.75$ and $\delta\mu$=0.1 for $\mu < 0.75$), are taken from the best-quality images that contain such a range of positions. However, for a given μ, images obtained at different solar latitude and longitude are treated separately to prevent the possibility of the properties of the granulation differing as a function of the coordinates on the solar disk. Following Cooley et al. (1970), the boxes are overlapped by 50%. In that way, the number of power spectra included in the average is substantially increased thus reducing the "noise" in PS stemming from the statistical differences between single boxes.

To fit a smooth model to PS, isocontours at various power levels are obtained and ellipses are fitted to them. The power levels have been chosen to have an isocontour representation as dense as possible, only being limited by the overlapping of the ellipse sampled points due to the small box size. No a priori assumptions about ellipticity or orientation of the fitted ellipses are made. The fitting method consist in taking many sets (typically 100000 for the outer iso-contours, and 700 for the inner ones) of three well separated points in an isocountour. Each set of three points defines an ellipse centered on the origin of frequencies. The parameters describing each of these ellipses are represented in the parameter space A-B-Θ, where $A \equiv$ major axis, $B \equiv$ minor axis and $\Theta \equiv$ ellipse orientation. The coordinates of the centroid (A_c, B_c, Θ_c) of the resulting cloud of points, defines the ellipse that is chosen as the best fit to the isocontour. This process has been applied to every isocontour obtaining a series of fitted ellipses at different power levels, all of them centered at the origin of spatial frequencies. Representation of these ellipses in the $12.''4\times12.''4$ box—i.e. sampling their analytical expressions—gives the ellipse data points describing PS at those power levels where isocontouring has been performed.

The overall slope of the power spectrum flattens in the highfrequency range and the peaks due to the remaining statistical uncertainties cause the isocontours in this region to be disconnected so that an elliptical shape cannot be recognized. Thus, an information gap concerning the PS's shape is found between the lowest isocontour elliptically shaped and the noise level, where the horizontal section of PS is a circumference of radius equal to the telescope cut-off spatial frequency. We assume an exponential decay of PS in this power range, described in every radial direction by the expression:

$$p = a \cdot e^{-b \cdot \nu} + c\,, \tag{4}$$

where ν is the spatial frequency, a and b are determined by forcing the decay to pass through the lowest contour and the cut-off, and c is the free parameter that rules the slope of the decay. The ensemble of data points compounded by the sampling of the ellipses and of the exponential decay, form an irregular grid describing the entire PS. To interpolate these data values a Delunay triangulation of the set of data points has been performed to obtain a triangular grid structure (cf. Dahlquist & Björck, 1974). Then, linear interpolation has been applied to obtain a regular grid of interpolated surface values. Interpolated values are only computed from nearby points since the circumcircle of any triangle of a Delunay triangulation has the property of containing no other vertices in its interior. A logarithmic scale has been used to perform interpolation of the surface values in order to obtain a much realistic steeper decay between isocontours.

Fig. 1a shows typical contours of PS at $\mu = 0.5$ and their fitted ellipses. The corresponding global smooth model fitting PS is presented in Fig. 1b.

The constant power level of the white noise is computed by averaging the values of PS at frequencies higher than the telescope cutoff. Substituting $|I(u_x, u_y)|^2$ in (3) by the fit to PS described above and $|N(u_x, u_y)|^2$ by the fit to the noise, the optimum noise filter is obtained. Fig.1c shows a typical example of a noise filter corresponding to the case also represented in Figs.1a,b.

3.3. Image restoration

According to expression (2) the quotient of $\Phi(u_x, u_y)$ and the OTF—$M(\nu)$ in our case—rules the restoration process and will be termed the "restoration filter". Hence, the construction of the restoration filter has been accomplished by dividing the noise filter, calculated in the preceding section, by the specific $M(\nu)$ derived from each particular image to be restored. Note that the optimum noise filter is constructed by averaging power spectra from different frames, affected by different $M(\nu)$s, whereas in the restoration filter the specific $M(\nu)$ for each frame is used. From numerical simulations of observed spectra, based on real measurements of $M(\nu)$, we conclude that the resulting restoration filters are mainly sensitive to changes of the PS's shape in the high-frequency range ($\nu > 2$ arcsec^{-1}). However, this sensitivity is only noticeable when the $M(\nu)$ values are very close to zero (poor image quality) or when they differ significantly (ample spread of image qualities) in this frequency range. Accordingly, we have selected for restoration only those images with good and similar quality in this frequency range by analyzing and comparing the corresponding $M(\nu)$ functions. The averaging of power spectra favors the decrease of the above-mentioned information gap in the power range close to the noise level. The more power spectra are averaged the larger is the statistical stability in PS and consequently the closer are the isocontours that can be determined to the noise level.

A compromise between image quality and the minimum number of boxes to be considered has been reached to obtain PS. The boxsize determines the number of boxes included in a given μ-range and consequently, this parameter has also to be considered to define the strategy for calculating PS. Based on the stability of parameters such as PS shape, centroid of the 1D azimuthally integrated spectrum and total integrated power, the final decision for the boxsize is 100×100 pixels ($12''.4 \times 12''.4$).

It is important to note that this filter is based on the information contained in the PS of granulation and that the only free parameter is c, which determines the exponential decay from the lowest isocontour elliptically shaped to the noise level. This affects only the highest frequency part of the filter, beyond the maximum of restoration, so most of the restoration filter is completely data-based. The normalized rms contrast of the restored field (i.e. the integrated power) is only slightly sensitive to changes in c because the largest contribution to the power comes from the region of large and medium spatial structures.

As can be seen in Fig.1d, the restoration filters are non-symmetric at positions other than the disk center due to the asymmetry in PS.

Once the restoration filter is ready for a given μ-range and a particular frame, individual boxes of $12''.4 \times 12''.4$ included in this range and frame are restored following expression (2). In this formula $I(u_x, u_y)$ represents the Fourier transform of a particular box. A cosine apodization has been applied to the extreme 10% pixels to avoid oscillatory behavior at the borders of the restored boxes.

The efficiency of the restoration filter that we use is evidenced in Fig. 2 where pairs of non-restored and restored granular fields for different μ-positions on the solar disk are shown for comparison.

4. CONCLUSIONS

Partial solar eclipses provide an opportunity to perform an estimate of the PSF representing the degradation originating in the terrestrial atmosphere and the telescope. Here we describe the sequential steps for constructing a restoration filter. Apart from the procedure followed for the PSF determination, we put special emphasis on the description of a novel method for the implementation of a noise filter based on the observed data themselves. This method is presently being applied to a statistical analysis of the center-to-limb variation of solar granulation that will be published elsewhere (Sánchez Cuberes et al., 1999a,b).

Acknowledgments. The support provided by R. Kever, G. Hosinsky and M. Sobotka during the observations is gratefully acknowledged. We thank Terry Mahoney for revising the English. The SVST is operated on the island of La Palma by the Royal Swedish Academy of Sciences in the Spanish Observatorio del Roque de los Muchachos of the Instituto de Astrofísica de Canarias. This work was partially funded by the Spanish DGES project 95 - 0028.

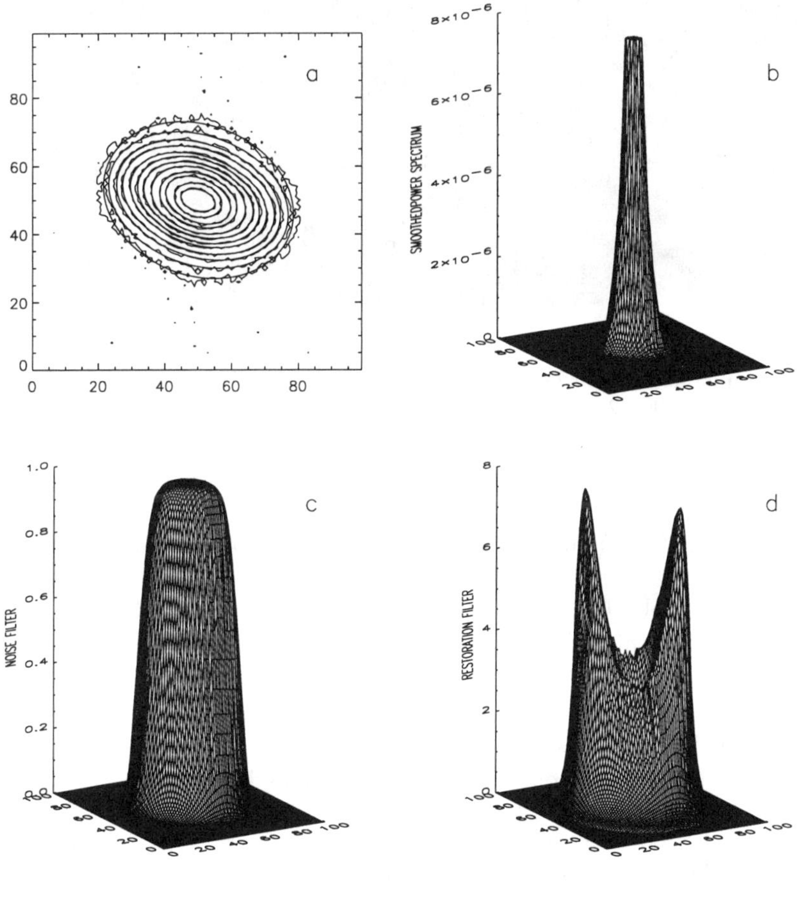

Figure 1. a) PS isocontours and the corresponding fitted ellipses; b) smooth model of PS; c) noise filter, and d) restoration filter. All cases correspond to PS at $\mu = 0.5$. The abscissa values are expressed in pixels.

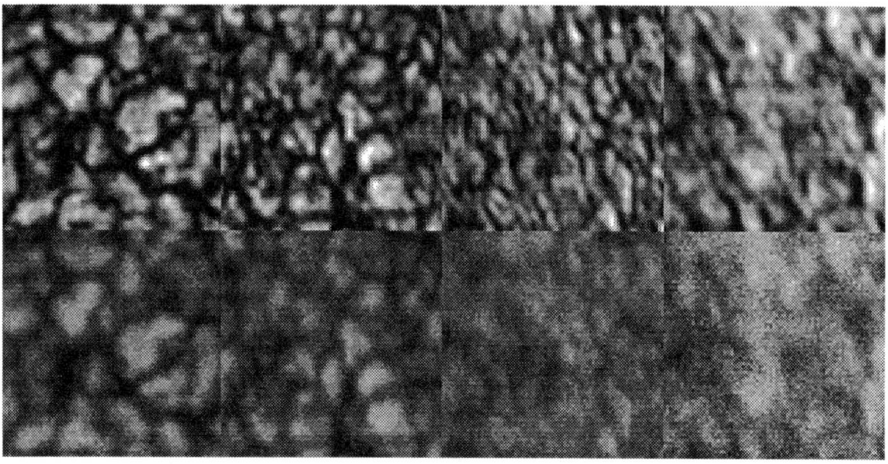

Figure 2. Pairs of reconstructed (top panel) and original images (bottom panel) at four μ-positions on the solar disk. The boxes cover a field of $9''.92\times9''.92$

References

Bonet, J.A., 1998, in *Motions in the Solar Atmosphere*, ed. A. Hanslmeier & M. Messerotti, (Dordrecht: Kluwer), (in press)

Bonet, J.A., Sobotka, M., and Vázquez, M., 1995, A&A, 296, 241

Brault, J.W., White, O.R.: 1971, A&A, 13, 169

Cooley, J.W., Lewis, P.A.W., and Welch, P.D., 1970, J. Sound Vib., 12 (3), 339

Dahlquist G., Björck, Å, 1974, *Numerical Methods.* Ed Prentice–Hall, Inc., Englewood Cliffs, N.J.

Sánchez Cuberes, M., Bonet, J.A., Vázquez, M. and Wittmann, A. D., 1999a, in *III Scientific Meeting of the Spanish Astronomical Society*, Astrophysics and Space Science, (in press)

Sánchez Cuberes, M., Bonet, J.A., Vázquez, M. and Wittmann, A., 1999b, (in preparation)

Scharmer, G.B., Brown, D.S., Pettersson, L., Rehn, J., 1985, Appl.Opt., 24, 2558

High Resolution Solar Physics: Theory, Observations, and Techniques
ASP Conference Series, Vol. 183, 1999
T.R. Rimmele, K. S. Balasubramaniam, R.R. Radick, eds.

Complex H_α Loop Activity in a Long Duration Flare

Debi Prasad C, G. Allen Gary

Space Science Laboratory, ES82, Marshall Space Flight Center/NASA, Huntsville, AL, 35812.

Ashok K. Ambastha

Udaipur Solar Observatory, Physical Research Laboratory, Udaipur, 313001, India.

Abstract. A long duration flare was observed in the active region NOAA 6555 on March 23, 1991 at a location of low magnetic shear. This flare displayed complex H_α loop activity during the decay phase. Initially, there were a few long loops orientated at an angle of about 45 degree with the east-west axis followed by the appearence of east-west axis. This configuration was aligned shorter loops. By using different magnetic field models, we have extrapolated the photospheric magnetograms to the chromospheric heights. The magnetic field lines computed by using the potential field model correspond to most of the observed H_α loops. In this paper, we report the analysis of high resolution chromospheric observations of the event and describe a possible senario for the post-flare loop activity.

1. Introduction

The reconnection of magnetic fields at the coronal height is the basic ingredient of the solar flare mechanism (Bruzek, 1964). According to the reconnection flare model (RFM), hot loops with temperatures of $\sim 3 \times 10^7$K are formed soon after the flare onset. When the loop plasma cools to $\sim 10^4$K, they become visible in the optical wavelengths. The regions joined by these loops are brightened due to heating by the loop plasma. The two-ribbon flares with post-flare loop activity are believed to be the classical examples of the RFM. In the standard picture the foot-point separation of the observed post-flare loops increase with time as these loops grow higher. Although this scenario of the RFM gives a qualitatively consistent description of solar flares, several observations do not comply with this general picture (Feldman, Seely and Yuhong, 1995). A major inconsistency is that in many cases the flare loops do not seem to follow the characteristics predicted by RFM in which the gradual appearance of new higher loops are expected while the old loops are fading. Since the observational manifestation of the RFM is the loop structure, their study would be helpful in better understanding the flares.

In this paper we present the analysis of a dynamic loop event associated with a long duration flare on 23 March 1991 in the active region NOAA 6555

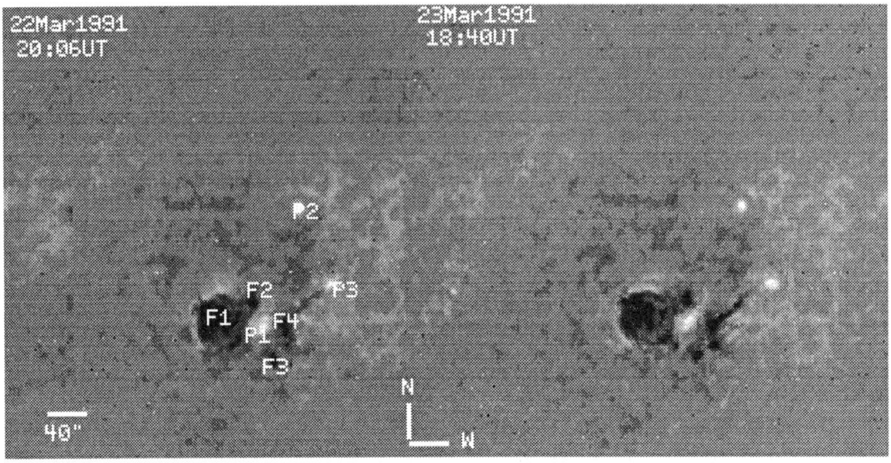

Figure 1. Kitt Peak Longitudinal magnetograms before and after the long duration flare.

located at latitude 24° S, 5° west of CMD and helio-longitude 188° at 01:00 UT. Most of the loops can be identified with the magnetic field lines of the potential model except one along the highly sheared neutral line. The temporal evolution of the H_α loop activity is compared with the computed field lines to infer the magnetic field structure of the active region at the decay phase of the flare. Also, a possible mechanism to describe the observed long duration flare is outlined.

2. Observations

The observations of the active region NOAA 6555 in the chromospheric H_α line were obtained from Udaipur Solar Observatory (USO), India. The H_α filtergrams have high spatial resolution of ~ 1 arcsecond and field of view 8×6 arcmin, which are taken through a 0.5 Å bandpass birefringent filter. The photospheric vector magnetograph in FeI line at 5324.19 Å and the chromospheric Dopplergrams in H_β line are obtained at Huairou Solar Station (HSS) of Beijing Astronomical Observatory, Beijing, China. These magnetograms have a field of view of 4 × 3 arcmin and spatial resolution of 2 arcsec. In order to perform the magnetic field line extrapolations, we have used data from the KPNO full disk magnetograms. These magnetograms are obtained in 8688Å FeI line with a spatial resolution of 1 arcsec before and after the flare at 20:06 UT and 18:40 UT respectively.

3. NOAA 6555 and H_α Loop Activity

NOAA 6555 produced around 150 flares, including five X-class flares, during the disk transit on March 17-31, 1991. The pre- and post-flare photospheric longitudinal magnetic field of the region are given in Fig. 1. The preceding and following spots are designated as **P** and **F** respectively. The magnetic shear was

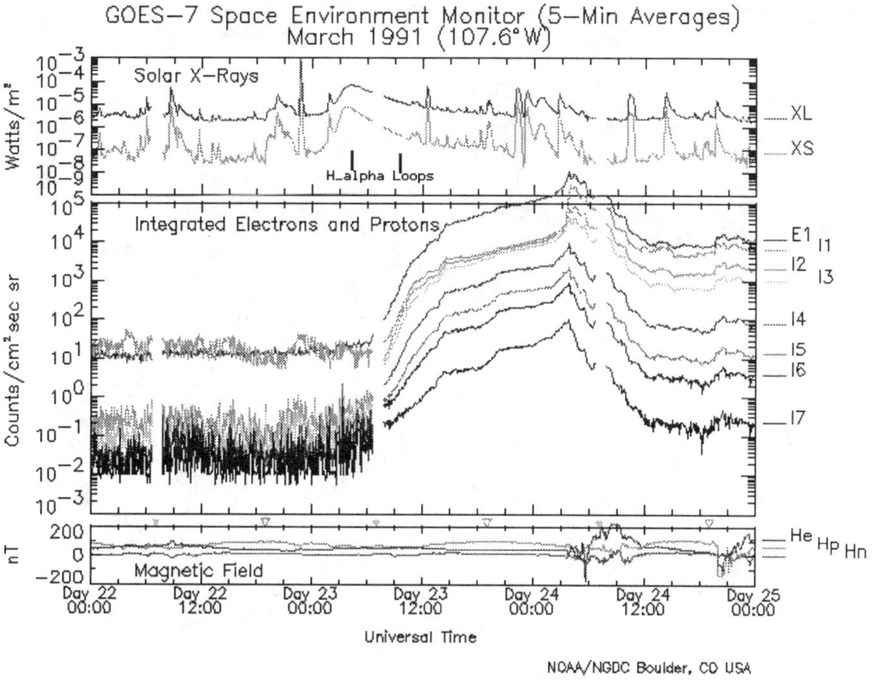

Figure 2. The intensity of the 1991 March 23 Long Duration Flare obtained by SXR detector on the GOES satellite in the wavelength band 0.5-4 Å. The duration of the H_α activity is indicated by two bars.

high at several locations of the active region, but all these locations were not flare productive in proportion to the amount of shear (Ambastha, Hagyard and West, 1993). Most of the flares occurred near the rapidly evolving neutral lines. The locations involving the spots **P2, P3, F3 and F4** was mostly inactive, where the Long Duration Event (LDE) discussed in this paper occurred during 01:15 UT and 15:10 UT. This LDE was responsible for the large CME observed by GOES as shown in Fig. 2.

The high spatial resolution H_α filtergrams during the decay phase of the LDE are shown in Fig. 3. They show two strides of the H_α ribbons and few bright points located between them. One of these ribbons, which is in between the spot **F1** and **P1**, is compact and the other extended along the neutral line, which were stationary during our observations. The LDE displayed a spectacular H_α loop activity shortly after its maximum phase (05:37 UT - 08:30 UT). The individual loops, visible in H_α, were short lived with an average lifetime of 10 minutes. The loops appearing in the beginning phase were long and oriented at an angle of 45° with the east-west axis and did not join the two main bright ribbons. Some of the prominent loops appeared from the sunspots and ended near the plage regions which were brightened. Following the temporal evolution of the prominent loop, it may be seen that after its disappearance, no morphologically

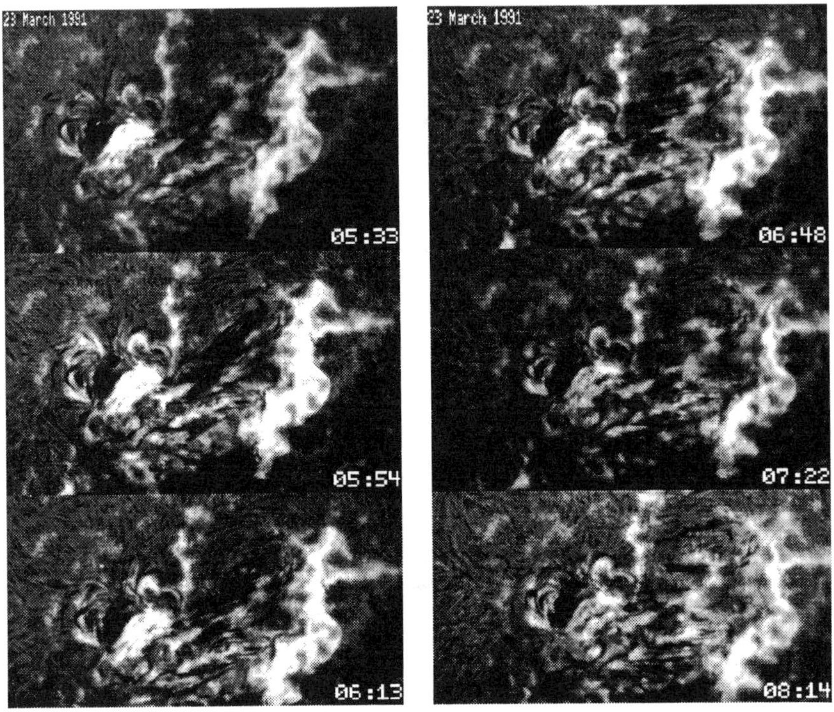

Figure 3. H_α filtergrams of the active region NOAA 6555 showing the temporal evolution of the loop activity.

similar loop of higher altitude was found. The short loops which were almost parallel to the east-west axis appeared towards the end of the activity.

According to their connectivity, the loop system can be designated into three groups as shown in Fig. 4. The group I are the loops from the spot **P1**, which appeared in the beginning of the activity. These loops were long and inclined by about 45° with east-west axis and connected with the nearby negative plage region. The group II loops were connecting the spot **F2** and **P2**. Most of the times these loops were seen to originate from one of the sunspots and end up at the plage regions. These loops also had similar appearance as group I loops. Group III loops connected the spots **F3** and **F4** with **P3** and nearby plage region. These loops were short and oriented along the east-west axis. The loop activity observed during this LDE is used to infer the magnetic field configuration of the active region by comparing them with the computed field lines.

4. Comparison of Model Field Lines and H_α Loops

In order to study the magnetic field configuration of the active region, the H_α loops are compared with the field lines computed by extrapolating the photospheric magnetograms. In the solar atmosphere above the active region, when

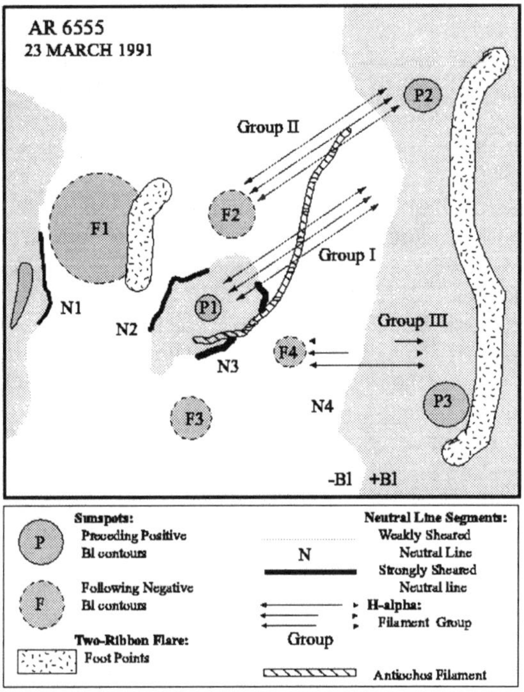

Figure 4. Schematic cartoon of the active region NOAA 6555 illustrating its configuration on 23 March 1991. **Bl**: Longitudinal magnetic field.

the current **J** is parallel to the magnetic field **B**, the field configuration is represented by

$$\nabla \times \mathbf{B} = \alpha \mathbf{B} \qquad (1)$$

(Gary, 1989, Alissandrakis, 1981). For zero value of α, the field is called current-free or potential, for a constant value of α, the configuration is known to be a linear force-free field. Although it is not expected that, in complex field configurations of active regions, α is either equal to zero or constant, these are the cases which can be easily computed numerically from the observational data. Furthermore, in the case of many active regions, it is seen that these simple computed fields in general agree well with the observed field configurations. Therefore, here we attempt to compare the H_α loop structure with the field lines computed by using both current-free and force-free models.

The detailed comparison of field lines computed using different models with the H_α loops showed that the potential model produces most of the field lines. Fig. 5 shows the computed magnetic field configuration using the Sakurai program (Sakurai, 1981). We identify the long inclined H_α loop originating from the spot **F2** (frame 05:33 UT) with the field line number 62. Similarly the field lines 61, 54, 67, 112, 107 can be identified with the group II loops. Group III loops can be identified with the field lines 64, 65, 66 and 105. The only unidentified loop is the one originating from **P1** and connecting northward. This loop is situated along the region with the sheared magnetic field. This filament might be tracing the sheared field and not reproduced by the potential or force-free model.

The main result of the comparison is that the longer and higher loops appear before the shorter and low-lying loops which is contrary to the prediction of RFM. This sequence of loop formation is, however, consistent with the mechanism in which the hot material is supplied through the foot points of the magnetic field lines to form the loops. In fact the alternate view of the formation of the loops as a result of the condensation of the coronal material following the reconnection at coronal heights was found to be inconsistent (Kleczek, 1957). In that case it was realized that there is not enough material in the pre-flare corona to build the loops.

5. Discussion and Conclusions

It is observed that there is a striking similarity between the magnetic field lines computed using the potential model and most of the H_α loops, which started at the decay phase of the LDE. This shows that at the time of the appearance of the H_α loop activity, no strong currents were present in the upper magnetic field configuration to make the field lines depart significantly from a potential field configuration (Levine and Altschuler, 1974). The overall magnetic field configuration of the active region was mostly potential in nature. Along with the potential field lines, there existed few low lying non potential field lines, which are designated as "*Antiochos Filaments*" due to the similarity with the prominance field lines described by Antiochos et al. (Antiochos, Dahlburg, and Klimchuk 1994).

We propose that at the time of onset of the H_α loop activity, a magnetic field configuration with low lying non-potential and higher potential field lines was present above the AR. Followed by the large impulsive 3B/X9.4 class flare, which occurred as a result of flux emergence near the south-east end of the spot **F1** before one hour of the onset of LDE (Wang and Tang, 1996), the magnetic field configuration of the active region was destabilized leading to the reconnection of the magnetic field lines at the coronal height. A part of this particle flux, drained down and heated the chromosphere producing the two ribbons and few bright points located between the two ribbons. As the chromospheric material was heated there was an enhancement of the local pressure at these locations. Consequently, the hot material was injected into the magnetic field lines from below, in the form of non-thermal particles accelerated during the flare process (Jefferies and Orrall, 1965). These trapped particles gradually thermalized and cooled to become visible in the form of H_α loops. The upward velocities observed

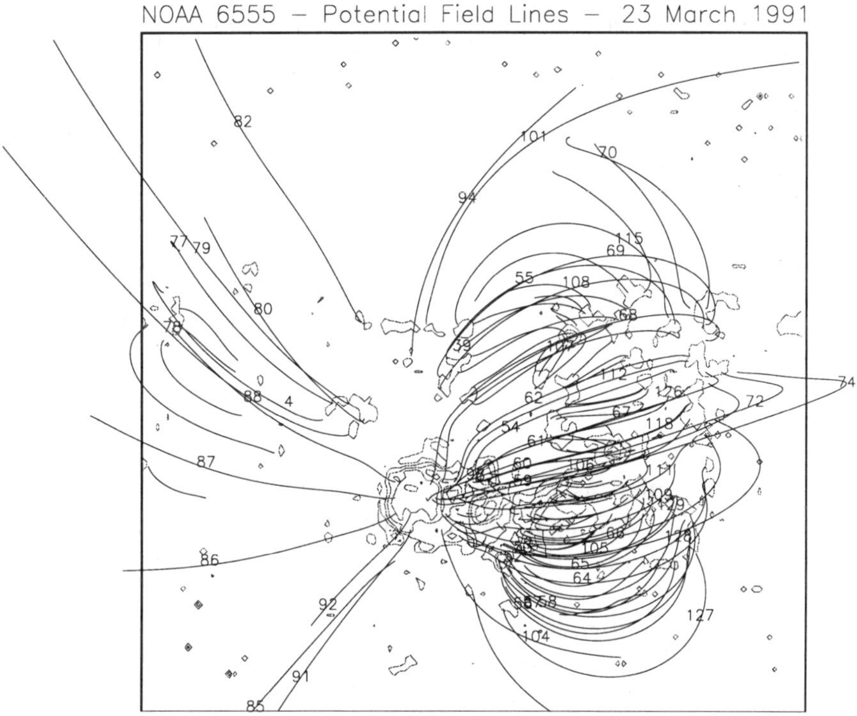

Figure 5. Computed magnetic field lines joining different polarity locations calculated by using potential field models with spherical geometry.

from the blue shifts of the H_β lines, near the loop foot points might indicate that the material injection process was continuing even at the time when it had attained lower temperatures. As a result of this process, first two long prominent loops shown in the 05:33 UT frame were formed. One of these loops, originating from **P1** (*Antiochos Loop*) was below the loop originating from **F2** which was a potential loop. As the Antiochos Loop expanded it interacted with the set of potential field lines like 62, 55, 108 and 39. As a result of this interaction the foot point of the *Antiochos filament* (near the spot **P1**) was heated and the material was filled into the field lines connecting these locations. Following the heating near **F4** area, due to the rise in pressure of this location, the material was filled into the field lines connecting the southern part of the neutral line generating the group III loops. In this way we believe that, these loops are not produced due to the physical process described by the standard RFM, but due to the interaction of a potential filed line with a non potential field line and consequently filling the material into the potential field lines.

The main conclusions derived from the observations and the analysis of the long duration event of 23 March 1998 are listed below. (1) The overall magnetic field configuration of the active region NOAA 6555 was consistent with the potential field configuration at the time of onset of H_α loop activity. (2) The LDE occurred involving a vast area of the active region, where the magnetic shear was low. The LDE area excluded the location of high magnetic shear and flux emergence, which were also the location of the energetic flares. (3) The low lying field lines originating from **P1** was sheared. (4) During the decay phase of the LDE, the flare energetic particles heated the chromosphere. As a result of injection of the hot chromospheric material through the foot points of the magnetic field lines, H_α loops were generated.

Acknowledgments. This work was performed while one of the authors (DP) held a National Research Council-MSFC Research Associateship. We thank Dr. Ron L. Moore and Dr. Mona J. Hagyard and Dr. S. K. Antiochos for very useful discussions.

References

Alissandrakis, C. E.:, 1981, Astron. Astrophys. 100, 197

Ambastha A., Hagyard M. J. and West, E.:, 1993, Solar Phys., 148, 277

Antiochos S. K., Dahlburg, R. B. and Klimchuk, J. A.:, 1994, Astrophys. J., 420, L41

Bruzek, A.:, 1964, Astrophys. J. 140, 746

Feldman U., Seely J. F. and Yuhong F.:, 1995, Astrophys. J. 450, 90

Gary G. A.:, 1989, Astrophys. J. Suppl. Ser. 69, 323

Jefferies, J. T. and Orrall, F. Q.:, 1965, Astrophys. J. 141, 519

Kleczek, J.:, 1957, Bull. Astron. Inst. Czeck. 8, 120

Levine and R. H., Altschuler, M: 1974, Solar. Phys. 36, 345

Sakurai, T.:, 1981, Solar Phys., 76, 313

Wang, H. and Tang, F.:, 1993, Astrophys. J., 76, 313

High Resolution Solar Physics: Theory, Observations, and Techniques
ASP Conference Series, Vol. 183, 1999
T.R. Rimmele, K. S. Balasubramaniam, R.R. Radick, eds.

H_α Intensity Oscillations in Solar Flares

Rajmal Jain, S. C. Tripathy, Lokesh Bharti and Brajesh Kumar

Udaipur Solar Observatory
(Physical Research Laboratory, Dept. of Space, Govt. of India),
Off Bari Road, Dewali, P. B. No. 198, Udaipur - 313001, India

Abstract. Based on the analysis of high spatial resolution H_α filtergrams, we report here the detection of H_α intensity oscillations in two extended flares of 15 November 1989 and 20 April 1991. The relative intensities measured with time at 18 different flare and chromospheric locations were analyzed to obtain the oscillation modes. The analysis shows prominent 5 and 3 min modes in flares in addition to their existence in the chromosphere. However, there exists a frequency difference between the flare and chromospheric modes. This frequency deviation of about 300 microhertz is proposed as an influence of higher magnetic field, location of measurements (height) in the chromosphere and high temperature in the flare. We extended our study to a recently observed solar flare on 12 May 1997, and a large sunspot active region on 14 March 1998 with a CCD camera to achieve high time resolution (2 - 10s) in contrast to earlier flares. The relative intensity is measured with the motion of 6 unambiguously identified flare kernels and 15 steady points inside the flare. Our analysis not only confirms the presence of 5 and 3 min oscillation modes but surprisingly shows the prominent modes near 7, 8.3 and 10 mHz in the flare. The similar analysis of filtergrams of 14 March 1998 further confirms the chromosphere oscillates predominantly with the 3 min mode while the plage shows shorter periods of oscillations observed in the same area under study. This indicates that high spatial and time resolution observations are extremely important. Further, the presence of short period oscillations in the flares indicates that the loops and hence flux tubes are oscillating in the active region very fast. We ask whether short period oscillations in flare kernels (foot points of flux tubes) may provide continuously enough energy in the fluxtubes from beneath the convection zone so as to accumulate at the top of the loop in the corona. When the situation of saturation takes place, then a minor instability or a disturbance may cause imbalance between magnetic field and plasma forces in the corona and hence the flare takes place. In this kind of scenario energy for the flare is pumped from beneath in pockets in a magnetic helical form by several oscillation modes, most prominently, perhaps, by $>$ 8.3 mHz modes.

1. Introduction

It has been known since the observations of Jensen and Orrall (1963) and Elliott (1969) that the solar chromosphere oscillates with a 120 - 300 s period. Liu (1974), Duvall, Livingston and Mahaffey (1980) and Kneer and Uexkull (1983) observed 3-min oscillations in both intensity and velocity in the Ca II K and H_α line profiles. However, some Doppler measurements made by Woodard & Libbrecht (1991) in calcium near disk center of the quiet sun and Fernandes et al. (1992) in sodium line in active region could not detect 3 - min oscillations. On the other hand Duvall et al. (1991) reported frequencies up to 6.7 mHz based on Ca II K full disk data taken at South Pole in November, 1987. Harvey et al. (1993) also detected 3 min oscillation modes in the H_α line but not in the photospheric line (6768Å) using the GONG prototype instrument in Tucson. Further, the observations taken from the South Pole by Harvey et al. (1993) and from Kitt Peak by Tripathy and Hill (1995) in the Ca II K line could measure 3 min oscillations as a background feature. Thus the experimental identification of 3 - min modes in solar atmosphere and its explanation is still debated.

The motivation of this investigation is twofold. First, we wish to confirm the existence of 3-min oscillations in the solar chromosphere. Second, we are interested to find the periods of oscillations, if any, in active region and in solar flares and their deviation from the chromospheric modes.

2. Observations and Analysis

The observational material for the proposed study consists of four time sequences of H_α filtergrams of solar flares taken through a Halle filter of $0.5\mathring{A}$ passband mounted on the 25cm refractor at Udaipur Solar Observatory. The first two time sequences of flares under study are two - ribbon flares of 15 November 1989 (N11 W28) and 20 April 1991 (N09 W50). Both flares were of 3B importance in H_α and X3.2 and X1.0, respectively, in X-ray. The observational details of these flares and their analysis have been reported recently by Jain and Tripathy (1998).

The H_α photographic observations on the film of both time sequences were digitized using a CCD camera and the Image Digitization and Analysis System (IDAS) at USO. During digitization, IDAS converts densities (D) on the film to intensities (I) using characteristic curves. However, the intensities of the filtergrams were kept within the dynamic range of the image grabber which has an 8 bit analog to digital converter of 0 to 255 gray levels. This enabled us to measure the intensity of H_α filtergrams precisely because the conversion of H_α film density to relative intensity is almost linear in the dynamic range of IDAS. These new intensity images were registered on a frame with 512 $\times$ 512 pixels. The registered intensity images were corrected for non-solar effects as described below.

(a) For correcting periodic guiding errors and image motion, we re-registered the new intensity images with the maximum accuracy of one pixel; which is the resolution limit of the telescope. During re-registration the sunspot position was assumed to be fixed for the observational interval under study.

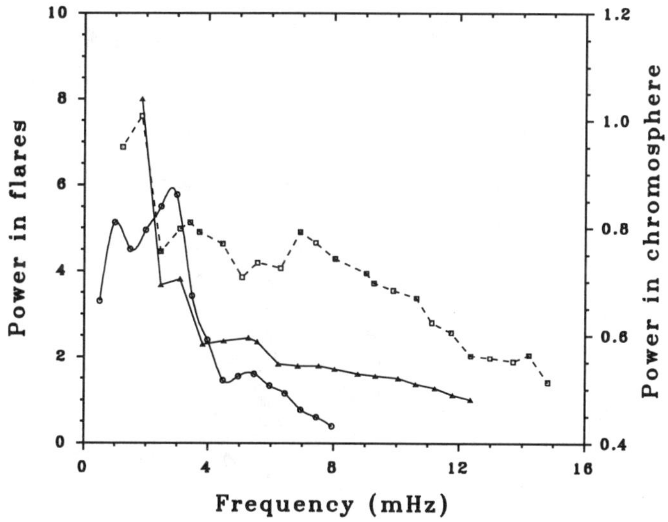

Figure 1. The average power spectrum for flares and chromospheric points. Solid line connecting the circles and triangles represent the spectrum for 15 November 1989 and 20 April 1991 respectively. The dashed line connecting squares represent the average power spectrum of quiet chromospheric points on 20 April, 1991.

(b) For precise measurement of relative intensity, many different locations for flare and for chromosphere were identified on the re-registered filtergrams as discussed earlier. The intensity measurements (I) were taken from the rms of a matrix consisting of 2×2 pixels with the selected position at its center. This improved the signal to noise ratio.

(c) To deal with the problem of non-uniformity on each frame, we measured rms intensity at six different chromospheric points. These rms measurements were averaged to find the normalized mean chromospheric intensity (Ich) for that frame.

(d) For correction of the different exposure time and sky transparency changes from frame to frame, we divided the rms intensity, measured for each point (flare and chromosphere), by Ich as described above.

(e) To account for blurring, we first manually examined the H_α filtergrams and selected only those that have seeing better than 2". Further, the technique of rms measurements of intensities at each identified location suppressed any minor blurring in the selected filtergrams.

The relative intensity of selected flare and chromospheric points with respect to the background chromosphere in each frame was finally measured as (I-Ich)/Ich.

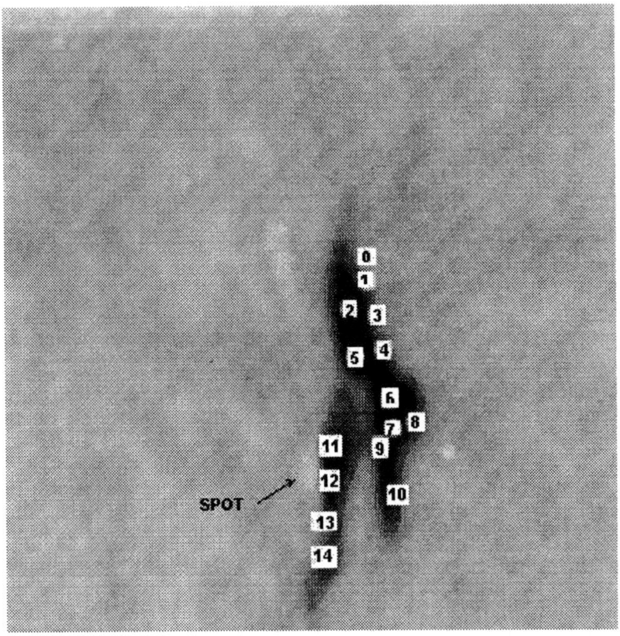

Figure 2. The negative of the flare of 12 May , 1997. The identified 15 different locations (0-14) for which relative intensity measurement is carried out are marked on the flare.

3. Results and Discussions

The time series of variation of H_α relative intensity, (I - Ich)/Ich, for 20 April 1991 and 15 November 1989, were subjected to Fourier transform to obtain the power spectrum of predominant oscillations. The average power spectra of all chromospheric and flare points on 20 April 1991 and 15 November 1989, is shown in Fig. 1. This indicates the prominent chromospheric oscillation modes at 1.85, 3.4, 5.6 and 6.92 mHz. In the case of flares, the predominant modes on 15 November are seen at 1.0, 2.98 and 5.46 mHz and on 20 April at 3.09 and 5.22 mHz.

These results of Jain and Tripathy (1998) mark the first detection of the prominent 5 and 3 min modes in H_α flares and further clearly show the presence of 3 min oscillations in the chromosphere. The present chromospheric results are in agreement with earlier results of Elliott (1969); Duvall et al. (1980, 1991); Kneer & Uexkull (1983) and Harvey et al. (1993). The interesting feature of our results is that the 5 min modes are clearly visible in H_α which could not be earlier detected by Kneer & Uexkell (1983) and by Harvey et al. (1993) using the same spectral line. Further, the power is slightly less but appreciable for

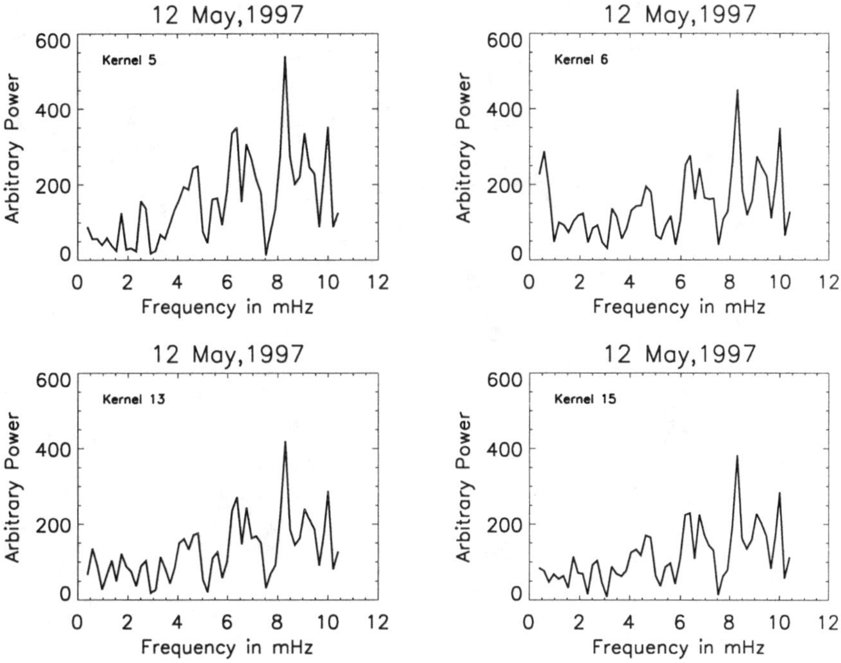

Figure 3. The power spectrum of the relative intensity variation in a few identified locations, 5, 6, 13 and 15 inside the flare.

100 and 60 s oscillation periods which also perhaps exist in the chromosphere as suggested earlier by Kneer & Uexkell (1983). But these modes have to be distinguished from quasi-periodic seeing and minute (<1") image motions in our data.

The prominent chromospheric mode frequencies of 3.4 and 5.6 mHz are close to those obtained by Elliot (1969) and Kneer and Uexkull (1983). We find the frequencies determined in the flaring region on 20 April 1991 are reduced approximately by 300 μHz as compared to the background chromosphere. Similarly, the frequencies of 5 and 3 - min modes in 15 November 1989 flare are reduced by 400 and 200 μHz, respectively, compared to the chromospheric modes of 20 April. The reduction in oscillation frequencies in flares compared to the nearby chromosphere may be attributed to the magnetic field gradient and high temperature (Evans & Roberts, 1990) in flares. However, the decrease in frequency is one order higher than their predicted values.

We extended this study for two more time sequences obtained recently with a CCD camera on 12 May 1997 and 14 March 1998 to achieve high time resolution in the range 2-20 s. Shown in Fig. 2 is a filtergram of 12 May 1997 on which the locations are marked where the relative intensity is measured.

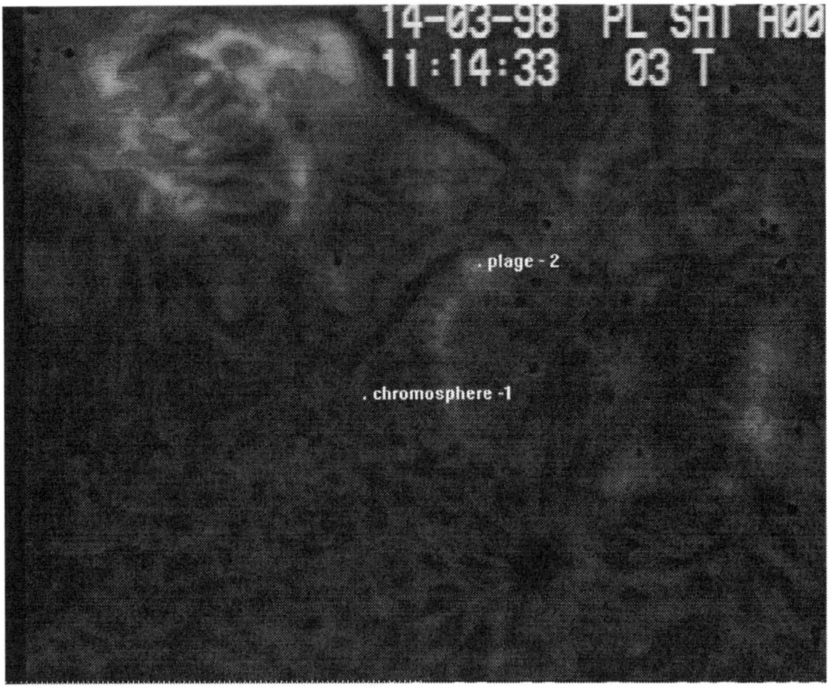

Figure 4. The computer processed image of H_α filtergram of an active region taken on 14 March 1998 through a CCD camera. The locations of relative intensity measurement in chromosphere and plage are marked on the image.

The details of the flare activity on 12 May 1997 are given by Jain et al (1997). The relative intensity measurement at these identified locations were carried out. The time sequences of these relative intensity variation were subjected to Fourier Spectra analysis to obtain the power spectrum. Shown in Fig. 3 are the power in a few identified locations. It may be noted, in contrast to earlier flares of Jain and Tripathy (1998), that the shorter periods (>7 mHz) modes are more predominant in this flare as compared to well known 5 and 3 min modes. This significant result could be due to high time resolution observations made through the CCD. Our result reveals that the shorter periods modes are prominent modes inside the flare. Further, though the 5 and 3 min modes are also present, their amplitude is comparatively weak and their frequencies are found deviate by few hundred microhertz compared to the two earlier cases studied by Jain and Tripathy (1998).

Similarly, we also carried out the relative intensity measurement on the H_α filtergrams taken on 14 March 1998 for an active region shown in Fig. 4. The power spectrum of relative intensity variation in chromospheric and plage points, marked on the filtergarm, are shown in Fig. 5. This figure again confirms that

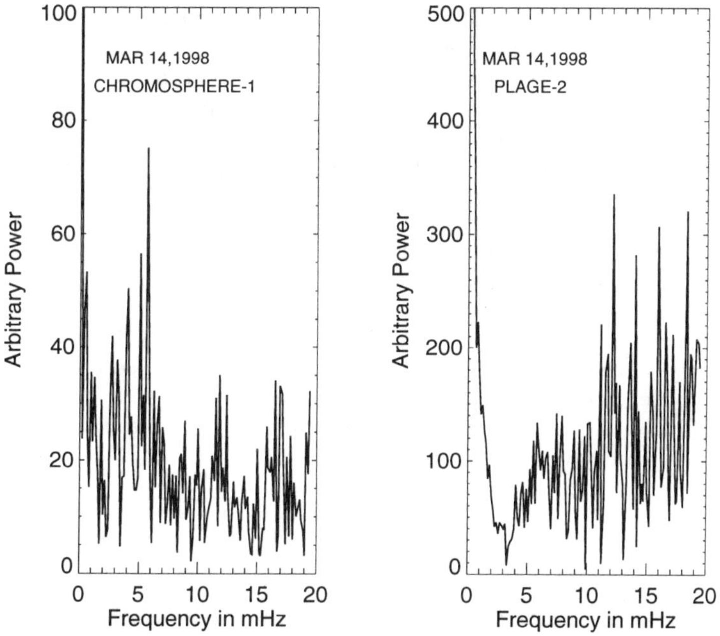

Figure 5. Power spectrum of relative intensity variation in chromospheric and plage points in an active region observed on 14 March, 1998.

in the quiet chromosphere the predominant oscillations are in the range of 3 -5 min while in the plage region (high temperature plasma) shorter periods are prominent. This shows that the 5 and 3 min modes which are seriously damped beyond the photosphere due to a drastic change in density are again dissipated by magnetic field and high temperature when they interact with the sunspot/flaring region during their propagation in the chromosphere. Our results show that the frequencies of oscillation modes increase with the height above photosphere.

To summarize, the H_α filtergrams have been analyzed to measure the variation of intensity of flare, plage and chromospheric points to determine various oscillation modes in the chromosphere. The present study clearly identified 5 and 3 - min modes in solar flares. The detection of these modes further confirms their existence in the chromosphere. It appears that the magnetic field gradient, location in the chromosphere and ambient temperature influence the oscillation frequencies. It appears that the shorter period modes (>8 mHz)

detected in flare kernels are in fact the modes at the foot points of the magnetic loops. This indicates that the loops are oscillating with these frequencies or still higher. If these detected frequencies and the scenario are correct then the oscillation of loops with wide frequencies may be interpreted as the source energy pumping not only to heat the chromosphere and corona but also to store enough energy for flares. The energy pumping in the loops through oscillation modes is possible if these modes are excited in the convection zone and propagate in the highly twisted/ helical loops. The magnetic field strength is extremely large ($10^5 G$) at the foot point of the loop in the convection zone. Thus the evanescent prominent p-modes during their propagation in these highly twisted, helical and strong magnetic field loops are modulated in their frequency and appear as shorter period modes in the chromosphere. If it were possible to measure these modes in the corona, perhaps their periodicity would be even smaller. In this light it is now easier to understand how these modes play an important role in heating the corona and perhaps produce micro and nano flares seen in soft X-rays. These small flares may periodically occur when the balance between the magnetic field configuration of the loop in the corona is disturbed by the plasma in the corona. These periodic micro/nano flares then transfer their energy to the corona to maintain its heating and hence the temperature. Recently, observations made by the TRACE satellite in 171$\AA$ have confirmed our results that the flux from the footpoints of the loop rises upward discretely and the loops oscillate continuously.

References

Duvall, T. L. Jr., Livingston, W., & Mahaffey, C. 1980, in "Nonradial and non-linear stellar oscillations", eds. Hill and Dziembowski, Springer Verlag, Berlin, p.237.

Duvall, T. L. Jr., Harvey, J. W., Jaferries, S. M., & Pomerantz, M. A. 1991, Astrophys. J. 373, 308.

Elliott, I. 1969, Solar Phys. 6, 28.

Evans, D. J. ,& Robert, B. 1990, Astrophys. J. 356, 704.

Fernandes, D. N., Scherrer, P. H., Tarbell, T. D. ,& Title, A. M. 1992, Astrophys. J. 392, 736.

Harvey, J. W., Duvall, T. L. Jr., Jaferries, S. M. ,& Pomerantz, M. A. 1993, in GONG 1992: "Seismic Investigation of the Sun and the Stars", ed. T. M. Brown, ASP conference Series, 42, p. 111.

Jain, R ,& 14 other authors, 1997, Proc. 31st ESLAB Symp., ESA SP-415.

Jain, R ,& Tripathy, S. C. 1998, Solar Phys. 181, 113.

Jensen, E. ,& Orrall, F. Q. 1963, Astrophys. J. 138, 252.

Kneer, F. ,& Uexkull, M. von 1983, Astron. Astrophys. 119, 124.

Liu, S. Y. 1974, Astrophys. J. 189, 359.

Tripathy, S. C., & Hill, F. 1995, in GONG ' 94 : Helio and Astroseismology from the Earth and Space, eds. R. K. Ulrich, E. J. Rhodes, Jr., & W. Dappen, ASP Conference Series, 76, p. 334.

Woodard, M. F., & Libbrecht, K. G. 1991, Astrophys. J. 374, L61.

Group Discussion

v.d. Luhe : What is the slope of the characteristic curve of the filter? Is there any coherence in the oscillations of the six points?

Jain : For the flare events of 15 November 1989 and 20 April 1991 only film was used. The film was Kodak Technical Pan 2415 and its characteristic curve was fitted in IDAS for D to I conversion. However, the grabbed images were kept within the dynamic range of the image grabber which has a 8 bit analog-digital converter producing 0-255 gray levels. This enabled us to measure the intensity of H-alpha filtergrams precisely because H-alpha film D to I is almost linear in the dynamic range of IDAS. The six flare kernels identified unambiguously on 12 May 1997 flare were observed with a CCD and not using film. Thus I think instrumental or systematic errors can not be introduced. The other observations from 14 March 1998 were also carried out with a CCD camera.

High Resolution Solar Physics: Theory, Observations, and Techniques
ASP Conference Series, Vol. 183, 1999
T.R. Rimmele, K. S. Balasubramaniam, R.R. Radick, eds.

Dynamical Motions as Precursors to Activity

S. L. Keil
Air Force Research Laboratory, Space Hazards Branch
Sunspot, NM 88349, USA

K. S. Balasubramaniam, L. J. Milano, A. Bayliss, J. Jones & J. Clark
National Solar Observatory/Sacramento Peak
National Optical Astronomy Observatories Sunspot, NM 88349, USA

Abstract. We have observed the evolution of several active regions prior to flares, Hα brightening, and CMEs. The observations include full-disk Hα images taken at a 10 second cadence with the NSO/SP flare patrol and high-resolution images of the selected active regions taken with the NSO/SP Dunn Solar Telescope. The high-resolution data includes 30 second cadence Doppler-grams and magnetograms and 5 second cadence Hα and G-band images. Using local correlation tracking techniques, we derive horizontal surface flows reflecting photospheric and chromospheric heights. These are compared with the line-of-sight magnetic flux and velocity field and with structures seen in the Hα and G-band images. We investigate these data for pre-activity signatures.

Our preliminary analysis of flares on 1997 April 7 and 1998 May 5 indicates that strong vortices in the surface flow often precede strong Hα brightening and filament eruptions. The full-disk Hα campaigns are being conducted to develop activity prediction techniques in anticipation of the high cadence imaging that will be provided by the Air Force's Improved Observing Optical Network (ISOON).

1. Introduction

The evolution of solar active regions prior to activity events such as flares and mass ejection has been intensely studied in hopes of finding observational signatures leading to activity prediction. Several predictors have been suggested including active region complexity, magnetic shear, intense Hα fibril activation, surface velocities, and pre-flare brightening. However, statistical studies (Neidig, 1990) show that even when taken together, prediction using these factors have been of limited accuracy and usefulness.

Longcope and Cowley (1996) demonstrated that stresses applied by photospheric motions can lead to current ribbons and Longcope (1996) showed these ribbons can store magnetic energy and facilitate flares. Recent work has shown that flux can emerge from the sun in pre-stressed states (Leka et al., 1996) that flux can emerge from the sun in pre-stressed states (Leka et al., 1996) and that surface flows are not necessary to have twisted field lines. However, the emergence of twisted field may induce twisting motions in the atmospheric plasma and Hα fibrils. In some cases, twisting motions have been clearly seen before the occurrence of a flare (e.g. Keil et al, 1994, Paper I). Although flare energy release is primarily a coronal phenomena, the source of energy is surely in the lower atmosphere or sub-atmospheric layers. The magnetic field lines involved in solar activity are deeply rooted through the photosphere into the convective zone. Thus, one expects to find observable manifestations in the photosphere and intervening chromosphere of the phenomena driving the active region toward instability.

Magnetic shear is important for flares, but several authors have shown that shear alone is not sufficient to predict the location, timing or frequency of flares (e.g. Ambasta, 1998, Choudhary, Ambasta and AI, 1998). Emergent flux and rapid motion of the field also play a role. Zirin and Wang (1990), have reported seeing motions associated with pre-flare sunspot regions and emerging flux.

In Paper I we discussed the roles of vertical and horizontal flows in flares that occurred on 1992 Jan 24 in NOAA 7016 and on 1993 Feb 6 in NOAA 7420. In both cases, strong vorticity in the horizontal flow and shear in the vertical flow were seen at the locations of the primary flare kernels as seen in Hα. In both cases only a small amount of data was obtained before the flare began, and it was not possible to say if the velocity shear and vorticity could serve as a predictor. In a second paper (Balasubramaniam, Milano, & Keil, 1998, Paper II) we used the 1-minute cadence Hα patrol at NSO/SP to investigate horizontal motions in the chromosphere prior to the 1997 Apr 7 flare in NOAA 8027. By tracking horizontal motions in Hα we can follow filament activations as well as net flows in the chromosphere. Strong vorticity were seen in association with the Hα flare kernels. Similar twisting motions have been reported by Canfield et. al. (1996) in connections with Hα surges and associated x-ray jets. Shimojo, Shibata, & Harvey (1998) show that the x-ray jets occur in region of rapidly changing flux and converging bipoles.

In this paper we report a preliminary attempt to improve the results of Papers I & II by increasing both the observational cadence and the length of time observed before the flare. Section 2 describes the observations and the following sections give results for flares observed in active regions NOAA 8210 and 8214 on 1998 May 5.

2. Observations & Reduction

The observations were obtained with the Richard B. Dunn Solar Telescope (DST) and the Hα flare patrol at Hilltop Facility at NSO/Sacramento Peak. Three camera's were used in the DST. A 403x256 pixel RCA 504 CCD was placed behind the Universal Birefringent Filter (UBF) resulting in a 2' x 1' field-of-view (FOV) with 0.3"x0.23" pixels. The UBF was used in 1/8th angstrom mode and tuned to two wavelengths in the blue wing of Ca I 6122 separated by 20 mÅ where Stokes I+V and I-V were recorded and then to the red and blue wing of Fe I 5434 Å to obtain a Doppler pair. All six measurements were obtained every 30 seconds. A 1024 x 1024 pixel Thomson camera was placed behind a 10 Å G-band filter at 4305 Å with a 3'x3' FOV. A second 1024 x 1024 Thomson camera was placed behind a 200 mÅ Zeiss Hα filter which was tuned on-band. The pixel size is 0.16"x0.16" on both Thomson cameras and they were operated at a 5-second cadence.

At the Hilltop Facility, the standard Hα film camera was replaced with a 2K x 2K Xedar CCD camera. The Hα patrol uses a 10 cm lens, stopped down to 6 cm to feed a 0.25 Å tunable Halle Hα filter. The resultant pixel size is 1.3"x1.3". In its standard patrol mode the filter is tuned to line center, Hα $\pm$ 0.5 Å, over exposed at line center for prominence, and to white light near Hα. For our observations, we only tuned to line center, permitting us to run at a 10-second cadence. The observations are summarized in Table 1.

Clouds permitting, we started observations near sunrise at both facilities and observed all day. The DST was pointed at the active region which seemed most likely to produce activity. Flat fields and dark calibrations were made periodically throughout the day.

Table 1. Observed Lines.

Richard B. Dunn Telescope

Filter	CCD	λ [Å]	Pol.	Cadence	FOV
				Seconds	FOV
UBF	403x256	6122.080	left	30	2'x1'
UBF	403x256	6122.080	right	30	2'x1'
UBF	403x256	6122.100	left	30	2'x1'
UBF	403x256	6122.100	right	30	2'x1'
UBF	403x256	5434.460	left	30	2'x1'
UBF	403x256	5434.590	left	30	3'x3'
G-Band	1Kx1K	4305.	n/a	5	3'x3'
Zeiss	1Kx1K	6562.808	n/a	5	3'x3'

Hilltop Patrol Facility

Filter	CCD	λ [Å]	Pol.	Cadence	FOV
Hα	2Kx2K	6562.8.	n/a	10	Full Disk

On 1998 May 5, 3918 images were obtained with the full disk Hα imager between UT 13:35 and 00:24. These were gain and dark corrected, circularized and rotated so heliospheric north was up. Fig. 1 shows a typical image and the location of the two active regions used in this study. After correcting each image for limb darkening, we selected an area containing NOAA 8210 (S12W50) and 8214 (N26W18) respectively and made a separate time sequence for each region from the full disk images. Foreshortening was removed to obtain a disk-center view, and each region was tracked to remove image motion. The corrected images were then divided into sub-images and local correlation tracking techniques (Keil et al. 1994) were applied to track the drift of these sub-images. For NOAA 8210, which is close to the limb, the sub-image size in the corrected images was 13"x13" and for NOAA 8214 the sub-image size was 9"x9". Rapidly varying displacements of sub-images due to seeing changes between images were separated

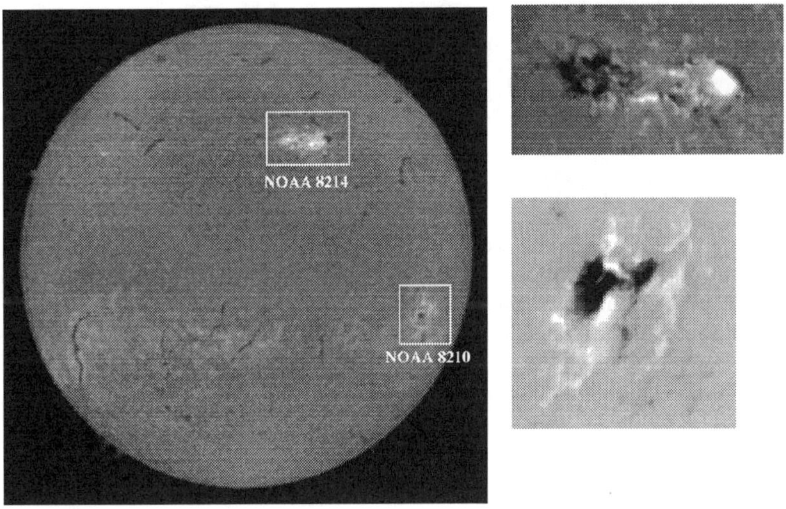

Figure 1. The Image on the left is a full disk Hα image from 1998 May 5, 13:35 UT with boxes drawn around the two active regions. From top to bottom the images on the right are magnetograms of NOAA 8214 and 8210 respectively from the NSO/KP daily patrol.

from slowly varying displacements due to actual motion of features on the sun using a low-pass filter. We then computed the derivative of the slowly varying component to obtain a horizontal surface velocity at each 10s time step.

For NOAA 8210, simultaneous observations were obtained with the DST. All DST images were first gain and dark corrected. The four polarization images in Ca I 6122 Å and the two Doppler in images in Fe 5434 Å from each UBF spectral scan were registered by destretching them with respect to the first Ca I image. The Ca I polarization pairs were then subtracted to obtain the Stokes V signal at each wavelength. The slope of the Ca I line was determined from the two wing wavelengths and the weak field approximation was used to compute the magnetic flux. The Fe I 5434 Å Doppler pairs were used to obtain an estimate of the line-of-sight velocity. Thus for each 30 s interval needed to scan the UBF through 6 wavelengths, we produce one magnetogram and one Doppler-gram. These were then corrected for foreshortening. After gain and dark correction, the G-band and Hα images were also corrected for foreshortening. Fig. 2 shows a G-band image along with a magnetogram and a Doppler-gram with foreshortening removed. As with the full disk Hα images, we then used local correlation tracking to obtain the time-dependence of the horizontal flow field for each set of images. Tracking was done on 5"x5" sub-images.

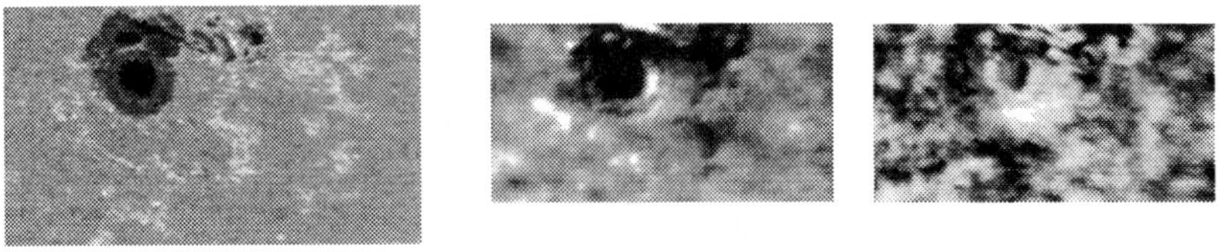

Figure 2. DST images of NOAA 8210 corrected for foreshortening. From left to right the images are a G-band image, a magnetogram, and a Doppler-gram from the DST time sequence at 22:15 UT. The magnetogram saturates in the sunspot umbra and the flux was set to a constant value. Positive flux is white in the magnetogram and redshift is white in the Dopplergram. The G-band images covers 281"x162" on the sun. The magnetogram and Doppler-gram cover a slightly smaller FOV (231"x118").

3. Results

3.1. NOAA 8210

NOAA 8210 was observed from 14:10 to 00:09 UT with the DST. A series of strong Hα brightening and small flares led up to a M2.5 flare at approximately 23:27 UT. From the NSO/KP magnetograms shown in Fig. 1, we see that NOAA 8210 is a complex δ-sunspot region. The spots exhibit strong negative polarity with positive flux intruding from above and below between the large spot on the left and the smaller spot on the right (Fig. 2). The G-band bright features just below the large spot correspond to negative polarity flux that is being transported away from the sunspot. This negative flux collides with the positive flux seen below and to the left of the large sunspot. The time sequence of magnetograms reveals flux disappearing at the interface between the two opposite polarity regions. The Doppler-gram clearly shows the Evershed flow and reveals an area of sheared line-of-sight flow directly below the sunspot.

Fig. 3 shows a sequence of Hα images, corrected for foreshortening. A small Hα brightening occurs at about 23:00 UT in the region to the right of the sunspot. The contours show levels of curl in the horizontal flow as measured by local correlation tracking of the Hα images. White represents positive curl (clockwise) and black negative and each contour level represents a doubling of the curl from the previous level. The curl is seen to increase in the region of the flare approximately one-hour prior to the Hα brightening.

Figs. 4 is similar to Figs 3, but shows the much stronger M-class flare that occurred approximately 30 minutes later. The net curl in the region begins to increase about 30 minutes prior to the flare and peaks about 10 minutes after the flare commences.

Fig. 6 shows the Hα brightness and the absolute value of the curl as functions of time, averaged over the boxes shown in Fig. 5. The net curl more than doubles in the region prior to the flare. The box labeled R-4 is in a region that remained fairly quiet throughout the period of observation. The average curl

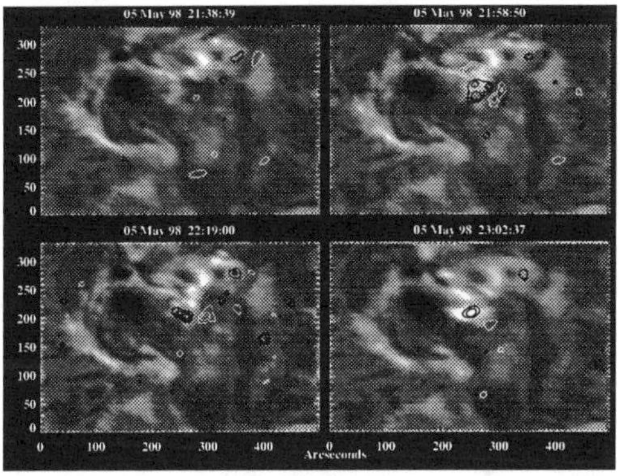

Figure 3. Curl Contours superposed on Hα images.

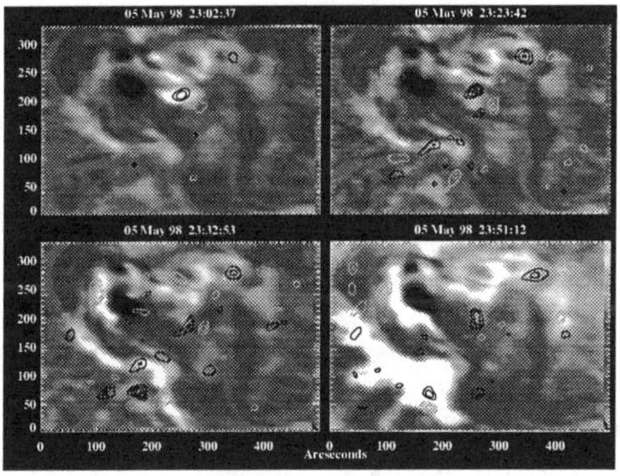

Figure 4. Curl Contours superposed on Hα images.

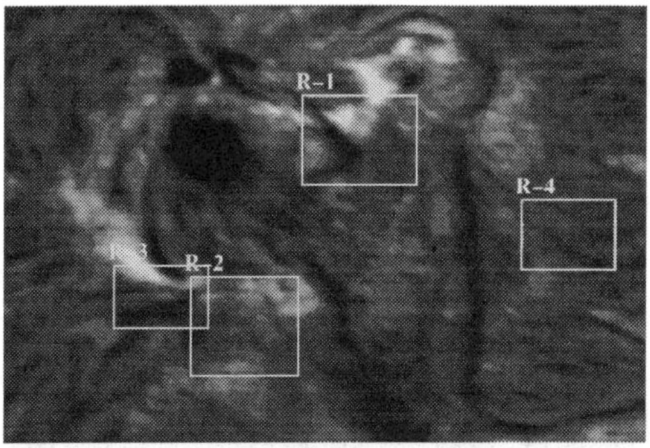

Figure 5. Absolute value of curl and Hα brightness vs time averaged over the boxes shown on curl image (upper right) and Hα image (lower right)

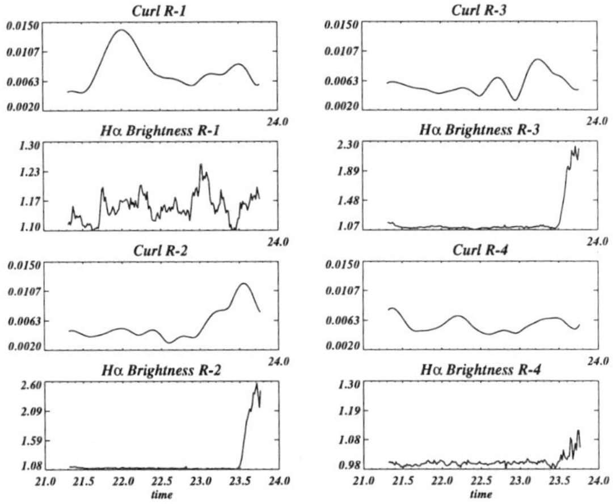

Figure 6. Absolute value of curl and Hα brightness vs time averaged over the boxes shown on curl image (upper right) and Hα image (lower right)

is seen to fluctuate by about 30% in this region. The region R-1 contained the small Hα brightening shown in Fig. 3. The curl is seen to reach a peak almost double its nominal value about 1 hour before the flare. R-2 contains the kernel of the much stronger M2.5 flare that began near 23:27 UT. The curl begins to increase about 30 minutes prior to the onset of the flare. R-3 contains some Hα filaments that also become involved in the flare. The filaments become activated before the flare, which is reflected in the curl shown in Fig. 6.

The much longer sequence of full-disk images was also used to investigate NOAA 8210. Fig. 7 shows the region after correction for foreshortening. The region labeled R-1 is above (North) of the FOV of the DST image shown in Fig. 5. The vertical component of the curl was again measured by tracking Hα surface flows. The larger sub-image size (13"x13") used to track the motions tends to smooth over and reduce the curl signals, but they are still clearly visible as was noted in Paper II. Fig. 8 plots the absolute value of the curl and Hα brightness vs time averaged over the four regions shown in Fig. 7. R-2 roughly corresponds to R-1 in Fig. 5 and R-3 corresponds to R-2 and R-3 in Fig. 5. R-4 in Fig. 7 is averaged over the entire active region.

For the region containing the small Hα brightening at 23:00 UT (R-2, Fig. 7) an increase in the curl begins near 21:00 UT and peak at approximately 22:15 UT, about 15 minutes later than the peak seen in the DST data (R-1, Fig. 5 and 6). The apparent time difference between the two data sets probably reflects the difference in spatial resolution. The increase in curl in R-3 (Fig. 7) is less

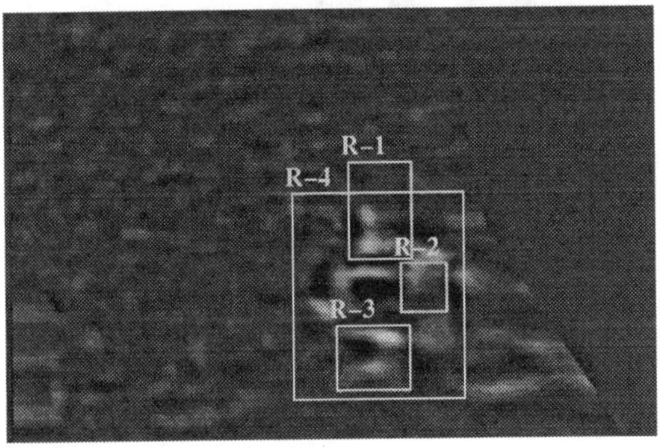

Figure 7. Hα image showing boxes where the curl was computed. Region R-4 contains the entire active region

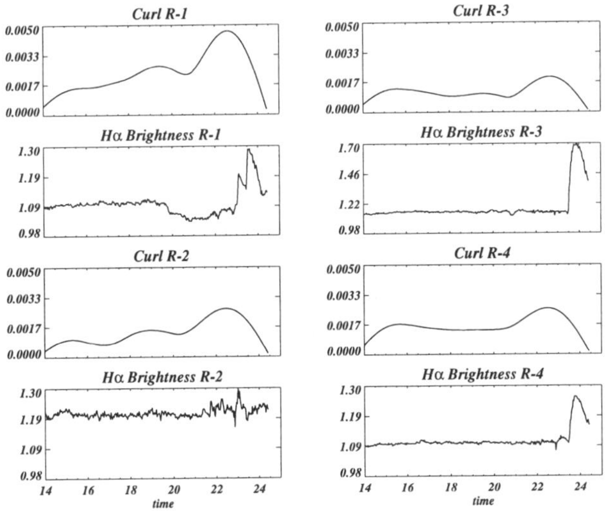

Figure 8. Absolute value of curl and Hα brightness vs time from NOAA 8210

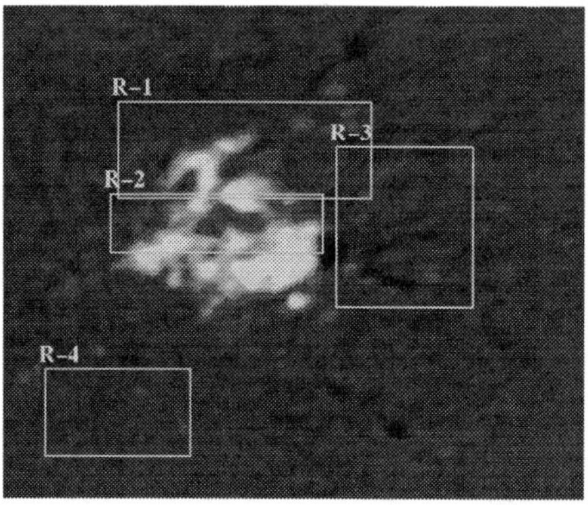

Figure 9. Hα image showing boxes where the curl was computed. Region R-4 is a quiet region for comparison

pronounced than seen in R-2 and R-3 (Fig. 5), but coincides well in time (about 30 minutes before flare onset). Again, the reduced spatial resolution probably accounts for the reduced curl signal. R-1 (Fig. 7), which also showed a strong flaring component, exhibits a steady increase in curl, peaking 30 minutes prior to the flare. Even when the curl is averaged over the entire active region (R-4, Fig. 7 and 8), an increase, although small, is still apparent before the flares.

3.2. NOAA 8214

NOAA 8214 was only observed with the full disk flare patrol. The data was processed in the same manner as NOAA 8210. The NSO/KP magnetogram shown in Fig. 1 shows that this region also has a weak δ-configuration. Regions labeled R-1 and R-2 in Fig. 9, contain an area that produced two C3.5 flares between 18:00 and 20:00 UT. R-1 contained several filaments and most of the strong Hα brightening took place in this region. The curl begins increasing about 45 minutes prior to the first flare and peaks between the two flares. Hα emission was already elevated in R-2, but it contained very little filamentary structure. A slight increase in the curl was visible before the first flare, but compared to the fluctuations observed in a quiet region (R-4, Fig. 9 and 10) it doesn't appear to be significant. R-3 exhibited several filament eruptions and finally a small flare near the end of the observed period. The curl remained slightly elevated in this region throughout the observing period, with a several fold increase before the flare.

4. Discussion

A wide variety of phenomena have been associated with the occurrence of solar flares. In his review, Wang (1998), lists: (1) strongly curved magnetic neutral lines ('S' or reversed 'S'-shaped), (2) steep gradients of the line-of-sight magnetic field, (3) filament activation, (4) new emerging flux, often being pre-twisted par-

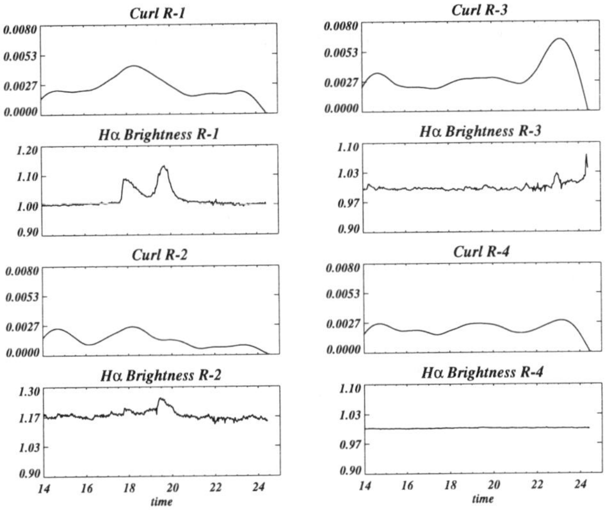

Figure 10. Absolute value of curl and Hα brightness vs time from NOAA 8214

ticularly in δ-sunspot regions, (5) highly sheared transverse field, (6) magnetic flux cancellation, and (7) vertical current concentration. The present work attempts to quantify (3) directly and (4) indirectly. Our primary results are:

- From Figs. 3–4 we see strong pairs of oppositely directed curl form near the location of subsequent flare kernels. In Paper II we speculated that these oppositely directed twists could result from the untwisting of magnetic flux.

- The amount of vorticity in the horizontal flow in flaring regions, as measured by summing the absolute value of the vertical component of the curl, increases up to an hour before the Hα brightening begins, as seen in Figs. 5, 8 and 10.

We have attempted to look closely at flows in the chromosphere associated with flaring regions and filament eruptions. In Paper I, photospheric motions associated with flares were investigated and strong vorticity were seen at the foot points of filaments that erupted during the flare. If these motions are caused by the relaxation of twisted fields, then one might expect to see similar flows in the chromosphere. Since we have only looked at the flows in this paper, we can not say whether or not they cause or are induced by changes in the magnetic field. Martres et al. (1977) noted changes in the twisted pattern of Hα fibril structures occurring up to an hour before the onset of flares. This work confirms and quantifies their observations.

Canfield and Pevtsov (1998) give a review of magnetic helicity of solar active-region magnetic fields arguing that the twist of active region fields observed in the photosphere and corona originate beneath the visible surface. On the small scale observed here, one expects turbulence below the surface to be a major factor in the local twist. The motions we observe could be due to the emergence of twisted field into the active region. The emergence of this flux into the pre-existing field of the active region could provide a trigger for the flare or provide the energy for the flare itself through reconnection. In any case, the increase in the vorticity of the surface motions seen in Hα appears to be a good indicator that an eruption is eminent.

We plan to extend our analysis of these regions to include the evolution of the line of sight magnetic field and the associated flows in the G-band images which reflect deeper layers in the photosphere.

Acknowledgments. We would like to thank the observing staff at Sacramento Peak for their help in acquiring the data. This work is supported by the Air Force Office of Scientific Research through their Task 2311G3 on Solar Environmental Disturbances. NSO/KP data used here are produce cooperatively by NSF/NOAO, NASA/GSFC, and NOAA/SEC.

References

Ambasta, A., 1998, Proc. XVIII National Solar Observatory/Sacramento Peak International Summer Workshop, 1997. Eds. K.S. Balasubramaniam, J. W. Harvey, & D.M. Rabin, ASP Conference Series, Vol. 140, pg 113

Balasubramaniam, K. S., Milano, L., & Keil, S. L., 1998, Proc. XVIII National Solar Observatory/Sacramento Peak International Summer Workshop, 1997. Eds. K.S. Balasubramaniam, J. W. Harvey, & D.M. Rabin, ASP Conference Series, Vol. 140, pg 189

Canfield, R. C., Reardon, K. P., Leka, K. D., Shibata, K., Yokoyama, T., & Shimojo, M., 1996, Astrophys. J. 464, 1016

Canfield, R. C. & Pevtsov, A. A., 1998, Proc. XVIII National Solar Observatory/Sacramento Peak International Summer Workshop, 1997. Eds. K.S. Balasubramaniam, J. W. Harvey, & D.M. Rabin, ASP Conference Series, Vol. 140, pg 131

Choudhary, D. P., Ambasta, A. & Ai, G., 1998, Solar Phys. 179, 133

Keil, S. L., Balasubramaniam, K. S., Bernasconi, P., Smaldone, L. A., & Cauzzi, G. 1994, Proc. XIV National Solar Observatory/Sacramento Peak International Summer Workshop, 1993. Eds. K. S. Balasubramaniam & G. W. Simon, ASP Conference Series, Vol. 68, pg 265

Leka, K. D., Canfield, R. C., McClymont, A. N., & van Driel-Gesztelyi, L., 1996, Astrophys. J. 462, 547

Longscope, D. W., 1996, Solar Phys. 169, 91

Longscope, D. W. & Cowley, S. C., 1996, Phys. Plasmas 3 (8), 2885

Martres, M-J., Soru-Escaut, I., Nakagawa, Y., 1977, Astron. Astrophys. 59, 255

Shimojo, M., Shibata, K., & Harvey, K. L., 1998, Solar Phys. 178, 379

Wang, J., 1998, Proc. XVIII National Solar Observatory/Sacramento Peak International Summer Workshop, 1997. Eds. K.S. Balasubramaniam, J. W. Harvey, & D.M. Rabin, ASP Conference Series, Vol. 140, pg 121

Zirin, H., and Wang, H. 1990, Solar Phys. 125, 45

High Resolution Solar Physics: Theory, Observations, and Techniques
ASP Conference Series, Vol. 183, 1999
T.R. Rimmele, K. S. Balasubramaniam, R.R. Radick, eds.

Average Properties of Flux Emergence in Young Active Region 5617

Louis H. Strous

Lockheed-Martin Advanced Technology Center, Solar & Astrophysics Laboratory; O/H1-12, B/252; 3251 Hanover Street, Palo Alto, CA 94304

Abstract. The average magnetic flux emergence event in young active region NOAA 5617, when it is about 8 hours old, is characterized (based on a sample of 111 emergence events) by coincident transient darkenings (of about 2 Mm and 10 minutes) in the 557.6 nm continuum and Fe I 557.6 nm line center intensity and an upflow in the Fe I 557.6 nm dopplergrams, followed by the appearance (when not obscured by the presence of pre-existing flux) of up to two new bright points flanking the line-center darkening, which coincide with magnetic flux elements in the Fe I 630.2 nm magnetograms and downflows in the dopplergrams. The footpoints then move apart at on average 1.4 km/s. The flux emergence happens all over the active region, and does not fit the notion of a well-defined magnetic neutral line. The flux emergence happens with a preferred orientation of the emerging magnetic loops which fits Hale's polarity law.

1. Introduction

Strous et al. (1996) studied the motion of granulation and magnetic elements (faculae, pores, sunspots) in young active region NOAA 5617, and found faculae of either polarity scattered throughout the active region, moving toward the end of the active region that has a magnetic polarity similar to their own. Magnetic flux emergence must therefore occur all over the active region, too.

Almost all visual spectral lines are observed to be weakened in faculae: to emit more flux in the line center in faculae than in nearby non-magnetic photosphere (Sheeley 1967, Chapman and Sheeley 1968). This behavior is reproduced by models (review by Solanki 1993). The appearance of bright points in line-center intensity can therefore be used as a proxy for the emergence of magnetic flux.

Just before the appearance of a small magnetic bipole, transient photospheric upflows are observed (e.g., Bruzek 1967, Frazier 1972, Brants 1985, and Tarbell et al. 1989). Downflows of the order of 1 – 2 km/s follow the upflow phase (Kawaguchi and Kitai 1976, Zwaan 1985, Brants 1985, Tarbell et al. 1989). Tarbell et al. see the upflow only in the dark central part of the emerging bipole, and downflows only in the bright endpoints.

Strous (1994) found qualitative results similar to those of Tarbell et al., in active region NOAA 5617, linking transient continuum and line-center darken-

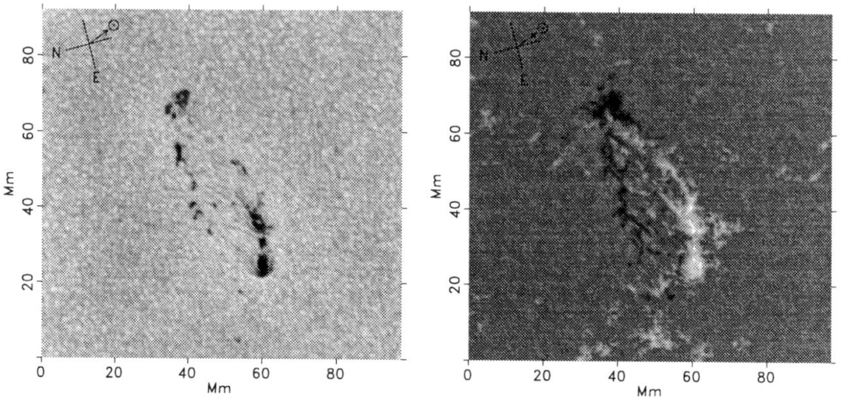

Figure 1. AR 5617 at 11:01 UT in the continuum (left) and a magnetogram (right). The orientation of the images and the direction to disk center are indicated in the upper left-hand corner.

ings as well as newly emerged nearby line-center bright points and downflows to magnetic flux emergence. Here I derive quantitative results for a statistical sample of flux emergence events in the same data set.

2. Data

The studied active region is NOAA 5617, observed on 29 July 1989 between 10:42 and 12:14 UT with the Lockheed SOUP Tunable Filter at the Swedish Vacuum Solar Telescope in La Palma, Spain. Observations were taken at a number of different wavelengths and filter positions in 150 cycles of on average 36.9 seconds in length. From the observations were derived the continuum intensity near 557.6 nm, magnetograms from the Fe I line at 630.2 nm, and fitted line-center intensity and doppler velocity from the magnetically insensitive Fe I line at 557.6 nm. A spatial-temporal filter was applied to each of these quantities to remove p-modes. After data processing, the field of view was 464×438 px in size at 0.29 arcsec/px. The active region was about 7.5 hours old and about 20 degrees from disk center at the beginning of the observations.

The magnetograms show circular polarization density, which is a proxy for magnetic flux density. For the current data, the ratio is about 30 G (3 mT) per percentage point of circular polarization for magnetic field strengths up to about 1200 G (120 mT). For field strengths exceeding 1500 G, the polarization density in the current observations *decreases* with increasing field strength (see Strous 1994, Ch. 6).

3. Qualitative Investigation

Fig.1 shows the active region in the full field of view, and Fig. 2 shows a detail of the field of view at greater magnification. The magnetograms show that in this active region at this time the concept of a magnetic neutral line is useless

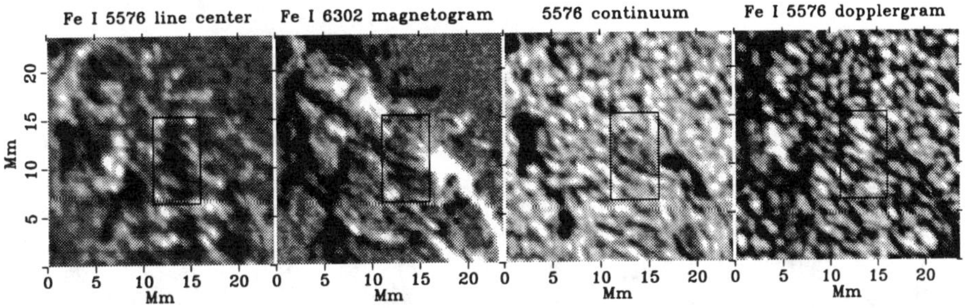

Figure 2. A detail of AR 5617 at 11:01 UT in (a) the line center, (b) the magnetogram (black indicates leading polarity and white following), (c) the continuum, and (d) the dopplergram (white indicates upflow, black downflow). The small rectangle encompasses a number of flux emergence events.

in the inner parts of the active region: magnetic flux of both polarities can be found in close proximity throughout the inner parts of the active region.

Regard the elongated line-center darkenings visible in the small rectangle in Fig. 2a. When the data is watched as a movie, then these darkenings last only about 15 minutes or so. Often, a bright point appears at one or both ends of such a darkening shortly after it reaches its greatest extent. Comparison with movies of the quantities displayed in the other panels of Fig. 2 shows that the transient line-center darkenings correspond to transient continuum darkenings, transient upflows, and nothing in particular in the magnetograms. The association of continuum darkness with upflows is opposite from what one finds in granulation: there continuum darkness is associated with downflows. Line-center bright points are associated with magnetic flux and with downflows.

These observations fit the emergence of magnetic flux into the photosphere: the bright points and downflows indicate the footpoints of the emerged magnetic flux loops, and the transient darkenings and upflows correspond to the rising top part of the loop, wherein the material has not yet had the opportunity to drain down through the footpoints.

The identification of line-center bright points with magnetic flux tubes suggests that a new bright point should appear at both ends of each transient line-center darkening, but I do not often see that. My impression is that new bright points become visible only if there are no pre-existing bright points (flux tubes) near the end of the transient line-center darkening where the new point is expected. Possible explanations for this absence of some line-center bright points are: (1) the newly emerged magnetic flux is immediately canceled before it could become visible as a line-center bright point; (2) the newly emerged flux is associated with a new bright point but the bright point just does not stand out between the pre-existing ones; (3) fast, conspicuous line-center brightening is suppressed or inhibited near pre-existing magnetic flux.

Fig.3 shows snapshots and time slices related to a single conspicuous flux emergence event in which a newly appearing line-center bright point remains isolated for 40 minutes. The transient line-center darkening lasts about 17 minutes, reaches about 2.5 Mm^2 and 0.65 of average line-center intensity. The coincident

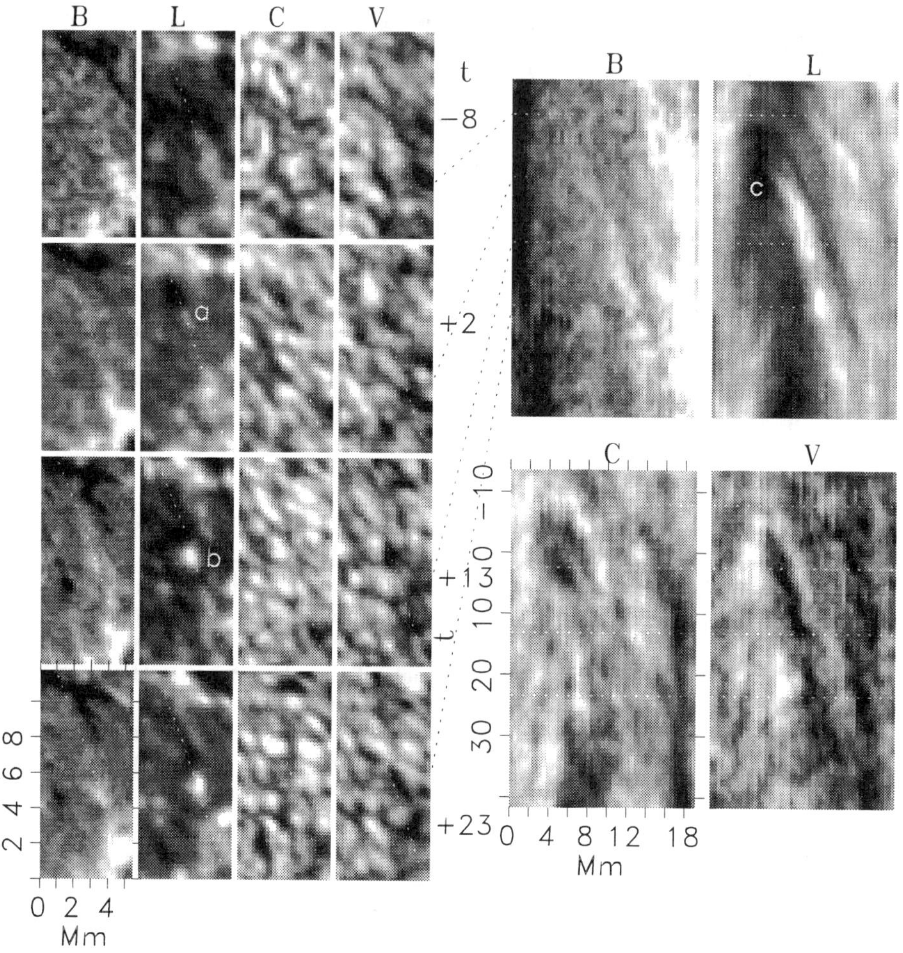

Figure 3. Snapshots and time slices of a conspicuous flux emergence event. The displayed quantities are: polarization density (B), line-center intensity (L), continuum intensity (C), and doppler speed (V). The left-hand sixteen snapshots show these quantities as a function of two spatial coordinates at four observation times: $t = -8$, $t = +2$, $t = +13$, and $t = +23$ minutes, relative to the time of least line-center intensity in the transient line-center darkening. The right-hand four time slices show quantities as a function of one spatial coordinate and time. The spatial coordinate runs along the dotted line visible in the snapshots; the coordinate increases from the upper left to the lower right. Time increases downward in the time slices, as indicated by the coordinate labels on the lower left-hand time slice, and runs from 6 to 62 minutes after the start of observations ($t = -14$ until $t = 42$). The four times at which snapshots are shown are indicated by the horizontal dotted lines in the time slices.

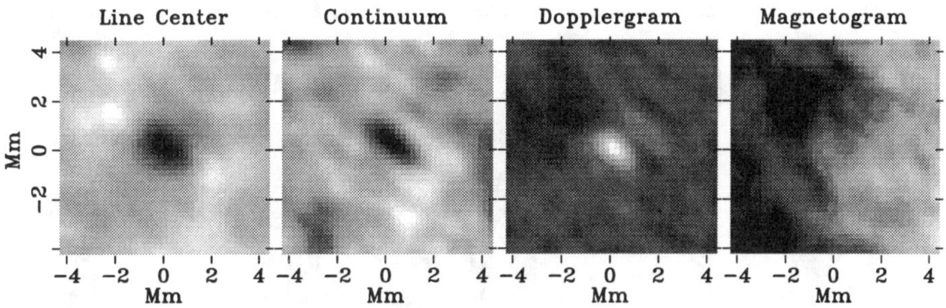

Figure 4. Average quantities vs. position at the time of maximum line-center darkening, for 111 events. Extreme values: line-center 0.736 – 1.119 times median; continuum 0.948 – 1.036 times median; doppler velocity −175 – +706 m/s; magnetogram −81 – +49 gauss (30 gauss per one percent polarization). The full contrast range is shown for all panels, except for the magnetogram, for which the range between −60 and +60 gauss is shown.

transient continuum darkening lasts about 13 minutes, reaches about 1 Mm^2 and 0.9 of average continuum intensity. The coincident updraft lasts about 16 minutes and reaches 0.86 km/s.

The peak line-center intensity (1.6 times the average) in the new bright point peaks at $t = +7$ minutes (relative to the peak line-center darkness in the transient darkening). The bright point moves away from the position of the darkening at 2.2 km/s. The coincident downdraft reaches −1.26 km/s at $t = +12$ minutes, but is probably not resolved in the observations. The coincident polarization density (magnetogram values) goes from polarity-balanced at $t = -7$ minutes to almost exclusively following polarity at $t = -3$ minutes, rising from 0.007 at $t = +6$ minutes to 0.03 at $t = +17$ minutes.

4. Quantitative Investigation

I determined the average properties of flux emergence events, as defined by the occurrence of transient line-center darkenings. I constructed a statistical sample of events by seeking local minima in three dimensions (two spatial and one temporal) in line-center intensity after smoothing in all dimensions with a gaussian smoothing kernel with a FWHM of 0.6 Mm spatial and 2 minutes temporal so that spurious multiple detections due to noise were suppressed. Only those 111 minima were accepted that were in the active region but outside the sunspots and pores and that had their minimum line-center intensity between 0.59 and 0.79 times the median line-center intensity over the whole field of view. The lower boundary is another measure to keep sunspots and pores out, and the upper boundary rejects local minima that occur all over the field of view without any apparent connection to flux emergence.

For all events in the sample, I calculated the four quantities discussed before as a function of coordinates relative to the time and place of the local minimum

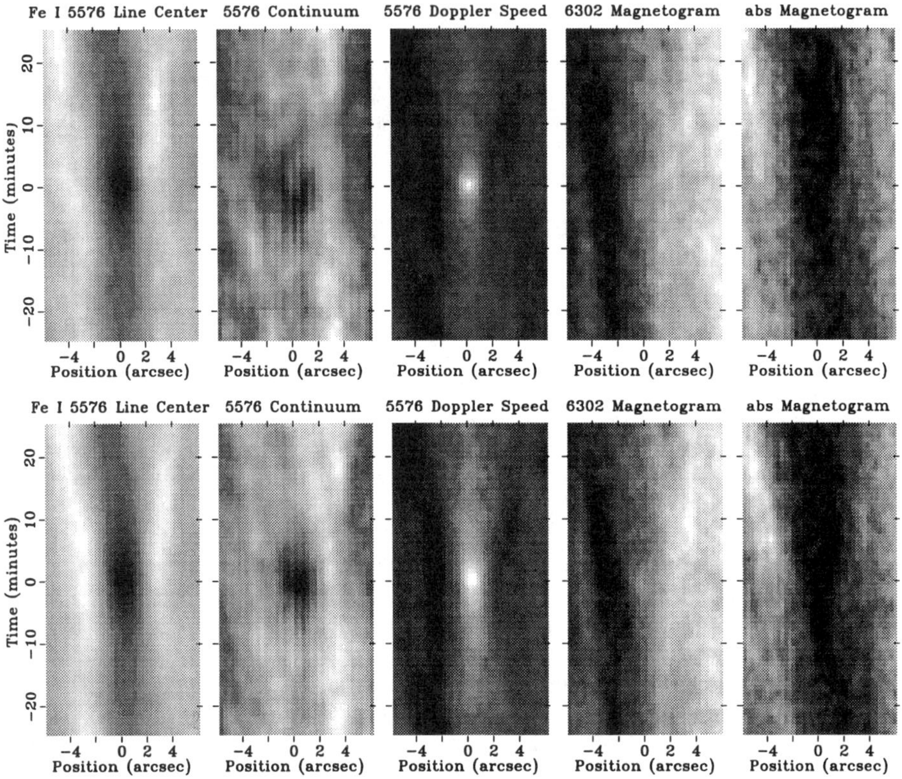

Figure 5. Average quantities vs. position along a line running from the northwest to the southeast through the main line-center bright points and the event centers (horizontal) and time since the event center (vertical) for 111 events. Columns: average Fe I 557.6 nm line-center intensity, average 557.6 nm continuum intensity, average Fe I 557.6 nm upward doppler velocity, average Fe I 630.2 nm magnetogram value ($\langle B \rangle$), and the average of the absolute value of the magnetogram value ($\langle |B| \rangle$). Top row: with the time and place of the peak upward doppler velocity at zero coordinates; extreme values: 0.716 – 1.100 times median, 0.960 – 1.040 times median, −206 – +728 m/s, −67 – +47 gauss (30 gauss per percent polarization), and 31 – 109 gauss, respectively. Bottom row: with the time and place of the minimum line-center intensity at zero coordinates; extreme values: 0.673 – 1.111 times median, 0.938 – 1.039 times median, −205 – +504 m/s, −72 – +47 gauss, and 32 – 107 gauss.

in line-center intensity, and then averaged those quantities over all events in the sample.

Snapshots of the averages are shown in Fig. 4. Average time slices are displayed in Fig. 5. The method of their construction does not favor sharply defined line-center bright points in the results, because no accomodation is made

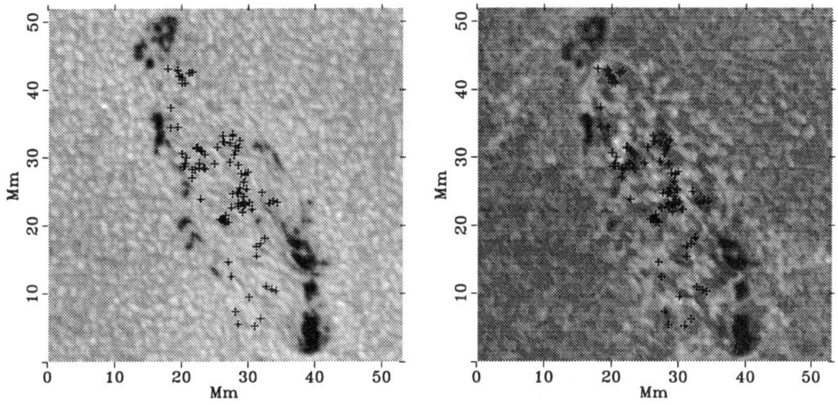

Figure 6. Locations (+) of magnetic flux emergence superimposed on part of the continuum image (left) and line-center image (right) at 11:02 UT.

for variation in the exact location of the bright points relative to the assumed zero coordinates. Indeed, the maximum line-center intensity contrast of the bright points in the average (10 percent of the median value) is much smaller than the maximum contrast of the single event studied before (of 60 percent).

From Figs. 4 and 5 I conclude that the average event has a preferred orientation in the transient line-center and continuum darkenings and in the associated upflow, with the major axis of its transient darkening oriented approximately along the line connecting the major sunspots of opposite polarity.

If there were a lot of spread around the average magnetogram values, then the maximum of $\langle|B|\rangle$ (5th column in Fig. 5) would be much greater than the maximum of $|\langle B\rangle|$ (4th column in Fig. 5). In our data, these quantities are of the same order of magnitude (107 G vs. 72 G), which indicates that the spread around the average is not substantial, and that the distribution of polarities in the average reflects the distribution of polarities in most if not all of the events. I conclude that the average event occurs just east of pre-existing leading flux, and that the magnetic flux emerges with the order of polarities aligned with the global distribution of magnetic flux in the active region. Because the new leading flux emerges next to pre-existing leading flux, immediate cancellation of flux is not the reason for the absence of a conspicuous new line-center bright point there (see Section 3).

The average event has a FWHM extent of about 10 minutes and about 2 Mm (3 arcsec) in line-center and continuum intensity. The pre-existing leading point is stable before the event but moves west at 1.1 km/s after the event. After the event, the following and leading points move apart at 1.4 km/s from an initial separation of 4.0 Mm (5.5 arcsec). In the time slices pinned on line-center darkness, the peak darkness is 0.675 times the median line-center intensity; the peak continuum darkness is 0.938 times the median continuum intensity; the peak upflow is 504 m/s; and the peak absolute magnetic flux density is 107 gauss. Pinning on the upflow rather than the line-center darkening sharpens the upflow and widens the leading bright point but leaves the essentials unchanged.

The distribution of the sample flux emergence events over time appears to be random, but their spatial distribution is not, as is shown in Fig. 6.

Acknowledgments. This research was supported by the Nederlandse organisatie voor wetenschappelijk onderzoek (NWO), the Lockheed Independent Research Program (NASA contracts NAS5-30386 [MDI], NASW-4612 [SR&T], and NAS8-39746 [SR&T]), the US National Science Foundation (ATM 92-13879 and ATM 89-12841), and NATO (Collaborative Research Grant 910997).

References

Brants, J. J. 1985, Sol. Phys., 95, 15

Bruzek, A. 1967, Sol. Phys., 2, 451

Chapman, G. A., & Sheeley Jr. N. R. 1968, Sol. Phys., 5, 442

Frazier, E. N. 1972, Sol. Phys., 24, 98

Kawaguchi, I., & Kitai, R. 1976, Sol. Phys., 46, 125

Sheeley, Jr. N. R. 1967, Sol. Phys., 1, 171

Solanki, S. K. 1993, Space Sci.Rev., 63, 1

Strous, L. H. 1994, Ph.D. Thesis, Utrecht University, The Netherlands

Strous, L. H., Scharmer, G., Tarbell, T. D., Title, A. M., & Zwaa, C. 1996, A&A, 306, 947

Tarbell, T., Topka, K., Ferguson, S., Frank, Z, & Title, A. 1989, in High Spatial Resolution Solar Observations (ed. O. von der Lühe), National Solar Observatory/Sacramento Peak, p. 506

Zwaan, C. 1985, Sol. Phys., 100, 397

High Resolution Solar Physics: Theory, Observations, and Techniques
ASP Conference Series, Vol. 183, 1999
T.R. Rimmele, K. S. Balasubramaniam, R.R. Radick, eds.

Thermal Structure in Sunspots and Dynamic Solar Infrared Granulation Imaged at 4.8, 12.4, and 18.1 Microns

Dan Gezari

NASA/Goddard Space Flight Center, Infrared Astrophysics Branch - Code 685, Greenbelt, MD 20771

William Livingston

NOAO/National Solar Observatory, 950 N. Cherry Ave., Tucson, AZ 85726

Frank Varosi

Ratheon STX Corp. Lanham, MD

Abstract. We have carried out a program of mid-infrared (4.8, 12.4, and 18.1 μ) imaging of the Sun at the McMath/Pierce Solar Telescope. Observations at these wavelengths provide several unique advantages, including the ability to make direct measurements of the gas temperature at well defined altitudes above the photosphere. Images in this new wavelength band have revealed new structure surrounding sunspots, and changing "infrared granulation" that differs significantly in cell structure and time scale of variation compared to granulation seen in visible light. The infrared also holds the potential for high spatial resolution imaging by minimizing the effects of atmospheric seeing.

1. Introduction

The mid-infrared (5 - 30 μ) is a new wavelength regime in solar astronomy, with several unique and potentially useful diagnostic capabilities. Compared to solar images in visible light, 12.4 μ continuum images reveal significantly different information, making the infrared a potentially useful new observational tool. These capabilities and differences are:

1. 1) Mid-infrared continuum observations provide a direct measurement of temperature at selectable altitudes within the lower solar atmosphere since they are not complicated by Doppler effects, such as line broadening and microturbulence. Relative temperatures can be measured directly without using spectral lines or radiative transfer models which complicate interpretation of visible light data.

   2) New brightness features are seen immediately surrounding sunspot penumbrae. These bright, plage-like regions are 50 - 100 Kelvin hotter than the surrounding quiet Sun.

3) Dynamic "infrared granulation" is observed, which differs markedly from visible light granulation in cell size and time scale of motion.

4) Scattered light contamination is reduced in the mid-infrared (Pierce 1991), and the effect of scattering is further reduced in sunspot images because sunspots are relatively much brighter in the infrared than in the visible (lower contrast compared to surrounding quiet Sun regions).

5) The infrared also holds the potential, in principle, for achieving sub-arcsecond spatial resolution since atmospheric seeing effects decrease with wavelength.

We have made 4.8, 12.4 and 18 μ images of selected regions of the Sun using our 58 x 62 array camera system (Gezari et al. 1992). These include sunspots, the surrounding plage-like regions, the quiet Sun near center disk, and the solar limb. These are the first mid-infrared observations of the Sun reported since the work of Turon and Lena (1970) using a single detector scanning photometer, and the first program of observations using a two-dimensional mid-infrared array camera imaging system.

2. Mid-Infrared Imaging of the Sun

2.1. Infrared emission from the solar atmosphere

In the solar photosphere and chromosphere the dominant source of the continuum opacity at visible wavelengths is H^-bound-free absorption, with a lesser contribution from free-free absorption. Electrons in the solar atmosphere are very nearly in LTE with the gas, so they emit radiation with a black body source spectrum which is characteristic of the gas temperature. The Rayleigh-Jeans limit to the Planck source spectrum applies, hence the infrared brightness is linearly proportional to the temperature of the emitting gas, linear to within about 10% at 5 μ and improving to within 2% at 12.4 μ. As would be expected for a simple blackbody source, solar limb darkening is negligible at mid-infrared wavelengths; the observed brightness remains at nearly its maximum center disk value out to within ~20 arcsec of the solar limb.

The solar continuum opacity at wavelengths longer than about 1.6 μ is essentially proportional to λ^2, so $\tau = 1$ is encountered at progressively higher altitudes as the wavelength of the observation is increased. The opacity also results in narrow weighting functions, allowing good altitude resolution. Continuum weighting functions for H- free-free opacity at mid-infrared wavelengths calculated from the VAL C solar model (Vernazza, Avrett & Loeser 1981) by J. Jefferies and C. Lindsey (private communication) indicate that observations at 4.8 μ probe an altitude $h = 70$ km above $\tau(0.5\ \mu) = 1$, where the corresponding temperature is $T = 5700$ K, 12.4 μ probes $h = 180$ km altitude, where $T = 5100$ K, and 18.1 μ probes $h = 250$ km, where $T = 4800$ K.

2.2. Special procedures for mid-infrared imaging

The array camera system we used was originally designed for nighttime, broad-band ($\Delta\lambda/\lambda$ ~0.1) imaging of faint sources, where the thermal background from the telescope, camera and sky dominates the detected flux (the background is typically 2 - 3 orders of magnitude larger than astronomical targets observed). In this high background application real-time background subtraction is required using the chopping secondary mirror on an infrared telescope. A mid-infrared observation of the Sun using this instrument in its normal operating configuration would saturate the array detector by a factor of roughly ~1000. Therefore a cold 0.001 neutral density filter is used to reduce the solar signal enough that it only half-fills the detector "well" in the shortest practical integration time (1/30 sec). This reduces the thermal background by the same factor, or about 10^{-3} of the full well. Since the background is negligible compared to the solar flux, a broadband mid-infrared imaging observation of the Sun is a low background measurement and background subtraction is not required. However the images must be flat field to correct for gain variations among the array detector pixels. To do this each image is divided by a gain matrix, which is obtained by normalizing an image a uniform, featureless target of about the same brightness as the source observed. The most convenient way to make this observation is to co-add a series of exposures taken while scanning across a small area of the quiet Sun containing no obvious active regions, thus averaging out any small brightness variations in the scene.

3. Observations and Results

3.1. Observations

We obtained mid-infrared images of the Sun during 16 - 29 September 1992 and 27 - 31 July 1998 at the National Solar Observatory 1.5-m McMath-Pierce Solar Telescope at Kitt Peak. At Kitt Peak (altitude 6000 feet, daytime ambient temperature ~300K) the 12 μ atmospheric/observatory thermal background is roughly double the level at Mauna Kea (altitude 14,000 feet, nighttime ambient temperature ~270K). At 20 μ the background level can easily be an order of magnitude larger. The all-reflecting design of the McMath-Pierce telescope and its low scattered light properties (Pierce 1991) make it well-suited for infrared observations. The diffraction limit at the 1.5-m main beam is 0.7 arcsec FWHM at 4.8 μ, 1.9 arcsec at 12.4 μ, and 2.8 arcsec at 18.1 μ. The array plate scale was 0.38 arcsec/pixel at the 1.5-m main focus allowing full sampling of the image data at wavelengths ≤ 5 μ, and the array camera field of view was 23 arcsec. Only the 4.8 μ images were degraded by seeing, as discussed below.

3.2. New emission structure around sunspots

One of the most dramatic new phenomena observed in the mid-infrared is significant ring-like bright structure along the outside edge of sunspot penumbrae (Fig.1). This brightening is not apparent in visible light images, but resembles structure seen in Ca K images. It represents atmospheric gas which is 50 - 100 K hotter than the temperature of the quiet Sun immediately surrounding it. These brighter (hotter) regions are likely to be analogous to faculae seen in the visible (which are observed in the continuum only near the limb due to altitude-dependent opacity effects).

Continuum images of sunspots at 4.8, 12.4 and 18 μ show a variety of new, unambiguous temperature features as image brightness structure. Images

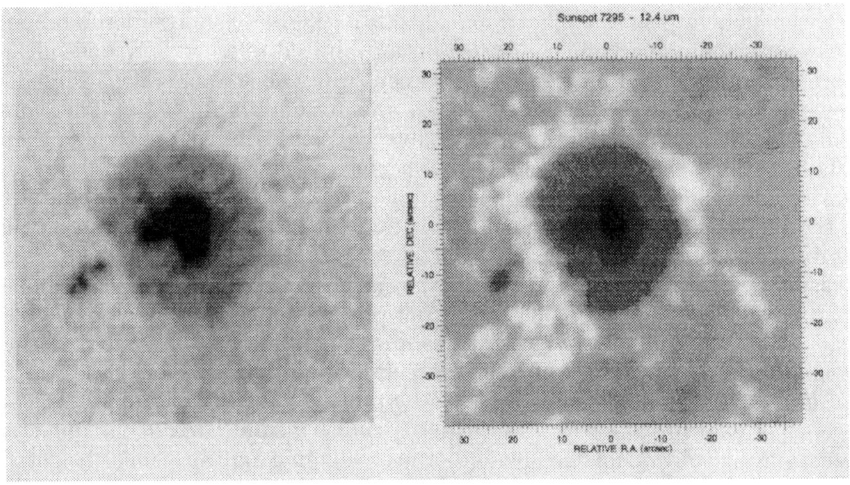

Figure 1. Comparison of sunspot NOAA 7295 in the visible and at 12.4 μ . The infrared image is essentially a "thermal map", linearly displaying temperature proportional to brightness (white is hot). The coolest point in the smooth central umbra is about 21% cooler than the surrounding 5000 K quiet Sun (orange) at the altitude sampled at 12.4 μ, while the penumbra is only 3% cooler. The infrared image clearly shows hotter regions around the periphery of the penumbra which are not evident in the visible light image. It is not possible to display all of the information contained in these images in a single figure since their dynamic range is large (~10^4), corresponding to temperature resolution of better than 1K. Ordinary images of high contrast regions (sunspots for example) can show temperature differences only as small as ~100K. However, if a low contrast region (quiet Sun) is isolated (Fig. 2) and the image color table stretched to correspond to the local minimum and maximum intensity values, the full temperature resolution of the data can be displayed .

of sunspot NOAA 7295 at 12.4 μ (Fig.1) show that the coolest point in the umbra is 21% cooler than the surrounding quiet photosphere, with the minimum temperature of 4000 K occurring at an altitude of 180 km as sampled by the 12.4 μ continuum. The penumbra is 3% cooler than the quiet Sun, and shows radial structure with thermal variations of roughly 2% (100 K).

Thermal structure is evident on spatial scales down to the diffraction limit. Very fine radial filamentary detail can be seen in the penumbrae of sunspots at 4.8 μ, which corresponds to the fine structure seen in the highest resolution visible light images, suggesting that these formations have significant temperature structure. There is also evidence that additional sub-arcsecond brightness details exist which would be further resolved by a larger solar telescopes.

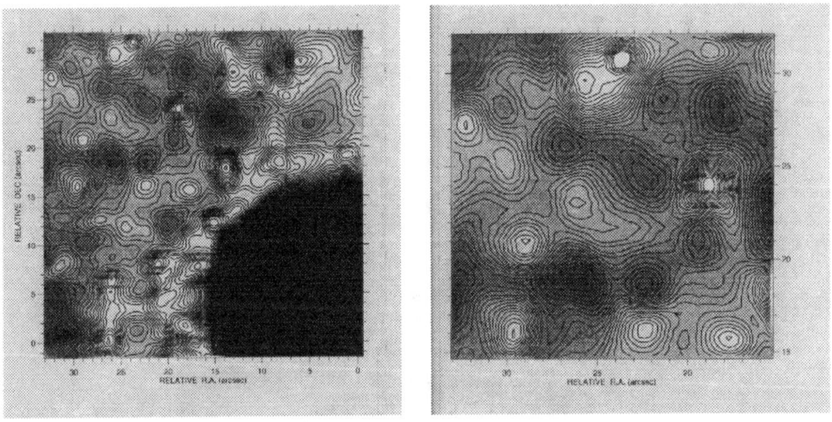

Figure 2. Four different contrast depictions of sunspot image data showing the dynamic range of temperature information in the image data. a) 12.4 μ continuum image of Sunspot NOAA 7295, displayed from a minimum (min) temperature of ~4000K at the center of the umbra to the maximum (max) of ~5000K outside the penumbra. b) The same image as Fig. 2a, but displayed with 100K range from min to max, showing small temperature variations in the region around the sunspot. c) A portion of the sunspot image, with temperature contours plotted at 5K intervals. d) Upper left quadrant of Fig. 2b, displayed with a min-max range of 75K and contour interval of 2.5K, showing spatial structure on the scale of the diffraction limit and temperature resolution of 1K in the image data.

3.3. Dynamic phenomena and "infrared granulation"

We observed "infrared granulation" in quiet Sun regions at 12.4 μ which differs markedly from photospheric visible light granulation in two ways: 1) Infrared granulation shows none of the cell-like character of visible granulation (distinct sharp-edged polygon shapes characteristic of well-developed convective cells). 2) Infrared granulation temperature structure changes on a time-scale roughly an order of magnitude faster than visible granulation. Mid-infrared granules are characterized by compact, point-like bright spots (and occasionally dark spots) which are randomly located but evolve in an orderly way, that is, they retain their shape but change in brightness and move in a "boiling" motion, gradually appearing and fading away.

Time-lapse 12.4 μ imaging observations of quiet Sun regions, made with images taken at 20 sec intervals over a period of 30 min and assembled into a "movie", show changing thermal structure at the 2% (100K) level on a diffraction-limited (~2 arcsec) spatial scale. A first order signal level drift can be introduced by the changing airmass or instrumental response during the observation since real-time background subtraction is not employed. When the linear drift is corrected for in data reduction by interpolating between average end values, more subtle brightness details can be seen (Fig. 3).

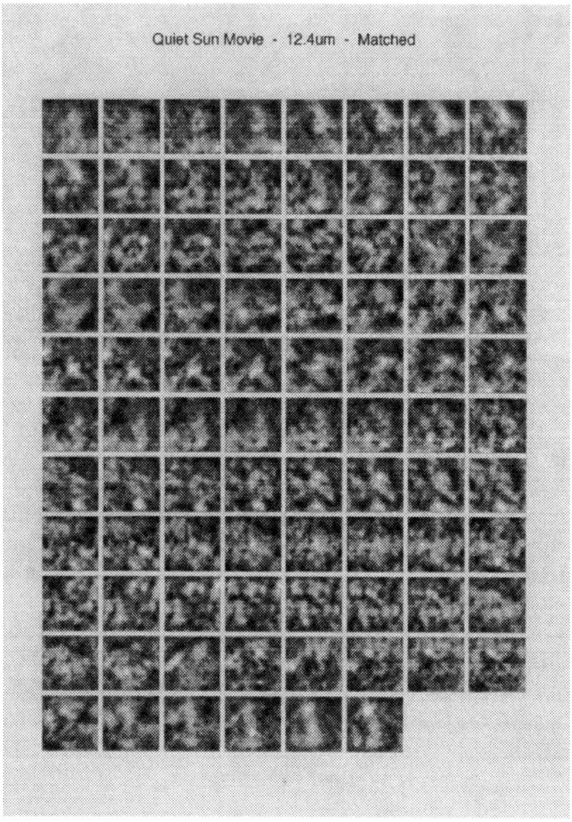

Figure 3. Infrared granulation seen in time-lapse images of quiet Sun disk center, taken at 20 sec intervals over a period of 34 min and assembled into a "movie". The sequence starts in the upper-left corner and progresses with time increasing to the right, continuing to the next row, etc. The image sequence shows random, nearly diffraction-limited (2 arcsec) point-like, granular structure with brightness (temperature) variations of about 2% (100 K) which change significantly on a 1 min time scale. Drifts of typically a few percent due to variations in atmospheric transmission (comparable in magnitude to the temperature variations observed in the images) have been removed in data reduction, to more clearly reveal the changing spatial details.

4. Discussion

4.1. Dynamic phenomena

The infrared granulation can be clearly seen to change in time if the second order drifts are corrected for (Fig.4). The bright infrared granules appear at distinct

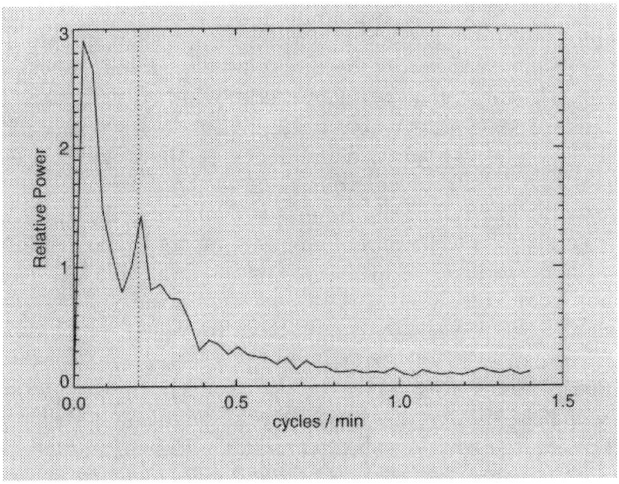

Figure 4. Spectral analysis of 95 images obtained over a period of 33 min. The frame interval is 20 sec. The power spectrum shows a line at $f = 3.5$ mHz (period = 4.8 min) at all spatial frequencies, which may be the signature of the "5 minute" oscillation.

sites 10 arcsec apart. There are rarely more than 3 such bright spots observed at any one time within the 23 arcsec field-of-view of the array. They are compact and often not fully resolved at the ~2 arcsec diffraction limit. The characteristics of mid-infrared granulation structure seem to differ at 4.8 and 12.4 μ. Some of the difference can be attributed to seeing effects (which are significant only in the 4.8 μ images) and diffraction differences (0.7 arcsec resolution at 4.8 μ vs. 1.9 arcsec at 12.4 μ). Other differences could be attributed to different properties at the solar altitude sampled above the photosphere (70 km at 4.8 μ vs. 180 km at 12.4 μ).

The bright, possibly unresolved emission features develop and fade on a ~1 min time scale. In contrast, typical lifetimes for visible granulation structures are ~15 minutes (Zirin 1988). These infrared features may be due to unresolved granulation, to temperature changes caused by compression from solar oscillations, or by thermal "overshoot" due to mass motions in the photosphere below.

Second order drifts (which were processed out in Fig.3) can be caused in part by real dynamic solar brightness effects. For example, spectral analysis of 95 quiet Sun images observed at equal time intervals over a period of 33 min shows the power spectrum of thermal variations, typically about $\pm 1\%$ of the average temperature. The spatial wavenumbers are averaged over constant $k_r = (k_x^2 + k_y^2)^{1/2}$ yielding a space and time power spectrum of quiet-Sun thermal oscillations in the 23 arcsec field of view analyzed. The power in spatial wavenumbers is summed to obtain the total temporal power spectrum, showing a line corresponding to a 5-minute oscillation (Fig. 4).

4.2. Maximizing Spatial Resolution in the Infrared

The potential for obtaining sub-arcsecond resolution solar images in the infrared is often discussed because seeing becomes weaker at longer wavelengths. We find that seeing is a small effect at ~12 μ under good conditions at the McMath-Pierce Solar Telescope, typically amounting to only a few tenths of an arcsecond of image wobble. The canonical dependence of the angular seeing size θ on wavelength is $\theta \propto \lambda^{-1/5}$, derived from the $\theta \propto \lambda/r_o$ dependence of the seeing on the coherence length r_o of atmospheric turbulence cells, and the wavelength dependence $r_o \propto \lambda^{6/5}$ in the visible and near infrared (Hardy 1998). A typical value of $r_o = 10cm$ produces ~1 arcsec seeing in the visible. At 12 μ, $r_o = 10cm$ (in the visible) would result in $r_o \approx 450cm$ or a seeing of ~0.55 arcsec, i.e., a telescope with an aperture of 4.5m would be diffraction limited for typical seeing conditions. At 3 μ $r_o \approx 86cm$ corresponding to seeing of 0.7 arcsec. According to Hardy (1998) values of $r_o = 1m$ may frequently be encountered at longer infrared wavelengths. There has been little observational verification of these relationships over large wavelength intervals or at longer infrared wavelengths.

Obtaining the best mid-infrared image quality using a camera capable of high spatial resolution and very high relative astrometric precision requires precise pointing and guiding. Automatic solar limb guiding is used to correct for tracking errors and to position the image for mosaic observations. Short exposures ($\geq$ 1 sec), correlation techniques, and shift-and-add image combining techniques can be used to obtain the best resolution when telescope wind-shake or seeing are a factor. Tip-tilt seeing correction with real-time active optics would be an important improvement in observing techniques. Considerable care must be taken in assembling mosaic images from individual exposures of extended, low contrast infrared surface detail, which may not have well defined features that could be used for registering overlapping data.

Acknowledgments. We thank Greg Kopp for his collaboration in the early stages of the mid-infrared solar imaging program. Mary Hewitt of Hughes/Santa Barbara Research Center (SBRC) provided expert guidance during the development of the array camera system at Goddard. We thank Dick Joyce at NOAO/KPNO for generously providing the 18.1 μ interference filter used in this project. This research is funded by NASA/OSSA (RTOP 188-44-23-08) with support from NOAO/National Solar Observatory.

References

Gezari, D. Y., Folz, W. C., Woods, L. A. and Varosi, F. 1992, P.A.S.P., 104, 191.

Hardy, J. W. 1998, "Adaptive Optics for Astronomical Telescopes", Oxford University Press (New York), 46.

Pierce, K. 1991, Solar Phys., 133, 215.

Turon, P. J. and Lena, P. J. 1970, Solar Phys., 14, 112.

Vernazza, J. E., Avrett, E. H., and Loeser, R., 1981, ApJ Supp. Series, 45, 635.

Zirin, H. 1988, Astrophysics of the Sun, Cambridge Univ. Press.

AUTHOR INDEX